INTERACTIONS OF ENERGY AND CLIMATE

Interactions of Energy and Climate

Proceedings of an International Workshop held in Münster, Germany, March 3-6, 1980

Edited by

W. BACH

Center for Applied Climatology and Environmental Studies,
The University of Münster, Germany

J. PANKRATH

Umweltbundesamt, Berlin (West), Germany

and

J. WILLIAMS

Climatological Consultant, Karlsruhe, Germany

D. Reidel Publishing Company

Dordrecht : Holland / Boston : U.S.A. / London : England

Library of Congress Cataloging in Publication Data

CIP

Main entry under title:

Interactions of energy and climate.

Proceedings of an international conference held at Münster, Germany, in March 1980.
Includes index.
1. Climatology–Congresses. 2. Energy development–Environmental aspects–Congresses. I. Bach, Wilfrid. II. Pankrath, Jürgen. III. Williams, Jill, 1949–
QC980.I56 551.5 80–23220
ISBN 90–277–1179–8
ISBN 90–277–1177–1 (pbk.)

Published by D. Reidel Publishing Company,
P.O. Box 17, 3300 AA Dordrecht, Holland.

Sold and distributed in the U.S.A. and Canada
by Kluwer Boston Inc.,
190 Old Derby Street, Hingham, MA 02043, U.S.A.

In all other countries, sold and distributed
by Kluwer Academic Publishers Group,
P.O. Box 322, 3300 AH Dordrecht, Holland.

D. Reidel Publishing Company is a member of the Kluwer Group.

Printed in The Netherlands

TABLE OF CONTENTS

FOREWORD vii–viii

PREFACE ix–xi

WORKING GROUP REPORTS AND RECOMMENDATIONS xiii–xxxiii

LIST OF AUTHORS AND PARTICIPANTS xxxv–xl

PART I / GLOBAL ENERGY DEMAND AND SUPPLY – PRESENT AND FUTURE

A. B. LOVINS / Economically Efficient Energy Futures 1
H.-H. ROGNER and W. SASSIN / High Energy Demand and Supply Scenario 33
G. A. McKAY and T. ALLSOPP / The Role of Climate in Affecting Energy Demand / Supply 53
J. K. PAGE / Climate Considerations and Energy Conservation 73
W. J. QUIRK and J. E. MORIARTY / Prospects for Using Improved Climate Information to Better Manage Energy Systems 89
Discussion 100

PART II / CLIMATIC IMPACTS OF CARBON DIOXIDE

H. OESCHGER, U SIEGENTHALER and U. HEIMANN / The Carbon Cycle and its Perturbation by Man 107
C. D. KEELING / The Oceans and Biosphere as Future Sinks for Fossil Fuel Carbon Dioxide 129
U. HAMPICKE / The Role of the Biosphere 149
W. L. GATES / Modeling the Surface Temperature Changes Due to Increased Atmospheric CO_2 169
R. M. ROTTY and G. MARLAND / Constraints on Fossil Fuel Use 191
D. H. SLADE / The U.S. Department of Energy and the Carbon Dioxide Issue 213
Discussion 235

PART III / CLIMATIC IMPACTS OF OTHER GASES AND AEROSOLS

D. H. EHHALT / The Effects of Chlorofluoromethanes on Climate 243
I. S. A. ISAKSEN / The Impact of Nitrogen Fertilization 257

V. RAMANATHAN / Climatic Effects of Anthropogenic Trace Gases 269
W. W. KELLOGG / Aerosols and Climate 281
Discussion 297

PART IV / CLIMATIC IMPACTS OF LAND USE CHANGES

A. BAUMGARTNER and M. KIRCHNER / Impacts Due to Deforestation 305
G. L. POTTER, H. W. ELLSAESSER, M. C. MacCRACKEN, J. S. ELLIS and F. M. LUTHER / Climate Change Due to Anthropogenic Surface Albedo Modification 317
J. WILLIAMS / Impacts of Changes Due to Large-Scale Deployment of Energy Conversion Systems 327
T. R. OKE / Climatic Impacts of Urbanization 339
Discussion 357

PART V / CLIMATIC IMPACTS OF HEAT EMISSION

J. PANKRATH / Impact of Waste Heat Emissions in the Upper-Rhine Region 363
H. G. FORTAK / Local and Regional Climatic Impacts of Heat Emission 383
R. M. CHERVIN / Computer Simulation Studies of the Regional and Global Climatic Impacts of Waste Heat Emission 399
Discussion 416

PART VI / RISK-BENEFIT AND COST-BENEFIT ASSESSMENT

S. C. BLACK and F. NIEHAUS / Comparison of Risks and Benefits Among Different Energy Systems 421
J. A. LAURMANN / Assessing the Importance of CO_2-induced Climatic Changes Using Risk-Benefit Analysis 437
W. O. ROBERTS / The Costs of Climatic Impacts 461
Discussion 469

PART VII / POSSIBLE PREVENTIVE MEASURES

G. EGBERTS and A. VOSS / Reduction of Fossil Fuel Use and Adoption of Alternative Energy Sources 473
H. SCHAEFER / Conservation Practices and Increased Efficiency of Energy Conversion and Usage 483
C. F. BEAS Jr., S. E. BEALL, D.W. LEE and G. MARLAND / The Collection, Disposal, and Storage of Carbon Dioxide 495
M. STEINBERG and A. S. ALBANESE / Environmental Control Technology for Atmospheric Carbon Dioxide 521
Discussion 552

INDEX 556

FOREWORD

Since the beginning of industrialization in the last century, a steady increase in energy consumption can be observed. At the same time, energy generation switched from wood and coal to predominantly oil, coal and natural gas.

Soon, many countries became aware of the fact that the resources of fossil fuels, especially of oil and natural gas are finite. Diversification of energy sources became a requirement for the future.

Governments expressed their concern by setting up natural energy programmes while international organisations undertook assessments of the global energy resources and possible rates of supply and substitution.

When it comes to setting up energy policies, the following factors must be taken into consideration: population growth, level and nature of socio-economic activity, the costs of energy, the adequacy and reliability of supply, the availability of technology and supporting infrastructure, the success of energy conservation programmes and concern about the environment, safety aspects of production and use of energy as well as educational efforts toward a rational use of energy.

When we express our most urgent concern, the long-term global energy provision, experts offer four interrelated partial strategies:

- the strategy of rational use and conservation of energy
- the strategy of using renewable energy sources
- the coal strategy including coal gasification and liquefaction
- the nuclear power strategy.

Any strategy, however, for securing future energy supply has, from my point of view, to be thoroughly examined as to its impact on the environment.

In the past, the environment discussion tended to focus more attention on short-term aspects than on long-term environmental, climatic and socio-economic consequences. But long-term developments, because of their irreversibility, need even more attention.

One important part of this risk assessment process is the attempt to quantify the impact of the various energy generation options

W. Bach, J. Pankrath, and J. Williams (eds.), Interactions of Energy and Climate, vii–viii.

on global and regional climate. And I am glad that the workshop, the proceedings of which I am pleased to introduce herewith, comes to scientific recommendations regarding energy use and production causing the least impact on climate and environment.

The outcome of this workshop is a fruitful contribution to the interdisciplinary Governmental Research Programme on Climatology of the Federal Republic of Germany. With this programme we are joining the relevant multinational activities of the WMO and the Commission of the European Community, which is to give the basis for political decisions to protect man and his environment from harmful effects.

Main points of interest are

- Research on the climate system
- Acquisition, management and availability of climate data
- Research on sources and sinks of air pollutants potentially affecting climate
- Research on the impact of climatic changes on socio-economic conditions
- Research on the impact of biospheric changes as well as impacts of air pollutants on the regional climate.

Within this programme it is one of the particular concerns of my agency, which on behalf of the Federal Minister of the Interior sponsored this workshop, to investigate the interactions between energy and climate. We therefore set up a research and development programme on the environmental compatability of specific developments and options in future energy generation. And, in this respect, I am grateful for the contribution of the workshop on energy/climate interactions to our programme.

The Federal Environmental Agency acknowledges, with gratitude, the efforts of Professor Wilfrid Bach in developing and organizing, and of the University of Münster, in hosting this workshop.

Dr. Heinrich von Lersner

President

Umweltbundesamt

PREFACE

Over the past decades climatic events in many regions of the world have made us aware of our vulnerability to climatic change and variability. We increasingly realise that not only is man vulnerable to variations in climate but also that climate is vulnerable to the actions of man. This is of particular concern in view of the continued world population growth and the world's increasing appetite for energy and food.

This concern is shared by a number of international organizations as well as a growing number of national governments. The Federal Republic of Germany, through the Department of the Interior and its Federal Environmental Agency, is sponsoring a pilot program of the effects of man's activities on the global climate. An important part of the overall program involves the organization of a series of international conferences. The first meeting, entitled "Man's Impact on Climate", which was held at the Federal Environmental Agency's headquarters in Berlin in 1978, was designed to give a technical review of the various aspects of climatic change. The second meeting, entitled "Energy/Climate Interactions" was held at Münster, Germany, in March 1980, and is documented in this book. This conference dealt with both the effects of energy systems on climate and with the effects of climate on energy use. A third meeting will be held at the Aspen Institute of Berlin in December 1980 to deal with the climate/food interactions and their implications for world food security. All of these conferences are planned and organized by the Center for Applied Climatology and Environmental Studies at the University of Münster.

The "Energy/Climate Interactions" meeting included 29 formal presentations to an audience of about 100 participants, representing 10 different nations. The lectures were grouped into 7 different sessions and are reflected by the sections of this book. Some of the important questions addressed at this meeting were:

- What climatic impacts can be expected from the various energy, economic and population growth policies?

- What energy strategies are feasible that would cause acceptable climatic impacts?

- What are the relative contributions of the various forcing functions to climatic change?

W. Bach, J. Pankrath, and J. Williams (eds.), Interactions of Energy and Climate, ix–xi.

- What are the relative risks, costs and/or benefits of climatic impacts?

- What possible countervailing measures could reduce or even avert impacts on climate?

An edited version of the conference discussions follows each group of papers.

The last two days of the meeting were devoted to three simultaneous working group sessions. Statements prepared by the working group chairmen were discussed with emphasis on the topics: 'Energy demand and supply - the preferred energy strategy causing the least impact on climate', 'Identification and assessment of the various climatic impacts', and 'Objectives of a climatic impact study program'.

Based on the papers, the discussions and the working group reports the following observations can be made. In the decades ahead decisions have to be made to reduce or avert the impacts of climatic change before all the answers have been obtained. Although a climatic impact assessment program is faced with many uncertainties, it nevertheless has to be started now, because society cannot afford to wait until all variables are quantified to the satisfaction of all the parties involved. Additionally, as a precautionary measure, society should follow a low risk-climate-energy-land use policy, which would

- promote the more efficient end use of energy,

- secure the expeditious introduction of energy sources that release little or no carbon dioxide to the atmosphere, and

- discourage deforestation and encourage reforestation and soil conservation.

The major advantage of such a low risk policy is that in the best case it may prevent climatic impacts altogether, and that in the worst case valuable time would be gained to obtain better information to redirect policy decisions. Since this policy is based on measures that make sense for other than climatic reasons, it is worthy of consideration in any case.

In conclusion, the convenors of this meeting wish to thank all colleagues who gave valuable advice during the planning stage of the conference and who contributed their papers to this book. Additionally, we are grateful to all conference participants who shared their ideas with us.

We further wish to acknowledge the financial support from our sponsor, the Department of the Interior, who through the Federal Environmental Agency, made available a grant to carry out this climate/energy project. We are also grateful to the Commission for the European Communities and the United Nations University for co-sponsoring this meeting. The conference host, the University of Münster, deserves special thanks for supplying the facilities which helped to make this a successful conference. The reception given by the City of Münster is also gratefully acknowledged.

Finally, we especially wish to recognise the assistance given by the staff of the Center for Applied Climatology and Environmental Studies at the University of Münster in the preparation of both the conference and the conference publication.

Wilfrid Bach
Jürgen Pankrath
Jill Williams

WORKING GROUP I:

ENERGY DEMAND AND SUPPLY – THE PREFERRED ENERGY STRATEGY CAUSING THE LEAST IMPACT ON CLIMATE

Chairman/Co-Chairman: F. Niehaus/A. Voss

W. Bach, P. Biscain, U. Hampicke
G. Hekstra, C. Keeling, F. Krause
H. Oeschger, W. Quirk, T. Waltz,
G. Zimmermeyer

1. THE ISSUE

Provided energy growth continues at the current rate and with the present mix of sources, there is a definite prospect of a significant climatic change. Along with the expected overall warming of the lower atmosphere there will be shifts in the atmospheric circulation patterns resulting in regional and seasonal temperature and precipitation anomalies. The major effect will not be the creation of a new type of climate, but rather a different distribution of climate. This can have an impact on energy requirements, food production, fisheries, water resources, forestry, land use, transportation, tourism, and human health and well-being. In short, the total global socio-economic structure is vulnerable in varying degrees to a naturally and/or anthropogenically induced climate change and variability.

The influence of energy use on climate and the environment, and thus on society, therefore cannot be evaluated in isolation, but has to be viewed together with economic, ecological and social developments. Thus the minimization of the influence on climate and environment can only be a sub-set of the overall objectives of policy decisions which need to be defined carefully. A broad systems approach is required that considers all aspects including those which are difficult to quantify.

The overall objective of policy decisions guiding energy strategies is the improvement of the well-being of all mankind now and in the future. Such decisions have a subjective component so that different conclusions can be reached by different individuals and different nations with different value systems. It is, therefore, important that an assessment of the various national and global energy scenarios takes fully into consideration both the technical potential and the economic, social and political realities.

It is also important to recognise that any optimization procedure

W. Bach, J. Pankrath, and J. Williams (eds.), Interactions of Energy and Climate, xiii–xxxiii.

regarding a desirable energy mix may lead to very different conclusions depending on the spatial and temporal scales considered. Thus it is necessary to consider not only what is desirable on a local and regional basis but also what is desirable on a global scale. Attention must also be paid to the short-term as well as the long-term solutions and their consequences.

Considering the problems of future energy demand and supply, it is desirable to have a wide variety of energy options at our disposal in order to increase the resilience of the energy economy. This can best be achieved by a more efficient use of the traditional energy sources and by the more expeditious development of alternative energy resources.

All options must be carefully scrutinized for their inherent risks, costs and benefits prior to their adoption. Such comparisons should consider the total fuel cycle of the various systems including construction and dismantling of plants as well as waste disposal. Moreover, comparisons must be made among systems that provide the same product to the end-user. Current risk-cost-benefit analyses require considerable improvement. Finally, it is important that the short-term benefits be carefully weighed against the medium-to long-term risks.

2. CLIMATIC INFLUENCE OF ENERGY USE

All energy use results in the emission of heat into the atmosphere. Model studies show that there are apparently no effects on global climate from heat emission when energy scenarios of less than 100 TW of primary energy consumption as compared to the present 9 TW are considered. Heat emission may, however, pose problems on a local or regional scale. The results of model and analogue studies and comparisons with natural phenomena suggest that the major effects of power plants clustered in energy centers of 10,000-50,000 MWe capacity could produce significant changes in cloudiness and precipitation with an increase in the probability of severe weather. Research on effects of heat emission should therefore concentrate on local-scale meteorological models and on the investigation of the impacts of large power centers.

Human activities, especially those related to industrial processes and the practice of agricultural burning and soil management, result in the release of particles into the atmosphere. In addition to direct particle emission, aerosols are also formed from the gaseous products of combustion. All aerosols scatter and absorb both solar and infrared radiation, and, by influencing the atmospheric heat balance, may also change the climate. There may be warming in some regions and cooling in others. The net global effect will probably be one of atmospheric warming.

The increase of atmospheric CO_2 poses a problem of major concern. This increase which is worldwide has to be studied in international co-operation and given high priority.

If energy consumption follows current projections it seems probable, based on present knowledge of the carbon cycle, that atmospheric CO_2 will increase to a level of about 380 ppm by volume by the end of the century and reach twice the preindustrial level around 2050 A.D. Still further increases could occur before limitations in fossil fuel supply will force a decline in production, so that a peak concentration 4 to 8 times the preindustrial level may occur within 2 or 3 centuries. High levels once reached will only slowly decline so that a level over twice the preindustrial is likely to persist for over 1,000 years.

Since the beginning of the industrial era, the atmospheric CO_2 level has increased by about 15%. The present rise in atmospheric CO_2 is equivalent to 50 to 60% of the CO_2 from fossil fuels remaining airborne. The year to year fraction is, however, variable. Accurate measurements of atmospheric CO_2 are therefore desirable to furnish a more reliable baseline for projecting the airborne fraction into the future.

Although present knowledge of the carbon cycle is adequate to predict future CO_2 fractions remaining airborne, efforts must continue and be expanded to determine more accurately the uptake by the oceans, and to assess both the contribution and the response of the biosphere to atmospheric CO_2 increases. These factors influence the airborne fraction and should be established to the highest possible accuracy.

Present estimates center around a global average value of 2-3°C surface air temperature increase per doubling of atmospheric CO_2 concentration with a 3-4 fold temperature increase in northern polar regions. Due to the inertia of the oceanic response, temperature increases are expected to follow the CO_2 increase with a lag of 10-20 years. Based on studies of the exploitable oil, gas and coal resources, carbon cycle models suggest a maximum possible atmospheric CO_2 increase by a factor of about 8. Since the temperature increases approximately logarithmically with increases in atmospheric CO_2 concentration (see Working Group Report No. II), the maximum global average rise could be as high as 8°C.

The most likely ultimate temperature increase, based on present knowledge, is considerably lower than this but nevertheless higher than that predicted by a single doubling of the preindustrial CO_2 level. If the exploitation of fossil fuels is not restrained and no large storage of carbon in the biosphere occurs, the most likely global average temperature rise could be of the order of 4 to 6°C.

If, on the other hand, the fossil fuel CO_2 production could be kept at the present level, a 50% CO_2 increase, corresponding to an average temperature increase of 1 to 1.5°C, would result in 2100 A.D. Over such a long time period this global average temperature increase is perhaps tolerable. If not tolerable, time is gained to take further remedial action.

In this context it is well to recall that there are other gases and probably also aerosols that will reinforce the CO_2-greenhouse effect making an eventual remedial action more difficult. At any rate scrubbing CO_2 from the stack gases or taking it out of the atmosphere and disposing of it somewhere are considered to be last resort measures in the prevention of a CO_2 build-up in the atmosphere because they are very expensive and require large additional energy expenditure.

The Panel looked also into the possibility of whether coal gasification or liquefaction would aggravate the CO_2 problem. These processes require both heat and hydrogen. If tnese are supplied by burning part of the coal (autothermal gasification) the total CO_2 emissions are some 40% higher than using coal directly. If, on the other hand, the energy is provided by a non-fossil energy source (allothermal gasification), the total CO_2 emissions are still some 6% higher than burning coal directly, because carbon is still needed to split the water for making hydrogen. Moreover, if the energy is supplied by an external energy source and the hydrogen production does not involve carbon (e.g. electrolysis, thermal water splitting) the total CO_2 emissions could be reduced by about 40%. Of course, CO_2 emissions would be prevented altogether, if hydrogen was produced from a non fossil fuel source and used as a direct energy carrier.

3. EVALUATION OF ENERGY SCENARIOS AND RECOMMENDATIONS

In the following emphasis is placed on the increasing atmospheric CO_2 levels due to fossil fuel use because they were identified as the most likely contributing cause to a global climatic change. Projections of CO_2 emission levels into the future are based on aggregated regional energy scenarios. A number of such scenarios were considered notably those developed at the International Institute for Applied Systems Analysis (IIASA) in Laxenburg in Austria, the International Energy Agency in Paris, France, the Institute for Energy Analysis (IEA) in Oak Ridge, USA, and the Massachusetts Institute of Technology in Boston, USA.

It was seen that all scenarios considered a continued growth in energy demand of between 20 and 40 TW by 2030 A.D., mainly because of the continued population growth and the obligation to reduce the gross inequities in energy consumption between the developed and the developing nations. The fossil fuel share in 1980 is about 90% of the total primary energy supply. Both the IEA scenario for 2025 and the IIASA high and low scenarios for 2030 project a fossil fuel share of 65-70%. It was also noted and taken into consideration that the increase in energy demand may not be as large as the above figures suggest in view of the fact that primary energy demand forecasts have been scaled down continually in recent years.

Based on this information the following strategies can be envisaged:

- to let fossil fuel use, and hence CO_2 emission, continue to grow at the historical exponential growth rate of 4.3%/yr.;
- to let fossil fuel use grow at a reduced rate;
- to keep fossil fuel use at the present level, or
- to reduce fossil fuel consumption.

Considering the fact that the world economy will continue to be strongly dependent upon fossil fuel use for at least a transitional period of several decades, the panel does not recommend a reduction in fossil fuel use at this time. However, the panel wants to make it equally clear that a continued growth of fossil fuel use at the historical rate is highly undesirable in the light of the magnitude and irreversibility of the potential effects.

After much deliberation, out of the remaining two strategies, the panel opted to recommend that the global fossil fuel consumption not be increased above the present level. We felt justified in making such a recommendation on account of recent energy studies prepared for a number of nations in the developed world and data presented at this and other workshops suggesting that

- the more efficient use of energy can result in substantial savings (initially as much as 30-50%) in several energy sectors without in any way jeopardizing prosperity; and that
- non-fossil energy sources have the technical potential to make a significant contribution given sufficient time and appropriate institutional support.

These strategies used in a complementary fashion could ensure growth of end use energy over a few decades until primary energy consumption equilibrium is reached. They can be viewed as a prudent insurance policy with a built-in flexibility sufficient to react to the potential impact of CO_2 and other agents on climate and environment. Further arguments which were put forth in support of our recommendation included

- the necessity of developed nations to curtail drastically their strong dependence on imported fossil fuels, thereby avoiding an imbalance of payments;
- the fact that the recovery and processing of unconventional fossil fuel resources will exert an extremely high strain both on the economy and on the environment; and
- the argument that the rapid combustion of fossil fuels to produce energy may be hard to justify to our progeny since fossil fuels represent non-renewable substances of high chemical value.

In making the recommendation to keep the global CO_2 emission from fossil fuel use at the present level the panel was fully aware that appropriate allowances must be made for the disparate fossil fuel needs of different countries on an equitable basis. Special care must be exercised that developed countries do not export their energy problems to developing nations i.e. by making available energy-intensive technologies that are unsuitable for their infrastructure. This could well lead to a situation where developing countries produce goods for the developed countries just for the purpose of paying for their energy bill (especially for crude oil imports and capital investments) and without much benefit to their own countries (such as raising their standard of living). In order to reduce North-South tensions it seems therefore highly desirable that developed nations use their high technological potential to reduce their share of global fossil fuel consumption (notably crude oil and natural gas) so that the developing nations can attain a correspondingly larger share. Developing countries will be especially sensitive toward climatic change since they often operate on a minimum of natural resources. These less resilient regions of the world need therefore particular attention and help.

All these actions may, however, be nullified if global population growth continues. Population increase in large regions of the world threatens to outweigh any material improvements. Clearly, the gap in energy use and prosperity between the developing and developed world can only be narrowed by a voluntary change of attitude. This requires an atmosphere of mutual trust and the realization that the problems of energy and population growth both in the developed and the developing nations are inseparably linked.

4. CONCLUSIONS

There is clear evidence of an impact of energy use on local climate. There may already be some perturbation of the global climate, but this may not be detectable in our measurements before the end of this century. Moreover, the interactive mechanisms involved are so complex and presently still so poorly understood that the impacts on climate, which may be quite serious and potentially irreversible, cannot be predicted with confidence at this time. The danger is that energy decisions based on such poor information may lock us into an energy development where we may lose flexibility once energy systems with long market penetration have been adopted.

In the face of present uncertainties prudence dictates a cautious and flexible energy strategy. We therefore recommend the following low climate-risk-energy policy, which would

- promote the more efficient end use of energy,
- secure the expeditious development of energy sources that add little or no CO_2 to the atmosphere, and
- keep global fossil fuel use, and hence CO_2 emission, at the present level.

This energy policy which should be followed anyway - CO_2 risk or not - has the advantage of both permitting energy growth and promoting energy quality. The major bonus is, however, that in the best case such a policy may prevent an impact of energy use on global climate altogether, and that in the worst case valuable time is gained to obtain better information to redirect our energy policy. The time gained can be used to reduce the uncertainties about the future CO_2 emissions, their climatic effects and their socio-economic implications.

WORKING GROUP II:

IDENTIFICATION AND ASSESSMENT OF THE VARIOUS CLIMATIC IMPACTS

Chairman/Co-Chairman: H. Flohn/R. Rotty

R. Chervin, D. Ehhalt, W. L. Gates, J. Hofmann, H.-J. Jung, H. Lieth, G. McKay, T. Oke, G. Potter, V. Ramanathan, G. Schlesser, C. C. Wallen

It is imperative that the climatological community provide guidance on the environmental consequences of growing global energy demand and the evolving mix of energy supply possibilities. The full understanding is very complex, involving physical, social and economic understanding of the entire global system.

There are fundamental techniques that can be employed in assessing the climatic impacts of anthropogenic uses of energy:

- analytical (numerical) models of varying degrees of complexity and resolution based on physical principles and,
- past climates that may have features similar to those that may come from increased CO_2 can be reconstructed using geological and historical evidence.

We believe that the state of development of each of these techniques is at present inadequate to give a picture of future climate with sufficient resolution and clarity to enable us to identify the specific changes necessary for a full assessment. We recommend that at present both techniques be used to complement each other - models to determine the large scale sensitivity of climate to man's activities and historical and geological reconstructions to obtain information with more detail and regional resolution.

Although the effects of atmospheric carbon dioxide are not the only ones associated with the energy used by man, we believe that they will prove to be the major factors that will give concern to mankind on a global scale within the foreseeable future.

Model studies of climate give general and global average results. History and geological knowledge can be exploited through case studies to acquire understanding of the probable consequences of large scale climatic change for the environment. However, before warm periods of the distant past (as reconstructed, for example, from historical and geological data) can be meaningfully interpreted as possible scenarios of a future CO_2-warmed climate, it is necessary to determine the extent to which regional anomalies resemble each other during periods of similar overall warmth. For this purpose it is useful to use recent instrumental climate data, classified or stratified according to overall warmth, from which temporally and physically consistent regional climate anomalies could be determined. The likelihood of a particular regional anomaly of temperature and precipitation occurring within an overall warm period could then be estimated. Such diagnostic research could be performed independently of paleoclimatic data and independently of climate models, but the process suggested would contribute to the interpretation of results of both procedures.

The physical processes and time lag involved with the likely increase of CO_2 need to be studied in more detail to help parameterization of the analytical models. We therefore have limited our discussions to the atmospheric CO_2 impacts as supplemented by other trace gases that function in a so-called greenhouse capacity. The change in the global mean surface temperature can be expressed as

$$\Delta T_s = \lambda n \ln(f)$$

where

λ is a feedback parameter in K
n is a non-dimensional factor to account for other trace gases (other than CO_2)
f is the fractional change in CO_2 (e.g. f = 2 for CO_2 doubling).

Present knowledge suggests that λ lies within the range 1.5/ln2 to 6/ln2, and n is probably between 1.0 and 1.6. We have assumed a mean value for λ = 3.0/ln2 and n of 1.3 but observe that strict regulations of CO and NO emissions and photo-chemical modelling uncertainties in N_2O nitrification and denitrification processes could eliminate the projected increases in CO, O_3, CH_4 and N_2O, and thus result in a value for n close to 1.0. In any case, atmospheric CO_2 could increase severalfold and thus a substantial temperature increase might eventually occur.

Atmospheric general circulation models have a particularly valuable future contribution to make to the estimation of the likely impacts of an increase of CO_2 on the climate. Although such models are known to contain systematic errors, they can provide a physically-consistent global distribution of the principle climatic variables as a function of the season of the year, as well as an estimate of the interannual climatic variability. For application to the CO_2-climate problem, such models should be run for a sufficient number of years with increased (i.e. doubled) CO_2 to enable the statistical significance of each CO_2 experiment's altered climatological statistics to be determined with respect to a parallel control integration. Such models should include an interactive ocean which provides at least an internally-determined sea-surface temperature. Ideally, such experiments should be performed with several models, and the degree of model consensus should be determined.

The use of case studies (scenarios) provides a "snapshot" of climate and the environment at a period in history. It should be remembered that climate evolves as a continuum. The progression of changing atmospheric states and their related human responses must be kept in mind in the interpretation of the scenarios. The scenarios do not imply a temporal response arising from progressively increasing atmospheric carbon dioxide. The uncertainties involved in the understanding of geophysical processes and their socio-economic feedbacks do not permit such an extrapolation. As examples of climate scenarios that could prove useful to those involved in assessing the socio-economic impacts of potential climatic warming, we offer the following, which are based on historical and geological analogues:

SCENARIO NO. 1 - global temperature increase of about 1°C.
Historical analogue: recent anomalous years, and the medieval warm period (ca. 900-1050 AD).

Regional climatological parameters for impact studies - such as length of vegetation period, or rainfall in selected seasons - can be quantified, as a first approximation, by using selected (5-10) warm seasons of the instrumental period. These data can be made more or less extreme by employing an appropriate amplification factor. Reasonable estimates could also be given of the interannual variability, but here extensive studies on the time variations of this parameter and its relation to large-scale climatic variations are needed. Using the medieval warm period, the following climatic consequences are indicated:

1. Seasonal disappearance of sea-ice in East Greenland Current and retreat towards Spitzbergen.
2. Frequent warm/dry summers in central and western Europe.
3. Warming and increase of vegetation period of 1-2 weeks in high latitudes.
4. Decreasing rainfall in subtropical winter rain belt, higher frequency of droughts in these areas.
5. Decreasing probability of drought in dry areas of tropical summer rain belt.

SCENARIO NO. 2 - global temperature increase of about 1.5°C.
Historical analogue: Holocene warm period (ca. 6000 BP) <u>except</u> humid period in Sahara/ Middle East.

The following climatic consequences are indicated:

1. Warming in the Subarctic, retreat of Arctic sea-ice during warm season.
2. Frequent summer droughts in mid-latitudes between 45 and 55°N.

3. Frequent droughts in northern winter rain belt.
4. Relatively wet tropical summer rain belt.
5. General weakening of tropical Hadley circulation:more frequent equatorial down-welling instead of up-welling.

In the following two scenarios no regional data can be given now since the boundary conditions differed much more from the present situation. For this purpose model studies on the role of these changing boundary conditions are necessary, also on their role on climatic variability.

SCENARIO NO. 3 - global temperature increase of 2 to 2.5°C.
Historical analogue:last interglacial
(ca. 125,000 BP).

In many places the precipitation could be higher. From the vegetation distribution in the arctic and subarctic latitudes it must be concluded that during the last interglacial the Arctic sea-ice had retreated (seasonally) to its central core between Greenland, Canada and eastern Siberia. In high northern latitudes permafrost had probably retreated several 100 km.

In tropical and subtropical latitudes the climate during the last interglacial was probably quite similar to that of Scenario 2. The observed rise of the sea-level by 5-7 m may have been caused by a partial disintegration of the Western Antarctic Ice Sheet during the latter half of this period. No details of the physical mechanism, the place and the time-scale can be given. Much more research is needed to put this event from speculation into the light of geophysical knowledge.

At any rate, the sea penetrated at this time far into western Siberia. A connection between the Baltic and the White Sea made Finland and Scandinavia islands. Climatic conditions in eastern Europe and western Siberia were more maritime than now. It should be emphasized, however, that although these events seem to have been typical during the last interglacial, it is not certain that they all would recur with a future temperature rise of 2 to 2.5°C.

SCENARIO NO. 4 - global temperature increase of about 4°C.
Historical analogue:Early and Mid-Pliocene
(5-3 million years BP).

In this period the Arctic Ocean was ice-free, surrounded by boreal or cool temperate forests up to the vicinty of the coasts;no evidence of tundra or permafrost could be found. The possibility of a seasonally ice-free Arctic Ocean has been suggested also from two recent model computations:Manabe-Stouffer (1979) and Parkinson-Kellogg (1979);such a stage could be expected at least as a transitory stage (of unknown length).

Simultaneously the Antarctic ice sheet - which had covered the continent at least since 15 millions years BP - had the same (or an even higher) ice volume than now. The atmospheric-oceanic circulation must have been quite asymmetric, even more than now, when the annual average troposphere temperature difference between the North Pole and the South Pole exceeds 11°C. Scanty evidence for this exists;rather complete evidence is available for a substantial northward shift of the northern subtropical arid belt which had reached (shortly before this time) Latitude 48°N (near Vienna). The southern arid zone seems to have extended to the vicinity of the equator, while the tropical rain-forests of Africa extended from the equator far into the Sahara. A major difference of the boundary conditions from to-day is the rudimentary stage of uplift of most mountains - the Alps and the Tibetan highlands had reached only a small portion of their present height. Another major difference was the existence of a narrow ocean connection through Panama, which was closed by volcanic activity only near the end of this period.

In some areas of the northern hemisphere a temperature increase of about 6°C seems to have occurred, but the climatic evolution during Oligocene/Mid-Miocene (38-15 million years ago) is not yet known with sufficient accuracy to allow the outline of a 6° warming scenario based on paleoclimatic evidence.

WORKING GROUP III:

OBJECTIVES OF A CLIMATIC IMPACT STUDY PROGRAM

Chairman/Co-Chairman: W.O. Roberts/W.W. Kellogg

C. Baes, G. Breuer, J.A. Laurmann,
H. Meinl, J. Page, I. Smith

1. PREFACE

This report of Panel III consists of an Introduction, prepared before the Workshop by the Panel Chairmen, Walter Orr Roberts and William W. Kellogg, followed by the discussion and recommendation of the members of the Panel. In the short time available the Panel could obviously not consider all the facets of such a complex set of problems as those involved with impacts of climate change, but there was strong agreement on the importance of those recommendations that were generated.

Attention should be called to the fact that this is the subject area of the Climatic Impact Studies Program, a part of the World Climate Program, for which the World Meteorological Organization (WMO) and the UN Environment Programme (UNEP) share responsibility. The main features of this Program are outlined in the Proceedings of the World Climate Conference, organized by the WMO and held in Geneva in February, 1979.

2. INTRODUCTION

The expectation that further increases in the CO_2 concentration of the atmosphere will cause a complex climatic change raises a host of other problems that are even more complex. These have to do with the implications of economic and societal impacts of such a climatic change, and the responses of individuals, governments, and the community of nations to these impacts - or in anticipation of them.

While most long-range decisions concerning energy, agriculture, transportation, urban and land use planning, and so forth have so far not seriously taken long-term climatic change into account, this will probably change in the years ahead. It will change as the scientific community gradually arrives at a better consensus about what to expect if we continue on our present course. Furthermore, it will almost surely be recognized as an important factor in long-range planning when the climatic change itself becomes evident.

The problem of assessing the socioeconomic impacts of CO_2-induced change can be thought of as a hierarchy of problems, a hierarchy imposed by the cascade of cause-and-effect that starts with the burning of fossil fuel. Thus, in its broadest conception one must consider the alternative courses that mankind may adopt to satisfy the world's energy needs, the consequent climatic change that is expected, the many ways in which this climate change impacts important economic activities, and, finally, the changes in the fabric of society that may ensue. These are sketched in Figure 1. This cascade of cause-and-effect can be seen to close back on itself and the resulting impacts will therefore be influenced by the actions of people.

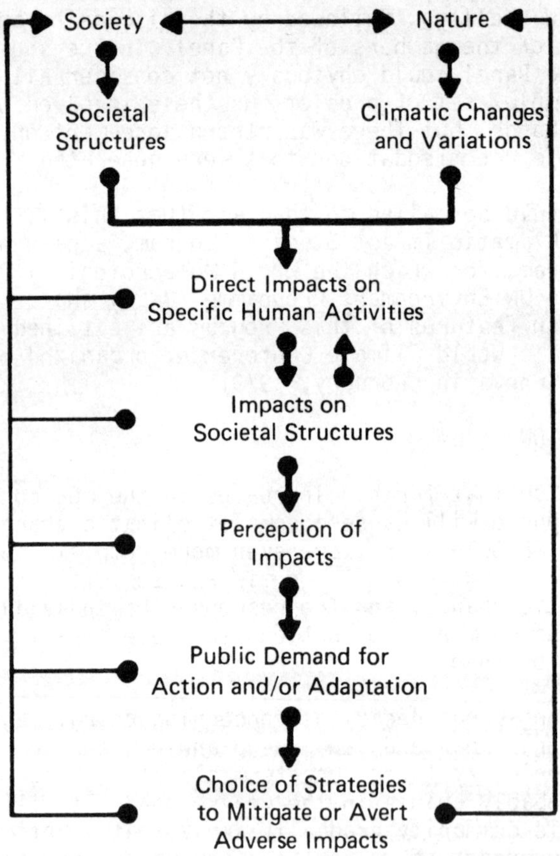

Figure 1. Components and linkages in Climatic Impact Studies

The development of the scenarios of likely climate change is thus an essential input to the research on economic and societal impacts, and there must therefore be continual guidance given to the geophysical research community in order that it produce the information needed for impact studies. (This is discussed in section 5).

The research needed for the last three steps is mainly in the areas of the social sciences and will inevitably involve interdisciplinary work as well. It is important to realize that the uncertainties in our understanding of climatic impacts will most likely increase as we move down the cascade of cause-and-effect from predictions of fossil fuel use, to development of scenarios, to foreseeing economic and social implications. This may be thought of figuratively as a "cascade of uncertainty", shown schematically in Figure 2. Here food and energy considerations are displayed as prime examples, but the same principles probably also apply to such activities as management of water resources, transportation, land use, and so forth. That is to say, uncertainty is likely to increase as we study the more and more indirect or secondary implications of climatic change, and as we move deeper into the possible development of international strategies to manage the global commons.

The following set of questions must now be asked: If there are so many uncertainties in our understanding of climatic impacts, can we make any useful predictions of their implications? Are there ways to build resilience to climatic changes, whether carbon dioxide-induced or natural? Will decision-makers be willing to rely on the results of this program? What incentives and disincentives exist for taking policy action on the carbon dioxide-climate issue? What is the long-term purpose of the research effort?

While there are no very clear answers to this set of questions, the following considerations may help to put any proposed carbon dioxide-climatic impact study program into better perspective. First, no predictions of the future can be made until the climatic change and its geophysical implications can be described in more detail and with more assurance than can be done now. Second, there appear to be a great many tools already at hand to assess the socioeconomic effects of a given climatic change on certain specific human activities. These tools and the methodologies involved need to be further developed and applied to a wider variety of activities, but the point is that they are to a certain extent at hand. Third, the public and the decision-makers need to be made aware of the alternatives open to them. Even though long-term predictions of the future behavior of mankind will probably be beyond our reach in the foreseeable future, impact studies can provide the information needed to comprehend

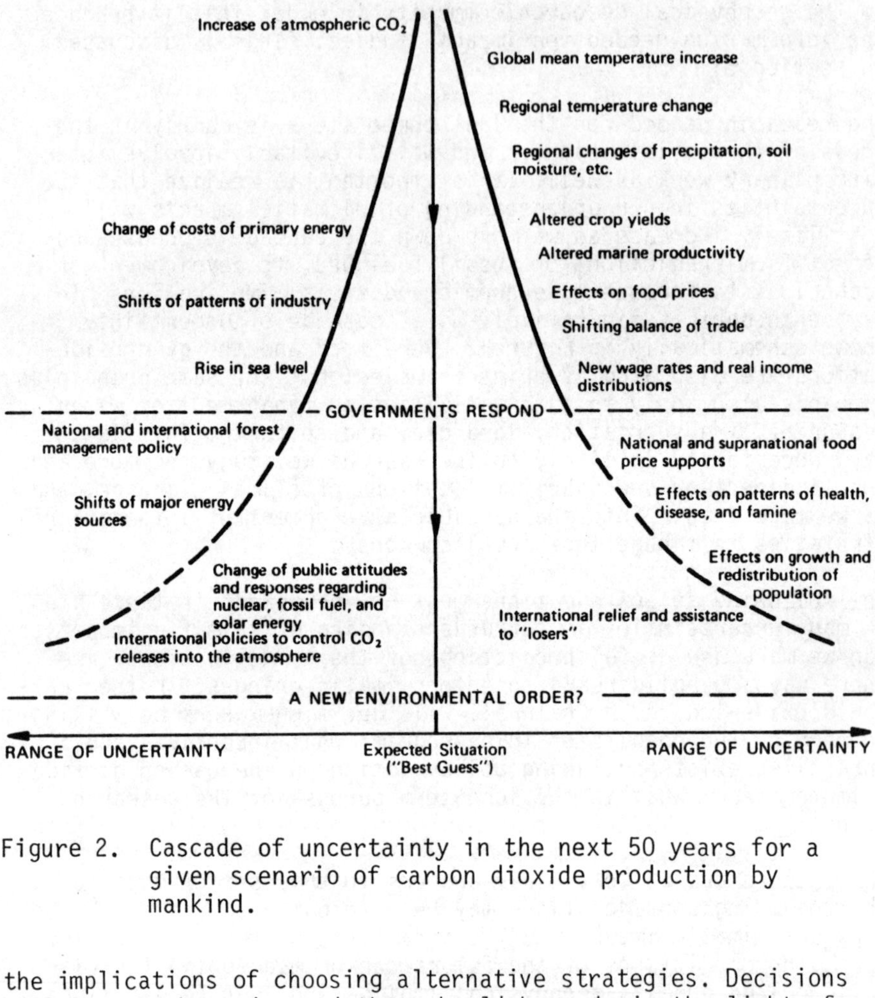

Figure 2. Cascade of uncertainty in the next 50 years for a given scenario of carbon dioxide production by mankind.

the implications of choosing alternative strategies. Decisions will have to be made, and they should be made in the light of the best perceptions of the likely consequences.

In Figure I the point was made that "perceptions of impacts" must precede any "choice of response" or decision. The provision of clearer and responsible perceptions of climatic impacts, then, can be considered as one of the major long-term goals of any program of research on impacts.

The ultimate objective of a sustained and continuing climatic impact assessment program is to provide policy-makers with timely information on the societal implications of climatic change induced by increasing carbon dioxide. It must therefore include the identification and evaluation of alternative national and international responses and strategies. This ultimate objective will shape the course of the entire program as it evolves through the years, and as the course of the climatic change and the options available become better defined. The course of research and its priorities can rarely be charted ahead of time, and the truly evolutionary character of the research plan must be built explicitly into the program itself.

3. SUGGESTED ATTRIBUTION OF RESPONSIBILITIES BETWEEN DISCIPLINES

It is important that appropriate techniques be developed to assess more accurately the impact of climatic variations and change. The hypothetical climate scenarios (see section 4) are deemed useful tools to assess the likely impact on various spheres of human activity. The rise in global temperatures and the associated changes in rainfall and soil moisture will not impact economics in the same way in different localities, and in the end some impacts may be beneficial. For example, space heating demands at middle and high latitudes in winter could decline significantly with the rise of temperature, whereas space cooling demands in summer would increase. It will thus be necessary to clearly identify the benefits and disbenefits of climatic warming in different regions and to weigh them against each other.

Taking account of the interactive nature of human activity at both the national and international levels, advanced economic analytical techniques such as cost-benefit analysis will have to be developed in modes that are suitable to the various climatic scenarios and the long time scale of the change in order to assess the consequent chain of probable impacts and the associated social consequences.

It should not be the task of climatologists to develop the necessary economic modelling techniques, but rather the task of social scientists. Sufficient research funds need to be allocated in this field, as well as in the climatological modelling field. Since the climatic change could lead to significant relocations of economic activity and consequently to shifts of presently accepted relationships between nations, international collaboration in this socio-economic study of the effects of climatic change is surely to prove essential.

A feedback cycle will need to be set up in such socio-economic analyses between the climatologists producing various future scenarios and the social scientists carrying out the socio-economic studies of the responses to these scenarios. The scenarios will continually need to be modified or made more detailed in response to the socio-economic analyses.

Climatologists however should clearly state all the environmental factors that are relevant to climate model building and experiments with those models, and should critically review all data input to models for internal consistency before utilizing it in their model building. Climatologists also should carefully examine environmental influences of various existing and proposed forms of human activity, over and above the production of CO_2, and should clearly state their reasons for viewing some types of economic activity as environmentally more adverse or more benign than others. They, together with economists, should give detailed attention to the relative merits of different approaches to energy and materials utilization with the objective of giving clear scientific guidance on the climatologically preferred socio-economic path to follow.

The fastest progress towards the reduction of adverse climatic impacts on economic activity is likely to come from the adoption of strategies or policies that not only reduce vulnerability to climatic variability and change but also confer direct economic benefits. Considerations of CO_2-induced climatic change will necessarily involve long-term planning and consequences that are hard to assess at the time the decisions must be made. Conservation of energy and materials is likely to offer the most immediate benefits. The ultimate advantages of developing renewable energy sources must be recognized, but the immediate economic advantages are likely to be less favorable than conservation.

The panel recommends that each country establish a group of experts in climate change, capable of applying the various existing techniques for the assessment of the probable impacts of alternative climatic scenarios on future human activity. Furthermore, efforts should be made to develop, preferably in international collaboration, new and improved techniques for climatic socio-economic assessment to enable the interactions between human activity and climatic change to be more accurately assessed in the future than is possible at the current relatively rudimentary state of knowledge in this area.

4. HYPOTHETICAL SCENARIOS FOR CLIMATE IMPACT ASSESSMENT

The Working Group recommends the commissioning of a small group of carefully designed hypothetical global Climate Scenarios suitable for use as research tools for assessment of economic, social, political and ethical impacts. The following features should guide the formulation of the scenarios:

a. It should be made clear that the scenarios are not intended as forecasts, but that they have been designed to describe plausible and physically consistent future possibilities.

b. Several scenarios should be designed, and the following are suggested:

   i. Moderate average warming compared to to-day

   ii. Extreme average warming

   iii. Moderate cooling.

c. All scenarios should have the mean values and their variability explicitly specified on a regional basis and as a function of season (monthly or seasonal). The weather variables considered should include temperature, rainfall, soil moisture and perhaps other parameters for each location.

d. The bases for the scenarios could be both climate models and observations of past climates, with plausible data filled in for missing regions. Examples of possible bases are:

   i. Holocene Altithermal or Hypsithermal (about 4500-8000 years ago, the warmest period since the last ice age).

   ii. The Eemian-Sangamon period (warm interglacial prior to last ice age).

   iii. The ten warmest years with good historical data.

   iv. Scenarios based on coupled ocean-atmosphere general circulation models, especially those that include the hydrologic cycle and that therefore give precipitation and soil moisture.

5. CONCLUSIONS AND RECOMMENDATIONS

As stated in the Introduction, the decisions that will have to be made in the decades ahead to prepare for or avert a carbon dioxide-induced climatic change will have to be made before all the answers have been obtained. Assessing the impacts of climatic changes is an assessment under uncertainties but nevertheless this assessment has to be started now, because we cannot wait until all variables and processes involved are quantified to satisfy all the parties involved.

The Panel makes the following recommendations to all nations with existing or potential climate programs, and to the international community that has already adopted the World Climate Programme:

a. Some methods of mitigating the CO_2 problem also have other beneficial effects on society, and can therefore be more easily justified than, for example, trying to prevent a further increase in coal consumption or turning to extensive use of nuclear energy to reduce the carbon dioxide build-up in the atmosphere. Some examples of such measures are:

i. Stopping deforestation and furthering reforestation, which would be advantageous for many reasons. Agricultural production would be increased in the long run by reducing the cultivated area and planting trees on mountain slopes, creating wind break forests, etc. (thus improving the water balance, preventing land slides, avalanches, erosion etc.). Reforestation cannot by itself solve the CO_2 problem, but a reasonable forest policy shifting the balance from deforestation to reforestation can reduce the size of the CO_2 problem and bring multiple ecologic and economic benefits as well.

ii. Soil conservation is important for agriculture and prevents oxydation of soil humus, which adds to CO_2. Soil conservation would also reduce the amount of aerosols from windblown soil.

iii. Conservation of energy realised by adequate insulation, energy storage, use of waste heat and so forth tends to reduce fossil fuels that are valuable resources for future generations, resources that will be required by mankind for a wide variety of non-energy uses.

iv. Use of renewable energy forms in developing countries as a substitute for wood or dung reduces the release of CO_2 from deforestation and releases dung for use as a fertilizer.

b. In assessing the risks of CO_2 build-up it should be borne in mind that non-fossil alternatives, especially a plutonium economy, have their risks, too, which are subject to considerable debate. While climatologists are experts in assessing climatological risks of CO_2 buildup, they are not qualified to compare the weight of these risks with the weight of nuclear risks which may or may not be considered as being "worse" - a judgement that depends on the scales of social and economic values used in the comparison.

c. Energy forecasts are continually coming down. In low scenarios the CO_2 problem becomes more tractable. While it does not make much difference whether CO_2 will increase seven times or eight times it makes a difference whether it goes up 80% or 100%. In low energy use scenarios marginal measures for reducing CO_2 only a small fraction may be valuable.

d. The disposal or sequestering of carbon dioxide from the use of fossil fuels, even under the most favorable circumstances, is judged to be economically a most unattractive option. Therefore it should be considered as virtually the last resort as a preventive or remedial measure in dealing with the impact of increasing atmospheric carbon dioxide, should that increase be considered unacceptable.

e. Comprehensive global socio-economic impact assessments present considerable challenges in testing the capabilities of the social sciences disciplines. Nevertheless, two extreme situations may be within the reach of analysis. In one the perturbation of the socio-economic system is small, and a complete, weakly interactive model, in quasi-equilibrium may be viable. By contrast, it may be possible to define a minimum value of CO_2-induced climatic change that can be judged to result in a fundamental (and thus large) disruption of the social, economic, or political structure, even though the eventual outcome of the event cannot be determined. It would be useful if such a critical level could be determined ahead of time. Such assessments could help in the definition of those climatic changes which would be "acceptable" and those that should be avoided if at all possible.

It is therefore recommended that an attempt be made to define (a) the limiting magnitude of climatic change, the socio-economic consequences of which are capable of analysis with present understanding of the field, and (b) the minimum climatic change that can clearly be seen to result in major economic or social disruption. In such analyses it is likely that rates of change of physical environmental conditions will be as crucial as their final mean states, and to the extent possible, the other changing factors in our social system, such as population, availability of food and land, changing balances of trade etc. should be taken into account.

LIST OF AUTHORS AND PARTICIPANTS

Albanese, A.
Process Sciences Division
Department of Energy and
Environment, Brookhaven
National Laboratory
Upton, New York 11973, USA.

Allsopp, T.
Canadian Climate Center (CCC)
4905 Dufferin Street
Downsview, Toronto
Ont. M3H 5T4, Canada.

Bach, W.
Forschungsstelle für Angewandte
Klimatologie und Umweltstudien
Westfälische Wilhelms-Universität
Robert-Koch-Straße 26
4400 Münster, Germany.

Baes, C.
Chemistry Division
Oak Ridge National Laboratory
Oak Ridge, TN. 37830, USA.

Bartholomäi, G.
Gesellschaft für Umweltüber-
wachung, Industriestraße
5173 Aldenhoren, Germany.

Baumgartner, A.
Lehrstuhl für Bioklimatologie
und Angewandte Meteorologie der
Universität München
Amalienstraße 52/111
8000 München 40, Germany.

Beall, S.
Chemistry Division
Oak Ridge National Laboratory
Oak Ridge, TN 37830, USA.

Bicsan, P.
Physikalisches Institut
Westfälische Wilhelms-Universität
Robert-Koch-Straße 26
4400 Münster, Germany.

Black S.
International Atomic Energy Agency
Kärtner Ring 11, P.O. Box 590
A-1011 Vienna, Austria.

Breuer, G.
Birnbaumgasse 4/3
A-1100 Wien
Austria.

Bröhl, H.
Institut für Physikalische Chemie
der Universität Frankfurt/M.
Feldbergstraße 47
6000 Frankfurt/M., Germany.

Chervin, R.
National Center for Atmospheric
Research
P.O. Box 3000,
Boulder, Col. 80307, USA.

Degens, E.
Geolog.-Paläontologisches
Institut der Universität Hamburg
Bundesstraße 55, Germany.

Döbritz, W.
Institut für Geographie
Westfälische Wilhelms-Universität
Robert-Koch-Straße 26
4400 Münster, Germany.

Ebel, A.
Meteorologisches Institut
Universität zu Köln
Albertus Magnus Platz
5000 Köln 41, Germany.

Egberts, G.
Systemforschung und Technologie
der Kernforschungsanlage GmbH
Postfach 1913, 5170 Jülich
Germany.

Ehhalt, D.
Institut für Chemie 3
Kernforschungsanlage Jülich
Postfach 1913, 5170 Jülich
Germany.

Ellis, J.
University of California
Lawrence Livermore Laboratory
P.O. Box 808, Livermore
California 94550, USA.

Ellsaesser, H.
University of California
Lawrence Livermore Laboratory
P.O. Box 808, Livermore
California 94550, USA.

Fantechi, R.
CEC Directorate General XII
Rue de la Loi 86
1046 Brussels, Belgium.

Fischbach, U.
Institut für Physikalische
Chemie der Universität Frankfurt/M.
Feldbergstraße 47
6000 Frankfurt/M., Germany.

Flohn, H.
Meteorologisches Institut der
Universität Bonn
Auf dem Hügel 20
5300 Bonn 1, Germany.

Fortak, H.
Institut für Geophys. Wissen-
schaften der Freien Universität
Berlin
Thielallee 49,
1000 Berlin 33, Germany.

Freise, W.
Mathematisches Institut
Westfälische Wilhelms-Universität
4400 Münster, Germany.

Gates, W. L.
Department of Atmospheric
Sciences
Oregon State University
Corvallis, Oregon 97331, USA.

Ghazi, A.
Institut für Biophysik und
Meteorologie der Universität Köln
Zülpicher Straße 49
5000 Köln - Lindenthal 41,
Germany.

Hampicke, U.
Arbeitsgruppe Umwelt, Gesellschaft
Energie (A.U.G.E.)
Universität Essen
Universitätsstraße 12
4300 Essen, Germany.

Heimann, M.
Physikalisches Institut
Universität Bern
CH-3006 Bern, Switzerland.

Hekestra, G.
Ministry of Public Health and
Environmental Protection
10-12 Doleter Reyersstraat
Leidschendam, Holland.

Hempelmann, R.
Universität Karlsruhe
Hertzstraße 16
7500 Karlsruhe 1, Germany.

Hermanns, P.
Bayerischer Rundfunk
Baaderplatz 1,
8000 München-5, Germany.

Hofmann, J.
Battelle Institut e.V.
Am Römerhof 35,
6000 Frankfurt/M., Germany.

Hüster, H.
Meteorologisches Institut
Universität Bonn
Auf dem Hügel 20
5300 Bonn 1, Germany.

Günter, J.
Deutscher Wetterdienst Zentralamt
Frankfurter Straße 135
6050 Offenbach/M., Germany.

Isaksen, I.
Institute of Geophysics
University of Oslo
P.O. Box 1022, Blindern
Oslo 3, Norway.

Keeling, C. D.
Scripps Institution of Oceanography
University of California
San Diego, California 92037, USA.

Kellogg, W. W.
National Center for Atmospheric
Research
P.O. Box 3000
Boulder, Colorado 80307, USA.

Kirchner, M.
Lehrstuhl für Bioklimatologie und
Angewandte Meteorologie der
Universität München
Amalienstraße 52/111
8000 München 40, Germany.

Köhnlein, W.
Institut für Strahlenbiologie
Westfälische Wilhelms-Universität
Mittorf Straße 17
4400 Münster, Germany.

Kratz, G.
Physikalisch-Chemisches Institut
Universität Frankfurt
Robert-Mayer-Straße 11
D-6000 Frankfurt, Germany.

Krause, F.
ÖKO-Institut
Freiburg/Breisgau
Germany.

Kress, B.
Institut für Markt und Planung
Universität Frankfurt/M.
6000 Frankfurt/M., Germany.

Laurmann, J.
Division of Applied Mechanics
School of Engineering
Stanford University
Stanford, Cal. 94305, USA.

Lee, D.
Chemistry Division
Oak Ridge National Laboratory
Oak Ridge, TN. 37830, USA.

Lieth, H.
Lehrstuhl für Ökologie
Universität Osnabrück
Postfach 4969
4500 Osnabrück, Germany.

Lovins, A. and H.
International Project for
Soft Energy Paths
124 Spear Street
San Francisco, CA. 94105, USA.

Luther, F.
University of California
Lawrence Livermore Laboratory
P.O. Box 808, Livermore
California 94550, USA.

MacCracken, M.
University of California
Lawrence Livermore Laboratory
P.O. Box, 808, Livermore
California 94550, USA.

Manshard, W.
The United Nations University
29th Floor, Toho Seimei Building
15-1 Shibuya 2-chome, Shibuya-ku
Tokyo 150, Japan.

Marland, G.
Institute for Energy Analysis
Oak Ridge Associated Universities
Oak Ridge, TN. 37830, USA.

McKay, G.
Canadian Climate Center (CCC)
4905 Dufferin Street, Downsview
Toronto, Ont. M3H 5T4, Canada.

Meinl, H.
Dornier System GmbH
7990 Friedrichshafen,
Germany.

Moriarty, J.
Federal Energy Regulatory
Commission
U.S. Department of Energy
Washington D.C. 20426, USA.

Müller-Warmuth, W.
Westfälische Wilhelms-Universität
Schlossplatz 2,
4400 Münster, Germany.

Niehaus, F.
International Atomic Energy
Agency
Kärtner Ring 11, P.O. Box 590
A-1011 Vienna, Austria.

Oeschger, H.
Physikalisches Institut
Universität Bern
CH-3006 Bern, Switzerland.

Oke, T.
Department of Geography
The University of British Columbia
Vancouver, B.C., Canada V6T 1W5.

Overdieck, D.
Universität Osnabrück
Albrechstraße 16
45 Osnabrück, Germany.

Page, J.
Department of Building Sciences
Sheffield University
Sheffield S10 2TN, England.

Pankrath, J.
Umweltbundesamt
1000 Berlin 33, Berlin, Germany.

Pestiaux, P.
Institut d'Astronomie et de
Géophysique George Lemaitre
Chemin du Cyclotron 2
B-1348 Louvain-la-Neuve,
Belgique.

Potter, G.
University of California
Lawrence Livermore Laboratory
P.O. Box 808, Livermore
Cal. 94550, USA.

Quirk, W.
Atmospheric and Geophysical
Sciences Division
University of California
Lawrence Livermore Laboratory
P.O. Box 808, Livermore
Cal. 94550, USA.

Ramanathan, V.
National Center for
Atmospheric Research
P.O. Box 3000
Boulder, Col. 80307, USA.

Roberts, W. O.
Aspen Institute for Humanistic
Studies
Boulder, Colorado 80302, USA.

Rogner, H.
International Institute for
Applied Systems Analysis
A-2361 Laxenburg, Austria.

Rotty, R.
Institute for Energy Analysis
Oak Ridge Associated Universities
Oak Ridge, TN. 37830, USA.

Sassin, W.
International Institute for
Applied Systems Analysis
A-2361 Laxenburg, Austria.

Schaefer, H.
Institut für Energietechnik,
Wirtschaftsenergie und
Kraftwerkstechnik
TU München, 8 München, Germany.

Schleser, G.
Systemforschung und Technologie
der Kernforschungsanlage GmbH
Postfach 1913
5170 Jülich, Germany.

Schmidt, W.
Systemforschung und Technologie
der Kernforschungsanlage GmbH
Postfach 1913
5170 Jülich, Germany.

Schnatz, G.
Hauptabteilung Physik
Battelle-Institut
Am Römerhof 35, Postfach 900 160
6000 Frankfurt 90, Germany.

Schunck, W.
Institut für Physikalische Chemie
der Universität Frankfurt/M.
Feldbergstraße 47
6000 Frankfurt/M., Germany.

Seeliger, J.
Gesamtverband des deutschen
Steinkohlenbergbaus
Abteilung Recht und Umweltschutz
Friedrichstraße 1
4300 Essen 1, Germany.

Siegenthaler, U.
Physikalisches Institut
Universität Bern
CH-3006, Bern, Switzerland.

Slade, D.
Office of Health and Environmental
Research
Department of Energy
Washington D.C. 20545, USA.

Smith, I.
IEA Coal Research
14-15 Lower Grosvenor Place
London SW1W OEX, England.

Steinberg, M.
Department of Energy and
Environment
Brookhaven National Laboratory
Upton, New York 11973, USA.

Tetzlaff, G.
Institut für Meteorologie und
Klimatologie der TU
Herrenhäuser Straße 2
3000 Hannover-Herrenhausen
Germany.

Voss, A.
Systemforschung und Technologie
der Kernforschungsanlage GmbH
Postfach 1913
5170 Jülich, Germany.

Wallen, C.
Global Environmental Monitoring
System, UNEP
P.O. Box 30552
Nairobi, Kenya.

Waltz, T. W.
U.S. National Climate
Program Office
6010 Executive Blvd.
Rockville, MD 20852, USA.

Wege, K.
Deutscher Wetterdienst Zentralamt
Frankfurter Straße 135
6050 Offenbach/M., Germany.

Weischet, W.
Geographisches Institut I
Albert-Ludwigs-Universität
Werderring 4
78 Freiburg, Germany.

Werner, J.
Institut für Geographie
Westfälische Wilhelms-Universität
4400 Münster, Germany.

Williams, J.
Reinhold-Schneider-Straße 6
7500 Karlsruhe-Rüppurr
Germany.

Zimen, K.
Hahn-Meitner-Institut für
Kernforschung
Glienicker Straße 100
1000 Berlin 39, Germany.

Zimmermeyer, G.
Gesamtverband des deutschen
Steinhohlenbergbaus
Friedrichstraße 1
43 Essen 1, Germany.

PART I

GLOBAL ENERGY DEMAND AND SUPPLY – PRESENT AND FUTURE

ECONOMICALLY EFFICIENT ENERGY FUTURES

Amory B. Lovins

Consultant physicist and British Representative,
Friends of the Earth, Inc., 124 Spear St.,
San Francisco, CA 94105, USA

ABSTRACT. The integrated burn of fossil fuel, and the associated risk of global climatic change, can be minimized by economically efficient energy policies based on very efficient energy use and rapid deployment of appropriate renewable energy sources. Such policies can stabilize the rate of burning fossil fuel and gradually, over a half-century or so, reduce it to approximately zero. Economically and technically sophisticated recent studies in many industrialized countries have shown that it is cheaper, faster, and easier to increase national energy productivity by severalfold than to increase energy supply. If such studies are taken as an existence proof, a worldwide Western European material standard of living for 8×10^9 people could be maintained with today's rate of world energy use (~ 8 TW) or less, even with unchanged life-styles in the developed countries and complete industrialization of the developing countries. At these cost-effective levels of energy productivity, virtually all long-term energy needs can be met by appropriate renewable sources that are already available and that are significantly cheaper, faster, and otherwise more attractive than competing power stations and synthetic-fuel plants. Only major efficiency improvements and, secondarily, appropriate renewable sources can substantially change the timing of, or reduce the risk of, CO_2 problems.

1. INTRODUCTION

Until the mid-1970s, energy demand in most countries was apparently forecast by armies of chimpanzees armed with semilogarithmic graph-paper and rulers. Today these have been largely succeeded by teams of systems analysts armed with computer models whose elaborate

W. Bach, J. Pankrath, and J. Williams (eds.), Interactions of Energy and Climate, 1–31.

opacity generally conceals primitive methodology and unrealistic assumptions. These extrapolative models (*1*) tend to predict large and rapidly rising demands for energy, especially in premium forms, hence a need to build huge new energy supply systems with increasingly nasty problems of many kinds, including risks of climatic change.

Before devoting much effort to studying these risks, we should carefully explore whether we need to incur them. How much energy *will* we need? In what forms? At what scales? and in consequence, Where from? How fast? Serious attention to these prior fundamentals reveals a very different result (and a realm of what Pogo called "insurmountable opportunities").

Greatly increased affluence, which this analysis assumes for all countries, does not require proportionate increases in energy use, because energy productivity--the amount of work wrung from each unit of energy--can increase even faster, at far lower cost than new energy supply. The next best buy is generally the appropriate renewable energy sources, which appear ample for virtually all long-term needs without any technical breakthroughs. Despite increasing wealth, then, the worldwide rate of burning fossil fuels can stay roughly constant in the short term and decline gradually, over about 50-75 years, to about zero, thus minimizing climatic risks of burning those fuels.

Such policies are both simpler to understand and technically more sophisticated than their predecessors. They can be relatively free of climatic risks in themselves, and at the same time offer faster displacement of oil and gas, longer-term sustainability, lower economic and political costs, lower risks of technical failure and of the proliferation of nuclear weapons, more tractable safety and environmental problems, and more favourable social side-effects than traditional, less economically efficient policies. These conclusions depend very little on geography, climate, culture, political system, level of economic activity or technical development, and density of population or industry.

2. HOW MUCH ENERGY?

The forecasts of growing future energy demand on which governments and international organizations have based their expansionist policies arise from the failure of nearly all forecasting models to account properly for ten important effects:

- *Price elasticity of demand* (*i.e.*, the more you charge people the less they buy). Many analyses ignore or minimize long-term adaptation to higher price. The IIASA projections, for example, have a very low implicit price elasticity of demand: that is, they assume that the very high energy prices needed to pay for the projected supply system will not lead energy users to substitute brains, capital, and other resources for much of their energy.

- *Inverse price elasticity of demand.* Many studies project from past energy demand in periods when real energy prices fell, as petrol prices did in the UK (by 14%) and the Netherlands (by 7%) during 1970-9, or as average US residential electricity prices did (by ∿80%) during 1940-70.

- *Subsidies.* Nearly all studies assume continued massive (and unevenly distributed) subsidies to make energy, especially additional energy, look cheaper to consumers than it really is to the society. The size of the historic and current subsidies via both tax and pricing policy is unknown except in the US, where subsidies on current account alone total over $\$1 \times 10^{11}$/y (*2*)--enough to reduce average energy price by over a third and nuclear electricity price by over half (*3*).

- *Saturations* may physically constrain growth in many traditional energy markets. Official UK projections have until recently assumed that space-heating demand is an unlimited linear function of income, but the rich may not wish to roast. Car traffic is ultimately limited by one's unwillingness to sit in a car all day and inability to drive more than one car at the same time--though some traffic forecasts, forgetting this, have implied old-age pensioners driving about for many hours each day. (Economic growth itself raises such problems--does the trebled GNP assumed below mean two cars, a boat, and a helicopter in every garage?--but this paper, for the sake of argument, tactfully ignores economic saturations.)

- *Promotion.* In most countries, rapid demand growth, especially in such fuel-intensive sectors as electric-resistance space heating, has been heavily promoted by advertising, concessionary tariffs, etc. which are now often being removed or reversed.

- *Distortions in basic data.* Many studies rely on outdated and excessive estimates of population, labour force, labour productivity, etc. Some countries have unique distortions: *e.g.*, the US annual energy growth rate in the 1960s was inflated 1-2 percentage points by the Vietnam war, leading to an enormous error if extrapolated for several more decades.

- *Structural changes.* Worldwide changes in terms of trade may shift production. In the Federal Republic of Germany (FRG), projected reductions in industrial energy use over the next 50 y are nearly as large from officially anticipated shifts in composition as from decreases in intensity per sector, but only part of the latter and none of the former is reflected in the government's energy forecasts (*4*). Further distortions are introduced by treating gross economic activity as if it measured welfare: in fact it omits important nonmarket economic activities but includes remedial and transaction costs, so that increasing GNP may for all we know *decrease* net welfare. To avoid these pitfalls, energy intensity must be measured by "technical coefficients"--energy used per *physical* unit of specific, tangible product or activity.

- *Primary energy projections* cause increasingly serious errors because, with the commonly assumed shift towards electrification and synthetic fuels, more than half of the officially projected growth in primary energy supply in the next few decades (*e.g.*, in France, FRG, UK, and US) will go to conversion and distribution losses, thus masking much slower growth in final consumption or "end-use energy". Forecasting should instead be done in end-use terms and primary energy treated as derivative.

- *Excessive aggregation* conceals crucial engineering details of potential savings. Because energy is used in milliards of devices of thousands of types, the efficiency improvements possible in each type are individually small on a national scale, but can add up rapidly. Large total savings are generally missed through failure to disaggregate demand sufficiently to detect most of the terms.

- *Technological progress*. Few forecasters, even in technically sophisticated institutes, appear to be aware of the extraordinarily rapid progress made in recent years, mainly in the private sector, in doing more work with less energy. (Such progress can be expected to yield further major improvements in the future, but as a conservatism, this paper will assume only present technologies that are already commercially available and technically mature.)

This paper reviews, with reference to fuller accounts cited elsewhere (*5,6*), recent efforts in many countries to overcome these problems through more sophisticated analyses (*7*) that are more disaggregated and detailed, take fuller account of the best state of the art, pay more attention to physics and engineering detail, and apply more rigorous economic tests than forecasts in the customary literature. These studies, mainly post-1978,

- are notable for their physical concreteness: they construct end-use needs "from the bottom up" and require the analyst to specify exactly what the energy will be used for;
- require only a hand calculator to compute;
- have the virtues of being simple, transparent, scrutable, and ratproof;
- are economically conservative, basing their decision rules not on political ideologies or thermodynamic idealism but on the orthodox market criterion of minimizing private internal cost;
- assume a greatly increased level of material affluence--a level unprecedented in the history of the world--rather than the agrarian utopia, primitive hardship, or asceticism often ascribed to such analyses by critics who have not read them;
- nonetheless show how to achieve a straightforward, economically attractive improvement in primary energy efficiency of *a factor of 3 to 6 or more* in some of the most advanced countries, at a more or less linear rate over 50-odd years; and
- achieve this entirely through "technical fixes": well-known, currently available technical measures that are now economic (at least against marginal-cost prices) and have no significant effect

on lifestyles. Technical fixes use energy more efficiently rather than curtailing the services it provides. The two ways of saving energy are quite different--insulating your roof does not mean freezing in the dark--though they are often confused.

These analyses of what I shall call "efficiency scenarios"--efficient in the use of both energy and money--do not imply low standards of living nor even a decrease in their rate of increase: the studies usually assume on the contrary that energy efficiency gains will be used mainly to fuel rapid, traditionally industrialized economic growth. They range in detail from mere sketches to perhaps the most precise and persuasive analyses published anywhere. Of the 23 efficiency scenarios presented by 32 analysts from 15 countries at a May 1979 conference (*8*), only the most detailed will be discussed here; some others are reviewed elsewhere (*5,6,9*). Yet *even one* such study, for a diverse, heavily industrialized economy in a rigorous climate (Northern Europe), serves as an existence proof--an *a fortiori* argument--for other countries. And the studies cited here, though often of remarkably high technical quality, are not the last word: they were generally done by individuals with little or no official help (some with official hindrance), so it may be safely assumed that with more time, money, and research facilities, still further savings could be discovered.

The first really detailed efficiency scenario, and still among the best, was sponsored during 1977-8 by the Ford Foundation and published in January 1979 by G. Leach *et al.* (*10*). It considered, often in greater detail than any prior official study, the systematic use of technical fixes, generally cheaper than recently delivered North Sea gas ($\sim$\$3.5/GJ), in over 400 sectors of the British economy. The resulting savings, singly often unimportant, collectively sufficed to increase UK primary energy efficiency (*11*) threefold: during 1976-2025, real GDP could increase 3.1×;* industrial production 2.2×, car travel 1.8×, and air travel 3×, while total primary energy use declined to 0.93×. Exhaustive review of these findings by British energy agencies and industries and by international bodies (including IIASA) failed to refute them: the study was, as claimed, technically conservative, often more so than UK government forecasts, and applied the weak economic test of near-present prices. Its aim, like that of the other studies cited here, was to say what could, not what would, happen: to be explicitly, not tacitly, normative. It began with a target date--2025, roughly the retirement date of a major energy facility ordered today, and far enough in the future that major capital stocks could be largely modified, replaced, or diluted--and then worked backwards to see what must be done when.

At first privately and then under UK government auspices, D. Olivier has carefully enlarged upon Leach's pioneering work (*12*). Olivier intended originally to see how far Leach's trebled efficiency could be increased if technical fixes were assumed which might not be

*"increase 3.1×" means "increase to initial value ×3.1", not ×4.1.

cheaper than today's North Sea gas but were at least cheaper than officially proposed long-run replacements for it, such as synthetic gas made from coal, or coal-fired or nuclear power plants (assuming prices for these well below official estimates). Though Olivier's assumptions about what can be done when are often less conservative than Leach's, they are defensible in detail as realistic and not pressing any technical or political limits. Olivier's explicit list of technical fixes suffices to double Leach's endpoint efficiency--*i.e.*, to improve present UK primary energy efficiency by sixfold, permitting real GNP to rise 2.9× and industrial production by 2.3× while total primary energy use decreases to 0.42× its 1976 value. He then found to his chagrin that the main energy-saving measures he had assumed would *still* compete with present North Sea gas. The saving achievable at prices below the long-run marginal cost of supply--the correct economic test for present investments--is thus larger than sixfold in Britain: according to preliminary estimates, probably nearer eightfold without nearly exhausting technical options.

F. Krause has prepared a similar analysis (*13*) for the FRG, relying mainly on official sources not previously assembled. He finds an economically attractive scope for technical fixes that improve FRG primary energy efficiency 3.3×: real GDP increases 2.3× (3.2× per capita) by 2030 and industrial production 1.7× (2.3×), but the 3.85× total saving (including that from structural shifts in industry(*4*)) decreases total primary energy use to 0.60× its 1973 level. The degree of technical conservatism is nearer Leach's than Olivier's.

J. Nørgård *et al.*, in the latest of a series of increasingly detailed studies (including a monumental survey of the dominant household sector (*14*)), have shown (*15*) how to save money by more than doubling Danish primary energy efficiency. An equally conservative study (*16*) showed how to decrease 1990 Swedish primary energy use to 0.87× its 1976 level despite increases of 1.45× in GNP and 1.51× in industrial production: a 40% drop in E/GNP (*11*) in 14 years.

A distinguished panel of the US National Academy of Sciences CONAES study has published US projections for 2010, assuming doubled real GNP. Despite the strong technical and economic conservatisms (*17*) required for consensus among a diverse group, they projected (*18*) "pure technical fix" changes of E/GNP (*11*) to 0.49-0.62× the 1978 level, depending on the vigour of policy actions, or 0.40× with modest changes in lifestyle (also, in the view of many members, achievable without those changes but with more realistic technical assumptions). Leach's technical coefficients are generally comparable in conservatism with the highest of these three US scenarios (*19*).

A study for the US Department of Energy (*20*) has found that with similarly conservative technical assumptions and no changes in lifestyle or values, California could treble its economic activity by 2025 with only a 20% increase in end-use energy. Making the technical assumptions more realistic and internally consistent (*21*) could readily turn this into a 22% decrease (a 3.85× E/GNP (*11*) gain).

Total end-use efficiency improvements of 2.53× for Canada derived (*22*) by three government analysts confirm a far more aggregated analysis done earlier for the Science Council of Canada (*23*). Both are technically very conservative.

Studies in France, Switzerland, and elsewhere (*5,6*), and more detailed continuations now in progress, have broadly confirmed the severalfold efficiency improvements found in the countries cited.

These efficiency scenarios, whether or not they are highly disaggregated, pay attention, not lip service, to the goal of supplying energy services at lowest cost. For example, an authoritative analysis (*24*) has shown that buying the cheapest United States energy system during the decade prior to 1978, in the light of actual 1978 prices, would have reduced 1978 use of oil by ∿28% (halving imports), reduced 1978 use of grid electricity by ∿43% (so that over a third of today's power stations would never have been built), and reduced 1978 payments for energy services actually received by ∿17%.

The efficiency scenarios also apply best present technology. For example, it is now straightforward to build a house so heat-tight that it requires little or no net space heating even in a frigid climate, at an extra capital cost ranging from a few thousand dollars (payback in a few years) to about zero (*25*). Most existing buildings can be retrofitted cost-effectively to very high standards (*10,14, 26*). A big, comfortable car using diesel-electric hybrid drive can readily average < 3 ℓ/100 km (> 78 mi/USgal), or 3× typical European efficiencies of today (*27*). The new generation of jet aircraft now being flight-tested is twice as fuel-efficient as the present fleet. Properly sizing and coupling industrial electric motors roughly doubles their practical efficiency (*28*). Properly designing household electric appliances roughly quadruples their average efficiency (*29*). Adding up such well-documented improvements throughout the economy yields severalfold total savings without great effort.

Gradual penetration of these insights into official planning circles has led to a remarkably consistent downward trend in most countries' energy demand forecasts. The following US example is a diagonal matrix with a two-year time constant for transitions towards the lower right in a direction of increasing sociological respectability. It accurately predicts several 1980 forecasts now in draft form. It has nowhere near hit bottom yet: for example, it does not reflect the recent finding (*30*) that an energy-conscious materials policy should roughly treble US energy efficiency. Olivier-style efficiency levels would lead, in a few more rows of the matrix, to long-run (post-2050) US values around 10-15 q/y--4-6× below present US energy use--even if real GNP approximately doubles.

Evolution of Approximate Forecasts of US Primary Energy Demand in the Year 2000 (q/y ≡ 10^{15} BTU/y ∿ EJ/y) (1978 rate ∿ 78 q/y)

Year of Forecast	Source of Forecast: Beyond the Pale	Heresy	Conventional Wisdom	Superstition
1972	125[a]	140[b]	160[c]	190[d]
1974	100[e]	124[f]	140[g]	160[h]
1976	75[a]	89[i] - 95[j]	124[g]	140[h]
1978	33[k]	63[m] - 77[m]	95[n] - 96[m] - 101[p]	123[q] - 124[r]

a Lovins speeches
b Sierra Club
c US Atomic Energy Commission
d other Federal agencies
e Energy Policy Project "ZEG"
f Energy Policy Project "TF"
g US Energy R&D Admin. (ERDA)
h Edison Electric Institute
i von Hippel & Williams (Princeton U
j Lovins, *Foreign Affairs*, 9/76
k Steinhart (U. Wisc.) (for 2050)
m CONAES (*18*) I-III (for 2010)
n USDOE, 9/79, \$32/bbl (1977 \$)
p Weinberg (IEA-Oak Ridge) "low"
q USDOE, 9/79, \$18 & \$25/bbl av.
r Lapp (believes E/GNP fixed)

An important analytic trend during 1979 was the increasing geographical disaggregation of national efficiency scenarios. Probably well over a thousand studies are underway in the US at state, county, and municipal level--where most of the institutional barriers to energy efficiency occur and are being attacked. Methodological tools for local (*7*) and county (*31*) studies are widely used, supplemented by national (*32*) and regional workshops. Efficiency scenarios, some rather detailed, are now available for eight of Canada's ten provinces (*33*) and for several regions of Italy (*8*). Conversely, reaggregation has revealed opportunities for sharing comparative advantages to smooth a transition to an efficient, renewable-based energy economy (*34*).

Efficiency scenarios are commonly misinterpreted (*35*) as assuming or entailing drastic social change for the worse. The studies described here offer, on the contrary, a means of *avoiding* such change through the intelligent application of proven technologies guided by sound market principles. They assume no significant change in where we live, how we live, or how we run our societies--save those changes unavoidably resulting from a doubling or trebling of GNP (an assumption which would lead to a per-capita National Product grosser than anyone else's, and which certainly requires some imagination). Many analysts of efficiency scenarios, including myself, consider such undifferentiated economic growth, in M. Goldberger's phrase, to be "spherically senseless"--it makes no sense no matter which way round one looks at it--but we have assumed it anyhow to save argument, keeping our personal preferences separate from our analytic assumptions, and arguing that if one wishes to Los Angelize the planet, efficiency scenarios can do it more cheaply. Should economic activity (or population) grow even faster than assumed here, energy-inefficient capital stocks can be diluted or replaced even faster, so energy needs will not rise proportionately more quickly. In fact, the economic growth rates assumed in most of the scenarios cited above are nearly twice those typical of the past decade. The conclusion is inescapable: that if we follow the principles of economic efficiency to which we profess allegiance, improved energy productivity should lead the industrialized countries to expect gradual energy *shrinkage* to the year 2000 and beyond, even if their most sanguine economic targets are reached. (Rates of implementation for energy savings will be discussed in §5.)

A review (*5*) of these concepts for the United Nations Economic Commission for Europe--whose area comprises North America, Western and Eastern Europe, and the USSR, 1.0 milliard people using 3/4 of the world's energy--has noted a useful thought-experiment. Suppose that by 2025 in this affluent region,

- population grows to 1.25 milliard (in line with official projections);
- gross economic activity per capita increases 82%;
- output ratio of goods to services (with an assumed 3× difference in average energy intensity) decreases 40%;
- ratio of primary to end-use energy drops from 1.35 to 1.05 through careful thermodynamic matching and total-energy systems;
- technical end-use efficiency improves by average factors of 4 in North America and 2.5 in Europe and the USSR (considerably less than the detailed studies cited earlier indicate).

The resulting primary (end-use) energy needs of the ECE region in 2025 will then be ∿107 (102) EJ/y or 3.4 (3.2) TW--40% (22%) *less* than in 1975 despite the 2.28× increase in gross economic activity.

This encouraging result does not yet take account of the aspirations of the other 3.3 milliard people in the world. High-energy forecasts commonly assume that developing countries must and will imitate the pattern of development which served OECD in a very different period, when OECD countries enjoyed ready access (via colonial control) to cheap raw materials at competitively depressed prices and sold their manufactures at monopoly rents. Today, with conditions rapidly reversing, it is hard to see how such a pattern could be reproduced. The emphasis of development economics is shifting from urban industry, based on capital-intensive, inequity-promoting, colonially supplied high technologies (hoped to produce "trickle-down" rural benefits) to a more indigenous rural development based on sustainable agriculture, capital-saving and equity-fostering technologies, and meeting basic human needs. Would this pattern, on which there is a consensus of responsible international bodies, require vastly more energy?

Developing countries tend to have poor energy statistics (*36*) which generally omit their 1-2 TW of noncommercial fuels. Much recent growth in commercial fuel use is mere substitution for dung and for the ∿1 T/cap·y of firewood (about half the world timber cut) now burnt by 80-90% of people in the poorer countries. Analyses of end-use patterns are sparse (*37*,*38*), but representative Indian village data, qualitatively characteristic for over 2 milliard people (*38*), are approximately: 218 W/cap domestic (96% cooking), 76 agriculture (mainly pumping and ploughing), 23 manufacturing, 15 lighting, 11 transport: total 344 W/cap, supplied as 226 noncommercial fuels, 48 animal and 33 human work, 26 oil, 5 coal, 6 electricity. Careful analyses (*38-43*) have consistently shown great scope for improved technical efficiency in these functions, starting with innovations as simple as replacing open cooking fires with cheap clay stoves (*43*). Such investments are generally agreed to be the best immediate buy. Without them, neither OECD-style energy supply nor the eco-development alternatives with their many attractions (*38-42*,*44*)--biogas, solar heat, wind, microhydro, pyrolysis (*45*)--can long succeed. What is at stake for development is neatly encapsulated in Reddy's comparison, using nominal empirical data from India, of two ways to make 2×10^8 kg/y of fixed nitrogen (*46*):

Characteristic	*Western-style coal-fed fertilizer plant*	*Indian-style gobar gas plants* (*42*)
number of plants	1	26,150
capital cost ($\$10^6$)	140	125 (∿60 with Chinese tech.
foreign exchange ($\$10^6$)	-70	0
direct employment (10^3)	1	131
energy balance (TW-h/y)	-0.1	+6.35

If dung is burnt for cooking, all the nitrogen and most of the heat go up in smoke (which blinds people). The biogas plants offer (*47*) an improved fertilizer, a more efficiently usable fuel, a substitute for dwindling firewood; the gas output can about cover all the cooking, lighting, and pumping needs of the village--about half India's total energy needs today. Similar incentives have led China to install over 9 million biogas plants since 1972 (*47*).

Some analysts, however, especially in the most industrialized countries, propose that every developing country should and will in the fullness of time be an imitation FRG, and that this will require enormous amounts of energy. Yet if such heavy industrialization occurs--even in addition to, rather than as a displacement of, its OECD counterpart--the same ultimate efficiency levels achievable in the FRG should also be achievable by its imitators, *but much faster*. The developing countries can do things right the first time. Building anew rather than retrofitting, they can become energy-efficient faster and cheaper. They are not burdened by OECD's ∿$\$2 \times 10^{13}$ worth of energy-inefficient capital stocks with 20+ y turnover time. Rapid population and economic growth would both mean faster accumulation of highly efficient energy-using devices to dilute the old, hence faster achievement of a highly competitive trading position.

Whether developing countries will buy truly modern (or modernized ancient (*48*)) passive solar buildings rather than sealed glass monuments to declining-block rate structures, best technologies rather than outmoded castoffs, depends on their originality and sophistication in buying, and on the example set by OECD countries. Rich, technically skilled countries with indigenous fuels, who treat energy efficiency and renewable sources as second-class and unworthy of prestige, must not be surprised if countries lacking those advantages come to similar conclusions. Industrialized countries' counsel to developing countries would be politically and morally more palatable if they got their own houses in order before offering gratuitous and apparently hypocritical advice.

Many observers who sympathize with the goal of world equity implied by forecasts of world industrialization (if not by Kahnian "two milliard Chinese driving Buicks" scenarios) nonetheless believe that reproducing resource-intensive OECD lifestyles on a world scale is improbable for reasons other than energy, and that achieving such distributional equity, if it can be done, will require a huge transfer of wealth from rich to poor countries (another opportunity for embodying the best energy-saving technologies). This argument only implies that energy is not as near or unavoidable a constraint on industrialism as water, soil fertility, ecological and social resilience, war, etc. (*49*). If for argument's sake we ignore these more proximate constraints, the energy needs of a heavily industrialized but economically efficient world can be roughly inferred from European "existence proofs" (subject to seven reservations below). For example, Olivier's sixfold efficiency improvement on a 5 kW/cap UK energy budget suggests that a standard of material affluence somewhat better than the present British average--undreamt-of wealth to most of the world's people--can be supported by a total primary energy budget of 1 kW per capita, equivalent in traditional (if repugnant) units to the food input of 5.7 slaves (each 170 W = 3500 kcal/d). For an asymptotic world population (again ignoring other constraints (*49*)) of 8 milliard, a 1 kW/cap·y budget implies a rate of world energy use $\sim$8 TW $\sim 8.7 \times 10^{12}$ kg coal/y (at 29 MJ/kg). This is slightly below today's rate, and far below other estimates that have rightly caused much climatic concern, *e.g.* IIASA's 30-50+ TW, Niehaus's 50 TW, Rotty's 32 TW, and even Columbo's 15 TW (*50*). The difference is efficiency.

This 8-TW thought-experiment is conservative in six ways and nonconservative (*i.e.* possibly an underestimate) in one:

- It assumes technical efficiency improvements well short of the economically worthwhile limit represented by the cost of conventional power plants and synfuel plants now being ordered.

- It assumes no significant changes in lifestyle--not even, for example, such reasonable adaptations as car-pooling, living near one's workplace, or dressing to suit the weather.

- It assumes no improvements in a fast-moving technology.

- It assumes that the world has a British climate. (Nearly half of European end-use needs are for heat $< 100^{\circ}C$.) As a counter-example, milder New Zealand, thought by some to have perhaps the highest quality of life in the world, got on very nicely with only 2.4 kW/cap in 1970, despite great inefficiency (including virtually uninsulated houses): this could easily be reduced well below 1 kW/cap by technical fixes. Most people live in even milder climates.

- It falls far short of practical limits (typically (*51*) $\sim$ 2-4 × Second Law theoretical limits (*52*)) for energy efficiency, because Olivier assumes sufficiently slow implementation that even by 2025, much of the stock of buildings and machines is still retro-fitted old items that perform worse than new ones would, *e.g.* in developing countries' new installations.

- It assumes that those social scientists who detect the start of a major and rapid transformation of values (*53*), toward a "Conserver Society" (*54*) seeking more meaningful work, are wrong: that nearly everyone's chief goal in life will continue to be, at least presumptively, the materialistic acquisition of more consumer ephemerals, produced increasingly by subordinating people to the imperatives of machines. This in turn assumes that societies will seek to maximize the flow of resources needed to maintain a desired stock of material artifacts, rather than seeking to maintain that stock with a minimum resource flow--a more rational procedure that would imply minimizing rather than maximizing most of GNP (which measures the flow rather than the stock (*55*)).

- Finally, tending potentially in the opposite direction, it ignores the increasing energy needed to win increasingly dispersed resources such as metals. This energy requirement rises very steeply with declining grade of ore and mineral grain size (*56-58*), and in large steps for the many less common metals with a discontinuous grade-tonnage distribution (*56,59*). In the long run, of course, we are all entropic (if not, as Keynes remarked with cheerful brutality, dead), and a time will come when high-grade (low-entropy) resources have unavoidably been so scattered that the technostructure needed to re-collect them cannot sustain itself. With sensible management (*60*) but without widespread use of more biologically based technologies, I estimate this time is at least 10^4 y hence--well beyond the bounds of current energy policy. Careful consideration of the relevant geochemistry and mining engineering suggests that: (a) over the next century or so, mineral constraints on economic growth will probably be those of land-use (*61*), water, geopolitics, and geologic availability; (b) during that period, despite technological progress (*62-64*), the energy needed to win each kg of most metals will increase substantially, but not enough in comparison to other energy needs to be important in the energy demand and supply budgets outlined here;

and (c) the energy and materials policies implied by the conservation-and-renewables strategy suggested here will postpone this problem as long as possible (*65*). For the time-scale of mere millenia, then, the substitutability of resources (*66-68*), albeit with some specialized inconveniences, can keep materials shortages from increasing energy needs (after efficiency improvements) by an amount sufficient to disturb the conclusions drawn here.

3. WHAT FORMS OF ENERGY?

Most analyses assume that the energy problem is where to get more energy of whatever kind to meet extrapolated *homogeneous* demands. But there are many different forms of energy whose different prices and thermodynamic qualities suit them to different applications. It is therefore sensible to redefine the energy problem to be how to meet *heterogeneous* end-use needs--how to do the many different tasks for which the energy is wanted--with an economically efficient minimum of energy, supplied in the cheapest form for each task (having regard to different forms' varying effectiveness). If energy of superfluously high quality is used for a low-grade task, the mismatch will cause poor Second Law efficiency (*52*) and poor economic efficiency. Least-cost solutions tend, to first order, to be thermodynamically elegant solutions too. For example, an extremely efficient heat pump powered by electricity from a newly ordered nuclear station reflects such a mismatch of source to task (degrading a 10^{12} ^{o}K reaction temperature via 10^{3} ^{o}K steam and ∞ ^{o}K work and electricity to provide 300 ^{o}K heat) that it is inherently uneconomic (*infra*).

Data on the physical categories of energy required to run national economies--a logical starting-point for designing energy supply systems--can be indirectly derived (*69*) from available statistics on end-uses. Typical approximate results (*6*) are:

Percentage of Total Delivered Energy (Heat Supplied Basis) Required in Various Forms in Selected Industrial Countries Around 1975 (70)

Form Required	*USA*	*Japan*	*Sweden*	*UK*	*France*	*FRG*	*Av.W.Eur.*
total heat	58	68	71	66	61	75	71
<100°C	35	22	48	55	36	50	45
100-600°C	15	31	14	6	14	12	13
>600°C	8	15	9	5	11	13	13
portable liquids	34	20	19	26	29	18	22
electr.-specific	8	10	10	8	10	7	7
indus. motors	5	7	6	4	6	4	4
other electr.*	3	5	4	4	4	3	3
(supplied as electricity)	(13)	(16)	(18)	(14)	(12)	(13)	(11)

*Lights, electronics & telecommunications, electrochemistry, electrometallurgy, household non-thermal appliances, el. rail, arc-welding.

Comparison of the electricity-specific requirements with the electrical supplies illustrates a principle to which no exception has yet been found: that the electricity now supplied substantially exceeds that required for the premium uses that can justify the very high cost (>$20/GJ) of this special form of energy. The excess is even greater--typically severalfold--if present supplies are compared with what electricity-specific demands would be after technical-fix efficiency improvements that are cheaper than new electricity (*6*). In contrast, most high-energy scenarios assume, without analytic foundation, greatly increased electrification, even though end-use demand both now and in the long term (*5*) is overwhelmingly--today over 90%--for heating and non-rail vehicles, uses in which new electricity is grossly uneconomic (*6*). From this perspective--doing each task in the cheapest way--new nuclear power stations are uneconomic that if you have just built one, you will probably save money by writing it off and never operating it (*71*). This suggests serious economic inconsistencies in scenarios involving new power stations of any kind, particularly central steam-raising plants (*72*).

Just as supplying energy in the right quality for each task can minimize the costs and losses of conversion, so appropriate scale can minimize those of distribution. This important issue is analyzed elsewhere (*73,74*) and is the subject of a growing literature (*75*) to the effect that gross economies of scale in supply, for uses that are chiefly small-scale and dispersed, are often more than counterbalanced by previously uncounted diseconomies of scale: what matters is the net, not the gross, economies, and the savings expected from centralizing energy systems often prove illusory. Matching supply to end-use in both quality and scale is an important reason for the economic advantages of the supply strategy proposed in §4.

4. WHERE FROM?

Renewable energy sources matched in energy quality and in scale to end-use needs are generally conceded to be a possible long-term option but are often said to require extensive and speculative development before they are available in significant numbers and diversity. Careful study of the best such technologies already in or entering commercial service, however, shows that this is no longer true (*73,76*). The rich array of "soft technologies" already available appears sufficient to meet virtually all long-term energy needs in every country so far studied, including *inter alia* the US, Canada, UK, FRG, France, Denmark, Sweden, and Japan (*5,6,8,9,77*).

This list of countries for which renewable energy supply scenarios have already been sketched in sufficient detail to make a respectable case is highly suggestive: it includes very densely populated and heavily industrialized countries with cold and cloudy climates, low biomass production, and little hydroelectricity. Nonetheless, while technical opportunities vary between and even within these countries, each has enough: countries that are poor in fuels, such as Japan, can be singularly rich in renewable energy of many kinds (*77*). This suggests similar inferences for other countries not yet studied in detail but clearly possessing a broadly similar range of opportunities owing to their varying combinations of insolation, biomass base, wind regime, hydro potential, and other energy flux resources.

A complementary analytic approach is to make approximate but instructive estimates for aggregated regions of the world. Sørensen (*78*) for example, has assumed a 2050 world population of 10 milliard, appropriately distributed among seven regions, and a worldwide level of material affluence twice that of Scandinavia today. He liberally supposes this to entail a per-capita energy budget of 4 kW--at least twice that indicated as economically efficient in §2. This 4 kW comprises 1 kW vehicular liquids (*79*), 1 kW electricity + mechanical work + high-temperature heat, and 2 kW low- and medium-temperature heat (split 1.5/0.5 and 0.5/1.5 for high and low latitudes respectively, with medium-temperature heat including cooling (*80*)). An illustrative balancing of renewable supply potential with demands shows, on these assumptions, self-sufficiency for each region in all categories, except for a modest deficit in Asia made up by exports of biofuels and hydrogen from regions with corresponding surpluses. With more realistic demand assumptions, no imbalances would arise and the renewable supply system would be far more attractive. Its main technical assumptions are:

- net solar heat supplies 20 TW, or 0.2% of mean insolation (land-use and structural constraints are taken into account);
- 25% of end-use energy (probably ∿1.3-2× the economically justifiable fraction and $\gtrsim$5× the economically justifiable level) is supplied as electricity--1.07 TW hydro (at all scales), 0.2 TW wind, 0.3 TW geothermal, 8.8 TW photovoltaic;
- biomass, including some mariculture, yields 9.7 TW.

These supplies are generally (except for hydro) a small fraction of technical potential and a very small fraction of gross resource base.

Taylor (*81*) has done an illuminating scoping calculation, noting that >99% of world population live in a solar regime between that of Stockholm and of Lake Rudolph (Kenya), which differ by 2.5× in mean insolation and ∿24× in seasonal fluctuation ratio on a horizontal surface (much less with optimal elevation angle). By showing that his conceptual designs for low-cost solar systems--based on ponds and air-inflated plastic-film collectors--are likely to achieve very attractive cost goals in these extreme climates, Taylor makes a powerful *a fortiori* case for intermediate climates (bolstered by his case-studies of nine developing countries (*82*)). His design philosophy appears compatible with low end-use efficiency, hence high demands; this is a conservatism.

The specific industrialized-country studies cited earlier, and my own analyses, generally assume that efficiency improvements cheaper than new supply are done first. They then consider a restricted list of presently available soft technologies (*73*):

- passive (preferably) or active low-temperature solar heating and cooling (*80*);
- active solar process heat (tracking and non-tracking concentrators or, in diffuse-light regimes, highly selective absorbers in hard vacua (*83*));
- efficient processes--including pyrolysis, thermogasification, fermentation and anaerobic digestion, acid and enzymatic hydrolysis, and advanced water-alcohol separation techniques--to convert municipal wastes and farm and forestry residues (*84*) to fluid fuels;
- present hydroelectricity, small-scale hydro, and wind, integrated into the existing grid (which makes storage problems far more manageable than in a central-electrified system (*85*));
- in some cases, other simple systems such as woodburners or solar ponds and Rankine engines.

Though further elaboration of end-use categories is possible--for example, there are often attractions to doing industrial drive with wind-hydraulic systems rather than with the electrical systems assumed here--the technologies listed suffice to meet virtually all long-term needs in all dozen-odd industrialized countries studied (*8*). (There may be trivial exceptions in certain applications of very-high-temperature industrial process heat; these can be minimized by suitable choice of materials and met indirectly by solar hydrogen systems if desired (*21*).) These analyses do not, in general, assume direct solar-electric conversion by photovoltaic devices, even though progress on many types, especially the potentially very cheap amorphous films, is so unexpectedly rapid that it is very likely that we shall have cheap solar cells in the next few years before we know what to do with them. (Some concentrator systems entering the market in 1980 are already competitive in many industrialized and most developing countries.) Photovoltaic technologies suffer mainly from the vast surplus of promising and rapidly evolving lines of development--which inhibit investment in particular ones for fear of rapid obsolescence--and by having hardware far ahead of analysis of how best to use it.

Six features of soft technologies deserve special emphasis because ignorance of them is widespread and underlies many prevalent misconceptions. First, the best present art is highly dispersed among thousands of workers in dozens of countries. Many have more ingenuity than technical credentials, and most are outside official R&D programmes, which are often the last to know. Information flows to the conventional energy-hardware community are slow and, in general, seriously deficient. (The public is often further ahead in suspecti the truth, but lacks particulars.) Only people with extensive informal networks and the freedom to travel a great deal to examine grassroots technical developments can expect to be aware of even a small fraction of what has been done.

Second, technical progress is dizzyingly fast: half the data in a recent survey (*73*) had to be revised within four months of the first draft. Most performance and cost goals are bettered monthly. Conventional channels of publication are years behind, and even informal networking newsletters (*86*) are up to a year behind communications on the telephone-and-typescript grapevine.

Third, there is no single panacea. The number of potentially interesting combinations of types of soft technologies could not be enumerated by a fast computer counting for the lifetime of the universe. The greatest strength of soft technologies is their immense *diversity* of type, integration opportunities, scale, suitable climate, appropriate application, level of technical sophistication, and accessibility to various actors. Soft technologies work best, technically and socially, when matched to a particular task, place, and user. The best designs have probably not been thought of yet, and fundamentally new and attractive concepts are still emerging frequently. It is a common and egregious error to dismiss as small the contribution of a particular one of a myriad soft technologies, or to fault it for its unattractiveness if it is irrationally used in the manner or at the scale of conventional centralized systems--*e.g.*, to imitate reactors.

Fourth, the cost of soft technologies, and of the energy services they deliver, is sensitive to their reliability, durability, and convenience (all assumed here to be equivalent to present conventional systems), but even more sensitive to their elegant simplicity of design and to their market structure. For example, equivalent active solar heating systems can be bought today in many parts of the US for a low price of \$100-150/m^2 collector if site-assembled (*87*) so that only one markup is paid, or, as assumed in most studies, for a high price of \$250-500/m^2 collector if packaged (three or four markups via manufacturer, wholesaler, retailer, and installer). Most governmentally funded prototype and demonstration projects are pathologically overpriced and useless for cost data.

Fifth, there are important and generally neglected technical synergisms between efficient energy use and cost-effective soft-technology design. For example, 100% coverage of space and water heating loads in a house in a harsh climate can be achieved with an active solar system 5-10× smaller and technically simpler than most official studies would conclude is needed for partial (say 50-70%) coverage--because the studies assume collectors are put on a sieve. Doing cost-effective conservation *first* changes the basic design physics: the space-heating load becomes a small addition to the water-heat baseload rather than the reverse, is smoothed by the enhanced thermal time constant of the house, and can be evenly met with no heat distribution system (because of convective and radiative near-equilibrium in the nearly sinkless house); the needed heat storage becomes so small that it is cheaper than backup heating (*88*). Such synergisms abound in integrating soft technologies with efficiency improvements, with each other, and with shelter, food, water, and other systems; but such integration, the single greatest key to low costs, is just starting to be seriously explored.

Sixth, the "supply curve" reflecting the cost, difficulty, or nastiness of soft technologies (including any needed storage devices) rises steeply and often discontinuously if they are used in large quantities. They are relatively benign and attractive only if used on the shallow-slope lower portion of the curve--the economically efficient region for supply-demand balance. For example, 8 TW of global soft-energy supply is straightforward: with a typical long-term end-use structure (*5,6,21*), the ∿1 TW of electricity needed (*89*) can be readily supplied by macro- and micro-hydro alone, but in practice would be extensively supplemented with wind and from photovoltaics if available; the ∿2-2½ TW of liquids needed--derived from ∿3-4 TW of farm and forest wastes--represent a small fraction of collectible biomass residues from present harvesting (*90*); and the remaining heat requires little or no extra land area for collection (and far less than for equivalent conventional systems). A 16-TW soft-supply scenario, though no doubt still feasible, may be less than half as attractive and probably begins to incur marginal land requirements (though probably still somewhat less than for "hard" technologies (*91*)). Higher energy demands would increasingly entail a spurious "need"--an artifact of economically inefficient energy use--for "hard" solar technologies, such as monocultural biomass plantations, power towers (*92*), solar space satellites, ocean-thermal-electric conversion, etc. In no case in which end-use efficiency has been properly studied have these systems proved necessary or met elementary economic tests.

Skeptics prepared to admit the existence of soft technologies often aver that such devices are hopelessly "uneconomic". This, too, is demonstrably incorrect (*73*), and generally results from an asymmetrical comparison in which soft technologies are expected to compete with dwindling, historically cheap, and often heavily subsidized fossil fuels while hard technologies (nuclear, synfuels, etc) are only expected to compete *with each other*: in fact they would fail the former test by a far wider margin (*71-73*). Some soft technologies are indeed cheaper today than oil and gas (*73*), and this may become generally true if Taylor's conceptual designs (*81*) or certain promising photovoltaic and photolytic developments succeed. My own analysis, however, asserts only the more modest--but still decisive--proposition that for each category of end-use tasks, the best presently available soft technologies, compared conservatively (*93*) with power stations and synfuel plants also available for ordering today, are generally several times cheaper in delivered energy price and nearly always cheaper even in capital cost (*73*). In other words, present soft technologies are seldom cheap, but they are cheaper than not having them. This proposition is corroborated in US conditions by an authoritative Harvard Business School study (*94*) and in European conditions by others (*95*). The traditional belief that soft technologies are much *more* costly than hard is confined increasingly to the uninformed.

This economic comparison omits the external advantages of a soft path, leaving them as a conservatism. These advantages, described elsewhere (*96*), include lower political costs (centrism, vulnerability, technocracy, inequitable allocation of costs and benefits (*97*)), better cash-flow (*72*), better employment (*98*), lower inflation, increased self-reliance and resilience (*99*), lower public and occupational hazards (*100*), greater suitability for development (*38-45*), leverage for nuclear nonproliferation (*101*), and lower risk of technical failure. The convergent economic and political attractions of a soft energy path are leading to its very rapid implementation at grassroots level in several countries, and appear to be the nucleus around which a new consensus on energy policy is coalescing. Implementation of a soft path, as of any policy, presents difficult social and political problems--chiefly in removing asymmetrical subsidies and pricing rules, and the "institutional barriers" (market imperfections) which inhibit response to price signals. But these problems look to many analysts (*94*) more tractable than those we would otherwise face. There is no free lunch--though some lunches are cheaper than others.

It is neither necessary nor possible to do everything at once. Some options are incompatible with others. The cultural conditioning, accretion of institutional barriers, and resource commitments of a high-energy, centralized, electrified, high-technology future could easily push major soft-technology implementation so far into the future that before we could get there, the fossil-fuel "bridge" to it would have been burnt. If we propose to shift rapidly to reliance on renewable sources, we must do so before the relatively cheap fossil fuels, and the cheap money made from them, are gone, because we shall not have another chance to get there (*35,96*).

A comparison between the difficulties of various energy strategies must also bear in mind that no analyst's view of what is important to a society can substitute for asking the people concerned (*97,102*). Not everyone shares my own view, for example, that living in a plutonium economy is a less pleasant change, and more difficult to achieve in a democracy (*103*), than living in a very well-insulated house. The disagreements about whether a soft energy path is feasible or desirable generally boil down to different political judgments about implementation, and in turn to different political philosophies:

the Jeffersonian (and market-economics) concept that people are pretty smart, and, given incentive and opportunity, can choose wisely for themselves, *vs.* the Hamiltonian concept that these issues should be centrally decided and enforced by a technocratic elite. Such disputes are not technical and cannot be resolved on technical merits. But under the latter philosophy, energy policy requires massive central planning and intervention; under the former, energy policy cannot tolerate it. In my own view, recent experience of what works (*94,97*) is empirically resolving the dispute in favour of the Jeffersonians.

5. HOW FAST?

The key question about any energy policy, whether for displacing oil or for minimizing the climatic effects of burning fossil fuels, is how fast it can be deployed at particular magnitudes. The formidable rate-and-magnitude constraints of new energy supply, especially (*104*) from technically or socially complex devices with long lead times, are seldom appreciated. Illustrative calculations about the potential role of nuclear power (applicable, *mutatis mutandis*, to coal-fired or hypothetical fusion power stations, synfuel plants, etc.) are summarized in the facing table, with details given elsewhere (*5,6* The inherently modest contribution shown even disregarding all financial (*72*), economic (*6*), and political constraints, is the reason that both IIASA (*105*) and the Institute for Energy Analysis (*106*) found that whether or not the world uses nuclear power does not make much difference to the date of attaining a particular global CO_2 level. Even the most generous nuclear projections, as the last column of the table confirms, are too little, too late, against the background of fossil-fuel use rising more slowly but from an enormously greater base. The only important variable in CO_2 timing is energy conservation (*105,106*).

The remarkable speed with which this can work is illustrated by the finding (*107*) that of all new US energy supply during 1973-8, 72% came from more efficient use, and only 28% from all supply expansions *combined*--more oil imports, Alaskan oil, more nuclear power, more coal-mining, etc. Thus US energy conservation, without much effort and before the National Energy Act 1978, has already yielded twice as much energy-"supplying" capacity, twice as fast, as synfuel advocates claim they can do (but their option, if it works, costs an order of magnitude more). Particular cases are even more impressive: *e.g.*, under a simple grant programme, Nova Scotians "weatherized" half their housing stock in one year, and the people of Fitchburg, Massachusetts devised a programme of door-to-door citizen action to do the same (saving $\gtrsim$25% of heating oil) in five *weeks*. (It actually took them seven weeks.) Institutional innovations used in previous retrofits of extensive capital stocks could be applied to improving energy productivity (*108*). Even without them, Leach

region	*target date*	*demand**	*J of fuel displaced by each J of elec.*	*nuclear goal at target date**	*nuclear plant construction time (y)*	*ordering interval (days/GWe nucl.)*	*approx. min. cost of new stns. & grid in 1976 currencies*	*final nuclear GWe (at 0.62 cap. fact.)*	*resulting fossil-fuel demand* at target date*
assumptions						*results*			
USA	2000	1978 PT§ +1.72%/y	2P	$\frac{1}{4}$(PT in 2000)	12	5	$>\frac{3}{4}\times\int$ NPDI†		
France	2000	1975 PT +3.8%/y					Fr 6×10^{11}	104	1975 O&G +70%
Japan	2000	PT = 41 EJ/y¶	2D		6	20	¥10^{14}	274	1978 O&G +73%
OECD Europe	2000	1975 DT +1%/y; 1975 DE +2%/y					\$$3\times10^{11}$	210	1975 + 13%
same	2025	same					\$$10^{12}$	614	1975 + 30%
same	2000	1975 DT +2%/y; 1975 DE +4.4%/y					\$$6\times10^{11}$	384	1975 + 33%
same	2025	same					\$$3\times10^{12}$	1890	1975 + 92%
non-Com. world	2000	1975 PT +3%/y					\$$2\times10^{12}$	1000	1975 + 89%
same	2025	...+1.6%/y					\$$4\times10^{12}$	2100	1975 + 150%
world	2000		2P	$\frac{1}{2}$(O&C in 1970)	8	3.5	\$$3\times10^{12}$	>1600	

*P = primary; D = delivered; T = total; E = electrical; O&G = oil & gas
§This yields 120 EJ/y in 2000, the 9/79 USDOE medium forecast.
†NPDI = net private domestic investment, assumed to remain 7% of a 1978-2000 GNP rising 3.5%/y. Omits costs of plants unfinished in 2000.
¶Assumes trebled GNP (below 1978 forecasts) and conservation 1/3 above MITI's 1978 "maximum possible" projection.

calculates that his trebled UK energy efficiency (*10*) could be achieved over about 50 years if the savings were implemented *more slowly* than they actually were during 1974-8. In fact, during 1973-8, the ratio of primary energy growth to real economic growth was negative in the UK, Netherlands, Belgium, and Luxembourg (*109*). In the US, it started the period at about 1 and ended below 0.5; in the nine EEC countries, none with a strong conservation programme, the highest ratios were 0.4 (Denmark) and 0.3 (FRG). For the whole EEC, primary energy efficiency improved during 1973-8 by ∿8% while primary supply increased 0.42% (*109*)--a ratio of ∿19, implying that ∿95% of effective new energy supply was efficiency improvements, compared with the US 72% (*107*). If an improvement on today's minimalist policies combined with diminishing returns to produce a more or less linear rate of long-term implementation, an 8% saving in 5 years implies an 80% (5-fold) saving in 50 years without anyone's really noticing. This makes Leach's 3-fold saving look almost staid.

Much evidence is emerging that the next fastest displacements of fossil fuel, after efficiency improvements, come from soft technologies. There are three fundamental reasons why they should be faster to deploy than competing power stations and synfuel plants (*5,6*):

- they take days, weeks, or months to build, not ten years;
- they diffuse spontaneously into huge consumer markets (like pocket calculators, CB radios, snowmobiles, colour TVs, etc) rather than requiring a tedious process of "technology delivery" to narrow, specialized utility markets (like reactors and basic oxygen furnaces)
- since the myriad types of soft technologies are held back by institutional problems largely *independent* of each other (as opposed to the generic constraints on monolithic technologies), dozens of individually slowly-growing types of soft technologies can independently add up, by strength of numbers, to very rapid total growth.

This potential is starting to be realized faster than anyone expected. In the US, for example, there are $\sim 2\times10^5$ solar buildings (of which half are passive and half of those are retrofits); in the most solar-conscious communities, from a quarter to all of the 1978 housing starts were passive solar; over 150 New England factories and over 40% of northern New England houses (>50% in rural Georgia) heat mainly with wood, entirely on private initiative; over half the state have active fuel alcohol programmes; over 40 manufacturers of wind machines share a market which in 1979 saw two large state-government contracts totalling $230 million, both fully commercial lowest bids. The size, dispersion, rate, and diversity of renewable energy activities are so great that Federal authorities are only dimly aware of their extent, and their statements correspondingly unreliable. In general, however, official projections of energy contributions are steadily rising for solar; steadily falling for nuclear and synfuels. If post-1973 trends persist--as their economic and political motivation seems bound to do--then the US will readily surpass its modest year-2000 official goal of 19.5 EJ/y of solar energy (*110*) or its officially noted opportunity of 21-32 EJ/y (*111*) (compared with its present primary solar contribution of 4.4 EJ/y (*112*)). There appears to be no reason save political inertia why other countries should not achieve similar proportions; some countries, like Sweden, have much of the needed infrastructure and consciousness in place (*95*), and others, like Japan, probably have the social opportunity to set new speed records if they so choose.

6. CONCLUSIONS

Through mechanisms that are complex, interactive, poorly understood, and possibly unpredictable even in principle, our species' energy conversion, especially our burning of fossil fuels, may already have perturbed global climate significantly, or may do so soon. If not, a continuing rapid rate of burning fossil fuels will very probably do so sometime in the next century; only the timing and exact effects are in question. Judging from chastening experience of intricate interactive systems, in which higher-order effects may dominate and which may be "almost-intransitive" (*113*) and subject to long lags, if the climate does unravel, it may well be in ways that we did not expect and are hard pressed to explain afterwards. The effects may be grave and irreversible. They are not predictable with confidence and may never become so. Prudence therefore requires forbearance in testing empirically the climatic limits of tolerance. If one's life may depend on the smooth functioning of a complex and mysterious machine, one is wise not to stick spanners into it. By the time we understood its basic workings--if we ever do--if by then we have already got into the habit of messing about with them, via enormous, long-lead-time energy systems, it may be too late to stop in time to prevent serious damage. Major changes in the world energy system take typically a half-century; with the inexorable force of exponential energy growth, major changes in climate may be set in motion in much less.

This paper argues that there is a technical, economic, and political possibility to stabilize and then steadily to reduce to approximately zero the world rate of burning fossil fuel--consistent with worldwide achievement of the most generously enthusiastic economic goals--by substituting efficiency improvements and renewable sources which are climatically quite benign *per se* (*100*). Such a policy would lead to a smaller integrated fossil-fuel burn than could be envisaged with any other policy (in the absence of catastrophic curtailments, *e.g.* by war, of either oil supply or economic activity). The time thus bought can be illustrated by making some crude simplifying assumptions which tend to make CO_2 problems look further in the future than they probably are:

- no CO_2 sources or sinks other than fossil-fuel combustion (*e.g.* no de- or reforestation), save the 45% sink assumed below;
- no feedback loops (*e.g.* marine CO_2 solubility) and no lags;
- each TW-y of heat from burning fossil fuel releases 0.605 + 0.0017/y Pg (10^{15}g) of carbon (*114*);
- 1 PgC emitted = 0.55 PgC in air = 0.26 ppmv (parts/10^6 by volume) increase in global CO_2 concentration.

Indicative projections for exponential growth, constancy, and linear decline of the rate of burning fossil fuel can then be calculated:

Assumed Fossil-Fuel Combustion				*Dates of Attaining CO_2 Levels*					
1975–2025 rate (1975 ≡ 8 TW)	*rate (TW) 2000*	*2025*	*TW-y ∫ 1975–2025*	*ppmv*	*350*	*400*	*450*	*500*	*550*
				1979×	*1.05*	*1.20*	*1.35*	*1.50*	*1.65*
				**1850×*	*1.25*	*1.43*	*1.61*	*1.79*	*1.96*
+ 4%/y	21	57	1278		1986	2003	2013	2020	2025
+ 3%/y	17	35	928		1987	2006	2018	2027	2033
+ 2%/y	13	22	687		1987	2010	2025	2037	2046
+ 1%/y	10	13	519		1988	2016	2037	2053	2067
constant	8	8	400		1989	2025	2059	2090	2119
constant to 1985, then −0.175 TW/y	5.4	1.	260		1989	2200			
same @−0.1975 TW/y	5.0	0.1	242		1989	3700			
same @−0.20 TW/y	5.	0.	240		1989	†			

*estimated pre-industrial concentration ∿280 ppmv
†converges asymptotically to ∿370 ppmv

The dates in the "350 ppmv" column show that if a CO_2 level $\lesssim$350 ppmv will cause major climatic change, it is too late already: except by drastic programmes of conservation, solar, and reforestation, decided upon immediately worldwide and aimed mainly at the largest fuel-burning nations, we could only slightly mitigate the degree of change. Conversely, the dates in the "+ 4%/y" row show that rapid demand growth is a recipe for events outrunning our ability to manage them. (And to provide the fuel: the integrated burn of 1278 TW-y would exhaust all the estimated ultimately recoverable world resources of oil and conventional gas plus nearly 90% of the coal now considered technically and economically recoverable; there is perhaps 16× more coal than that, but burning it, if half the resulting CO_2 stayed airborne, would increase CO_2 levels 7×, so coal is almost certainly climatically rather than resource-constrained.)

A 50-year time constant--say until 2025--is important for policy both because it is how long ordering a coal facility today locks us fairly firmly into burning coal in it and because it is probably about how long it would take to displace fossil fuels essentially completely with climatically benign energy sources of whatever kind. Taking this target date as an indicative planning horizon, the above table shows that:

- the rate of burning fossil fuel must not increase by more than 4%/y if the CO_2 level that causes major climatic change is 550 ppmv, nor ∿3.3%/y if 500, nor 2%/y if 450, nor 0%/y if 400: that is, as the CO_2 "alarm threshold" decreases, the urgency of reducing the rate of growth of the fossil-fuel burn rises nonlinearly, more and more steeply;
- it is not possible, may not soon be possible, and may never

be possible to predict with confidence which, if any, of these levels is of concern, so *a constant rate of fossil-fuel burning is the highest tolerable rate* if we wish to keep safely beyond our minimum planning horizon the lowest possibly unacceptable CO_2 level whose attainment is not already virtually inevitable;

- even at that lowest (most time-sensitive) CO_2 level, a comfortable several centuries' time for a supply transition can be bought by linearly *reducing* the rate of burning fossil fuel, over about 50 years, to about 0-1 TW;
- such stabilization and reduction can be won through any combination of efficiency improvements and alternative sources, but the latter are slower to build and can benefit from the $1/(1-x)$ longer time available to build them if a fraction x of the fossil-fuel reduction is done by efficiency improvements;
- because of *the disproportionate time-buying benefits of reductions* in the rate of burning fossil fuel, a strategy of efficiency improvements and renewable sources--argued in this paper to be the cheapest and fastest method--deserves a high priority for climatic protection in the face of uncertainty.

Contrary to what others may suggest in this Conference, a formally quantified balancing of costs, risks, and benefits for a decision on whether to follow an economically efficient energy path is neither theoretically nor practically possible, even in principle (*102,115*). A political decision is required--or rather, a very large number of political decisions at all levels of many societies. This in return requires information, as people cannot choose options they do not perceive. The key information here is that a minimum-climatic-risk energy policy, which phases out fossil fuels as fast as possible (*116*), is a real option. Energy-climate problems are thus probably an artifact of economically inefficient energy policies. A pragmatic, low-climatic-risk energy policy--an unusual piece of good news--would be worth the most urgent and detailed consideration on that ground alone. Such a policy which also appears desirable on other grounds deserves to be quickly refined and widely shared.

NOTES

1. A B Lovins, *Soft Energy Paths: Toward a Durable Peace*, Harper & Row (NY), 1979, and Pelican (UK), 1977, pp 63-66.
2. For details, see ref. (*6*), n. 7.
3. A B Lovins, I.79 supplement to 20-21.XI.79 testimony on Federal subsidies to nuclear power, available from IPSEP, 124 Spear St, San Francisco, CA 94105, USA, (415) 495-5210.
4. F Krause, "Energieversorgung der Bundesrepublik ohne Kernenergie und Erdöl", Öko-Institut (Schönauerstr 3, 7800 Freiburg i Br), XII.79; seedetails in "Wirtschaftswachstum bei sinkendem Energieverbrauch", *id*, 1980; summarized *Soft En Notes 2:*82-5 (1979).

5. A B Lovins, "Re-Examining the Nature of the ECE Energy Problem", ECE(XXXIII)/2.I.G., UN Economic Commission for Europe (Geneva), I.78, reprinted *En Policy 7:*178-198 (IX.79).
6. A B Lovins, "Is Nuclear Power Necessary?", Friends of the Earth Ltd (9 Poland St, London W1V 3DG, UK), V.79, £1.50.
7. For suggestions on methodology, see D Brooks & S Casey, *Alternatives 8(3/4):*10-23 (Summer/Fall 1979), Trent U (Peterborough, Ont K9J 7B8, Canada).
8. Supplemented proceedings are to be published by the sponsors, IPSEP (International Project for Soft Energy Paths)(*3*) and Amici della Terra (Piazza Sforza Cesarini 28, 00186 Roma), 1980, from whom the regional Italian studies may also be had. Summaries of many papers appear in *Soft En Notes* (*9*).
9. *Soft Energy Notes*, an approximately bimonthly periodical of IPSEP (*3*), I.78- ; a reprint of issues 1-7, DOE/PE-0016/1, is free from Office of Solar Policy, Office of Policy & Evaluation, USDOE, Mail Stop 7E-088, Forrestal Bldg, Washington, DC 20585.
10. G Leach *et al*, *A Low Energy Strategy for the United Kingdom*, Intl Inst for Envir & Develt (10 Percy St, London W1, UK) and Science Reviews Ltd, I.79, £7.50.
11. This paper uses aggregated E/GNP ratios only as shorthand for a far more complex set of detailed technical coefficients obtainable from the particular studies. The E/GNP ratios are *not* suitable for rigorous comparisons over time or between countries, for reasons explained in (*5*) at 190-1. Technical coefficients for individual energy services--J/kg steel, J/kg-km transport, $J/m^2 \cdot C^o \cdot d$ heating, etc--must be used for this purpose. E/GNP ratios are used here only as a handy summary benchmark.
12. Olivier's equally detailed work is to be submitted to its sponsor, the Energy Technology Support Unit (Harwell), at end II.80 and will be published in due course by Earth Resources Research Ltd (40 James St, London W1, UK). A brief interim summary of results is in *Soft En Notes 2:*50-52 (VII.79).
13. *Op cit* (*4*).
14. J Nørgård, *Husholdninger og Energi*, DEMO Projekt (Fysisk Lab. III, Danmarks Tekniske Højskole, 2800 Lyngby) and Polyteknisk Forlag (København), 1979.
15. Reports of the DEMO Project, *id.*
16. Miljörörelsens Alternative Energiplan (MALTE), "Huvudrapport och Bilagor", 3 vols, 1977 (Box 24023, 40022 Göteborg). See also T B Johansson & P Steen, "Solar Sweden", Secretariat for Future Studies (Fack, 10310 Stockholm), 1978.
17. *E.g.*, both the 77- and 96-q CONAES scenarios assumed that US houses built in 2010 would be less heat-tight than the average Swedish house is today: L Schipper & A Lichtenberg, *Science 194:*1001-13 (1976), a classic study of technical coefficients.
18. Demand & Conservation Panel of the Committee on Nuclear & Alternative Energy Systems (CONAES), *Science 200:*142-152 (1978).
19. P G Schipper & E J Tuininga, Annex 6 to J Saint-Geours *et al*, "In Favour of an Energy-Efficient Society", EEC (Brussels), 1979.

20. P Craig *et al*, eds, *Distributed Energy Systems in California's Future: Interim Report*, HCP/P7405-01 and -02, '78, from DOE (*9*).
21. A B Lovins, 3d Regents' Lecture, V.78, from IPSEP (*3*).
22. D B Brooks *et al*, "Some Scenarios of Energy Demand in Canada to the Year 2025", reprinted at 1718-1801 in US Senate, Small Business & Interior Comms, *Alternative Long-Range Energy Strategies: Additional Appendixes*, Int Comm 94-47/92-137, USGPO, 1977.
23. A B Lovins, "Exploring Energy-Efficient Futures for Canada", *Conserver Society Notes 1(4)*:5-16 (1976) & USS *op cit* 1405-16.
24. R Sant, "The Least-Cost Energy Strategy", Energy Productivity Ctr, Mellon Inst (Suite 1200, 1925 N Lynn St, Arlington, VA 22209), 1979. Sant has perhaps the best data base on costs.
25. *Procs Sask Conf on Low En Passive Solar Housing* (X.79), summ. in *Soft En Notes no 9* (1980); *Low Energy Passive Solar Housing Techniques*, U Sask Div of External & Community Relns (Saskatoon S7N OWO), 1979; A B Lovins (*88*); A H Rosenfeld *et al*, LBL-8912, Lawrence Berkeley Lab (Berkeley, CA), 1980; Project 2020, *How to Build a Superinsulated House*, Cold Weather Edition (Box 81961, College, AK 99708), 1979, $3; R Korda, "What About Windows?", Ctr for Community Technology (1121 U Ave, Madison, WI 53715), 1978; G Leger house described in *Soft En Notes no 9*.
26. *Procs*, Rosenfeld *et al*, & Korda, *op cit* (*25*); *Soft En Notes 2*:10-3 (1979); numerous reports of Swedish Bldg Res Counc (Stockholm) on exterior retrofit insulation; M Ross & R H Williams, "Drilling for Oil and Gas in Our Buildings", PU/CEES 87, Ctr for Envtl Studies, Princeton U (Princeton, NJ 08540), VII.79.
27. See official and industry citations by Leach (*10*), Olivier (*12*), and Krause (*4*). They do not assume diesel-electric hybrid drive, which, in recent German tests, yielded 2.8 ℓ/100 km the first time with off-the-shelf components in a 1550-kg car. At least one small US maker's cars use <2 ℓ/100 km. Even today's diesel Rabbit does 5.3, its successor model does 4, and available continuously-variable transmissions could yield 2.0-2.8.
28. W Murgatroyd & B C Wilkins, *Energy 1*:337-345 (1976). The FRG Fichtner-Studie draws similar conclusions but then ignores them.
29. J Nørgård (*14*), Ch 13, summarized in *En Policy 7*:43-56 (III.79).
30. R U Ayres & N Narkus-Kramer, 76-WA/TS-4, ASME, 1976.
31. A Okagaki & J Benson, *County Energy Plan Guidebook*, Inst for Ecological Policies (9208 Christopher St, Fairfax, VA), 1979.
32. First Annual Conf on Community Renewable Energy Systems (VIII. 79, Boulder); procs in preparation by SERI (Golden, CO 80401).
33. *Alternatives* (*7*) and subsequent issue discuss & summarize.
34. B Sørensen, "A Renewable Energy System for Scandinavia", paper to Rome conf (*8*); Niels Bohr Institute, Copenhagen, 1979.
35. A B Lovins & critics, *The Energy Controversy: Soft Path Questions & Answers*, FOE (124 Spear, San Francisco 94105), 1979.
36. See Intl Energy Agency (Paris), *Workshop on Energy Data of Developing Countries*, XII.78. But developed-country statistics omit noncommercial fuels too, *e.g.* the wood and wood wastes that supply ∿15% of primary energy in Finland, 9% Sweden, 2-3% US.

37. *Soft En Notes* (*9*) has published end-use data for Hong Kong, Bangladesh, Sudan, Mexico City, and an Indian village.
38. J Goldemberg, "Energy Strategies for Developed and Less Developed Countries", PU/CES 70 (*26*, II.78).
39. D Hayes, "Energy for Development: Third World Options", paper 15, Worldwatch Inst (1776 Mass Ave NW, Wash DC 20036), 1977.
40. A Makhijani & A Poole, *Energy and Agriculture in the Third World*, Ballinger (Cambridge MA), 1975; A Makhijani, "Energy Policy for the Rural Third World", IIED (*10*), 80p, 1976.
41. G Leach, "Report of the Energy Resources Wkg Gp", Intl Conf *Agricultural Production: R&D Strategies for the 1980s* (Bonn, X.79), from IIED (*10*).
42. A K N Reddy, *Bull Atom Scient*, 28-33 V.78 & 54-55 XII.78; — & K K Prasad, "Technological Alternatives and the Indian Energy Crisis", *Econ & Pol Wkly 12(33-34SN):*1465-1502 (VIII.77), Bombay.
43. *Soft En Notes 2:*89-90 (XII.79).
44. J W Howe, *Energy for the Villages of Africa*, Overseas Development Council (Washington DC), 1977; N L Brown & J W Howe, *Science 199:*651-7 (1978).
45. J W Tatom (4070 Ridge Rd, Smyrna, GA 30080) has published reports on low-technology pyrolyzers for Senegal, Ghana, and Indonesia. They seem highly effective.
46. A K N Reddy (Indian Inst of Science, 560012 Bangalore).
47. *Soft En Notes 1:*10-12, 45-6, 81-3 (1978), *2:*80-1 (1979). The 9 million figure is from *Peking Rev* XII.78, cited by T B Taylor
48. K Butti & J Perlin, *A Golden Thread: 2500 Years of Solar Architecture & Technology*, Cheshire/Van Nostrand Reinhold (NY), 1980.
49. A B Lovins, *Envir Conserv 3(1):*3-14 & USS (*22*) 1450-1461.
50. W Bach *et al*, eds, *Man's Impact on Climate*, Elsevier (Amsterdam) 1979. Umberto Columbo provided a 15-TW scenario in 1979 as part of his critical review of the IIASA Energy Project.
51. M Ross & R H Williams, "Energy & Economic Growth", US Congress Joint Ec Comm (Energy Subcomm), 1977.
52. Am Inst of Physics Conf Proc 25, *Efficient Use of Energy*, 1975, AIP (NY). This concept reflects only the process-independent energy needed for a change of state, not the opportunity to redefine initial or final states to provide the same function.
53. *E.g.* W Harman, *An Incomplete Guide to the Future*, Stanford [U] Alumni Assn, 1976; see also polls in USS (*22*), 2186-2196.
54. Science Council of Canada, *Canada as a Conserver Society*, 1977.
55. H Daly, *Steady-State Economics*, W H Freeman (San Francisco), '78.
56. A B Lovins, *Openpit Mining*, 1973, from FOE (*35*).
57. P F Chapman, *Metals & Materials* 107-111, II.74.
58. So important is the $\sim$50-80 kJ/kg requirement for comminution of typical ore to liberation size that the London Exchange price of >20 non-monetary metals can be accurately derived from a regression only on average grade now mined, Gibbs free energy of occurrence in common ore mineral, and entropy of mixing in ore.
59. NAS/NRC, *Resources and Man*, W H Freeman (San Francisco), 1969.
60. Including improved product longevity and low-entropy design to

facilitate repair, recycling, and remanufacture--which should all *reduce* total energy needs. See Ayres & Narkus-Kramer (*30*); T V Long II & L Schipper, LBL-5986 (*25*), 1977; P F Chapman, *Metals and Materials* 311-319, VI.74; D Hayes, "Repairs, Reuse, Recycling", paper 23, Worldwatch Inst (*39*), 1978.

61. D Brooks, "Minerals: an Expanding or a Dwindling Resource?", Energy Mines & Resources Canada (Ottawa), 1973.
62. Styrelsen för Teknisk Utveckling (Stockholm), "The Future Steelplant--A Study on Energy Consumption", rpt 59-1977.
63. P G Kihlstedt, *Scand J Metall 4:*145-9 (1975).
64. H H Kellogg, "Conservation & Metallurgical Process Design", 13th Wernher Mem Lect, Inst Min Metall (London), 1977.
65. The supply systems of §4 generally need less material and offer a far wider choice of it--especially of commoner, less energy-intensive options--than their high-technology competitors. Energy-saving measures do better still: take foam insulation out of your walls, burn it in the furnace instead of the gas that was invested in the foam, and see how long it heats you.
66. R U Ayres, "Substitution Possibilities & Problems in Regard to the Major Metals", TAD/RD/ENV/R.7, UNCTAD/UNEP, Geneva, 1974.
67. — *et al*, *Critical Materials: A Problem Assessment*, IRT-348-R, Intl Res & Technol Corp (1501 Wilson Blvd, Arlington, VA), 1974.
68. H E Goeller & A M Weinberg, "The Age of Substitutability", Inst for Energy Analysis (Oak Ridge) & ORNL, 18.IX.75; and see generally V K Smith, ed, *Scarcity & Growth Reconsidered*, RFF / Johns Hopkins U Press (Baltimore/London), 1979.
69. See Lovins (*1*), Ch 4.
70. Approximate %s *required* to run present processes at present rates, not necessarily in forms currently supplied. For sources and further details, see (*6*), notes 58 and 59.
71. This is because low-grade heating and cooling--all the marginal electricity can be used for--can be done with conservation and passive solar at a cost below the plant's *running cost alone* (fuel + O&M; no wastes or decommissioning). Further, if a new US plant were written off to tax funds, the capital-cost write-off would probably cost the taxpayers less than the unpaid tax subsidies and utility profits which they would pay if it ran.
72. A B Lovins, "Electric Utility Investments: *Excelsior* or Confetti?", III.79 E F Hutton paper, from IPSEP (*3*). Note also that if |long-run price elasticity of demand| >1, as it may be, then revenue elasticity <0: higher prices (entailed by higher marginal costs) *reduce* revenue (population and income held constant).
73. A B Lovins, *Annu Rev En 3:*477-517 (1978) & *Soft En Notes no 5*.
74. — , (*1*), Ch 5.
75. M F Cantley, *Options '79:3*, IIASA (Laxenburg), 4-5; J F Fisher (GE), "Optimum Size of Subcritical Fossil Fueled Electric Generating Units", in *The Scale in Production Systems*, IIASA, in press, 1980; A Ford & T Flaim, LA-8033-MS, Los Alamos Sci Lab, X.79; C Komanoff, "Escalation in Capital Costs of Nuclear & Coal-Fired Plants", submitted to *Science*, I.80; see also (*72*).

76. If indeed it ever was (*48*).
77. H Tsuchiya, "From Energy Hunting Civilization To Energy Cultivating Civilization", Res Inst for Systs Technol (Tokyo), 1979; A B Lovins, preface to Japanese edn of (*1*), Jiji (Tokyo), 1979.
78. B Sørensen (*34*), "Global Energy Policy & Develt Strategy", 1979.
79. Olivier (*21*) calculates 0.4 kW despite per capita increases on 1976 UK levels of 47% in car travel, 71% in inland freight, 382% in international air travel, etc.
80. A need for cooling is an artifact of bad architecture. I have seen traditional designs maintaining 20°C ΔT in the tropics.
81. T B Taylor (Dept of Aerospace & Mech Scis, Princeton U (*26*)), *Prospects for Worldwide Use of Solar Energy*, report to Rockefeller Fndn (NY), 1979, unpublished.
82. *Id*, Appendix II.
83. Such systems can yield $500\text{-}600^{o}$C under load on a cloudy winter day at high latitude. A lower-temperature nonevacuated system is developed: K Raetz (Gassnerstr 12, 3300 Braunschweig, FRG) reports achieving the needed $\alpha/\varepsilon \sim 50$, *PTB-Mitteilungen 89:*217-8.
84. Not crops (with possible rare exceptions, *e.g.* high-latitude forests), and carefully conserving nutrients and soil fertility. C Lewis (*En World* 2-8, I.80) reports the energy in wasted French straw is 10% of total French energy use.
85. E Kahn, *Annu Rev En 4:*313-352 (1979).
86. There are probably over 1000 US topical or regional ones alone.
87. *Procs 1st New England Site-Built Solar Collector Conf*, Mech Eng Dept, Worcester (MA) Polytechnic Inst, 1978.
88. A B Lovins, *Soft En Notes 2:*16-18 (1979) & *Sol Age 4(5):*21,24.
89. One TW is too high. Preliminary results from Olivier (I.80) suggest that a trebled UK GDP conservatively requires (§2) of order 220 W el/cap, implying 1.7 TWe for 8×10^9 people; but our 8 TW scenario here assumes $<\frac{1}{2}$ this economic activity/cap.
90. Annual-average world gross primary productivity on land is $\sim$76 TW ($\sim$0.2% insolation) and standing biomass $\sim$450 TW-y.
91. M Grenon at 42 in W Häfele *et al*, *2d Status Rpt on the IIASA Project on Energy Systems 1975*, IIASA (Laxenburg), & 4 (F3).
92. J M Weingart, "The Helios Strategy", IIASA, 8.II.78.
93. *E.g.* in the solar heat/nuclear heat-pump comparison (A B Lovins, *Science 201:*1077-8 (1978), *204:*124-9 (1979)): low nuclear costs, baseload = average price, peak COP = 2.5, high fixed charge rate, no cash-flow differences or differential escalation, no cheap solar designs (passive, integrated, pond, neighborhood, simple).
94. R Stobaugh & D Yergin, eds, *Energy Future*, Random House (NY), '79
95. See *e.g. Soft En Notes 2:*51 (1979) and the reversal of priorities in the 1978-81 Swedish R&D budget ((*6*), note 75).
96. See (*1*) generally.
97. L Gerlach, "Energy Wars & Social Change", Dept of Anthropology, U Minn (Minneapolis 55455), 1979; A B Lovins, "Democracy & the Energy Mobilization Board", 1979, from IPSEP (*3*).
98. See M Schachter, DOE/PE-0013, VII.79; CEP study of Long Island (summarized in *Soft En Notes no 9*, 1980); "Employment Impact of

the Solar Transition", US Congress Joint Ec Comm, 6.IV.79.

99. A B Lovins, "Energy, National Security, & Resilience", in preparation for submission to CEQ (Washington DC) IV.81.
100. J P Holdren, in (*20*), 2:1-63, plus '79 critique of Inhaber study.
101. See (*1*), Ch 11, and Lovins, *Foreign Affairs* VII.80, in prep.
102. A B Lovins, *45 Geo Wash L Rev* 911-943 (1977) & (*22*) 1417-49.
103. Justice (British Section, Intl Commission of Jurists), "Plutonium and Liberty", London, 1978.
104. A B Lovins, *World Energy Strategies*, FOE/Ballinger, '77, 49-50.
105. IEA, *Economic & Environmental Impacts of a U.S. Nuclear Moratorium 1985-2010*, MIT Press, 1979, at 59 (T3-1), 264-5, 271.
106. This is not an explicit IIASA conclusion, but is obvious from close comparison of the published IIASA scenarios.
107. V Taylor, "The Easy Path Energy Plan", UCS (1208 Mass Ave, Cambridge, MA 02138), 1979.
108. *E.g.* in converting the UK to North Sea gas, smokeless fuels, decimal coinage, and metric units; Holland to Groningen gas; Sweden to district heating and right-hand driving; and Toronto and Montréal from 25- to 60-Hz electricity after World War II.
109. J Saint-Geours (*19*), Annex 2, p 1 (*cf* p 2).
110. White House Fact Sheet on solar message, 20.VI.79, p 3.
111. President's Council on Environmental Quality, "Solar Energy: Progress and Promise", CEQ, Washington DC, IV.78.
112. *Domestic Policy Review of Solar Energy*, TID-22834, II.79, ii, iv.
113. E N Lorenz, *Tellus 17:*321-3 (1965), *Meterol Monog 8(30):*1-3 ('68).
114. This linear form conveniently represents a plausible shift in fuel mix, *e.g.:*

fuel share in:	*1975**	*2000*	*2025*	*PgC emitted/TW-y*†
oil	0.46	0.30	0.15	0.60
gas	0.24	0.20	0.20	0.43
direct coal	0.29	0.45	0.55	0.75
coal/shale synfuel	<0.01	0.05	0.10	1.02
composite PgC/TW-y	0.605	0.655	0.691	

*JASON data cited by Elliott & Machta at 72 in US Senate, Comm on Govtl Affairs, Symposium "Carbon Dioxide Accumulation in the Atmosphere, Synthetic Fuels & Energy Policy", 30.VII.79, GPO
†MacDonald data from Woodwell *et al*, *id* 116

Longer-term results of the linear approximation are reasonably consistent with depletion: W Bach (*id* 124-5) estimates world ultimately recoverable resources as ∿370 TW-y oil, 350 conventional gas, speculative unconventional gas, and ∿10 000 coal of which 640 is now considered technically and economically recoverable. A substantial burn thus entails a switch to coal.

115. P Junger, *27 Case Western Reserve L Rev* 3-335 (1976).
116. This is a necessary but not sufficient condition for minimum climatic risk: Beaufort oil spills, particulates, ^{85}Kr emissions, river diversions, albedo changes, deforestation, etc and non-energy-related perturbations would need control too.

HIGH ENERGY DEMAND AND SUPPLY SCENARIO

H.-H. Rogner and W. Sassin

International Institute for Applied Systems Analysis
A-2361 Laxenburg, Austria

ABSTRACT. An adequate energy supply system is a key issue in economic development and human welfare. The industrial development of the Northern part of the globe owes much to the abundant supply of cheap fossil fuels. It is the Southern countries' aspirations for industrialization that will call for a significantly larger energy supply. Sustaining economic growth in the industrialized regions will add considerably to the world's future energy consumption. Two scenarios labelled High and Low are described that indicate a plausible range of the long-term evolution of future economic and energy related activities in seven world regions. On this basis, the energy demand is calculated for each region. An aggregate view of the world's energy resources is followed by a detailed presentation of energy supply strategies. Of particular interest are the implications of a High energy demand and supply scenario on the stock of world energy resources.

INTRODUCTION

Industrialization, technological progress, and a steady rise in prosperity and human welfare--these characteristics of the last one hundred years are to a considerable extent based on an adequate energy supply structure. In the mid-19th century, wood was the predominant source of energy meeting about 70% of the energy demand. Subsequent technological innovation in industry, however, asked for another type of energy, one that has a higher specific energy density, and that can more easily be stored and shipped than wood. Coal, endowed with such characteristics, consequently replaced wood. This structural change in the energy system was one of the essential features of a development leading to the first industrial revolution. Other structural changes followed in due course.

W. Bach, J. Pankrath, and J. Williams (eds.), Interactions of Energy and Climate, 33–52.

Figure 1 shows the decline and rise of market shares of various types of primary energy during a rather long historical period. The substitution processes given appear to be governed by some remarkable regularity. One such regular feature is the slopes of the curves of the market shares. It suggests that the time required for an energy technology to become a substitute for another technology is about constant. On a world level differences in industrial maturity of particular economies are smoothed out. The case of the FRG indicates that any new energy carrier, if not retarded by external events, needs about 3-5 decades to increase its market share from 1% to 50%.

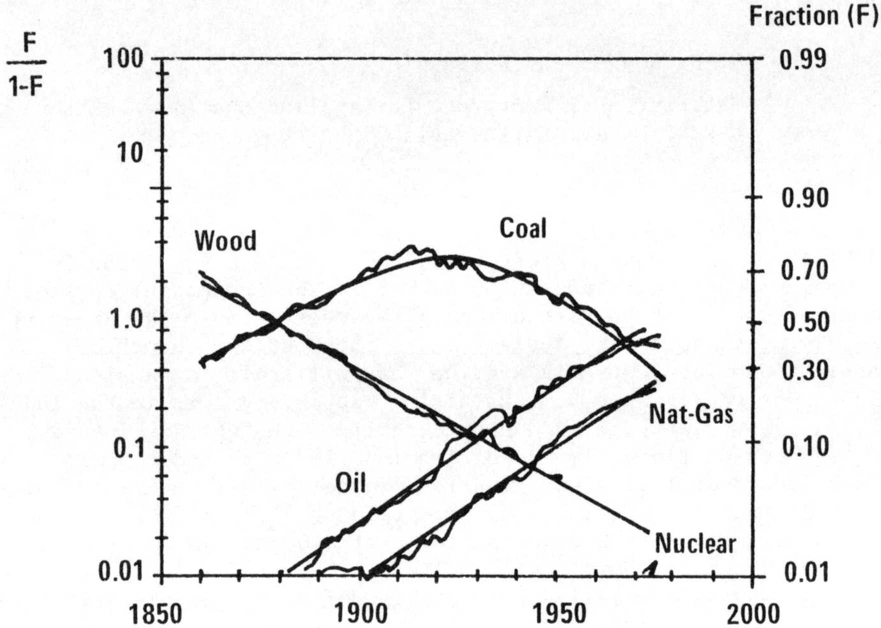

Figure 1. Primary energy substitution for the FGR (1).

Thus it appears meaningful for an analysis of future energy demand and supply in scenario form to span a time horizon of at least 50 years, capturing in this way any essential structural changes in an economy and in society at large. Furthermore, market penetration analysis (1) indicates that there are certain limits on the speed at which new and future energy production and conservation technologies can be introduced.

Of course, the interplay of energy demand and supply is also of global importance. Firstly, industrialization as it is aspired by the developing countries will take up an increasingly large share in the world's energy consumption. Since 1950, the increase in commercial energy consumption has been three-fold in the industrialized countries, and about eight-fold in the world's developing regions. But, percentage-wise, the gap in energy consumption has not been narrowed down at all. With 70% of the world's population, the developing countries hold only a 16% share in energy consumption (see Figure 2).

Secondly, roughly 25% of the world's energy consumption comes from only one region, i.e., the Middle East. The energy trade linking almost all parts of the world thus creates strong economic interdependencies, as well as difficulties in politics and balancing of trade.

Thirdly, national energy demand and supply studies tend to support the necessity of energy imports for meeting the domestic energy needs. But they rarely specify the countries that are supposed to make this energy available. One begins to wonder if in some cases the same barrel of oil may have been taken into account more than once.

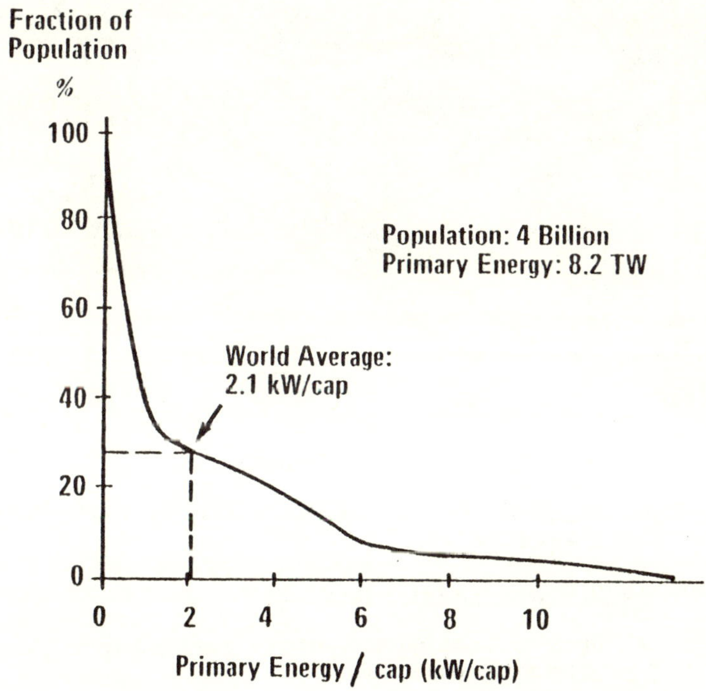

Figure 2. Per capita commercial energy consumption world, 1975.

It is not surprising then that global comprehensiveness and consistency were sought in preparation of a global energy study on the world's energy system, which has been carried out at the

International Institute for Applied Systems Analysis (2). An investigation as complex as this one, covering a planning horizon of 50 years, 1980–2030, has problems of its own, however.

On the other hand, the world's energy demand, its resources and its technological know-how are not uniformly distributed. Moreover, the evolution of energy demand is expected to be quite dissimilar in world regions that differ largely in terms of their economic, social and demographic structures. For IIASA's study the world is then divided into seven regions , as is shown in Figure 3. The groupings of countries into regions is not necessarily based on geographical proximity but serves to reflect similarities in economic infrastructure, life-styles and energy resources.

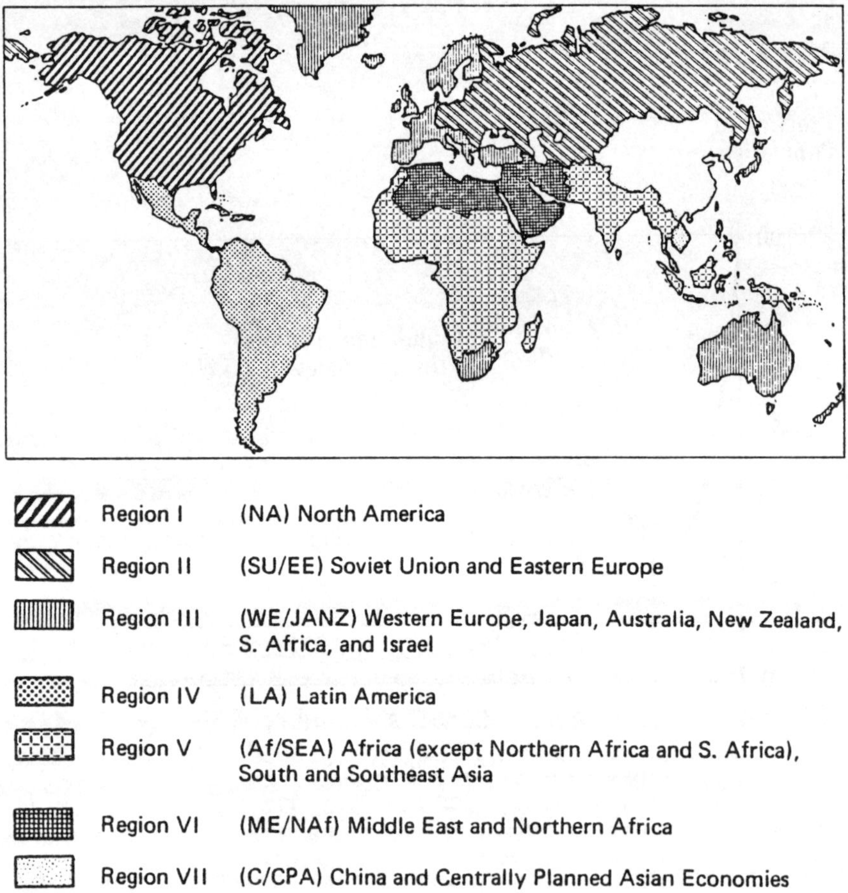

Figure 3. Composition of the seven world regions.

FUTURE GLOBAL ENERGY SCENARIOS

Long-term projections of energy demand and supply can be made only in the light of mutually consistent projections of population, economic growth, availability of energy, materials and other resources, with some perception of technological innovation and development, and in the wake of various physical, social and environmental constraints. In order to obtain a consistent picture, one has to look at all these factors both individually and collectively, trying by way of a reiterative procedure to eliminate internal inconsistencies, if any.

The analysis of IIASA's Energy Systems Program was facilitated by the use of a set of mathematical models serving as the main analytical tools. The schematized flow of information between the various models is shown in Figure 4. The demand for final energy in each world region is evaluated by way of the demand model MEDEE-2. This model is driven by the projected growth of population and economic development both of which are to be specified exogenously. The energy supply model MESSAGE leads to optimum supply strategies. These supply strategies are consistent with the availability of energy resources and subject to various constraints encompassing technological, environmental and other related issues. Consideration

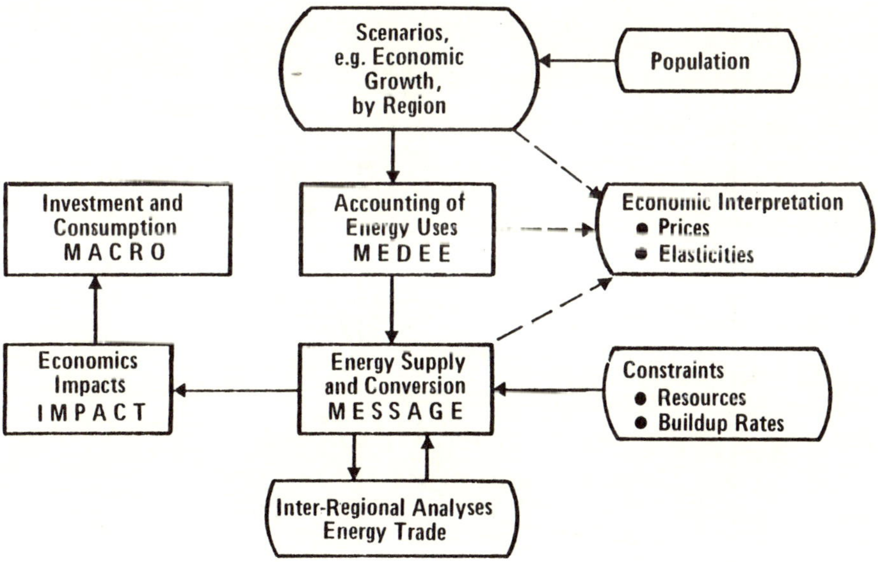

Figure 4. IIASA's set of models for energy strategies.

of interregional energy trade makes it necessary to reiterate the MESSAGE runs for various regions until a globally consistent picture emerges. The economic impacts of the regional energy supply strategies are analyzed with the help of the energy-economy interaction model IMPACT. Implications in terms of macroeconomic parameters are assessed in the aggregated economy model called MACRO. In addition, several consistency checks are carried out at various stages of each iteration step.

The analysis leads to two scenarios of global energy development until the year 2030. The scenarios, characterized as High and Low, are based on our perception of future population growth and correspond to two different rates of economic development in the world. Figure 5 shows the historical development as well as projections of the world's population growth (3). Accordingly, the world's population would increase from about 4.0 billion in 1975 to about 8.0 billion by the year 2030, showing only a modest increase thereafter.

It may be noted that the time horizon of this study, 1980-2030, also overlaps with the phase of transition in population growth. Figure 5 presents an aggregated picture; one expects, of course, marked variations in the current and projected growth of population

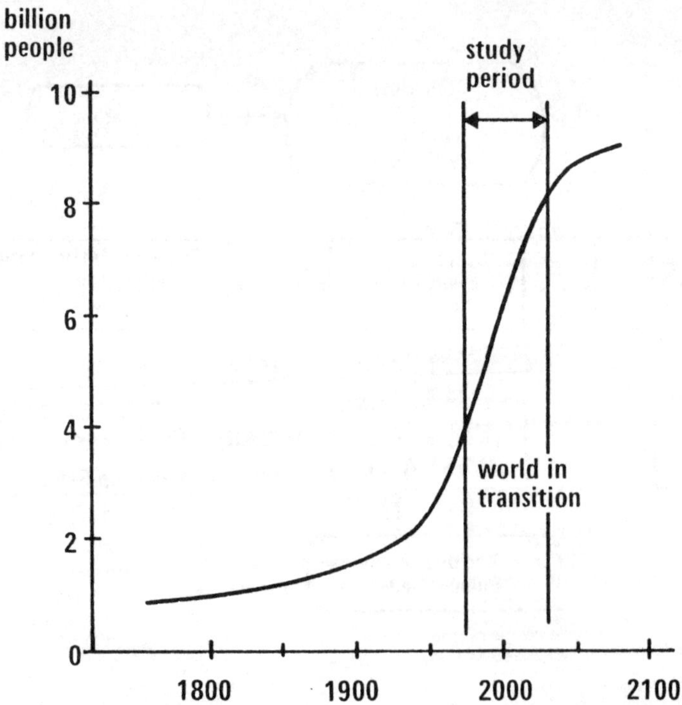

Figure 5. World population historical and projection (3).

in different world regions. As is shown in Figure 6, population growth rates are expected to decline in all world regions, but remain higher throughout in the developing regions than in the industrialized countries.

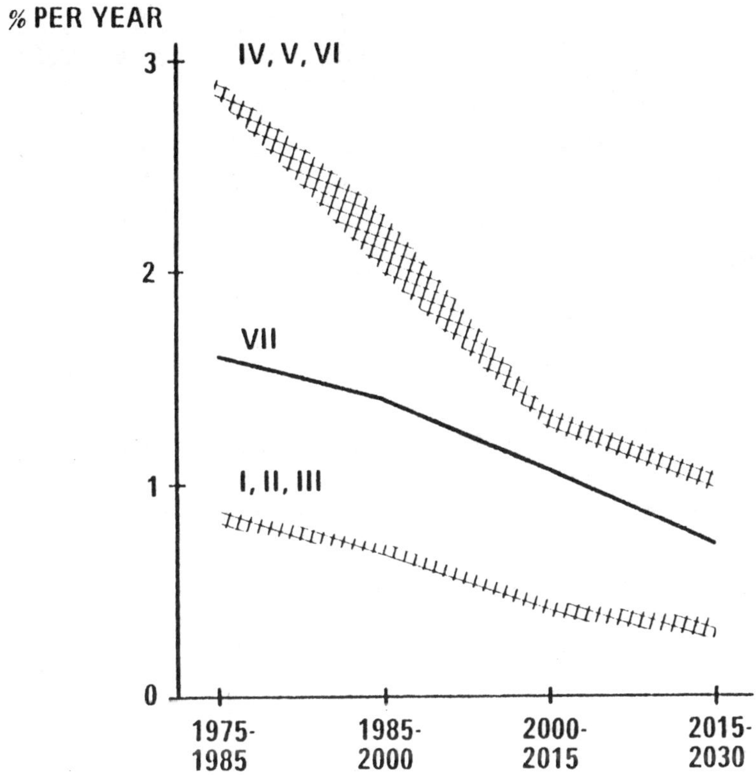

Figure 6. Population growth rates.

These population projections are common to both scenarios that are considered High and Low with respect to their differences in projected GDP growth. Again, the GDP growth rates in all world regions show a gradual decline with time that is considered with the reduction in population growth and the increasing scarcity of basic resources. Figure 7 shows the projected GDP growth rates in different world regions, disaggregated over certain periods. These projections implicitly assume that the economic growth in the developing regions will be dependent on that in the developed regions, i.e., on account of trade relationships (4).

GLOBAL ENERGY DEMAND

In order to properly appreciate the two global scenarios of primary energy consumption one has to look first of all into the

	HISTORICAL		HIGH SCENARIO			
	1950-1960	1960-1975	1975-1985	1985-2000	2000-2015	2015-2030
I (NA)	3.3	3.4	4.3	3.3	2.4	2.0
II (SU/EE)	10.4	6.5	5.0	4.0	3.5	3.5
III (WE/JANZ)	5.0	5.2	4.3	3.4	2.5	2.0
IV (LA)	5.0	6.1	6.2	4.9	3.7	3.3
V (Af/SEA)	3.9	5.5	5.8	4.8	3.8	3.4
VI (ME/NAf)	7.0	9.8	7.2	5.9	4.2	3.8
VII (C/CPA)	8.0	6.1	5.0	4.0	3.5	3.0
WORLD	5.0	5.0	4.7	3.8	3.0	2.7
I + III	4.2	4.4	4.3	3.4	2.5	2.0
IV + V + VI	4.7	6.5	6.3	5.1	3.9	3.5

	HISTORICAL		LOW SCENARIO			
	1950-1960	1960-1975	1975-1985	1985-2000	2000-2015	2015-2020
I (NA)	3.3	3.4	3.1	2.0	1.1	1.0
II (SU/EE)	10.4	6.5	4.5	3.5	2.5	2.0
III (WE/JANZ)	5.0	5.2	3.2	2.1	1.5	1.2
IV (LA)	5.0	6.1	4.7	3.6	3.0	3.0
V (Af/SEA)	3.9	5.5	4.8	3.6	2.8	2.4
VI (ME/NAf)	7.0	9.8	5.6	4.6	2.7	2.1
VII (C/CPA)	8.0	6.1	3.3	3.0	2.5	2.0
WORLD	5.0	5.0	3.6	2.7	1.9	1.7
I + III	4.2	4.4	3.1	2.1	1.3	1.1
IV + V + VI	4.7	6.5	5.0	3.8	2.9	2.6

Figure 7. Real growth rate of GDP (%/year).

nature of energy demand as it arises at various levels and in different world regions. Note that primary energy is only a means to obtain energy for use and not an end in itself. Primary energy consumption thus serves to indicate how much of an energy resource is needed to meet the ultimate objective of providing energy services.

Before it reaches the final user, energy passes through various stages of conversion (see Figure 8). It is converted from primary energy to secondary energy and to final energy and useful energy, and at each stage conversion and/or transmission losses occur. The ultimate goal is energy services, such as the generation of value added in industry, transportation of passengers and goods, provision of amenities, etc. Energy services are useful energy in combination with capital, labor and know-how. An evaluation of energy demand should not be distorted by big differences in losses. Thus it is desirable to evaluate demand in terms of useful and final energy, especially if major changes in the primary fuel mix are to be expected.

Final energy demand was studied for each region. It was derived from a wide range of activities all relating in a sense to energy conservation. Such activities must be based on scenario parameters, projecting any important changes that can be foreseen in terms of economic structure, social needs, life-styles, and technological progress. A parameter illustrating the mutual dependence of economic activity, structural change, and final energy demand over time is the ratio of final energy-to-GDP underlying the concept of energy intensiveness.

Figure 9 shows energy intensiveness versus per capita income for the High scenario. Of interest is the dissimilarity of the slopes of energy intensiveness in the various regions. A considerable improvement over the last two or three decades, both in energy intensiveness and in income, is seen for developed regions I, II, and III. Of course, the input factors to the production process in these regions are used rather efficiently, on account of a developed industrial infrastructure and technological progress.

The situation in the developing countries is converse. These regions are faced with the need to industrialize and to build up their infrastructures--which requires an overproportionately greater amount of energy than the operation of an existing infrastructure. Figure 9 suggests that the buildup of an adequate infrastructure is completed when an income of about \$2,000 per capita is reached. This range of per capita income corresponds to the maximum values of the final energy-GDP ratios.

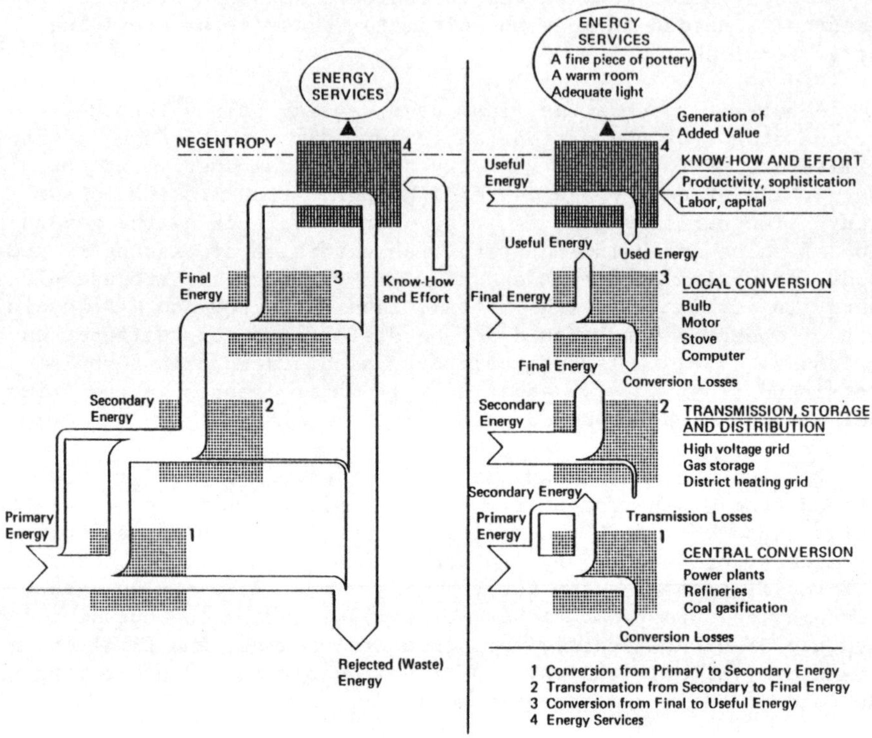

Figure 8. Energy consumption and use.

The respective final energy-GDP elasticities for each region are given in Figure 10. These elasticities, serving as an output--rather than as an input--in the IIASA approach, are a means of monitoring the evolution of energy demand and supply. Figure 11 summarizes the per capita final energy demand in each region for the High and the Low scenarios.

ENERGY RESOURCES

In the foregoing, the magnitude and the nature of the world's energy demand in the two scenarios were identified. In order to assess the global energy situation over the long-run, one also has to look into the potential of the world's energy supply system. For

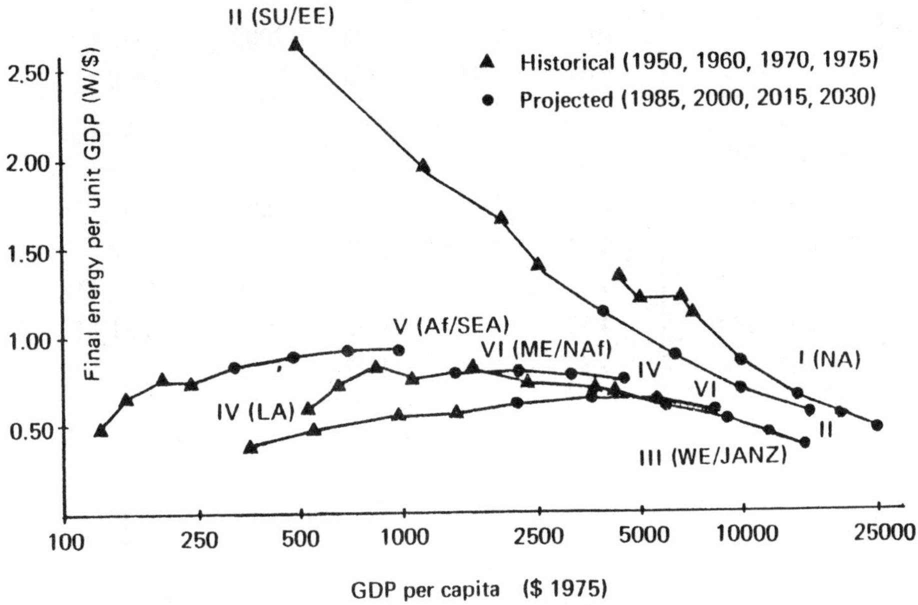

Figure 9. Energy intensiveness versus per capita income - High scenario.

a period of as many as fifty years, one may indeed anticipate considerable additions to the reserves and resources of the fuels commonly used today--even more so since technological developments are likely to open up new avenues. Thus it is a hazardous undertaking to predict the future supply potential of conventional and non-conventional energy sources and their production costs. Yet such predictions are a necessary step in understanding the long-term evolution of the energy supply system. At IIASA, considerable effort has been made to obtain realistic estimates of this development on the basis of published information, opinions of knowledgeable persons and analysis of the related technical data. Such estimates of the ultimately recoverable fossil fuels in all seven regions are summarized in Figure 12.

The economic feasibility of recovering coal, oil and gas leads to distinctive cost categories. The aggregates of the resources estimated are about two or three times higher than the numbers usually reported. This is readily understandable as the traditional assessments of fossil energy reserves and resources refer to the present energy price level of typically 12$/barrel of crude oil. The estimates given here include more expensive fuels up to 25$/barrel in 1975 terms. It should be noted that exploitation of

	HIGH				
	1950-1975	1975-1985	1985-2000	2000-2015	2015-2030
I (NA)	0.84	0.31	0.43	0.53	0.48
II (SU/EE)	0.68	0.59	0.58	0.52	0.53
III (WE/JANZ)	0.84	0.77	0.65	0.58	0.51
IV (LA)	1.21	1.07	1.01	0.97	0.90
V (Af/SEA)	1.42	1.20	1.08	1.05	1.01
VI (ME/NAf)	1.17	1.12	1.07	0.95	0.81
VII (C/CPA)	1.53	1.10	1.02	1.02	0.96
WORLD	0.87	0.69	0.73	0.78	0.77

	LOW				
	1950-1975	1975-1985	1985-2000	2000-2015	2015-2030
I (NA)	0.84	0.24	0.38	0.53	0.46
II (NSU/EE)	0.68	0.54	0.57	0.50	0.41
III (WE/JANZ)	0.84	0.67	0.64	0.60	0.49
IV (LA)	1.21	1.10	1.03	0.95	0.88
V (Af/SEA)	1.42	1.19	1.12	1.14	1.06
VI (ME/NAf)	1.17	1.21	1.11	1.01	0.93
VII (C/CPA)	1.53	1.02	0.98	0.99	0.90
WORLD	0.87	0.64	0.73	0.79	0.74

Figure 10. Final energy-GDP elasticities, ε_f .

the more expensive categories of fossil resources do not only mean larger costs but also enormous ecological and environmental impacts that will warrant limits on production rates. Oil of category III, for example, such as shale oil or tar sand, may call for mining operations at a scale similar to that required for coal. This is to say that even to continue with fossil fuel use will result in a considerable transition.

REGION	1975	HIGH SCENARIO 2000	HIGH SCENARIO 2030	LOW SCENARIO 2000	LOW SCENARIO 2030
I (NA)	7.89	9.25	11.63	7.95	8.37
II (SU/EE)	3.52	5.47	8.57	4.98	6.15
III (WE/JANZ)	2.84	4.46	5.70	3.52	3.90
IV (LA)	0.80	1.75	3.31	1.28	2.08
V (Af/SEA)	0.18	0.42	0.89	0.32	0.53
VI (ME/NAf)	0.80	2.34	4.64	1.76	2.46
VII (C/CPA)	0.43	0.93	1.87	0.64	0.33
WORLD	1.46	1.96	2.86	1.58	1.83

Figure 11. Per capita final energy demand (kW/cap).

There are, of course, several other energy supply possibilities such as nuclear energy, solar energy and renewable resources that may contribute to the future supply system. Figure 13 summarizes, on a global basis, the estimated magnitudes of all commercial energy resources and their production potentials. It also lists the most important factors constraining their exploitation beyond certain levels.

Although the renewable forms of energy (wood plantation, hydropower, wind power, geothermal energy, ocean thermal gradient, etc.) have infinite resource potentials, all that can be exploited technologically speaking is probably about 14 TWyr/yr. The number is reduced further to about half the value in terms of a practically realizable production under given social, environmental, economic, and other constraints. There are schools of thought that consider the actual potential of renewable resources to be 50-70% higher than the numbers used here. In fact, much more analysis and in-depth study are still required to settle the controversy.

Another source with an infinite resource potential is solar energy. The soft solar technologies may perhaps be called upon to meet as much as 30% of all the hot water requirements in buildings, in addition to supplying space heat in up to 50% of the newly constructed single-family homes. If so, the use of soft solar in 2030 will be around 1 TWyr/yr. The large-scale use of hard solar technologies, on the other hand, will be largely constrained by high material requirements. For example, at a density of 50 kg per m^2

RESOURCE	COAL (TWyr)		OIL (TWyr)			GAS (TWyr)			URANIUM (10^3 tU)	
COST * CATEGORY	I	II	I	II	III	I	II	III	I	II
I (NA)	174	232	23	26	125	34	40	29	1920	1500
II (SU/EE)	136	448	37	45	69	66	51	31	n.e.	4140
III (WE/JANZ)	93	151	17	3	21	19	5	14	793	2087
IV (LA)	10	11	19	81	110	17	12	14	56	3544
V (AF/SEA)	55	52	25	5	33	16	10	14	311	5269
VI (ME/NAF)	<1	<1	132	27	n.e.	108	10	14	78	1524
VII (C/CPA)	92	124	11	13	15	7	13	14	n.e.	1980
WORLD	560	1019	264	200	373	267	141	130	3158	20044

* Cost categories represent estimates of costs either at or below the stated volume of recoverable resources (in constant 1975 $).

For oil and natural gas: cat. 1: 12$/boe
cat. 2: 12–20$/boe
cat. 3: 20–25$/boe
For coal: cat. 1: 25$/ton
cat. 2: 25–50$/ton
For uranium: cat. 1: 80$/kgU
cat. 2: 80–130$/kgU

Figure 12. Ultimately recoverable energy resources.

	PRODUCTION (TWyr/yr)	RESOURCE (TWyr)	CONSTRAINTS
RENEWABLES			
WOOD	2.5	∞	ECONOMY – ENVIRONMENT
HYDRO	1–1.5	∞	ECONOMY – ENVIRONMENT
TOTAL	6–(14)	∞	ECONOMY – (NATURE)
OIL AND GAS	8–12(?)	1000	ECONOMY – ENVIRONMENT – RESOURCES
COAL	10–14(??)	2000(?)	SOCIETY – ENVIRONMENT – ECONOMY
NUCLEAR			
BURNER	12 for 2020	300	RESOURCES
BREEDERS	≤17 by 2030	300 000	BUILDUP RATES – RESOURCES
FUSION	2–3 by 2030	300 000	TECHNOLOGY – BUILDUP RATES
SOLAR			
SOFT	1–2	∞	ECONOMY – LAND – INFRASTRUCTURE
HARD	2–3 by 2030	∞	BUILDUP RATES – MATERIALS

Figure 13. Summary of energy resources, production potentials and constraints.

of land, the material requirements for hard solar to provide 1 TWyr/yr will lie between one and three billion tons, as against current world production levels of about 700 million tons each of iron ore and steel.

Nuclear power is limited to 300 TWyr resource-wise, unless reprocessing, advanced converters and fast breeder reactors are introduced which can multiply the resource availability by a factor of one thousand. (The maximum production potentials of nuclear power indicated in Figure 13 refer to resource consumption. If it were all to be used for electricity generation only, the production potentials of burner and breeder reactors would be about 5-7 TW(e)yr/year, respectively.) 300 TWyr of resources, corresponds to 20 million tons of uranium. This is about five times the estimates tranditionally referred to by OECD/IAEA, and includes ores with a concentration up to 50 times lower than the present 2000 ppm.

In the case of oil and gas, the production limits are estimated to be in the range of 8-12 Twyr/yr compared to the coal production potential of as much as 14 TWyr/yr. The production potentials shown in Figure 13 only give an idea of what may be physically and technologically possible under certain constraints. The production levels actually to be achieved by 2030 will largely depend on the new technologies used: their time and dynamics of introduction into the market, their cost economics relative to major technologies already in use, as well as on political considerations and societal acceptance.

ENERGY SUPPLY SCENARIOS

Final energy demand as it was described above must then be translated into energy supply strategies for the High and Low scenarios. This is done by the energy supply model MESSAGE. Its cost-minimization algorithm takes into account the regional allocation of different primary energy sources as a number of constraints on the time of availability and the buildup rates of new technologies, interregional energy trade flows, and so on. The resource estimates contained in Figure 12 are an orientation to the detailed and quantified energy supply scenarios made for each region. Here only the aggregate supply strategies are described.

Figure 14 shows the relative distribution of primary energy sources in the High scenario. One should note that the distribution is virtually identical with that in the Low scenario. Liquid fuels continue to retain their relative market share, but the general composition of the liquid fuel supply is changed considerably. The share of oil in primary energy supply decreases from about 45% in 1975 to about 18% in 2030. In absolute terms, this is 6.8 TWyr of oil, which is roughly comparable with the world's energy consumption in the year 1970. But the oil in the year 2030 is quite different from the cheap, clean and easily manageable fuel of the early 1970s. 6.8 TWyr/yr includes shale oil, tar sands, etc., whose extraction is costly involving enormous ecological and environmental impacts.

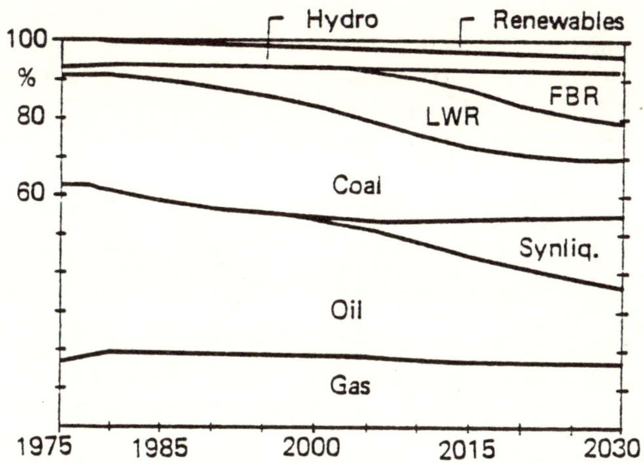

Figure 14. Global primary energy: High scenario, 1975-2030.

The relative decline of oil creates a serious gap in liquid fuel supply that must be filled. Partially, this can be done with coal liquefaction entering the market. But present-day autothermal liquefaction requires sizeable amounts of coal. These quantities can only be made available if another energy production technology takes over in electricity generation, where traditionally coal is used. In the IIASA scenarios, electricity is largely produced from nuclear energy, thus freeing coal for liquefaction purposes. Gas and hydropower maintain their present market shares, while renewables including solar do not influence the global energy supply situation before the year 2030.

Figure 15 summarizes the global energy supply strategies in absolute numbers. The High scenario totals 36 TWyr/yr, and the overall energy requirement of the Low scenario is 22 TWyr/yr. The distribution of the various primary energy sources is characterized by significant increases in the use of all available sources. In the High scenario, oil requirements nearly double over the next 50 years, but coal reaches the incredible amount of 12 TWyr/yr in 2030, compared to 2.26 TWyr/yr in 1975. One should bear in mind that a scenario of 12 billion tons of annual production appears feasible by definition, but does not consider environmental aspects. If coal production on such a scale does not seem advisable, one must embark upon alternatives; for instance, by employing allothermal coal liquefaction instead of autothermal coal liquefaction, which might cut down the coal input for liquefaction to about 30%. In such a configuration, nuclear power or hard solar would have to provide the necessary hydrogen, e.g., by way of electrolysis. According to Figure 13, the nuclear potential is about 17 TWyr/yr, but the High scenario allocates only 8 TWyr/yr. This is to say that there is still a certain amount of flexibility in the energy supply system as it is anticipated.

		HIGH		LOW	
	1975	2000	2030	2000	2030
I (NA)	2.65	3.89	6.02	3.31	4.37
II (SU/EE)	1.84	3.69	7.33	3.31	5.00
III (WE/JANZ)	2.26	4.29	7.14	3.39	4.54
IV (LA)	0.34	1.34	3.68	0.97	2.31
V (Af/SEA)	0.33	1.43	4.65	1.07	2.66
VI (ME/NAf)	0.13	0.77	2.38	0.56	1.23
VII (C/CPA)	0.46	1.44	4.45	0.98	2.29
	8.21	16.84	35.65	13.59	22.39

Figure 15. Global primary energy by region (TW).

CONCLUDING REMARKS

With such a plausible picture of the world's energy demand and supply over the next 50 years before one's eyes, one may want to take stock of the remaining fossil fuel. How much will supposedly be consumed by 2030, especially in the High scenario, and what is left for future use?

Figure 16 illustrates that, under the supply strategies suggested, the world will not run out of stock by 2030. Rather, the stock of conventional oil is reduced to 13% (categories 1 and 1A) of its 1975 value. As for coal, 43% of category 1 is still left. 61% of conventional gas is consumed. Unconventional gas, oil, and coal are barely touched. Thus more than 2000 TWyr/yr, out of about 3000 TWyr/yr of total fossil fuel resources

RESOURCE TYPE AND COST CATEGORY*	HIGH SCENARIO	ENHANCED NUCLEAR CASE
	PERCENT OF ORIGINAL RESOURCES REMAINING AT 2030	
COAL		
CAT. I	43	51
CAT. II	100	100
OIL		
CAT. I	11	11
CAT. IA	2	1
CAT. II	76	81
CAT. III	99	99
GAS		
CAT. I	39	36
CAT. II	84	81
CAT. III	100	100

*Fossil fuel by cost category

Figure 16. Fossil fuel consumption, enhanced nuclear case and High scenario.

in 1980, is still available for use after 2030. The phase of transition in energy supply and demand, which starts around the turn of the century, therefore continues beyond the year 2030. As a consequence, the phenomena of fossil fuels running out are expected to set in only in the second half of the 21st century.

In spite of the remaining considerable stock of fossil fuels, the future energy supply system does not appear to be very resilient to disturbances it may have to absorb. Of course, supply alternatives, such as more nuclear as well as solar power can be called in (see, i.e., the impact of an enhanced nuclear case on the overall resource situation in Figure 16). But this diversification would only partly alleviate the risk of fuel embargoes or hardly reduce societal and environmental pressures against a large-scale exploitation of the "dirty" fossil fuel resources. By 2030, no more than an additional 7 TWyr/yr from such sources may have entered the market in any case, given their presented supply potentials. The choice is therefore more restricted than was expected.

NOTES

[1]This paper is an interpretation by the authors of the extensive work undertaken by IIASA's Energy Systems Program on the long-term global energy problem. It largely draws upon an update of a contribution by W. Häfele, A. Khan, and H.-H. Rogner on "Geographic Diversity in Energy Supply and Demand", to the First Arab Energy Conference, Abu Dhabi, United Arab Emirates.

REFERENCES

(1) Marchetti, C., Nakicenovic, N., Peterka, V., and Fleck, F.: 1978, *The Dynamics of Energy Systems and the Logistic Substitution Model*, AR-78-1A/B/C, prepared for the Stiftung Volkswagenwerk, Hannover, F.R.G. Laxenburg, Austria: International Institute for Applied Systems Analysis.

(2) Energy Systems Program Group of IIASA: 1980, *Energy in a Finite World, A Global Energy Systems Analysis*. Laxenburg, Austria: International Institute for Applied Systems Analysis, (forthcoming).

(3) Keyfitz, N.: 1979, *Population of the World and Its Regions*, WP-79-119. Laxenburg, Austria: International Institute for Applied Systems Analysis.

(4) Hicks, N. L., et al.: 1976, *A Model of Trade and Growth for the Developing World*, Eur. Econ. Rev. 7, pp. 239-255.

THE ROLE OF CLIMATE IN AFFECTING ENERGY DEMAND/SUPPLY

G.A. McKay and T. Allsopp

Canadian Climate Centre
Atmospheric Environment Service
Toronto, Canada.

ABSTRACT. From an energy viewpoint climate is a liability and a resource. Energy is expended to offset cold, heat, snow, humidity and other attributes of climate. Space heating is the dominant demand accounting for probably, one third of the energy used in industrialized regions. Variations in this demand and societal change have found many energy systems wanting in recent years. The use of conventional fuels for space-heating can be significantly reduced by modern technology and using natural energy flows. The substantial energy-related costs of anomalous climate on agriculture, transportation and other sectors of activity can also be reduced by enhanced design and operational efficiencies. The potential of renewable energy exceeds Mankinds needs, but the sources are diffuse and variable. They are generally most abundant in the Tropics and Subtropics, decreasing poleward. Their variability necessitates storage or back-up energy sources for their systematic use. In some instances the supply from different renewable energy sources varies in a complementary manner that may be exploited to achieve stability of supply. The mounting cost of conventional fuels, the desire for energy independence, and the need for environmental integrity are persuasive arguments for their use.

INTRODUCTION

Climate is a major determinant of energy demand and is assuming increasing importance relative to supply. Energy is required to offset cold, heat, winds, excesses of precipitation, snow and ice. Defending against, exploiting, and, in particular, designing

W. Bach, J. Pankrath, and J. Williams (eds.), Interactions of Energy and Climate, 53–72.

to overcome the variable nature of climate are major activities in the planning of energy systems. An understanding of climate is necessary for the effective and efficient exploration, transportation and use of fossil fuels and for the transmission and distribution of electrical energy. Climate is also a control in the hydrological, biological and other processes that are central to the supply of renewable energy.

The vulnerability of energy systems to climatic change has increased as a result of population growth, altered attitudes and technology. This vulnerability is compounded by man's ability to alter climate on local to global scales. Since the climate system responds on a global basis, local energy systems are not immune to actions taken by the global community at large.

Good resource planning and management requires the ability to predict and adapt to climatic change in order to minimize losses and optimize on opportunities. This need has been underscored by recent climate-induced energy crises. Despite our present level of understanding and use of climatic knowledge, major climate-induced energy crises continue to occur. Losses to the United States' gross national product due to cold and drought during the winter of 1976-77 have been estimated at $20 billion, most of this energy related.

Such losses are frequently the result of changes in society rather than unprecedented climate. Their roots often lie in inadequate assessment of potential risks, growth, speculation and unforeseen occurrences such as strikes. A close examination of recent cold winters in the United Kingdom disclosed that climate was not the unique cause of energy shortages, and that equally cold winters had occurred before (7). Man continues to introduce new complications, and vulnerability will increase as more people compete for finite resources. Changes in climate due to human activities may exacerbate this condition since they alter the design bases for the demand and supply of energy.

The credible prediction of future climates would aid energy planners in their response to future demands. The present level of skill falls short (10). As a fall back, scenarios and risk statistics provide useful guidance, but they too, may fail to provide the quantitative impact information needed in planning.

Both climatic and energy futures are uncertain. However, the nature of climatic controls are known and these define the regions where heating is required, and where solar energy is available directly or indirectly in useful quantities. Climatic anomalies, their impacts on energy supply/demand and the ways in which climate can be exploited in responding to energy needs are also being defined more concretely. The exploitation of these known features of climate is explored in this report.

CLIMATE AND ENERGY DEMAND

Climate has a dual character, that of a hazard and of a resource. Man's major preoccupation has been with the hazardous aspects of climate. Much effort has been expended on technologies that counter cold, excessive heat, humidity, ice, snow, winds and other manifestations of climate. Over one-third of the energy consumed in industrialized North America (Table 1) and about 50% of that consumed in Europe (50% in Denmark (29) and Britain (5)) is used to overcome the direct or indirect consequences of climate. Marked regional differences in this fraction exist because of the spatial variations in vulnerabilities.

Direct Effects on Demand

The influence of climate on energy use is most evident in space heating. Significant other demands occur in agriculture where large amounts are consumed in field work, crop drying and irrigation, in transportation to overcome snow, ice and cold; and in other outdoor industries such as construction, forestry etc.; however, these are small by comparison to that for heating.

Table 1. Energy used in climate-sensitive activities expressed as per cent of total use. (13, 14, 15, 30)

	U.S.A.	Canada	Denmark
space heating	18	e 29	50
water heating	4		
air conditioning	3		
refrigeration	2		
transportation	26	26	17
agriculture	e 5	8	6

Values are approximate and for the past decade. They are from diverse sources and when prefixed by e, they are estimated from generalized statistics.

Space heating is greatest where there are concentrations of people in the pursuit of commerce, transportation and agriculture. The concentration of space-heating tends to conform with the 10C mean annual temperature isotherm and exists mainly between the 2 and 18C isotherms (Table 2 and Figure 1). The industrial areas of Europe, Asia and North America which lie within this belt close to the 10C isotherm account for about 85 per cent of global energy use. (22).

Table 2. Regional Energy Consumption in 1970 and Estimates for 2000 AD. (1).

Region (a)	1970			2000		
	Consumption (b)	Percent of market	kw per capita	Consumption (b)	Percent of market	kw per capita
Africa	109	1.6	0.29	564	1.6	0.63
Asia (c)	1,043	15.3	0.46	13,378	37.8	3.23
Central America	142	2.1	1.94	1,216	3.4	6.17
North America	2,472	36.3	9.90	7,934	22.4	21.76
South America	112	1.6	0.47	710	2.0	1.37
Western Europe	1,350	19.8	3.44	4,367	12.4	9.04
Eastern (d) Europe	460	6.7	4,04	1,625	4.6	11.69
Oceania	78	1.1	3.67	364	1.0	9.44
U.S.S.R.	1,055	15.5	3.97	5,242	14.8	14.51
Total	6,821	100		35,400	100	4.96

a) United Nations Statistical Year Book
b) Millions of tonnes coal equivalent
c) Excluding the Asian part of the U.S.S.R.
d) Excluding the European part of the U.S.S.R.

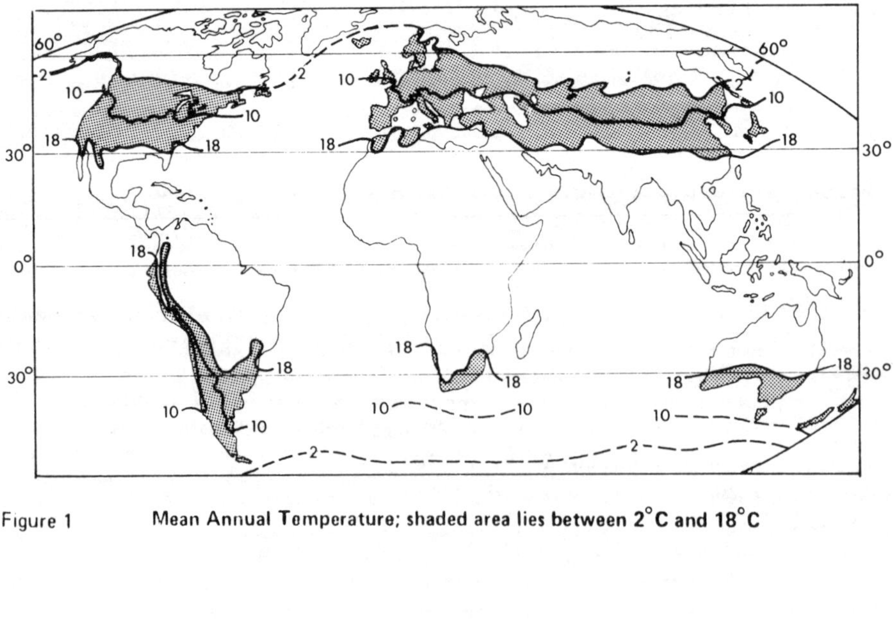

Figure 1 Mean Annual Temperature; shaded area lies between 2°C and 18°C

Both heating and cooling demand can be estimated by the use of degree-days. The base temperature for computation is subjective because of differences in insulation, construction practices, building exposure and management, and the actual consumption is also dependent on the efficiency of the heating (cooling) system. Furthermore, large differences in energy use per degree day occur even at the community level because of differing life-styles. Nevertheless the relationships are useful in treating with aggregated use. The use of heating degree-days assumes a normal condition of wind, humidity, radiation etc., but the introduction of these variables into the energy-use equation does not enhance its utility for purposes under discussion (24, 26). Base temperatures of 15, 18, 18.3, 19C etc. have been used; the value 18.3 is employed later in this paper for convenience. Annual accumulations are of particular interest since many climatic predictions relate to annual temperatures.

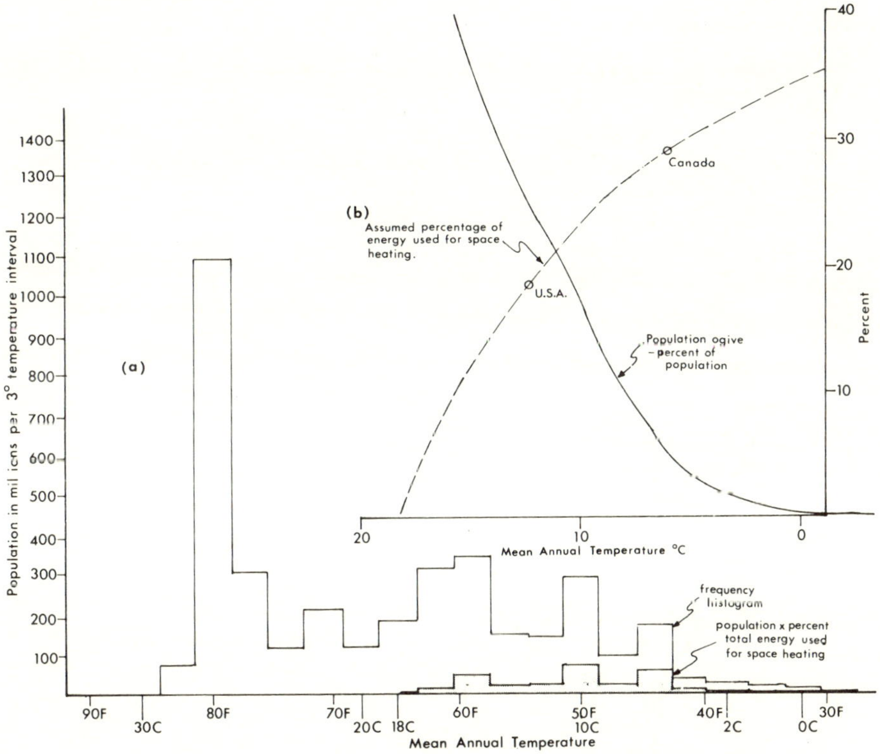

Figure 2 (a) Concentration of population according to mean annual temperature
(b) Population ogive and assumed percentage of energy used for space heating

Anomalous Periods

Figure 2 depicts the relationship between mean annual air temperature and population. Assuming consumption for heating approaches zero at 18C, and 35 per cent at 0C (based on U.S. and Canadian rates) and equal per capita energy use in the space-heating belt, greatest energy use can be shown to occur between 5C and 15C, peaking near 10C. Were rates more similar to that for Denmark, then the modal value of the energy-use histogram would be shifted toward 6C, but there would still be relatively little energy used in areas with mean annual temperatures below 2C. Climatic warming or cooling is equivalent to shifting the abscissa in Figure 2, to the right or left respectively. A one degree shift would alter space heating energy demand for people living in the present intensive-energy-use area by roughly 10 per cent. (Figure 3).

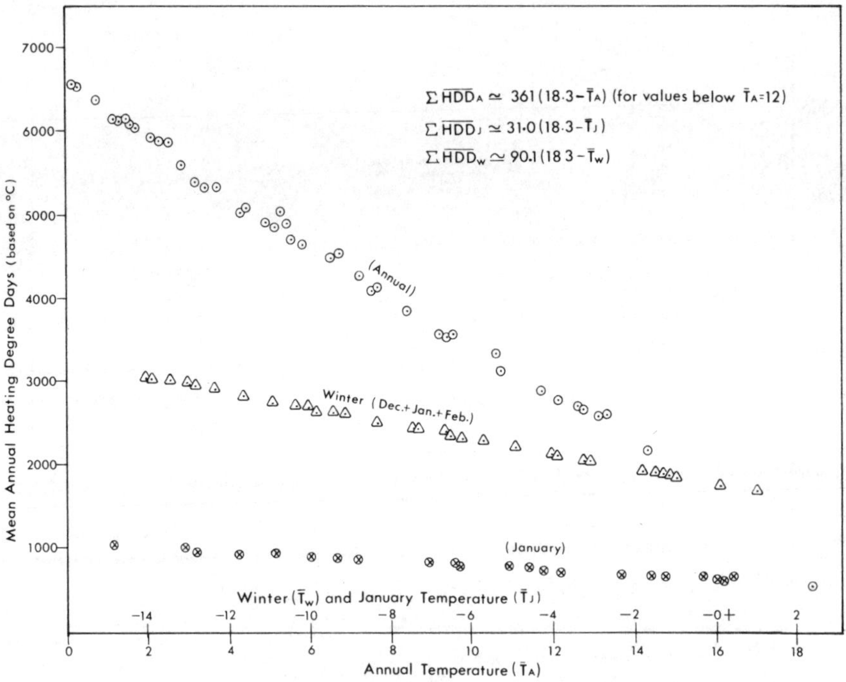

Figure 3 **Relationships between heating degree-days, and mean temperature**

Events during recent cold winters can be used to further assess the impact of climate on energy demand. In the winter of 1976-77 the temperature anomaly south of the Great Lakes was in excess of -4C and reached -6C in the Ohio Valley (36). January was extremely cold with heating degree-day anomalies reaching +58% in Alabama, +59% in Georgia and +96% in Florida (33) (Figure 4). Locally, winter heating-fuel-use was increased 50 per cent or more, the higher values being in areas where heating demands are normally low and buildings not well insulated. An American periodical, Business Week, (Feb. 14, 1977) claimed that the resulting "diversion of purchasing power to higher fuel consumption and higher prices for goods is estimated at 2 billion to 5 billion dollars". The 45-day "freezing-in" of river barges that transported oil, and stoppages in unloading of fuel from railway cars etc. compounded the problem so that consumption was probably less than climatology would indicate.

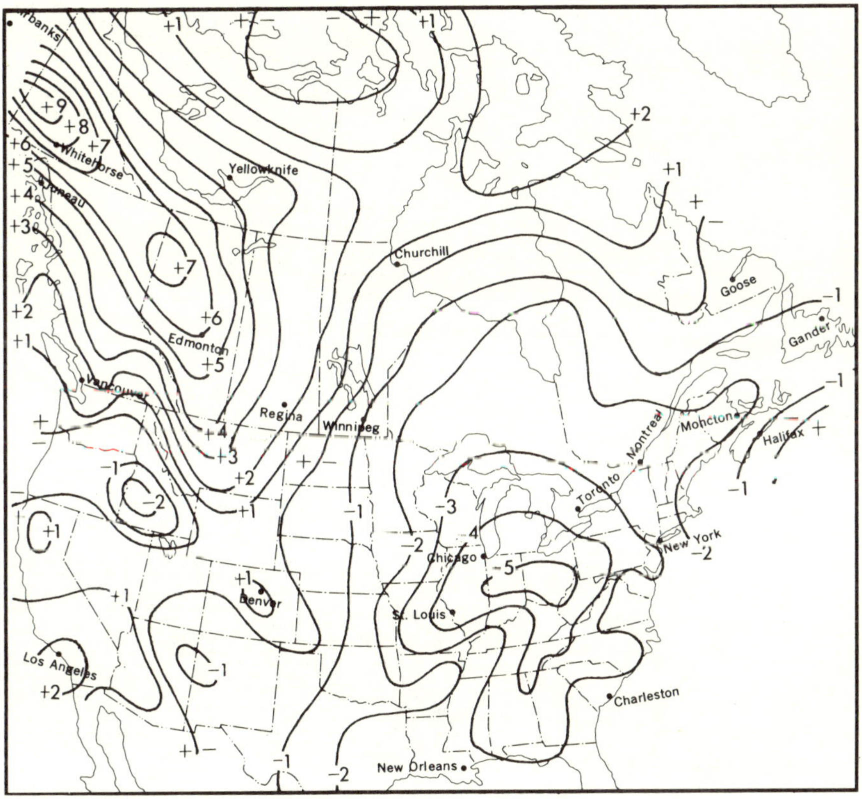

Figure 4 **Departure from 1941-70 Average Temperature (°C) Winter 1976-77** (Dec.-Jan.-Feb.)

Other "design" winters or epochs may be evaluated in a similar manner to assess the effectiveness of today's technology under changed social and climatic conditions and to develop strategic plans. The same approach can be used for cooling energy, (27, 31).

Man's Impact

Changes in climate wrought by man are presently evident on the local to regional scale. The urban heat island of large temperate-zone cities is well known. The space-time patterns for Toronto, Canada illustrate that over the past century the mean temperature at Toronto has risen 1.6C as a result of urbanization (Figure 5). A similar increase (1.2C over the period 1891-1977)

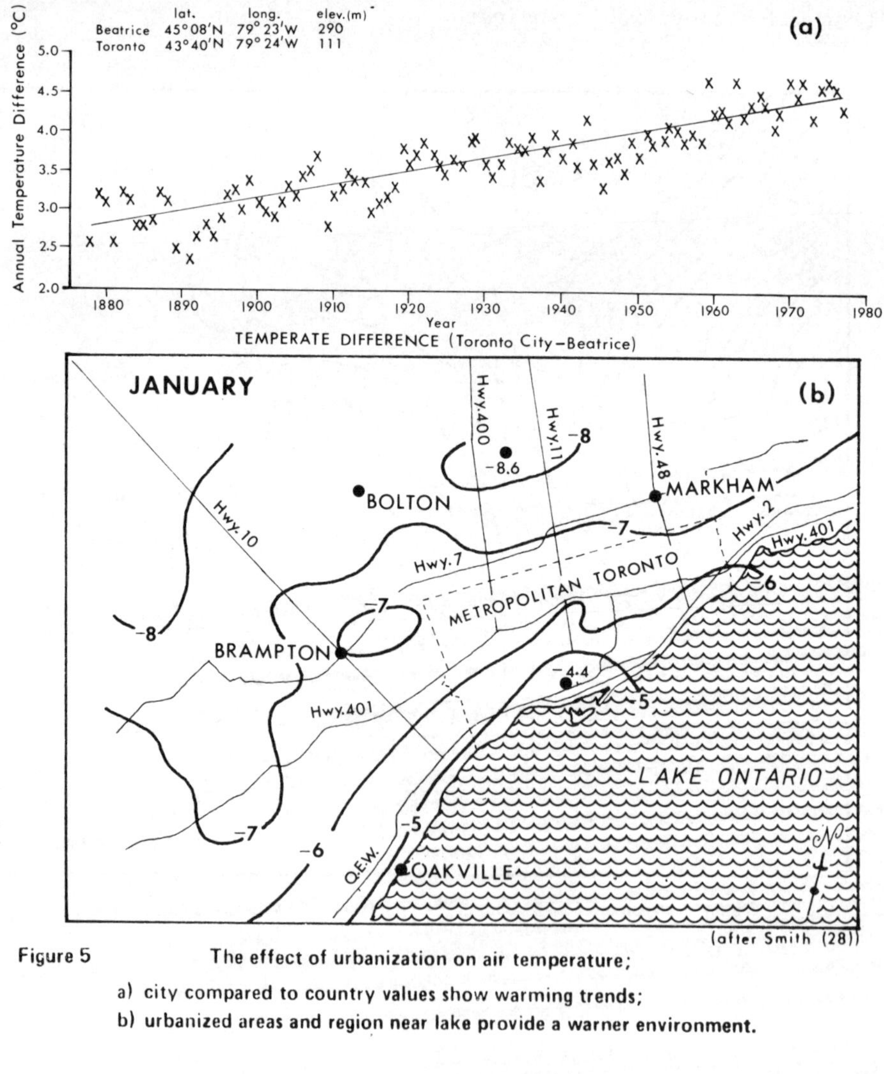

(after Smith (28))

Figure 5 The effect of urbanization on air temperature;

a) city compared to country values show warming trends;

b) urbanized areas and region near lake provide a warner environment.

has been observed for Paris, France (11). Such changes in temperature have major implications for heating demand. As a city grows the unit-heating-demand is reduced within the city, and downwind in the urban plume.

A 1.6C temperature increase in Toronto would reduce space heating demands about 12 per cent. This benefit may be partly offset by increased demands for air conditioning in the summer months. In warmer, more humid climates urbanization has contributed significantly to heat stress and created a surging demand for air-conditioning. Urban and building architecture that exploits shade and ventilation can be used to reduce air-conditioning energy demands.

Of equal significance is the opportunity to use land form and cover, using airshed concepts, to minimize air quality problems over the vulnerable areas, and to exploit not just building and heat source areas but also the opportunities offered by the natural mesoclimate. Exposures which are warmer, sunnier and less exposed to wind can be identified and exploited. Energy savings of 1.7 per cent of total use in the U.S.A. may be attainable in this manner (25). In northern regions urban and rural designs which limit the energy expended to overcome wind and snow, and which take full advantage of sunlight and waste energy, may have even greater significance.

These considerations apply mainly to new construction, and in some instances the additional capital costs have made them unattractive to the developer.

Operational Efficiency

The effects of climate on operational energy requirements are diverse and difficult to aggregate. Changing costs for energy have made it advantageous to consider climate more critically in achieving efficiency in energy use. Because of their sensitivity to climate, agriculture and transportation provide good examples.

Agriculture

Climatic variability can affect crop selection, field work, irrigation, drainage, shipping and market opportunities. Unexpected energy demands may result such as in grain drying, working in wet soils or even in replacing damaged crops. In the production of corn, 50 per cent of the energy use is related to fertilizing, 25 per cent to mechanical operations, 20 per cent to corn drying and 3.5 to pesticide use (17). Climate is a factor in each of these activities, and climate variations increase or decrease the energy consumption levels.

Possibly 15 per cent of nitrogeneous fertilizers are washed away as a result of untimely application (25). Studies by the United Kingdom Meteorological Office show potential accumulated savings of £105 million by buying tractors whose power is determined by farm tractionability, a function of soil and climate, and of £20 million in the use of climatic information in vegetable storage. These would reduce energy demand a) in tractor production, b) in the lower consumption rate of tractor fuels, and c) in the energy expended to grow vegetables to replace those that were damaged. Similar examples can be given for drainage, crop spraying, and other capital or energy intensive activities.

Much on-the-shelf technology exists in agriculture. A major task is to put it to practical use. Progress is being made in this domain continually with the appearance of "climate watches", and the development of models that provide farmers with timely decision information rather than climatic abstractions. This type of service has the potential to effect major energy economies in agriculture, from seeding through grain drying, to marketing. Losses in the United States for which protection was possible given adequate warning, have been estimated to be $3.5 billion (32).

Gradual climate change will slowly change the ratio of good-to-bad years. If some lands become uneconomic they will be abandoned while others are developed. Major costs are incurred during transition and if the lower class lands are substituted. It is possible to assess the effects on land-use capability using modelling techniques and by the use of isolines of climatic threshold or index values, but computation of the net effect on energy demand requires much more understanding of technology, lands and markets than is presently available.

Transportation

A transportation system moves goods from where they are plentiful to where supplies are deficient. Climate variations create local-to-regional surpluses and deficiencies. For example, coal or oil may have to be shipped to a drought-stricken area to replace hydropower. The marshalling of energy supplies to accommodate the transportation can be facilitated by the use of climatic monitoring and prediction.

In northern latitudes the understanding of factors such as sea ice, permafrost, cold and snow have assumed new importance in planning and operations. The use of climatology for the selection of the safe and economic air and sea routes, and reliable ports can result in major savings over the long run. A 3 per cent fuel economy has been estimated for ocean travel using meteorological information for route selection (25). The blockage of harbour facilities and routes by ice or fog results in costly delays, high risks, and possibly undesirable storage of offshore energy resources. The disappearance of ice from inland and Arctic waterways would extend shipping seasons, reduce the fleet size and eliminate the cost of fighting ice in terms of ship design as well as fuel.

Snowfalls are major obstacles to transportation. Large cities now find removal costs for individual storms frequently in excess of a million U.S. dollars, and additional energy consumed in transportation probably equals this amount. Anomalous winters such as 1971-72 produce record snow cover that can paralyze transportation over long periods of time. Microclimates and snow control structures are being used increasingly in cities and rural highways and railroads to reduce the mounting energy bills for snow removal illustrated by Montreal's snow removal budget (Figure 6).

ENERGY SUPPLY

Climate dictates the potential supply of renewable energy resources. It is also a factor to be contended with in most phases of conventional energy system activities.

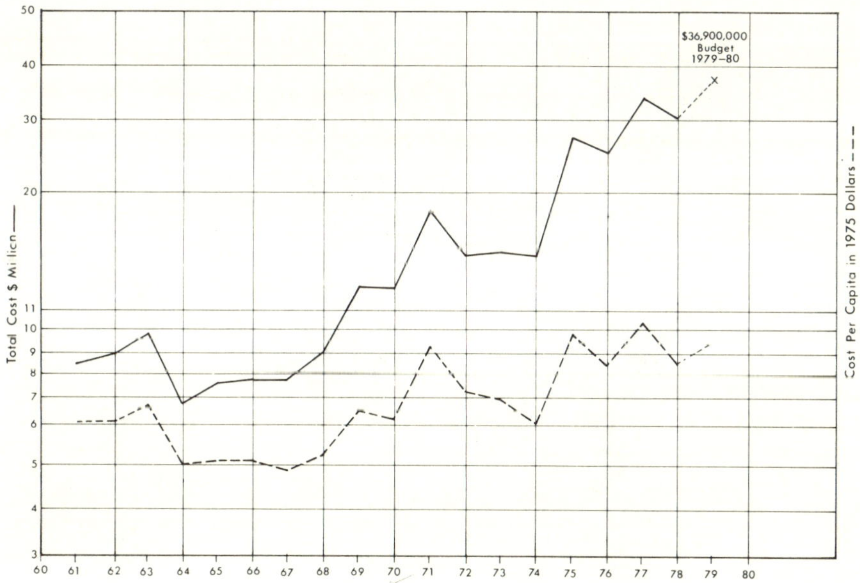

Figure 6 Costs of snow removal Montreal, Canada

Renewable Energy

Renewable resources are not expected to contribute a major portion of the world's energy for many years. Nevertheless, there are very persuasive arguments, economic and environmental, for adapting to the natural energy flows which originate with the sun and which are manifest as light, sensible heat, in living things, and in motion or other physical states of our environment. The supply is large (4, 35) and there is an increasing array of technologies to exploit renewable energy, and the diversity of sources compensates in part for their dilute nature. The mounting costs of conventional fuels and national desires for energy independence are making the renewable technologies increasingly attractive (21, 29).

In general supply is related to geography (Figure 7). Solar energy systems are most promising between 35 degrees North and South latitude, and of value as a supplemental source to 55 degrees. Significant hydropower and biomass production also occurs in these zones, the greatest productivity occurring in the humid tropics. The ocean thermal gradients are most favourable in tropical latitudes, wave energy in ice-free waters, while wind energy is ubiquitous. Changes in climate reduce or increase the overall supplies, and must be considered in the assessment of economic viability and the need to develop substitute or back-up systems.

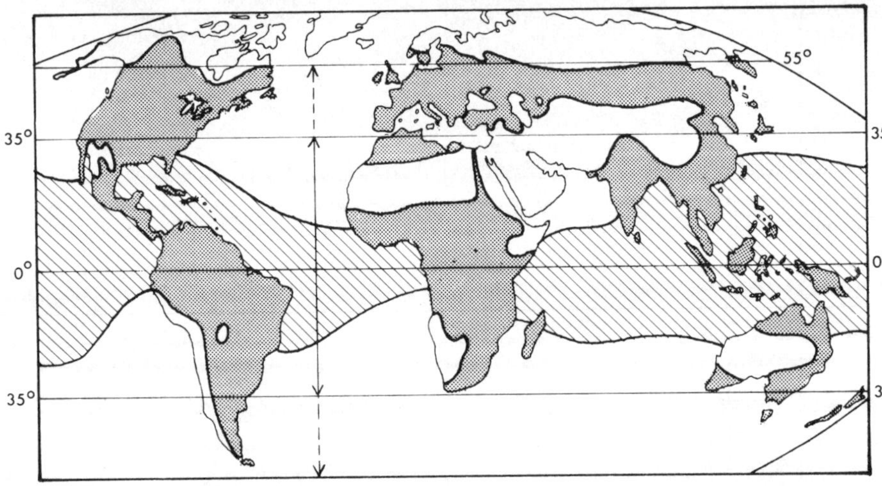

area of significant potential biomass ⟍⟍⟍ ≥ 20°C gradient - surface to 1000 m depth

⟶ latitudes between which solar is viable as main source

- - → latitudes between which direct passive solar is viable as a supplementary source

Figure 7 Areas favorable for exploitation of solar, biomass and ocean-thermal energy

Hydropower

Hydropower presently supplies about 3 per cent of the world's energy needs. Its potentials have been estimated as:

	kW hours/yr.	% of global total
Africa	358,300	16
Asia except the USSR	610,000	28
Europe except the USSR	163,000	7
North America	356,400	16
Oceanic	45,000	2
South America	431,900	20
USSR	250,000	11

Major development is expected over the next quarter century (3). As with all renewable systems, the possibility of climate change creates uncertainty as to future viability, return on invested capital, and bases for their engineering design.

Storage must be provided to overcome climatic variability. Over the United States the annual precipitation ranged from 4937 to 6691 Km^3 during the 1931-71 period, the two extremes occurring in consecutive years (L. Heindl, personal communication). Changes at Barnaul, Siberia (Figure 8) convincingly show the need for climatic information in system planning. Systems designed on

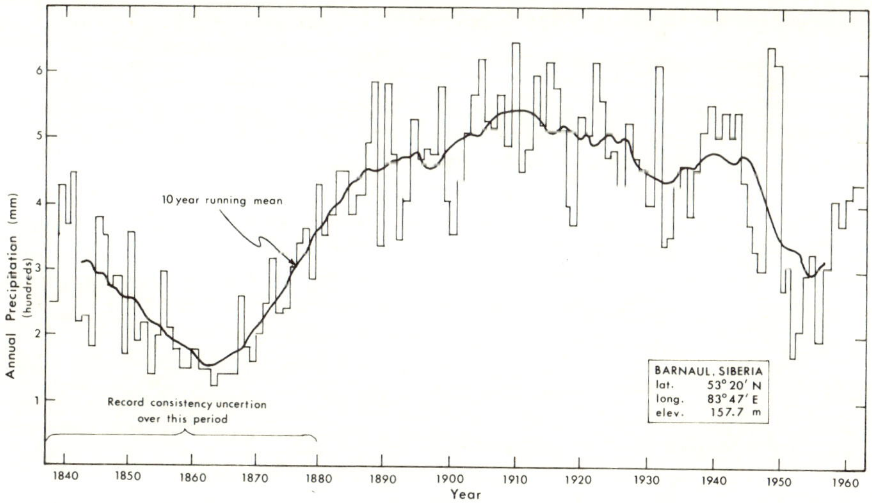

Figure 8 Annual precipitation at Barnaul, Siberia.

(Source Annales de l'OBservitoire Physique Central – Russie (2) ; World Weather Records (34))

the record prior to 1880 would be quite deficient in terms of the subsequent climate. Changes in evaporation, (the major demand on most reservoirs) and the reliability of future rainfall must be known to estimate optimum reservoir capacity and for planning operational strategies.

The resource potential is greatly diminished in drought and increased in wet periods. For example, the power generation of the Niagara River System in 1964 was reduced 26 per cent by drought conditions. Generation at points downstream was also reduced, e.g. by 19 per cent at Cornwall, Ontario.

Complications introduced by societal change were evident in the drought of 1976-77 in California. Precipitation was 65 per cent of normal in 1976 and only 45 per cent in the 1977 water year. By October 1, 1977, stored water was 38 per cent of average and 23 per cent of capacity of the States' reservoirs. Mandatory curtailment of use followed in 150 communities. Hydroelectric energy generation (20% of total on average) was reduced from about 33 billion to 16 billion in 1976 and 13 billion kWh in 1977. The additional power required to pump ground water was 1 billion kWh (6). Obviously knowledge of drought is of great relevance in planning for and responding to such occurrences.

The hydrological balance is being altered by urbanization and industrialization. Increases of 20 to 30 per cent in summer rainfall and of 15 per cent in stream flow were reported to have occurred near St. Louis, Missouri (8). Readjustments in rainfall patterns of this order and larger have clearly economic significance for the design and operation of hydroelectric systems.

Global precipitation patterns have been quite different throughout history. Changes in the atmospheric circulation and moisture loading could greatly alter the volume and timing of orographic, monsoonal and other precipitation, thereby changing the opportunity for hydropower generation. In some areas snow is a major storage medium and its reduction would reduce manageable supplies. The past is not necessarily a blue print for the future, but recognition of history and of how society is altering its vulnerability to climate, is necessary for the design of more resilient systems.

Active and Passive Solar Energy

Solar energy is now extensively used for water heating, and its capture for space heating is a growing practice. On-site systems are expected to be competitive with other central utility power in 10 to 15 years. There is speculation that solar energy will be a prime source given appropriate levels of development funding.

Much attention has been given to technology for the direct conversion of this energy to heat or electricity. More recently there has been increased interest in the potential of exploiting all energy flows in both a direct and passive manner. With inexpensive energy, architects tended to ignore the opportunities provided by the natural environment. Changes in building codes in the Soviet Union have been recommended (9) to enable the capture of solar energy, estimating that energy savings of 20 to 30 per cent are possible by this method. Throughout history, shade, the absence of windows, and the capture of the wind have been exploited to obtain an acceptable indoor climate. The natural energy sources and sinks can be exploited to replace other energy demands. By exploiting climate in this manner, Vancouver, Canada housing designers estimate they can reduce conventional energy costs by 80 per cent for an increase of 10 per cent of the capital cost of new housing. Retrofit of old housing has also been demonstrated to save energy at costs that are competitive with conventional fuel alternatives (Bryenton, personal communication).

Back-up systems must be designed to cope with periods of low insolation. Longer-term change is unlikely to be of major concern as additional capital costs will be recovered within a few decades. Climatic shifts such as that from wet to dry as in the Sahara (from humid pluvial in 5500-2350 BC to extreme aridity in 2350-800 BC (16)) would greatly alter patterns of opportunity, but the time element and compensating features make this an item of interesting speculation rather than a concern.

Biomass Conversion

Wood provides fuel for possibly over 2 billion people in developing countries (4). Its popularity elsewhere is increasing and merchantable supply is diminishing as costs of fossil fuels rise. The opportunity for its exploitation is highly related to climate. Biomass productivity is limited by temperature and aridity; productivity tending to double for every 10 degree C increase in mean annual temperature between the values -10 and 20 (20). This is reasonably consistent with agricultural productivity which in Canada and the USSR is only 50 to 60 per cent of that in warmer agricultural areas of the USA, (12).

The postulated trend to warming and increased atmospheric carbon dioxide concentration favours biomass production, but some regions would be disadvantaged by unfavourable water balances. Since many forests require decades to mature, the matter of change is of relevance in forest regeneration for optimum biomass production. Short-term changes are of concern in the control of fire and pestilence as well as in the harvesting of forests. Climatic monitoring, risk statistics and prediction are essential for mitigation and marshalling resources for these purposes.

Wind and Wave Energy

Wind power is very practical in select locations. These are generally along sea coasts, on mountain crests where the free air flow is relatively unimpeded, and in locations where orography funnels or accelerates air flow. Favoured countries may satisfy 10 to 30 per cent of their energy demand from this source, without storage (30).

Wind is competitive with other energy sources in localities where speeds are strong (extractable power α $\frac{1}{2}$ (wind speed)^3). Wind systems are superior energy extractors (currently 35% efficiency vs 5 to 15% for solar) and can produce power at costs that are competitive with fossil fuels. Multi-megawatt turbines are in the development stage. The spatial patterns of power and reliability need for estimation of the source potential storage requirements, and network operation are revealed by the interpretation of climatic records for an area.

Wave energy, too, is great but highly distributed. Exploitation opportunities are analogous to coastal wind-power generation. Both may require back-up systems for periods when winds are light (23).

Ocean Thermal Gradients

Ocean thermal gradients of about 20C or greater over a depth of 1000 m are considered viable energy sources. Realizable efficiency for a 20-degree gradient and a warm source of 22C is about 2.2% (18). A 2-degree reduction in thermal gradient would reduce the theoretical efficiency by 0.6 per cent, i.e. 27 per cent of the achievable potential. The sustained cold winter of 1976-77 caused a mixed layer of 18C water to a depth of 550 m in the Sargasso Sea (35N) (19). Areas where sustained cold outbreaks can occur would appear vulnerable. General warming or cooling would expand or contract substantially the large area now considered viable (Figure 7).

Conventional Systems

Since climate is a factor in virtually every phase of conventional fuel cycles, it has a direct relationship with conventional energy supplies.

Exploration and transportation are climatically-vulnerable phases at sea and in high latitudes. Climate must be considered in the selection of optimum shipping routes, ports and harbours from the viewpoint of economy and safety. Strategies, economics, engineering and environmental aspects of offshore drilling, and the development of coastal facilities require similar attention to climate and climate-induced sea states. This is particularly true in extreme cold and ice. The extent, thickness and ridging of ice relates to climate as does freezing spray, the brittleness of metals and the performance of "temperate zone" technology. Pipelining presents similar challenges, including concerns over expansion and contraction due to temperature changes.

Storage and Use

Air temperature, humidity, solar radiation, wind, and precipitation have relevance to the design and operation of storage systems. Since future energy systems may exploit a variety of energy sources a diversity of climatic statistics and scenarios will be needed to help avoid future crises due to inadequate storage and delivery strategies. The better use of climatic scenarios and statistics in the planning and regulation of these systems would have mitigated recent losses.

The climatologist should be involved in many other aspects of design such as the study of the releases of waste heat to cooling ponds or the atmosphere which may alter cloudiness, precipitation, fog frequencies, evaporation, and in certain instances, local atmospheric circulations. Climatology can also serve in planning the use of waste energy, such as for agricultural purposes. The proposed development of smaller, highly dispersed reservoirs and centrals will necessitate more detailed understanding of meso-scale climate. Siting of thermal plants and their safe, reliable and economic operation requires understanding of atmospheric dispersion and transport and deposition of emissions, of tornadoes, strong winds, lightning strikes and of ice accretion that may damage peripheral structures, transmission systems, and powergrid performances.

Climatology is needed in the understanding, predicting and minimizing the social and environmental costs. Its role in the control of air quality is well recognized as is its value in the avoidance or mitigation of adverse impacts from energy centres, offshore exploration, etc. Many meteorologically based strategies have already been proven. For example, the pollution created by the use of low quality fuels in individual heating systems may be overcome by the use of higher-combustion-efficiency central heating plants. Natural-hazard type analyses or scenarios are being used to evolve other strategic and tactical plans for minimizing environmental disasters that might result from spills of oil or the release of hazardous materials. Land use policies needed to minimize environmental damage by energy systems must be made with due recognition of the many manifestations of climate and the opportunities and limitations they impose.

CONCLUSIONS

Climate is intimately linked with energy demand and supply.

The association is most evident in the demand for space heating. The bulk of the demand is created by the industrialized countries in temperate latitudes. Within the heavily industrialized zone a one-degree change in mean annual temperature would alter demand by about 10 per cent. The amount would vary highly within countries as a result of differing levels of industrialization and life-styles.

The regional energy crises which occurred during recent cold winters are evidence of the increasing vulnerability of society to climatic change. Although climate is a factor in their occurrence the problem is usually more properly attributable to a societal change which has altered the consequences of an accepted level of risk. On the other hand the creation of cities and industrial areas is warming climate and reducing the per capita need for space heating.

Cooling is the greater need for over half of the world's population, however the requirement is less critical, and the 'conventional-fuel' energy demand for this purpose is relatively small.

Climate controls the quantity and variability of renewable energy sources. At present these supply only a small percentage of global energy consumption. Nationally and regionally they will become much more important for economic, environmental and political reasons. Their rational development require adequate knowledge of climate, its many manifestations of variability, and of related consequences. Generally the renewable energy supply is great and will not be used to natural limits, so that small variations have small overall impacts.

Climate impacts on the supply of conventional energy, affecting exploration, transportation, storage and use. Environmental concerns as well as economics make the increased understanding of climate and its use in design essential. This need is increased by trends to smaller, dispersed production units, and the need to diversify fuels.

REFERENCES

(1) Ahti, K.: 1975, "Application of Meteorology to Consumption and Transport of Energy," World Meteorological Organization, rapporteur report, unpublished, 25 pp.

(2) Annales de l'Observatoire Physique Central Nicolas: 1838-1920, St. Petersburg, Obscestvennaja Poliza.

(3) Armstrong, E.L.: 1978,"Hydraulic Resources," World Energy Resources 1985-2020, World Energy Conference, 1 PC Science and Technology Press.

(4) Auer, P.L. et al: 1978, "Unconventional Energy Resources," World Energy Resources 1985-2020. World Energy Conference 1 PC Science Tech. Press.

(5) Benn, A.: 1978, RIBA Journal, London, 85, pp. 218, London.

(6) Brown, A.J.: 1978, "The California Two-Year Drought,"Proc. 46th Western Snow Conference, pp. 85-91.

(7) Burroughs, W.: 1978, New Scientist, 77, pp. 146-148.

(8) Changnon, S.A. 1977, "Climatic Change and Potential Impacts on Water Resources," Living With Climatic Change Phase II, MITRE Corporation, McLean Virginia U.S.A, pp. 85-94.

(9) Davidson, B.M.: 1978, Meteorologiya i Gedrologiya, 8, pp. 109-112.

(10) Davis, W.K.: 1977, "Energy and Climate," Living With Climatic Change Phase II, MITRE Corporation, McLean Virginia U.S.A., pp. 95-100.

(11) Detwiller, J.: 1978, La météorologie, VI, pp. 95-138.

(12) Field, N.C.: 1968, Canadian Geographer, XII, pp. 1-14.

(13) Ford Foundation: 1974, Exploring Energy Crises, Washington, D.C, 141 pp.

(14) Gander, J.E. and Belaire, F.W.: 1978, "Energy Futures for Canadians," Canadian Government Publishing Centre, Hull, Quebec, 353 pp.

(15) Hirst, E. and Moyers, J.C.: 1973, Science, 179, pp. 1299-1304.

(16) Kovda, Victor A.: 1979, "Soil Reclamation and Food Production," Food Climate and Man, John Wiley and Sons, New York, pp. 159-186.

(17) Laflen, J.M.: 1977, "Energy Reduction," Research Progress and Needs; Conservation Tillage, Agriculture Research Service, U.S. Dept. of Agriculture ARS-NC57, pp. 64-67.

(18) Lavi, A.: 1979, "Ocean Thermal Energy Conversion (OTEC). A Worldwide Solar Energy Resource." UNITAR Conference on Long-Term Energy Resources, Montreal, 31 pp.

(19) Leetma, A.: 1977, Science, 198, pp. 188-189.

(20) Lieth, H.: 1973, Human ecology, 1, pp. 303-332.

(21) Lovins, A.B.: 1978, "Soft energy paths," Fourth Annual Conference, Solar Energy Society of Canada, Inc., 30 pp.

(22) Massachussets Institute of Technology: 1971, Inadvertant Climate Modification, MIT Press, Cambridge, Mass., 291 pp.

(23) Merriam, M.F.: 1978, "Wind, Waves and Tides." Annual Review of Energy, 3:29. Annual Reviews Inc., Palo Alto, California.

(24) Murphy, A.H.: 1954, "An analysis on the influence of the wind on the relationship between degree-days and fuel consumption," B.Sc. thesis, Mass. Institute of Technology, 46 pp.

(25) Murphy, B.L., Schulman, L.L. and Mahoney, J.R.: 1977, Bull. Amer. Met. Soc., 58, pp. 304-317.

(26) Pallot, A.C.: 1940-41, Jour. Institution of Heating and Ventilating Engineers, 8, pp. 249-277.

(27) Robertson, P.: 1974, The Building Service Engineer, 42, pp. 195-198.

(28) Smith, R.: 1976, "Climate," Central waterfront planning committee information base. Toronto, 65 pp.

(29) Sørensen, B.: 1975; Science, 189, pp. 255-260.

(30) Sørensen, B.: 1979, "Wind energy," UNITAR Conference on Long-Term Energy Resources, Montreal, 27 pp.

(31) Thom, H.C.S.: 1954, Monthly Weather Review, 82, pp. 1-6.

(32) Thompson, J.C.: 1977, "The economic value of improved long-range weather forecasts." Living With Climatic Change, Phase II, MITRE Corporation, McLean Virginia, U.S.A., pp. 67-76

(33) United States Dept. of Commerce: 1979, State Regional and National Monthly and Seasonal Heating Degree-Days, weighted by population. National Climate Center, Asheville, 67 pp.

(34) United States Dept. of Commerce: 1960, World Weather Records, (1921-60). Supt. of Documents, Washington.

(35) Von Arx, W.S.: 1974, GEOS, 55, pp. 828-832.

(36) Wagner, A.J.: 1978, Weatherwise, 31, pp. 53-59.

CLIMATE CONSIDERATIONS AND ENERGY CONSERVATION

J. K. Page

Department of Building Science, University of Sheffield, Western Bank, Sheffield, S10 2TN, U.K.

ABSTRACT. Global climate impacts are the consequence of billions of individual impacts arising mainly within the settlement patterns. Settlements consume a large proportion of global energy resources in their construction and running. There is, therefore, the need to think in terms of the "material-energy conservation city". Climate can play an important role in the conservation of energy. Correct urban design can reduce the adverse climatic impacts leading to lower energy losses and simultaneously contribute through alternative energy inputs to the solution of the energy problem in an ecologically acceptable way. Urbanism needs a sounder energy base, enlivened by proper climatological awareness.

INTRODUCTION

The currently developing adverse modifications of the global climate are the consequence of the integrated impacts of billions of small activities taking place over decades in the individual home and at the individual workplace, either agricultural or industrial. As only the individual can be the ultimate conserver, in the final analysis, the future of the global climate is essentially in the hands of the billions of individuals, who make most, but not all, of their climatic impacts through the various energy chains with which their lifestyles are interlinked. While conservation, pollution and lifestyle are very strongly interwoven, the links between them are currently very poorly perceived. One reason is that the global feedback develops so slowly that the feedback is totally separated temporally from the particular individual's actions, with the consequence that few people can

W. Bach, J. Pankrath, and J. Williams (eds.), Interactions of Energy and Climate, 73–88.

perceive any real reason to change their personal actions. Besides, living in a world, which exists in mounting tension between large power blocs, we, as individuals, have acquired a sense of powerlessness, within the global political scene, that we easily tend to carry over into the global environmental scene. The argument frequently runs - "If I deny myself something, my neighbour will not, so what is the point in my doing anything ?" Clearly the creation of a better global sense of environmental purpose is urgently necessary, but unfortunately the background arguments involved are very complex. If these environmental goals can be tackled in ways that simultaneously confer positive economic benefits, the quality of human life can be improved both on the material plane and on the environmental plane. It is for this reason that renewable energy is potentially so important. The sun is our only reliable entropy pump for creating more order from potentially greater disorder. The natural world has evolved to export the entropy increases associated with the creation of greater local natural order to space. Fundamentally we need to consider the entropy balance of the world.

CLIMATE AS A RESOURCE AND HAZARD FOR HABITATION

The demands for energy in habitation are strongly influenced by climate for the natural climate is usually only acceptable for human life indoors for part of the year. While we have become habituated in recent years to achieving the indoor regulation through fossil fuels without proper regard for climate in building design, this is not a necessary attitude, nor even a historic attitude, because buildings used to be much better conceived climatically in the past. Human settlements in turn must be considered in terms of their impacts on the climate at the microclimatic, mesoclimatic and macroclimatic levels (1) (2). In terms of human settlements, climate must be considered both as a renewable resource and a periodic hazard (3). We need to increase the climatic contribution as a resource, for example making wider use of the renewable energy for space and water heating and for cooking, especially in the developing world. We also need to reduce climatic hazards, for example controlling the potential destruction produced by storms and floods which represent short term conditions of excessive energy and mass flow leading to the serious disordering of an ordered settlement system.

HUMAN DIFFICULTIES IN MAKING PROGRESS

While the various components of the problem are strongly interactive, the precise interactions are hard to unravel, particularly for ordinary people, due to temporal and spatial separation effects. It is not obvious, for example, that the cutting down and burning of the extensive rain forest jungles in Brazil for industrial and urban purposes could eventually lead to a significant melting of the ice caps thousands of miles away, and be associated with dramatic adverse long term changes of world climate. It is virtually impossible for an individual peasant cutting down a particular tree in Brazil to perceive this change to be the eventual outcome of the integration of his own activities with those of millions of others like him. Any approach based on high intellectual appraisal of cause and effect at a complicated academic level will clearly make no impact on actual world activity. If we believe the future of climate is vital to the future of man, we must communicate the concepts clearly at the primary school level, which means for a start we must re-educate all the primary school teachers of the world, which is no small task.

STATUS OF CLIMATOLOGISTS IN WORLD PLANNING

However, if we believe this to be the problem, do we yet have enough status as climatologists to be held in proper world respect for global planning purposes ? As long as the same global input phenomena are predicted to lead to totally different long term climatic outcomes by different groups of experts, we are unlikely to convince anyone of the need for immediate action. All we can do in such circumstances is express scientific misgivings and plead for greater resources to help resolve the fundamental scientific issues more quickly. It is vital that we, as specialists, work together for an agreed global consensus on which to found future policy. Scientific ignorance can lead to the wrong actions being taken with the best of intentions.

I think, unfortunately, it is firmly agreed that the magnitude of the global impacts will increase with the magnitude of the global atmospheric inputs. Therefore the main challenge of conservation is to reduce overall consumption which will, in turn, reduce the global scale of any pollution outputs, and so minimise potential climatic change. It is less clear what the precise changes in atmospheric behaviour will occur at what point of time with any given pattern of change.

LOCAL VERSUS DISTANT IMPACTS

The impacts of settlements on the environment may occur at a local level, for example the combustion of firewood within a home, or the impacts may be far more distant, for example the consumption of electricity from a distant fossil fuel power station. In the first case the user may perceive some relationship between his activity and pollution of the environment while in the second case he may not. The outputs to the environment depend on the scale of the demand and the efficiency with which that demand is met in terms of primary energy supply. As electric utilities typically reject two thirds of the input energy to the environment, they must be considered in terms of carbon dioxide or sulphur dioxide pollution as more polluting globally than any existing direct combustion system with an efficiency of over 30%, a goal which is easy to achieve even with primitive technology. Only when we use combined heat and power systems for electricity, are we likely to come anywhere near the right combustion efficiency levels to make electricity less polluting at a global level. The goal in the design of human settlements must, therefore, be to reduce energy demands by appropriate design of the settlement, and to supply energy in the form that will give the highest primary fuel efficiency for the purpose being considered. I do not wish to take sides on the current nuclear debate, but it is clear that heat rejection to the environment will be a key problem in any successful major development of nuclear energy, either by fission or fusion. The large generating scale already used, with even larger scales contemplated, produce local concentrations of dumped energy which may exceed the capacity of the local environment to absorb them without excessive rises of temperature. The local entropy pump has to operate at a higher temperature to export the necessary consequent disorder.

COMPONENTS OF URBAN ENERGY USE

Typically 50% of energy is used for building heating and cooling purposes, about 25% goes towards manufacturing purposes and about 25% towards transport. If we are to reduce energy demands of human settlements, we have simultaneously to tackle the impact of urban form on building energy demand and on transportation demand, as well as scrutinise the energy content of the building materials we use to assemble our habitats. In high cold latitudes the energy running costs are dominant even in countries where insulation standards are high, for example Höglund (4) suggests that, for Sweden, 2.5% of energy utilisation is for construction, 0.2% for haulage and 3.8% for materials compared with 93.5% for heating for a building with a life span of approximately 40 years. The primary drive in the area of human settlements must, therefore, be to reduce energy demands for heating and cooling. There are two clear and complimentary goals to be pursued, improved conservation and more effective use of renewable energy sources. The significance of the use of renewable energy sources is not just one of energy supply, but also one of preserving existing source sink relationships in the atmosphere. Nuclear energy does not add to the CO_2 problem, but its thermal impacts are very considerable. We may leave on one side the very real additional hazards of a radioactive nature when discussing global climate interactions because at the global scale, the thermal climatic impacts are likely to be far greater.

In the developing world, where the real building problem lies, because most people in the developed world have reasonable habitations by world standards, the population is expanding very fast and adequate habitation is rare. Here the immediate energy demands for building materials represent a much larger proportion of the total settlement energy demand (5).

THE CONCEPT OF THE MATERIALS AND ENERGY CONSERVATION CITY

Architecture and planning are inextricably crosslinked. It is important, therefore, to ask how we might plan to produce a city effectively designed for materials and energy conservation. How then does city design actually affect material resources consumption ? This question clearly resolves itself into two parts:

i) The effects of urban design on the material and energy resources needed to actually construct the city and its facilities.

ii) The effects of urban design on the consumption of material and energy resources resulting from the various activities carried on in the city.

The first part is easier to answer than the second, though there are little properly evaluated data at present. Clearly, more compact building designs with appropriately related areas of wall, roof and window, constructed to appropriate thermal insulation standards, would conserve both energy and materials. Durable long-life building materials would have to be incorporated into suitably flexible building designs. In housing, compact terraced development would offer conservation by wall sharing, thus reducing materials and energy demands, but such buildings would have to be designed to high environmental standards, especially if noise was not to be troublesome. High rise buildings which are resource extravagant both in materials and energy would be rejected in favour of low rise compact development. The present suburban solution of detached houses and bungalows, which have adverse perimeter floor area ratios and consume excessive energy resources, would become increasingly unacceptable. A fundamental problem would be to select an appropriate urban density that balanced the various factors correctly to give optimal resource conservation, i.e. not too dense and not too diffuse. If solar energy is not used, account would need to be taken of the lost photosynthetic potential. The actual construction resource demands depend not only on the resources needed for the building structure themselves, but also on all the resources needed for the service infrastructure, which can be very extensive and costly, including roads, surface water drainage, foul water drainage, electricity and gas energy supply and telecommunication systems. It follows that sensible design evaluation would have to be based on an overall systems approach.

Compactness would favour economic service distribution, give economic public transportation opportunities and reduce the demand for private transportation and the scale of the roads needed for it. Detailed study is likely to show substantial energy conservation advantages for compactness, as well as a saving in materials resources. The diffusiveness of suburbia represents a high resource response, because of the extremely high infrastructure costs to which it leads. It should be noted that infrastructure costs are usually not properly costed at present, and there is a need to use an overall urban systems approach, so the various pay-offs and penalties are correctly identified. Infrastructure costs are usually treated as unavoidable overheads, insensitive to the basic building design decisions and are costed by average oncost value techniques. The infrastructure resource demands of different types of urban form thus need to be far more systematically explored on a collaborative intersystem basis to secure appropriate economy. Design for the bicycle would be given high priority, because of its very great energy and resource conservation advantages.

URBAN SHAPE

Critical attention would have to be given to the shape of the city. The circular city represents the most compact form, the linear city the most resource extravagant form. Both the travelling distances and infrastructure costs are high for the linear city. The circular city form is, of course, the traditional form, and it retains its traditional advantages. High priority would need to be given to environmental planning and recreational facilities. The necessary inter-urban open land could conveniently be provided by wedge systems of open space running in from the countryside. The green wedge open space is the inverse concept to ribbon extension of urban development into the countryside. Such a policy retains both high urban compactness and economic servicing potential, and also convenient access to recreation facilities and to nature. Some open space at the centre of cities is most critical, as it is the most accessible. The linear city would be rejected as resource extravagant, and the isolated house fed from centralised services even more so. Highly serviced isolation is a very materials and resource extravagant mode of living, if combined with city centre work places.

The energy conservation techniques used would not only involve greatly improved insulation standards, but would take advantage of total energy systems, using compound cycles as a normal part of the structure of urban energy servicing for example.

RECYCLING IN URBANISM

Integrated energy supply waste disposal systems would be evolved, linked with district heating schemes, energy and materials recovery. Appropriate use would be made of natural energy resource systems like solar energy and wind power as discussed later.

Waste recycling systems would be fully integrated and their design would form a central part of the planning process. Adequate facilities for citizen participation in waste recycling would be incorporated by providing intelligently designed waste collection facilities openly available to the public. The profits from these citizen participation waste recycling units could be devoted to social purposes, like building old people's facilities, subsidising theatres and youth enterprises, etc.

The processing of solid waste products of utilities and industry would be closely integrated with the overall urban building construction processes. For example, use of pozzolanic cements based on P.F.A. additives from electricity generation stations, building blocks made from slags, etc., would form part of an integrated approach to resource conservation and waste elimination that was crosslinked with economic industrial production. The aim would be to eliminate the majority of wastes by a systems approach that interlinked the production processes with the total urban system, so tackling environmental pollution due to wastes simultaneously with economic production and resource conservation by waste recycling.

EFFICIENT BUILDING SPACE UTILISATION

Individual buildings would be designed to be flexible on a day to day basis, to enable high occupancy factors to be achieved. They would be occupied for different uses at different times of day by different groups to achieve economy of use, resource and energy, so reducing the overall amount of public building needed. The actual public buildings produced would have to be designed to higher standards, with facilities to make flexibility possible. With proper management, these could be operated outside the constraints of the petty control of present divisive committee structure of the past local government system, which results in low load factors. Public buildings would thus be made to serve the whole community, rather than sections of it, and substantial resources would be saved. Agents of sub-sections in local government, like school caretakers, would be reorganised into a new general public service concerned with high building utilisation, and there would be appropriate pay incentives for achieving a high degree of building utilisation to replace the policy of shutting the buildings as soon as possible after their specialist use is completed. This approach would save vast sums of public money, as well as substantial amounts of materials resources and energy.

OUTDOOR SPACE UTILISATION

Outdoor space associated with public buildings would become truly communal space, accessible to the whole community, and available at all seasons of the year on a properly managed basis. Land resources would be managed with a true sense of their value. Space sharing of public open space, at present sectionalised into separate school playing fields, parks, industrial sports complexes, etc., would enhance environmental facilities and give a much better return for the high level of investment of land than the present sectional policies allow, and would contribute to urban compactness, so substantially reducing the resources needed for the distribution infrastructure. The value of the school playing fields for a single urban school can run into millions of pounds. We need to remember again that as land is not available for productive photosynthesis work residence systems would be much closely integrated for environmentally acceptable industries,so reducing long distance journeys to work. Certain industrial facilities might be shared with the community. Abolition of the rigid distinction between work place and residence would favour social interaction and encourage resource sharing.

THE END OF THE HIGH CONSUMPTION CITY

All this is, of course, a rapidly conceived personal viewpoint, but one thing is clear to many people now. Our resource consumptive way of living cannot continue for long in its present direction without total disaster. A key feature of the materials and energy conservation city is, therefore, that it should offer a quality of life that provides social satisfaction without excessive resource consumption. It will be the quality of the social, cultural and environmental design which will determine whether people have a full and interesting life style, that does not need to be propped up by escaping into the fantasy of more goods for their own sake, and more travelling about for its own sake. The basic policy will be to use resources to secure a high quality of life, and not as a substitute for the quality of life. This involves deriving a credible concept of the aims of industry of the future in a resource limited world. The industrial pride of achievement which is based on sheer volume of production, linked with productivity and avoidance of unemployment, will have to be replaced by more effective alternative industrial concepts that accept the resource limitations, which we have yet to evolve. The choice of viewing the industrial future as an exponential projection of the industrial past is simply not open to us. This is the central problem which will not go away. So, crosslinked with the evolution of the materials conservation city, we have the problem of the redesign of industry to meet new and achievable goals. It presents simultaneously a sociological and technological challenge.

ENERGY STORAGE AS A CONSERVATION TOOL

Another key area of future conservation policy must be to develop more effective techniques of heat storage. Storage techniques are particularly crucial in renewable energy design, because all natural energy supply sources fluctuate with time of day, weather and season. However, storage is also important for base load supply systems, like nuclear power. Fundamentally, in scientific terms, energy stores provide capacitance that enable one to deal with the phase lags between supply and demand. Stores allow surpluses to be laid down at the time they can be economically produced and to be consequently recovered at will at times of energy shortage. The storage may be long term or short term (6). Storage, of course, will contribute directly to energy conservation only if it enables heat energy which would otherwise be dumped to the environment to be conserved for practical use. System storage, however, has an important indirect effect on conservation because it enables peak loads to be met from past surpluses. It can thus significantly contribute to the reduction of the scale of the energy supply systems needed, either conventional or renewable. If the plant needed is smaller, the materials needed to construct it will be less and so materials conservation is interlinked with storage strategies.

ROLE OF CLIMATE RESOURCES IN MATERIALS AND ENERGY CONSERVATION IN HUMAN SETTLEMENTS

Having set up the general objectives of the materials and energy conservation city, we can now ask what contributions could our climatic resources make to the solution of the problems involved. If we accept that typically 50% of energy used in any country is devoted to building heating and cooling purposes, we must ask what climatic measures could be taken to change the situation. One has to view climate simultaneously as an energy demand creator and a potential energy supplier. Conservation is concerned with the reduction of demands. In heating terms this means both improved insulation and better control of natural ventilation. It also means making full use of wild heat gains from people and their activities.

THE ROLE OF WIND

Building performance is significantly affected by wind and the aerodynamic characteristics of human settlements are important both for outdoor comfort in different seasons, for natural ventilation and for surface heat transfer. Wind, of course, is also critical in pollution dispersion and its consideration of vital importance in the layout of towns from the pollution control viewpoint. A fundamental problem is that aerodynamics remains an experimental rather than a predictive science. Theoretical prediction remains difficult and modelling in wind tunnels provides the only convenient experimental route forward for studying the detailed effects of building layout. We consider it a key feature of our work to maintain and operate a first class boundary layer aerodynamic facility because in no other way can we assess the impact of wind flows on energy conservation satisfactorily.

The effects of terrain on wind flow are, of course, considerable. Built form, layout and associated landscape strongly influence the actual local flows and the consequent pressure drops across buildings (7). Shelter should become a more carefully considered factor in future urban design. For example, shelter belts can play the double role of wind protection and photosynthetic energy/materials production. The fuel plantation can thus confer environmental benefits, if properly considered in its location and layout in relation to urban form.

Wind plays a rather different role in hot weather, and its flows to the interior have to be encouraged in those circumstances where the air temperature is below skin temperature, especially in regions where the air is humid. The role of natural ventilation is critical in the humid tropics for human comfort. In desert areas the wind may be above skin temperature and then the wind becomes a hazard rather than an asset and shelter from wind becomes a necessity. People who live in high latitudes understand what it is to be blown cold, but not what it is to be blown hot.

The use of air conditioning is highly energy intensive, but, by appropriate architectural design of natural ventilation systems a great deal may be done to make the use of air conditioning unnecessary with great saving of energy and money.

WINDOW SYSTEMS AND ENERGY CONSERVATION

Window systems form a key pathway for energy exchanges between the interior and exterior and their successful design is a vital aspect of practical energy conservation in buildings. The detailed energy balance of any building has to be studied on an inter-seasonal basis. There are times of the year when the window energy fluxes may be valuable and other times when they are excessive. Traditional solutions like the venetian shutter treated the problem as a problem of thermal radiative valve design. Unfortunately, much of this traditional understanding of shading system performance has been lost in modern architecture. The consequence has been that air conditioning has had to be more widely used than is necessary, simply because most architects no longer understand how to shade windows efficiently.

ROLE OF BUILDING MASS AND COLOUR

The role of building thermal mass is also critical for temperature regulation. Thermal storage in the fabric helps damp diurnal temperature variations and allows transfer of peak solar energy inputs from day to night when the air is often cool. Mass damps both internal temperature variations and in air conditioned buildings the peak energy loads. External building colour is also important because it strongly influences the radiative balance of the external surface. The precise understanding of building thermal performance also demands an assessment of the long wave energy exchanges.

IMPORTANCE OF COMPUTER AIDED DESIGN TECHNIQUES

The study of such dynamic energy exchanges is complex and mathematically time consuming. One has to cope with variations in the solar radiation, long wave radiation, wind speed and temperature, as well as allowing for surface properties and storage within the building fabric. In my own laboratory we have devoted an extensive effort to use the development of computer aided design techniques for buildings applicable anywhere in the world, that enable a full scientific assessment to be made of the dynamically complex problems involved (8). Such tools are essential if one is to explore building energy conservation in an advanced scientific way. The computer system covering the meteorological aspects of our system is given in Figure 1.

CLOSING THE ARCHITECTURAL EDUCATIONAL GAP

Once again we have a huge educational problem on our hands, because the majority of world architects have been educated within educational establishments which have paid scant disregard to the role of climate in energy conservation. At last the world tide seems to be changing and recognition is being given to the climatic sophistication of many historical solutions. International organisations like HABITAT are coming to recognise the key role of climatically appropriate design in helping solve some of the world's energy problems. It is the developing world which could benefit most immediately. The UN HABITAT organisation is hoping to prepare some key background papers on this area for the UN conference on New and Renewable Sources of Energy on this topic. W.M.O. has also been active,working through specialised technical groups on topics like building climatology, urban climatology and applied solar radiation studies.

OTHER SIGNIFICANT WORLD HABITAT PROBLEMS - COOKING

The problem of cooking has become one of the world's most important problems, because a very significant proportion of food on a world scale is cooked by wood or charcoal. The scavaging for wood is having particularly serious effects in the areas of the world close to the great deserts, leading on to rapidly increasing desertification. It is for very good reason that the UN set up a technical panel of Fuelwood and Charcoal, as well as one on Biomass to help in the preparation of the technical agenda for the UN Conference on New and Renewable Sources of Energy (9) planned for 1981.

The conservation of wood by improved stoves is clearly important, but any solutions have to take account of the other functions of fires in rural settlements like the provision of heat to offset the cool of the evening, the provision of light within the home and the deterrent effect of smoke on insects which abound in many hot rural areas of the world and also other animals. Simplistic technological solutions may not meet the real social needs and may be rejected as a consequence. Adequate socially acceptable solar cooking technologies have yet to be evolved, so at the current moment of time one must view the cooking technologies of many rural and urban settlements as destroying the wider habitats in which they are placed. Therefore, cooking must be viewed as a key issue in world conservation policy.

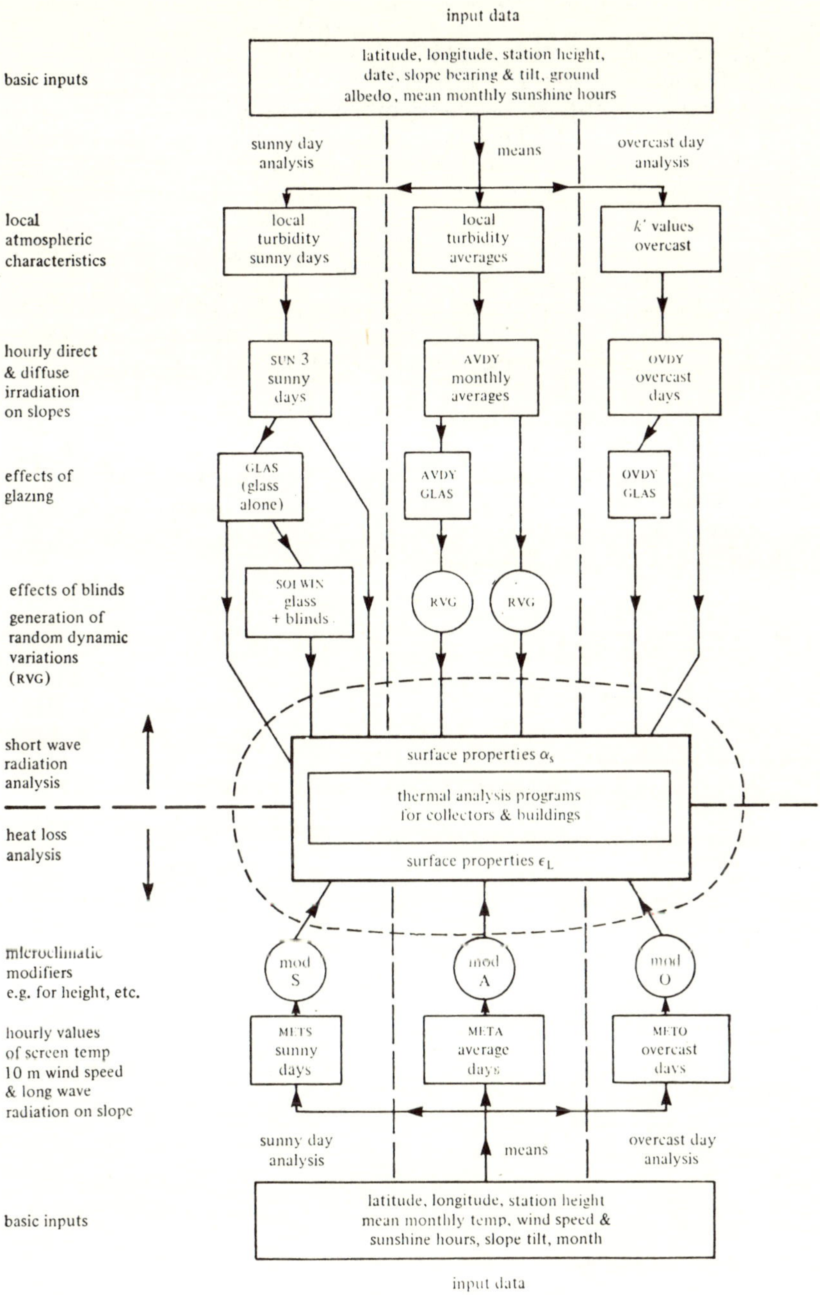

FIGURE 1. System diagram showing flow of climatological information through University of Sheffield Department of Building Science computing system into thermal programs.

COOLING AND RURAL FOOD STORAGE

Another key issue of great climatic significance is the storage of food in remote communities, especially in hot climates, where the wastage due to putrefaction, insect attack, etc. may lead to a loss of half the basic agricultural produce. Ideally, conservation in such climates involves refrigeration, but without the capital resources and the necessary energy supplies, it is difficult to make progress. Active solar cooling is not yet an economic technique. Passive cooling techniques, however, have a long world history especially in the Middle East. Some of these traditional methods have been described by Bahadori (10). Ice making techniques based on the selective use of outgoing long wave radiation have existed for at least a thousand years.

The cooling-food cycle interacts with other issues in this conference because, if half the agricultural produce is lost, any adverse climatic impacts of the agricultural system will be doubled if the same level of food production is maintained. Hence biomass conservation is also a key issue for human settlements and their climatic impacts.

RENEWABLE SOURCES OF ENERGY

Renewable sources of energy could have a key role to play in the energy economy of habitat, but the low energy density makes it difficult to achieve economic energy collection. Complex solutions are likely to become too expensive and there has recently been a marked shift of emphasis from active systems to passive systems in the architectural sphere. In passive systems the movement of energy is entirely by natural means. Passive systems, however, have important limitations. In particular, their capacity to store energy is frequently too low in relation to the large climatic variations in the energy fluxes which occur. The need to increase storage capacity often implies the development of hybrid systems which collect the energy passively and store it actively, for example with the use of fans, etc. in rock stores or latent heat stores. There is a considerable economic penalty if too big a proportion of energy is sought from solar at current levels of fossil fuel prices. Enthusiasts of renewable energy tend to over-exaggerate the current state of the art, and the economic role that solar can presently play. Big active systems are inherently more cost effective than small systems, thus while small may be beautiful, medium sized is likely to be better. As this field has become a major field of international R&D involving global expenditure at a level of around \$1½ billion per annum, there is no way such a vast field can be summarised in a single paragraph. I will merely stress the climatic advantages of freedom from chemical pollution and absence of further thermal burdens on the atmospheric environment other than those already encountered in the natural evolution of that environment, and solved in the normal meteorological processes.

Obviously if solar energy is used on a large scale, there will be climatological impacts at the micro and meso-climatological scale and some of these may be locally adverse. These relatively local adverse climatic changes must be balanced against the much more serious long term impacts of fossil fuels and nuclear energy on the global environment. There will, nevertheless, be some important pieces of research to carry out on climatic aspects of renewable energy and its impacts.

CONCLUSION

I have attempted to review a vast field in a short space, so it has been impossible to go really deeply into any one aspect of the subject. Climatological aspects of conservation are extremely important and there is considerable scope for the development of greater climatological awareness in the overall design of human habitats. Application of sound climatic experience will reduce energy consumption, promote great human comfort, and help control adverse global impacts of human settlements. There is a huge educational task ahead to change existing attitudes. The evidence of history of human settlements also indicates that we need to relearn much that has been forgotten in recent years in order that we may survive in human settlements into the long term future with an acceptable global climate.

REFERENCES

1. Page, J.K.: 1976: Application of building climatology to the problems of housing and building for human settlements, W.M.O. Tech. Note No. 150, W.M.O., Geneva.

2. Chandler, T.J.: 1976: Urban climatology and its relevance to urban design, W.M.O. Tech. Note No. 149, W.M.O., Geneva.

3. Page, J.K.: 1973: Use of climatological data in building design with respect to safety, Proc. Conf. Teaching the Teachers on Building Climatology, CIB Proceedings No. 25, Doc. D20, The National Swedish Institute for Building Research, pp. 97-115.

4. Höglund, B.I. and Johnsson, B.: 1976: Energy conservation measures taken in the thermal insulation and installation fields for existing buildings, in Energy Conservation in the Built Environment, Construction Press/CIB, pp. 412-435.

5. Page, J.K.: 1974: Conservation of building materials and the future of the built environment, in Proc. Conf. Conservation of Materials, Harwell, March 1974, Harwell, Oxfordshire, OX11 ORA, pp. 169-185.
The following section is drawn from that paper.

6. Bellecci, C., Visentin, R., Mancini, N.A. and Quercia, I.F.: 1978: Solar energy storage, International Centre of Theoretical Physics, P.O. Box 586, Trieste, Italy.

7. Lee, B.E., Hussain, M. and Soliman, B.S.: 1980, A method for the assessment of the wind-induced natural ventilation forces acting on low rise building arrays, Building Services Engineer, Research and Technology, 1, No. 1, pp. 35-49.

8. Page, J.K., Rodgers, G.G. and Souster, C.G.: 1980, Systematic design assessment techniques for solar buildings, Phil. Trans. R. Soc. Lond. A 295, pp. 379-401.

9. UN General Assembly: 1979: United Nations Conference on New and Renewable Sources of Energy, Report of the Secretary General, 34th Session, Agenda Item 71, UN Document A/34/585, 29th Oct., 1979.

10. Bahadori, M.N.: 1979: Natural cooling in hot arid regions, in Solar Energy Application in Buildings, Ed. A.A.M. Sayigh, Academic Press, N.Y., pp. 195-225.

PROSPECTS FOR USING IMPROVED CLIMATE INFORMATION TO BETTER MANAGE ENERGY SYSTEMS

William J. Quirk

Lawrence Livermore Laboratory, University of California,
Livermore, CA 94550, USA

John E. Moriarty

Federal Energy Regulatory Commission,
U. S. Department of Energy
Washington, D. C. 20426, USA

ABSTRACT. Given the rising costs of energy and the tight supply of key fuels, climate variations can have a major impact on society by affecting energy production and demand. The winter of 1976-77 brought continued drought to the western United States and cold weather to the eastern United States. This lifted U. S. energy demand for fossil fuels by the equivalent of 200 million barrels of oil. The weather pattern for that winter was an extreme version of a common pattern that is associated with colder than normal temperatures for most of the heavily populated north temperate zone. If such an extreme climate event had occurred during the winter of 1979-80, it would have increased the world oil shortfall by several percent making a bad situation worse. An international effort would be needed to determine the frequency and nature of such extreme climate variations, and to discover if these events are predictable. Monitoring of global temperature and precipitation data could provide early warning that world-wide energy consumption is unusually high. Use of such information could help ameliorate the impact of an extreme climate event on the global economy.

W. Bach, J. Pankrath, and J. Williams (eds.), Interactions of Energy and Climate, 89–99.

1. RECENT CHANGES IN ENERGY ECONOMICS

As little as ten years ago crude oil was cheap and appeared abundant. Rapid growth in energy consumption has put the world in a position where supply and demand for oil are precariously balanced (1). Twice in the last seven years events in the Middle East have resulted in slight shortfalls in energy supply. These minor shortfalls have permitted a catastrophic rise in the price of a barrel of oil from $2 to $30. Even as the present crisis abates, decisions by major producing countries not to rapidly increase (or even to decrease) production mean that further supply disruptions and attendent price increases are likely.

Given tight supplies and the resulting high prices, climate variations can have a major impact on society. An unusually cold winter can raise heating bills beyond the ability of some people to pay. Domestic energy production can be severely disrupted. Frozen rivers impede the barge transportation of coal. Unusually cold weather can stress mechanical systems such as natural gas pumping facilities past their design limits resulting in curtailments of delivery. Power production can be reduced when freezing weather solidifies piles of coal. Severe weather conditions can also substantially impede oil production, particularly offshore. Drought can significantly diminish the supply of hydroelectric power and increase the use of fossil fuels for pumping irrigation water. Production reductions coupled with increased demand will result in increased oil imports. This will weaken the national currency. Extreme climate events can even cause large enough energy shortfalls to result in widespread unemployment.

This paper will describe the impacts that climate anamolies have had on energy systems in the United States, and the resulting impacts on society. It will also point out how better use of climate information could alleviate some of the adverse consequences of these anomalies. Finally, it will point out the possibility of international cooperation in using climate information to avoid potentially worsening conditions during a global energy shortage.

2. IMPACT OF CLIMATIC VARIATIONS ON ENERGY SYSTEMS

Historically we have seen climate variations on every time and space scale. For example:

- o In the United States (2) the November to March heating season of 1976-77 had 22% more heating degree days than the same period of 1975-76. This corresponded to a 1.8°C difference in average temperatures (3).

- o Nationally the heating seasons of 1976-79 have had 15% more heating degree days than those of 1973-76 (2).

- o In California (4) the rainy seasons of 1975-76 and 1976-77 were 60% drier than the 1931-77 long-term mean.
- o In the Mid-Atlantic states (4) the years 1962-66 had 18% less rain than the long-term mean.
- o World-wide (5) the five years from 1962 to 1966 were .7°C cooler than the preceding three years.
- o In Europe (6) the 1690's were 1.5°C colder than the average temperature for 1580-1790.
- o The whole World (7) was cold from 1812 to 1817. During the summer of 1816 frosts were reported every month in New York and New England. During that year, Western Europe had similarly severe weather.
- o According to Flohn (8) there may have been a number of abrupt climate events in the last 700,000 years where temperatures have dropped 5°C in 50 years over large areas. Such events may occur as often as once every 1,000 to 10,000 years.

Extreme climate events lasting several years seem to occur every decade. One can speculate on the effect a world-wide cold period would have on the international market for petroleum; or how an abrupt temperature decrease of 5°C over a large area would affect the local economy. In this section we will describe two climate events that had a substantial impact on energy systems in the United States.

In 1975-77 a severe drought in the Western United States had a major impact on the economy of that region. Power costs were raised considerably because hydropower had to be replaced by more expensive thermal generation. In California the increased cost was borne primarily by Pacific Gas and Electric (PG&E) customers since the utility is heavily hydropower dependent. In 1977 PG&E's operating expenses were about 30% greater than the 1976 operating expenses of about $1,275 million (9). That year (10) 50 million barrels of oil had to be burned statewide to make up the energy deficit. At the January 1980 world price level this would have increased the U. S. trade deficit by almost $1.5 billion. Ultimately, these costs were passed on to the California consumer.

Farmers in particular were hard hit by the additional energy costs. It is estimated that in a normal year, approximately 85% of the total water demand in California comes from agriculture. With surface water supplies severely restricted in 1977, water allotments to agriculture were reduced by over 60%. Agricultural users attempted to replace this deficit by pumping up more ground water. During 1977, over 6 million acre feet of water were brought to the surface for irrigation purposes.

Ground water is estimated to have accounted for over one-half the water used in California agriculture that year. Statewide the extra energy associated with the additional pumping was approximately 1 billion kilowatt-hours. This expenditure cost California farmers over $25 million (10).

In the Pacific Northwest, part of the expected energy deficiency was reduced by the curtailments of interruptible customers. This resulted in some unemployment, operational shutdowns, and additional costs for purchasing thermal energy. The greatest impacts were in the aluminum industry. The drought ended before the Bonneville Power Administration (BPA) was forced to curtail all its interruptible customers. According to the BPA (9) the total curtailment of all its interruptible customers would have resulted in the following economic impact:

- "370,000 tons per year cut in aluminum production (23.4% of 1,578,000 tons).
- "$355 million added to the nation's foreign trade deficit, if the lost production were replaced by imports.
- "$53 million annually in lost wages in the Pacific Northwest.
- "$21.6 million annually in lost freight payments.
- "$32.3 million lost annually in materials and supplies not purchased.
- "1,900 jobs lost (14.0% of 13,700) in direct Pacific Northwest in the electroprocess industry.
- "6,800 — total of direct and secondary jobs lost in Pacific Northwest."

The effect of a prolonged drought on energy generation alone would cost the Western United States billions of dollars. Resulting increases in the U. S. trade deficit would further weaken the dollar.

While the Western United States suffered energy shortfalls due to drought in 1976-77, the Eastern two-thirds of the country suffered one of the worst winters in history. The protracted period of record setting low temperatures raised the nation's energy bill by $4-8 billion (1). Oil imports had to be increased by about 150 million barrels. The natural gas industry was unprepared for the unpredicted increase in demand, much of their equipment could not operate at these record low temperatures. As a result, there was a natural gas shortage that caused curtailment of supplies to industrial customers and schools. Unemployment (10) caused by the resulting gas curtailments rose from 240,000 in the later part of January to a peak of 1,200,000 on February 4. By the end of

February about 160,000 were still unemployed because of the shortage. Approximately 20 states were affected by the curtailment. Ohio suffered over 45% of the total national shortfall, and New York about 20%. It has been estimated that the federal government may have paid up to $100 million in unemployment benefits (10). Because lost production was probably made up eventually, the true economic impact of this temporary unemployment is hard to estimate.

Congress reacted to the shortages of early-1977 by passing the Natural Gas Policy Act of 1978. This act gave absolute priority for natural gas usage to residential and small commercial customers. This is an example of a political reaction to a climate variation. This legislation will have a significant impact on the natural gas industry, for reasons given below.

Utilities and pipeline companies have high fixed costs. These fixed costs are figured into their rate structure based on normal demand. When severe weather increases demand, revenues are increased faster than total costs, so the companies may earn excess profits; on the other hand, if demand is down they may suffer decreased profits.

Residential and small commercial users of natural gas have been given absolute priority for natural gas use. Supplies to industry and large commercial users must be interrupted to avoid curtailing delivery to homes and small businesses. Since the small users are highly temperature sensitive, they would receive a larger share of the total gas supplies during cold weather. If increased demand were to cause a curtailment of natural gas, a utility with few high priority residential and commercial users might lose substantial revenue, while a neighboring utility with many uninterruptible customers would reap windfall profits. Thus, it would seem incumbent on utilities to raise their proportion of temperature sensitive uninterruptible customers. If such a trend were to develop nation-wide, it could eventually have severe consequences if supplies were for some reason to decrease.

Thus, a law intended to ameliorate the impact of climate variations might in the long run enhance this impact. The natural gas policy act is a good example of how climate impacts can affect society by perturbing the political process. It shows the need for careful study before implementing strategies to deal with climate variations or any effort to further regulate the energy market.

In summary, it can be said that recent climate events have cost U. S. citizens billions of dollars for additional energy. They have resulted in increased oil imports, and thus have weakened the dollar and the U. S. economy. They have resulted in a Congressional Act that might actually increase U. S. vulnerability to climate variations. The history of climate variations shows us that even larger climate anomalies can be expected in the future. Unless the energy situation improves radically, these anomalies will have important consequences for society.

3. AMELIORATION OF CLIMATE IMPACTS

Given that the impact of climate variations on the national economy has been large, an important question to ask is whether these impacts could have been ameliorated. The winter of 1976-77 is an example of a case where some simple precautions could have been taken to avoid a major societal disruption (10). There were two major reasons why the increased natural gas demand for that winter could not be met. First, one of the major pipeline companies actually sold its reserves before the January record cold weather. This gas was used by utilities and other industries who could have burned other more abundant supplies such as propane and heating oil. Second, the severe cold dropped the pressure in the natural gas storage areas so low that the companies had insufficient pumping capacity to retrieve the stored gas. Had the pipeline companies not sold off their reserves early in the winter and had more pumping capacity been available, much of the anxiety caused the nation could have been avoided.

In the aftermath of the national trauma caused by the January-February 1977 natural gas shortage, the utilities and pipeline companies have been adequately prepared for extremely cold winters. The heating season of 1976-77 had 11% more population weighted heating degree days (HDD) than normal. The heating season of 1977-78 was nearly as severe. It had 10% more degree days than normal. Yet there were virtually no shortages or weather produced unemployment. Clearly, American industry had the resources to deal with unusually cold winters, it was simply unprepared in 1976-77.

There was evidence to suggest that such cold weather was possible. While the cold weather suffered by the Eastern half of the country in October-January of 1976-77 was unusual, it was not unprecedented (12). Virtually all of the excess HDD's occurred in October through January. February was normal and March was above normal. October-January of 1917-18 and 1836-37 were about as cold as 1976-77. While January of 1977 set records in many places; sites with very long temperature records show that the Januarys of 1856 and 1857 were about as cold (12). Unfortunately, this information is still not generally available to the utility industry.

There was, however, at least one report published prior to 1976-77 suggesting that a winter as cold as 1976-77 was possible. In 1974, a report (13) to the Energy Policy Office of the President was made based on the 1931-73 temperature record. It stated that years with a 10% increase in HDD's would occur about every 100 years. Thus, weather conditions as extreme as 1976-77 were known to occur once every hundred years. Once the natural gas industry experienced these conditions, they were prepared for them the next year. Energy system planners should examine the cost and benefits of being prepared for such adverse conditions, decide how prepared they should be, and then take long-term action to carry out their decisions.

The winter of 1976-77 is an extreme example of why climate variability should be taken into account in managing energy systems. Gordon McKay of the Canadian Climate Center and John Page of the University of Sheffield have discussed other cases at this workshop. Another good example is the concern in the Western United States about the probability of the recurrence of a drought as severe as that of 1975-77. If such a drought is likely to recur, then these states must consider reexamining their policies on growth of water use. At the present time, ground water is being pumped out of the Central Valley of California water table faster than it is replaced. Unless the water table is allowed to recover, it will take larger and larger amounts of energy to pump this water. If a severe drought were to occur during an energy shortage it would be difficult and expensive to pump adequate water for irrigation.

The situation in the upper Colorado River Basin is even more critical. It is widely believed that water supplies are already over-committed in this area. A boom in energy development will put further stress on supplies. Hopefully, it will not take a disastrous drought to make people aware of the need to be prepared for climate variations.

So far we have stressed the need for defining the likelihood of extreme climate events, and of building or retrofitting systems to be prepared for such events. Of course accurate long-range forecasts would be extremely useful. Again the winter of 1976-77 is a good example. During that winter, cold temperatures were experienced unusually early. On some systems gas was being withdrawn from underground storage during the fall, a time of year when gas is typically being injected. About 12% of the nation's inventory of stored gas was consumed during November; the previous November almost no gas was withdrawn from storage. Had it been known in November that December would be very cold many large industrial customers would have been switched to alternative fuels, and the stored natural gas would have been conserved. This would have considerably reduced the level of curtailments necessary in January and February. The curtailments could also have been further reduced had it been known that March would be unusually warm (14).

It is interesting to note that in late November of 1976, several of the long-range forecasters most respected by professional meteorologists predicted a switch from warm to cold weather for the Eastern United States (14). One even stated that it would be the coldest winter in 20 years. Unfortunately, past experience has shown that present day forecasts have very limited skill so a special effort by the climatological community to put the nation on alert for cold weather would not have been justified. Until such time as long-range forecasting is substantially improved, the best strategy for ameliorating climate impacts is to build systems which can respond to extreme events in a flexible manner.

4. PROSPECTS FOR THE FUTURE

In this paper we have given a few examples of how better use of information about climate variations can improve the management of energy systems. There is a rising concern both in private industry and in governmental regulatory and information offices about the potentially disruptive effect extreme variations in the climate can have on energy and therefore on the economy and society. Energy managers and analysts agree that:

1. Present methods for relating climate variations to energy use or disruption of energy supply are not sufficiently accurate; the economic and societal effects of adverse weather are largely unknown.

2. The improvement of monthly and seasonal forecasts would be of help in the allocation of scarce fuels.

3. Data on past climate variations are not available on a timely basis or in the most useful formats; the formats of presently available forecasts could be improved.

4. Methods of using presently available climate data and forecasts could be greatly improved.

5. The climate effects of the various energy technologies are largely unknown.

As part of the United States Climate Program a multi-agency interdisciplinary program is being instituted to minimize the impact of climate on energy and energy on climate. The following projects are now being initiated; they are listed in order of priority:

1. Detailed data are being gathered on energy usage and supply; they are being correlated with weather data down to the county level. This will help resolve many of the uncertainties about the impact of climate variations.

2. Economic studies are being made about the effect of climate variations on industry and the consumer, particularly the poor and aged. Implications about societal impacts will be drawn from these studies.

3. These impact studies will serve as the basis for programs that will develop systems less vulnerable to climate variations from both a financial and engineering point of view; the studies will also aid in making better use of climate information.

4. Data bases of energy related climate variables, such as heating degree days, are being developed; they will contain

both forecasts and observations and will be available on a real-time basis to concerned parties.

5. Specific efforts will be made to improve climate forecasts of variables critical to energy systems.

6. More intense investigations of the effects of energy on climate will be made.

These programs require the cooperation of energy experts, climatologists, economists, decision analysts, and many other physical and social scientists. Communication between these groups has already begun. The United States Department of Energy will be primarily responsible for the impact and engineering studies; to perform these studies it will need input on the type of climate variations to be expected in the future, particularly extreme events. It is expected that NOAA, NSF, and NASA will attempt to devise improved climate forecast methods.

While the national efforts mentioned above will give a good start on climate impact assessment, the problem is ultimately an international one. Energy is scarce and expensive world-wide. There are international agreements on how much oil each consuming nation should import. However, if a nation has an exceptionally cold winter it may need more oil. Global weather patterns (15) are such that it is quite common for the eastern two-thirds of the United States, Europe, Korea and Japan all to have cold weather at the same time. When this happens, dry conditions in the Western United States are also likely. An extreme version of this pattern occurred in 1976-77. These weather conditions caused an increase of over one million barrels per day in U. S. imports (2,3). We have no data on how much non-U. S. imports changed, but they probably increased as well. Since only about 30 million barrels a day are available for import world-wide, this was a substantial increase in demand. The last two oil supply crises have involved shortfalls of only a few percent. If such a climate extreme were to occur when there was a crude oil shortfall, it would make a bad situation worse. The winter of 1976-77 is just one example of a climate event that could have adverse consequences for the international economy.

As climatologists we can make a contribution to society by making both national and international institutions aware of the likelihood of extreme climate events. An important aspect of the U. S. Department of Energy Climate Program has been to encourage climatologists to work with energy planners and managers to propose strategies for ameliorating climate impacts. An international effort would be needed to determine the frequency of world-wide adverse climate events and determine their effect on global energy demand. Monitoring of world-wide temperature and precipitation data could provide early warning that world-wide energy consumption is unusually high. Use of such information could help alleviate the adverse impacts of climate extremes on the global economy.

ACKNOWLEDGMENTS

We would like to thank Mary Schrot, Bill Rousseau, Michael MacCracken, Jerry Potter, and Hugh Ellsaesser of LLL for their comments; and Harry Moses of DOE Office of Health and Environmental Research and Joseph Knox of LLL for their efforts in bringing this topic to our attention.

This work was performed under the auspices of the U. S. Department of Energy by the Lawrence Livermore Laboratory under contract No. W-7405-Eng-48.

REFERENCES

1. Brown, H.: 1976, "Energy in Our Future," in Annual Review of Energy, Volume 1, J. M. Hollander and M. K. Simmons (eds.), Annual Reviews, Inc., Palo Also, CA, pp. 1-36.

2. National Climatic Center: 1979, "State, Regional, and National Monthly and Seasonal Heating Degree Days Weighted by Population," Asheville, NC.

3. National Climatic Center: 1979, "State, Regional, and National Monthly and Annual Temperatures Weighted by Area," Asheville, NC.

4. National Climatic Center: 1979, "State, Regional, and National Monthly and Annual Total Precipitation Weighted by Area," Asheville, NC.

5. Mass, C. and Schneider, S. H.: 1977, "Statistical Evidence on the Influence of Sunspots and Volcanic Dust," J. Atmos. Sci. 34, pp. 1995-2004.

6. von Rudloff, H.: 1967, "Die Schwankungen und Pendelungen des Klimas," in Europa seit dem Beginn der regelmässigen Instrumenten-Beobachtungen (1670), Braunschweig, Vieweg, 370 pp.

7. Hughes, P.: 1979, "The Year Without a Summer," Weatherwise 32, pp. 108-111.

8. Flohn, H.: 1979, "On Time Scales and Causes of Abrupt Paleoclimatic Events," Quartenary Res. 12, pp. 135-149.

9. Federal Power Commission, Federal Energy Administration: 1977, "Impacts of the Western Drought on the Regional Electricity Situation," Washington, D. C.

10. Task Force on Community and Physical Resources of the Committee on the Budget, House of Representatives: 1978, "Weather Impacts on the Budget and the Economy," U. S. Government Printing Office TF-7-95-36, Washington, D. C.

11. Bardin, D. J.: 1977, from "National Climate Program Act, Hearings before the Senate Subcommittee on Science, Technology and Space," Serial No. 95-33, U. S. Government Printing Office, Washington, D. C., pp. 137-152.

12. Diaz, H. F. and Quayle, R. G.: 1978, "The 1976-77 Winter in the Contiguous United States in Comparison with Past Records," Mon. Wea. Rev. 106, pp. 1393-1421.

13. Mitchell, J., Felch, R. E., Gilman, D. L., Quinlan, F. T., and Rotty, R. M.: 1974, "Variability of Seasonal Total Heating Demand in the United States," Environmental Data Service, pp. 4-9.

14. Namias, J.: 1977, "Forecasting Climate Fluctuations: The Winter of 1976-77," Science 196, pp. 1386-1387.

15. van Loon, H. and Rogers, J. C.: 1978, "The Seasaw in Winter Temperatures between Greenland and Northern Europe. Part I: General Description," Mon. Wea. Rev. 106, pp. 296-310.

DISCUSSION

Kellogg:
Your ideas have been widely discussed since your famous article on "Soft Energy Paths" was published in Foreign Affairs in 1977; and so many people have been sympathetic to those ideas that some now refer to "the Lovins Syndrome". Why then, has there been so little official recognition of the need for conservation of energy, or on soft energy strategy, in the formulations of official energy policy by the governments of the developed countries?

Lovins:
A soft energy path is already implementing itself in many countries, far faster than I would ever have thought possible, but it is happening from the bottom up, not from the top down and most central governments will be the last to know. With a problem like energy, with billions of diverse components in heterogeneous societies, central management is indeed part of the problem not of the solution, and only decentralized methods of implementation are likely to work. Thousands of towns, cities, and counties in the US (and one or two states, notably California) are already on a soft path because they can think of nothing else that makes political and economic sense. The role of central government in encouraging this development should be to encourage, facilitate, and get out of the way, especially by helping to clear away the institutional barriers (market imperfections) that its tax and pricing policies have introduced and that certain of its regulations enforce. The effect of these barriers is that people do not get correct price signals or are prevented from acting on them.

Rotty:
1. Is your 1KW/cap. a forecast, or simply a possibility if society adopts certain (perhaps unlikely) steps, say within 50 years? Will this be uniformly distributed or will we still have rich and poor?
2. The "free-lunch" idea leads me to wonder how much individual choice is to be retained. Can I pay more for lunch to meet my individual dietary preference, and how much of this type of freedom is possible within a 1KW/cap. total energy budget? I hear answers to the question "What will you do when gasoline costs $2/gal.?" in the vein of "I won't eat meat". This suggests that many affluent people will be slow to give up some of the nice things that their higher energy budget buys them.

W. Bach, J. Pankrath, and J. Williams (eds.), Interactions of Energy and Climate, 100–106.

Lovins:

1. It is a plausible future that could be practically achieved if societies chose the lowest cost (and, incidentally, lowest climatic risk) energy strategy. It would probably take some 50 years to achieve without special accelerating measures such as many societies have taken in the past to retrofit major capital stocks. I assumed uniform distribution, but that is neither necessary to obtain my analytic result nor a necessary consequence of a soft path.
2. I favor complete individual choice - and having individuals pay the full social cost. You can choose your lunch, but I don't want to buy it for you.

Quirk:

1. Would you please give us references so that we can all make our homes more energy efficient?
2. Hydropower in your scenario will be the major source of electricity, however hydropower is quite variable. The California drought of 1975-77 caused a 40% drop in hydroelectricity. How would you respond in such a situation if hydropower were the main electricity source?

Lovins:

1. Basic references are given in notes 25 and 26 of my written paper. Houses should be well ventilated but in a controllable way - not through the usual technique of having about $1m^2$ of random holes in the house - and preferably through a heat exchanger (recuperator). Besant, Dumont, and van Ee (Dept. of Mech. Eng., Univ. of Saskatchewan at Saskatoon) have published details of a very cheap (about $30 material cost) counterflow recuperator made from plastic sheets and giving an average efficiency around 80% in a cold, dry climate. Mitsubishi sells a "Lossnay" crossflow recuperator at only slightly higher installed prices, which uses not plastic sheets but cardboard as the barrier material and which can thus exchange latent as well as sensible heat. This is usually preferable in humid climates. A. H. Rosenfeld at Lawrence Berkeley Laboratory can supply comparative cost and performance data on air-to-air recuperators.
2. If the US, for example, used electricity in an economically efficient way, today's economic output could be provided with less than a third of today's electricity. Present hydro-, microhydro-, and a modest amount of windpower would then suffice with no thermal stations of any kind. These three ingredients would make roughly equal contributions, but the latter two would be highly dispersed geographically (for example, the state with the most small dams is Kansas) and wind and hydro are statistically complementary. Some expansion of reservoir storage as against hydro peaking capacity could be desirable, as Edward Kahn of Lawrence Berkeley Laboratory has shown for the case of California. In practice, I suspect there would be a large dispersed capacity of photovoltaics, which work

best in the conditions in which drought and slack winds are likely to occur. These four sources are collectively a surprisingly firm source, especially given the extensive interconnections available in the present grid, which already ships British Columbia hydropower to San Diego and Churchill Falls hydropower to the Southeast. Though Kahn, Bent Sørensen (Niels Bohr Institute), and others have shown the statistical complementarity of the various renewable electric sources, I agree that more detailed simulations are important. They will probably confirm that the "soft" sources yield a more resilient grid than the "hard" sources (see LBL-7877 and Kahn in Annual Review of Energy 4, 1979), but should help to design the cheapest and most resilient combinations of renewable components for each area and climate.

Lovins:
The cheapest way to buy resilience against climatic fluctuations or other perturbations is very high end-use efficiency. For example, a superinsulated house has a long thermal time constant and can be heated entirely by any small, improvised point source of heat; a national fleet of very efficient vehicles can be run entirely on alcohols or pyrolysis oil derived from a wide range of farm and forestry wastes, even at high latitudes. The variation with latitude both of total insolation and of its seasonal variation is a relatively small factor and can be readily accommodated by design, if the energy is used in an economically efficient way so that little storage is required.
It is also important to note that the thermodynamic quality of sunlight is very high (thousands of degrees) even on a completely cloudy winter day in Northern Europe. This quality can be captured by highly selective surfaces, such as the selectivity-50 foils which Dr. Karl-Heinz Raetz (Physikalisch Technische Bundesanstalt, Braunschweig) reports can be made for DM 4.50/m^2. Such surfaces, on a cloudy winter day in Münster can supply under load a temperature difference of tens of ^{o}C (ample for domestic water heating) if in an airfilled flat-plate collector and of hundreds of ^{o}C if in a hard vacuum. There is therefore essentially no inhabited part of the world which is not suitable for direct solar collection.

McKay:
This is a commentary which is confirmed by Canadian experience. The use of "super" insulation, and other design that exploits and stores climatic and intrinsic energy flows make energy independence possible with regard to space and water heating. This is possible with no increase in capital cost; furthermore, progress is being made in the development of inexpensive animal and other shelters using simple absorbers and heat exchangers. However, these developments are not likely to insulate society from negative impacts of climate. Retrofit will most likely be gradual and "process" energy from both conventional and alternative sources

will remain vulnerable to hazards such as drought, flood and cold. Vulnerability to climate fluctuations such as those in Europe and Northern America in 1976-77 therefore remains a major problem.

Flohn:
Recent studies of interannual variability of temperature, precipitation, and run-off have been made based on 150 to 200 years of European records. They indicate that I.A.V. varies much more with time than expected and was particularly high at the beginning of the 19th century and relatively low in mid 20th century. Much more research on variability is needed. Energy systems need a lot of resilience to cope with a variability higher than now.

McKay:
I heartily concur. The need is increasing. For example, the competition for limited water resources between energy, agriculture, and urbanization requirements make necessary a much improved understanding of variations in precipitation, evaporation, and air temperature, if major shortages of supply are to be avoided in the future. This understanding is needed for systems design and for scenarios used to assess the vulnerabilities of an altered future society.

Schikarski:
Superinsolation of buildings has a disadvantage; because of the usage of stones (at least in Central Europe), there is an increase of radiation level due to more radioactive noble gases released and kept in the building space. This would require more ventilation in private homes. There seems to be an optimum between energy saving by insulation and energy losses by more ventilation.

Page:
While radioactivity may set limits on ventilation rates, currently ventilation rates are very much above this lower limit. Using heat recovery techniques to recover the outgoing heat in a flow heat exchanger, rates of ventilation can be kept at safe levels without large losses of heat.

Hekstra:
Great emphasis should be given to increasing end-use efficiency of fuel wood, in particular in developing countries. The Technical University of Eindhoven (NL) has developed a new domestic cooker, using about one-fourth of the usual amount of fuel wood for the daily family food cooking.

Page:
This work is very important, but it must take account of social

factors in the different countries in which the work is to be applied. Improved use of fuelwood is one of the topics for the 1981 UN Conference on New and Renewable Sources of Energy, because it is a key world energy issue.

Page:
Energy use in the western part of North America was less than normal in 1976-77, and they therefore might have sent power to the Midwest and East.

Quirk:
During 1976-77, the Eastern and Midwestern US were very cold and had to use 150 million more barrels of oil than normal from September to March. Energy use in the Western USA was reduced by about 5% overall, but the hydropower was reduced by 40%. Because of the reduction of the hydropower, the demand for oil was increased by 50 million barrels.

Waltz:
There is an important distinction to be made between measuring the cost of a climatic disaster and determining the amount of losses which could be avoided by appropriate mitigating strategies. Do you know whether detailed estimates of avoidable losses (in the case of recent climatic anomalies in the US or elsewhere) have been made by the US Department of Energy or anyone else?

Quirk:
Most of the types of losses incurred in 1976-77 could be avoided in the future. Energy conservation efforts could vastly reduce the sensitivity of energy demand to climatic variations. By keeping large reserves we could avoid oil or gas shortages. Unfortunately no one has done a cost/benefit/risk study to determine whether society would benefit by taking the action necessary to avoid these losses.

McKay:
1. Climatic data are not "locked up in archives", most engineers are well informed on their existence. While there may be a need for advising on their existence and means of access, a greater need is on how they may be used and enhanced in decision making. One should not have to wait until climate performs the experiment. Most climate anomalies, which are rare in nature, can be transposed and assigned risk limits. Development and promotion of such techniques warrant urgent attention.
2. Climate monitoring and strategic plans based on climatic case histories provided another excellent means of aiding the decision makers to respond to climate stress/opportunities.

For example, the 1976-77 drought made the substitution of coal for hydropower a necessity in Ontario. The railways were not prepared. They could have been warned of the pending need for coal shipments by an effective monitoring and advisory system.

Quirk:
1. You are certainly right that much work needs to be done on how best to use climate data. I can't comment on the Canadian situation, but in the United States I can assure you that many engineers need climate data that is not available to them. The Federal Energy Regulatory Commission has been trying to arrive at a policy on how much stored natural gas should be kept through the late part of the heating season (late February and March). The data they need to make this policy is not now available. It is a simple task from a scientific point of view to make such data available, but there are non-scientific barriers that make this data difficult to obtain.
2. That is a very good example of the sort of communication problems that exist in getting data to potential users. However, a cost/benefit/risk study must also be made to determine if there is benefit to society in being prepared for such rare events.

Roberts:
Some measures of conservation, it appears to me, may increase our vulnerability to climate extremes. Large gasoline-wasting cars in the U.S.A. keep a large volume of gasoline flowing in the supply system, and thus available for diversion in a climate emergency. I don't advocate wasteful cars, but I do observe that people with tight belts certainly store less fat for a food emergency. Please comment.

Quirk:
You have a good point and it requires more study than we can give it during this question period. My guess is that while we may have more efficient cars in the future we will still drive many more miles than we really have to. If we have homes that use much less energy for heating we would probably be able to supply all the additional fuel needed due to cold weather by driving a few less miles.

Breuer:
If hydro- and wind power were to go down together, insolation most probably would increase. Thus, if solar panels or solar electricity are also used, the supply would probably balance out.

Quirk:
Yes. This is true. My point was that one should not rely on hydropower and wind alone for electricity. In designing a soft

energy" supply strategy one must consider climatic factors even more than for a "hard energy" strategy.

PART II

CLIMATIC IMPACTS OF CARBON DIOXIDE

THE CARBON CYCLE AND ITS PERTURBATION BY MAN

H. Oeschger, U. Siegenthaler and M. Heimann

Physics Institute, University of Bern, Sidlerstr. 5,
CH-3012 Bern/Switzerland

ABSTRACT. Calibration and validation of carbon cycle models is mainly based on isotope data of the steady state of the system as well as its responses to natural and anthropogenic disturbances. The progress in the use of carbon isotopes is reviewed. The discussion covers the following phenomena:

- Response of the system to the anthropogenic CO_2 input: rise of atmospheric CO_2, dilution of the atmospheric ^{14}C concentration
- Prenuclear ^{14}C distribution in the ocean
- ^{14}C rise in atmosphere and ocean due to bomb produced ^{14}C
- Modulation of natural ^{14}C production and its atmospheric response, as reflected in $^{14}C/C$ ratios recorded in tree-rings
- Past changes of the atmospheric CO_2 concentration inferred from ice core studies.

The information is used to derive the probable range of the CO_2 release, resp. uptake by the biosphere, and to check the general validity of carbon cycle models.

INTRODUCTION

The carbon cycle and its modelling have been subject to an intensive discussion. In the first half of the last decade, main emphasis in carbon cycle models has been given to the atmosphere - ocean exchange. Models were developed which allowed a satisfactory description of the atmospheric CO_2 increase and ^{14}C dilution (Suess effect) due to the fossil CO_2 input as well as of

W. Bach, J. Pankrath, and J. Williams (eds.), Interactions of Energy and Climate, 107–127.

the atmospheric and oceanic ^{14}C increase due to ^{14}C produced by nuclear weapon testing (1, 2). A few years ago, biologists pointed out the role of the biosphere and its man-made change. It was stated that cutting of living biomass could result in a net carbon release between 0.1 and several times the fossil CO_2 input (3, 4). These estimates cast serious doubt on the validity of existing carbon cycle models.

In this paper, we derive what we consider the most probable value for the net contribution by the biosphere, as well as an upper limit, based on the fossil fuel consumption, the observed atmospheric CO_2 increase and the estimated CO_2 uptake by the oceans. We discuss the so-called airborne fraction of the CO_2 input, a quantity fundamental for predictions of future atmospheric CO_2 concentrations.

During the last few years, several laboratories were able to unambiguously show the existence of secular variations in the atmospheric $^{14}C/C$ ratio as recorded in tree-rings (5, 6). These fluctuations are probably due to a modulation of the ^{14}C production rate by cosmic radiation. We will discuss here the information on the dynamics of the carbon cycle which can be derived from them.

Measurements of the CO_2 content of air included in ice cores from the polar ice caps dating back 30,000 yrs give indications of changes of the atmospheric CO_2 content. We will briefly consider these results and their implications regarding the CO_2 - climate relationship.

THE CO_2 INCREASE AND THE AIRBORNE FRACTION

The increase of the atmospheric CO_2 concentration has been continuously monitored by Keeling and co-workers (7, 8) at Mauna Loa, Hawaii, and at the South Pole. The Mauna Loa record, covering the period 1958 to 1978 is given in Fig. 1. The yearly average increased during that period from 315 ppm to 335 ppm. During the same period an amount of CO_2 corresponding to 36 ppm has been released into the atmosphere due to fossil fuel combustion and cement manufacture (9).

An apparent airborne fraction for the fossil CO_2 can be defined as

$$F_a = \frac{\text{atmospheric } CO_2 \text{ increase}}{\text{fossil } CO_2 \text{ input}} \qquad \{1\}$$

For the period 1958-78, the apparent airborne fraction is

$$F_a = \frac{20 \text{ ppm}}{36 \text{ ppm}} = 0.56$$

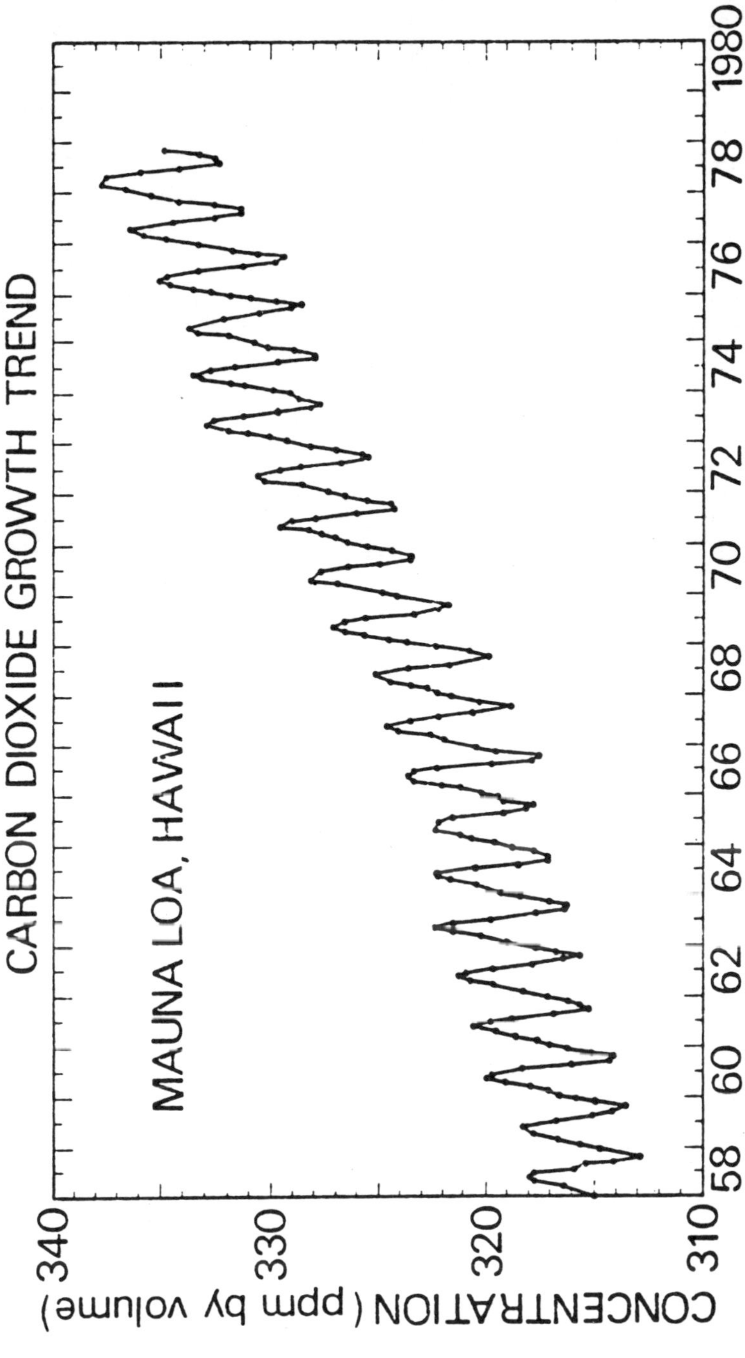

Fig. 1: CO_2 concentration observed at Mauna Loa by C.D. Keeling and co-workers

The observed CO_2 level does not increase regularly; superimposed on the secular increase there are fluctuations which are probably related to a natural meteorological phenomenon, the Southern Oscillation (10). The airborne fraction as calculated for different periods is therefore not constant, but varies by several %. Furthermore, the data for fossil fuel consumption are subject to an uncertainty of perhaps 10 percent. Therefore, the actual apparent airborne fraction can be estimated to be between about 0.50 and 0.62, for the period 1958-78.

The airborne fraction for the whole period since the beginning of industrialization can be determined with large error limits only, because the pre-industrial CO_2 level is not well known. The most probable range is 288 to 295 ppm, but values outside this range cannot be excluded. We therefore base our discussion on the period since 1958, for which reliable atmospheric observations are available.

The apparent airborne fraction, as defined above, only relates to the CO_2 produced from fossil fuel. It would be preferable to relate it to the total anthropogenic input, including the contribution from deforestation and soil modification. It is, however, not possible to indicate a reasonably accurate value for this actual airborne fraction, because the magnitude of the biospheric contribution is not known.

The excess CO_2 not remaining in the atmosphere must be taken up either by the ocean or the biosphere. A typical result of a model for the carbon cycle (2, 11) is a partitioning of the total CO_2 input between atmosphere, ocean and biosphere of 0.57 : 0.34 : 0.09 (see below). In order to discuss the CO_2 partitioning, we find it convenient to introduce a dilution factor $D = 1/F_a$, which gives the apparent total volume, in terms of atmospheric units, into which the anthropogenic CO_2 gets distributed. The dilution factor corresponding to the observed airborne fraction of 0.56 is $D = 1.79$, with the quoted uncertainty of about $\pm 10\%$. The dilution factor corresponding to the mentioned model result, $D = 1 + 0.34/0.57 + 0.09/0.57 = 1 + 0.60 + 0.15$, can be interpreted in a simple way: the three terms indicate the relative uptake capacities of atmosphere, ocean and biosphere, for the given CO_2 production history.

According to this model result, the biosphere would have to take up some fossil CO_2. However, the global biomass may be decreasing because of human influence. Bolin (3) evaluated FAO data on world forest resources and concluded that the net annual input of carbon into the atmosphere due to human influence on the land biosphere is 1.0 ± 0.6 gigatons per year, compared to 5 Gt of C per year of fossil input (1 ppm = 2.12 Gt). Woodwell et al. (4) even estimated that the annual release of carbon from the biota may be between 4 and 8 Gt, and possibly as high as 18 Gt. Their estimates are, however, apparently biased towards high values. An important argument of Woodwell et al. is a study by Veillon, which indicates a rate of forest clearing of about 1.3 percent per year in a region of Venezuela. Woodwell et al. apparently simply extrapolated from this and other studies of restricted areas to a worldwide rate of clearing of 1 to 2 percent per year. Loomis (12) presents evidence that the actual global rate of deforestation is probably significantly less, and that forest regrowth may in general be underestimated.

If there is a considerable CO_2 flux from the biosphere, the actual airborne fraction and dilution factor(related to fossil plus biospheric input) may be significantly different from their apparent values (related to fossil CO_2 only):

	D	F_a
Fossil only (1958-78):	1.79	0.56
Biospheric flux = 20% of fossil:	2.15	0.47
Biospheric flux = 100% of fossil:	3.58	0.28

In the following we discuss whether such large dilution factors are compatible with the present understanding of the CO_2 partitioning between atmosphere and ocean.

THE CO_2 SYSTEM AND ITS ISOTOPIC LABELLING

Figure 2 shows the main carbon reservoirs - atmosphere, biosphere and ocean - and the exchange fluxes between them. The exchange flux atmosphere - ocean and the vertical mixing rate of the ocean are derived from isotope studies (^{14}C, ^{222}Rn); this is discussed in a later section.

The exchange fluxes of the atmosphere with the biosphere and the ocean are of the order of 0.1 atmospheric CO_2 mass per year and therefore of relevance for the fate of excess CO_2. Organic and inorganic carbon get into the sediments and return to the atmosphere or ocean by degassing of the lithosphere and by weathering. These geological cycles have turnover times long compared to the time scales of $\lesssim 10^2$ years which are of interest when discussing the CO_2 problem, and need therefore not be considered in the present context.

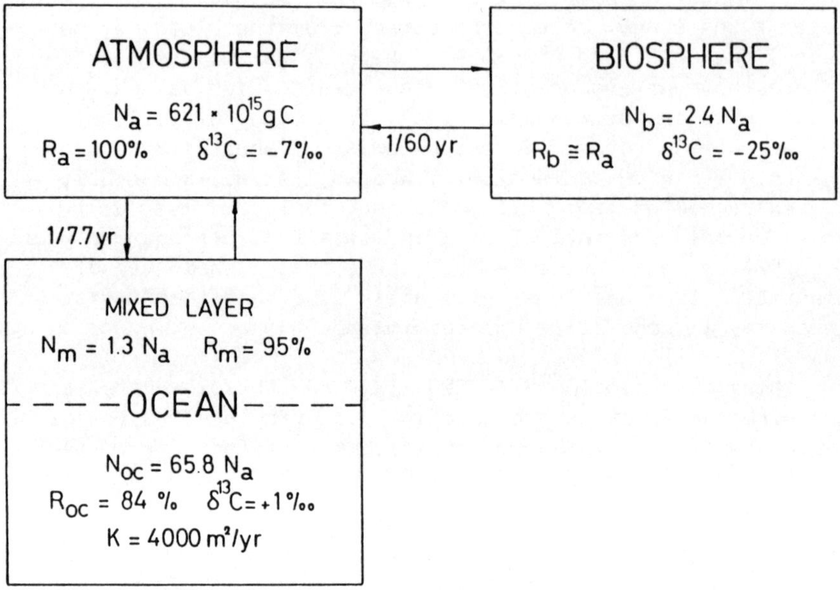

Fig. 2: Carbon reservoirs and fluxes, and isotopic composition (R_i = ^{14}C concentration in reservoir i)

Information on the time scales of the exchange processes and the mixing of the ocean can be obtained from the radioactive isotope ^{14}C (half-life 5730 years) which is produced by cosmic radiation in the atmosphere. Measurements of ocean water samples taken before the large nuclear weapon tests of 1961/62 and corrected for the dilution effect by fossil CO_2 (see below) showed that the pre-industrial, pre-nuclear ^{14}C concentration of surface water was 95% and the average value for the whole ocean about 84% of the atmospheric ^{14}C concentration. The lower ^{14}C activity in the ocean is due to radioactive decay and enables one to estimate the average time elapsed since the CO_2 in a given ocean layer left the atmosphere.

Photosynthesis leads to a depletion of ^{13}C ($\delta^{13}C$ in the biosphere $\sim$ -25‰, compared to the atmospheric value of $\sim$ -7‰), whereas the fractionation between atmospheric CO_2 and total CO_2 in the ocean leads to a ^{13}C enrichment in the ocean ($\delta^{13}C \sim$ 1‰).

THE STATIC CAPACITY OF THE SYSTEM FOR UPTAKE OF EXCESS CO_2

After a time period which is long compared to the mixing and exchange rates in and between the reservoirs, any CO_2 input into the system will be distributed among the different reservoirs. For physico-chemical and biological reasons the partitioning of the CO_2 excess will not be proportional to the carbon amounts in the different reservoirs.

An increase in atmospheric CO_2 and dissolved CO_2 gas in the oceans brings about a shift in the chemical equilibria between dissolved CO_2, HCO_3^- and $CO_3^=$, resulting in an increase of the total CO_2 concentration ($CO_2 + HCO_3^- + CO_3^=$) which is smaller than that of the dissolved CO_2 gas alone. This is taken into account by introducing a buffer factor ξ: if the dissolved CO_2 increases by p percent, the total CO_2 concentration of seawater increases only by p/ξ percent. For average surface seawater and a CO_2 level of $\sim$ 300 ppm, $\xi \simeq 10$; ξ increases for higher CO_2 pressure (1).

Since the biospheric growth rate depends, among other factors, on the CO_2 supply, it is probable that the CO_2 increase induces at least for part of the biosphere an increased growth rate ("CO_2 fertilization"). A simple concept to consider this is the introduction of a biota growth factor ε: if the atmospheric CO_2 pressure increases by p percent, the CO_2 flux to the biosphere increases by εp percent. Typically, values between 0 and 0.5 have been used in carbon cycle models (1, 2).

This now allows to calculate the fraction F_∞ of the total CO_2 input remaining in the atmosphere forever (neglecting interaction with the sediments). Taking into account the buffer factor and the biota growth factor, the static uptake capacities of atmosphere, ocean and biosphere are proportional to N_a, N_{oc}/ξ and εN_b, respectively, and the static dilution factor becomes

$$D_\infty = 1 + \frac{1}{\xi}\frac{N_{oc}}{N_a} + \varepsilon\frac{N_b}{N_a} \qquad \{2\}$$

According to Fig. 2, $N_{oc}/N_a = 65.8$ and $N_b/N_a = 2.4$. The static capacity of the whole system for $\varepsilon = 0$ (constant biomass) and $\varepsilon = 0.2$ (moderate CO_2 fertilization) is characterized by

$$D_\infty = 1 + 6.58 + \begin{cases} 0 & (\varepsilon = 0) \\ 0.48 & (\varepsilon = 0.2) \end{cases} \qquad \{3\}$$

The corresponding airborne fractions F_a are 0.132 ($\varepsilon = 0$) and 0.124 ($\varepsilon = 0.2$). These values hold for small CO_2 inputs. For large inputs, ξ increases and F_a is larger.

Obviously, the ocean is finally the most important sink for excess CO_2. At equilibrium, the ocean and biosphere will take up ∿ 7/8 of a given, relatively small CO_2 input. This uptake is, however, reached only after many centuries. For a continuing CO_2 production the CO_2 system is not in equilibrium as illustrated e.g. by the average age of deep ocean water of about 1000 years. Therefore, in order to estimate the dynamic CO_2 uptake capacity for different possible input histories, we need dynamic dilution factors.

THE DYNAMIC CO_2 UPTAKE CAPACITY

In the following we assume that the CO_2 production rate increased exponentially, $p = p_o\ e^{\mu t}$, which is approximately true for the past fossil fuel consumption. The e-folding time, μ^{-1}, of the fossil fuel increase was 41 yr prior to 1956, and 23 yr from 1956 to 1978 (13); the time constant of 28 yr given in ref. (14) seems, therefore, an appropriate overall value. When the production rate ceases to grow exponentially, the airborne fraction will change; it will also change when the CO_2 increase becomes comparable to the pre-industrial level, because then the buffer factor ξ is no longer constant.

For the model calculations, the box diffusion model is used (2, 11) (see Fig. 2), in which vertical mixing in the ocean is simulated by eddy diffusion, with a constant eddy diffusivity $K = 4000\ m^2/yr$. Atmosphere, biosphere and mixed layer (= surface ocean) are modelled as well-mixed reservoirs; the exchange fluxes between them obey first order kinetics. The depth of the ocean, H_{oc}, is 3800 m, that of the mixed layer, h_m, 75 m. The equivalent depth of the atmosphere, h_a, defined as thickness of an ocean layer containing the same amount of carbon as the pre-industrial atmosphere, is taken as 58 m. (Some numbers, especially h_a, do not correspond to the best possible estimates; they are chosen so as to keep consistency with refs. (2, 11)). The dynamic dilution factor can be expressed as follows:

$$D(\mu) = 1 + e(\mu) \cdot \frac{h_{oc}(\mu)}{H_{oc}} \cdot \frac{N_{oc}}{\xi N_a} + \varepsilon \frac{h_b(\mu)}{H_b} \cdot \frac{N_b}{N_a} \qquad \{4\}$$

$e(\mu)$ = degree of equilibration of ocean surface water with excess CO_2,

$h_{oc}(\mu)$ = equivalent CO_2 penetration depth in ocean,

$h_{oc}(\mu)/H_{oc}$ = fraction of ocean in equilibrium with ocean surface,

$h_b(\mu)/H_b$ = fraction of biosphere in equilibrium with atmosphere (symbols chosen in analogy to ocean).

The different terms can be calculated by means of the box diffusion model; see also eq. {35} in ref. (2). The penetration depth into the diffusive deep sea is given by $\sqrt{K/\mu}$, as long as $\sqrt{K/\mu} \ll H_{oc}$ i.e. the fossil CO_2 has not effectively penetrated to the ocean bottom. The total penetration depth is

$$h_{oc} = h_m + \sqrt{K/\mu} \qquad \{5\}$$

The degree of equilibration is given by

$$e(\mu) = \frac{\xi \; h_a \; k_{am}}{\xi \; h_a \; k_{am} + \mu h_{oc}} \qquad \{6\}$$

where k_{am} is the exchange coefficient between atmosphere and ocean for atmospheric CO_2, k_{am} = 1/7.7 yr.

The biosphere is simulated by an exponential box, with an exchange coefficient k_{ba} = 1/60 yr to the atmosphere. The fraction in equilibrium with the atmosphere then is

$$h_b/H_b = \frac{k_{ba}}{k_{ba} + \mu} \qquad \{7\}$$

Numerically we obtain, with ξ = 10, h_{oc} = 410 m, h_{oc}/H_{oc} = 0.108, e = 0.837, h_b/H_b = 0.318, and finally a dilution factor

$$D(\mu{=}1/28\text{yr}) = 1 + 0.60 + \begin{cases} 0 & (\varepsilon = 0) \\ 0.15 & (\varepsilon = 0.2) \end{cases}$$

For an exponential increase with μ = 1/28 yr, the dynamic is therefore by a factor of about 4.6 smaller than the static capacity. The model-predicted dilution factors are within the error limits of the observed value, D = 1.79 ± 10%, for both values of ε. For optimum agreement, the biota growth factor ε should be 0.25.

There is, however, a considerable discrepancy between the model-calculated dilution factor and the dilution factor required if the biospheric CO_2 input were of comparable size as the fossil CO_2 input. The question, therefore, arises if the model parameters can be changed so as to allow for a significant biospheric input without getting into conflict with the isotopic data used for model calibration. The parameters to be adopted are the atmosphere - ocean exchange coefficient k_{am} and the eddy diffusivity K; both are derived from the pre-industrial ^{14}C distribution in atmosphere and ocean. Considering the uncertainty and scarcity of pre-nuclear ^{14}C measurements in the ocean, a maximum error of ± 30% can be assigned to both parameters (Table 2 in ref. (2)). Assuming that K and k_{am} are actually by a factor 1.5 larger than our nominal values of

4000 m^2/yr and 1/7.7 yr is therefore an extreme upper limit. The corresponding oceanic mixing depth would increase to h_{oc} = 485 m, the degree of saturation to e = 0.867. The contribution of the ocean to the dilution factor, however, would only be increased from 0.60 to 0.73 which is relatively insignificant when compared to the high dilution factors postulated in the paragraph "CO_2 increase and airborne fraction".

OTHER MODEL AND PARAMETER CHECKS

The ^{14}C dilution (Suess-effect)

The fossil CO_2 brought into the atmosphere does not contain ^{14}C and leads to a ^{14}C dilution. Without exchange between the atmosphere and the other reservoirs a fossil CO_2 input amounting to 10% of the atmospheric CO_2 content (corresponding to 1950) would lead to a decrease of the $^{14}C/C$ ratio by 10%. Because of the exchange with the other reservoirs this effect is reduced by a dilution factor $D_{14C}(\mu)$ which can be calculated from the box-diffusion model:

$$D_{14C}(\mu) = 1 + e(\mu) \cdot \frac{h_{oc}(\mu)}{H_{oc}} \cdot \frac{N_{oc}}{N_a} + \frac{h_b(\mu)}{H_b} \frac{N_b}{N_a} \qquad \{8\}$$

$$\text{with } e(\mu) = \frac{h_a\, k_{am}}{h_a\, k_{am} + \mu h_{oc}}$$

Numerically, D_{14C}(1/28 yr) = 1 + 2.4 + 0.8 = 4.2.

In 1950 instead of a 10% decrease only a 2.4% decrease of $^{14}C/C$ is predicted. Using the actual CO_2 production history Oeschger et al. (2) calculated for the Suess effect in 1950 a value of -2.0%. This is essentially in agreement with the measurements (15), though ^{14}C fluctuations due to ^{14}C production rate variations make a precise determination of the Suess-effect difficult.

The ^{14}C fluctuations

Tree-rings preserve a record of the $^{14}C/C$ ratio in the lower atmosphere at the time of their formation. ^{14}C measurements on tree-rings covering approximately the last 7000 years show two types of variations. From 7000 B.P. (before present) to 2000 B.P. the average atmospheric ^{14}C activity decreased by about 10%. Superimposed on this general trend are secular variations ("Suess-wiggles") of the order of 1 to 2% (5, 6). Both effects can be explained principally either by variations of the ^{14}C production rate or by variations in the carbon exchange system, although the former mechanism appears to be more important (16). In periods of high (low) terrestrial and/or solar magnetism, shielding of the galactic component of cosmic radiation leads to relatively low (high) isotope production in the atmosphere. Palaeomagnetic data indicate that 7000 years ago the value of the earth's geomagnetic dipole moment was only one half of that of 2000 years ago, and estimates using a model for the geomagnetic modulation of radioisotope production and a carbon cycle model show that this increase in the magnetic field strength could well have resulted in the observed decrease in the atmospheric $^{14}C/C$ ratio.

Of direct interest regarding the response of the atmospheric CO_2 to disturbances with characteristic times of ∿ 30 years (e-folding time of CO_2 increase) are the secular variations superimposed on the general trend. During the last few years several laboratories (5, 6) showed that such characteristic ^{14}C deviations really exist. The saw-tooth-like variations have peak-to-peak amplitudes of up to 2%, with a basic period of about 200 yrs (17, 18). When comparing the system response to an exponential forcing with that to a cyclic forcing, we expect similarities for $\mu = \omega = 2\pi/_T$, where T is the period of the cyclic forcing. The period of 200 yrs (ω = 1/32 yrs) therefore happens to correspond to the e-folding time of the CO_2 increase. The relative amplitude of the atmospheric ^{14}C variations is by a damping factor $D_{^{14}C}(i\omega)$ smaller than the relative amplitude of the production rate variations (neglecting radioactive decay):

$$D_{^{14}C}(i\omega) = \frac{i\omega}{\lambda \cdot 58} \{1+e(i\omega)\cdot \frac{h_{oc}(i\omega)}{H_{oc}} \cdot \frac{N_{oc}}{N_a} + \frac{h_b(i\omega)}{H_b} \cdot \frac{N_b}{N_a}\} \qquad \{9\}$$

The number 58 gives the ratio of total ^{14}C in the system to ^{14}C in the atmosphere at steady state.

The damping factor is the product of a dilution factor for the atmospheric reservoir and the one for the system. The latter corresponds to the dilution factor $D_{^{14}C}(\mu)$ for the Suess-effect, with $i\omega$ replacing μ, $i = \sqrt{-1}$. It is a complex number expressing both the ratio of the relative amplitudes and the phase shift between ^{14}C production and observed ^{14}C concentration.

Equation {9} leads for $\omega = \frac{2\pi}{200 \text{ yrs}}$ to a total attenuation factor of 20 and a phase angle of 36°, corresponding to a lag of 20 yrs. During many years, this caclculated strong model attenuation was the reason why the existence of the secular

^{14}C fluctuations was doubted: variations in ^{14}C of 2% would correspond to ^{14}C production rate variations of ∿ 40%, compared to variations of the order of 10% to 20% as predicted from ^{14}C production models. Stuiver and Quay (6) used the box-diffusion model to calculate the ^{14}C production rate variations that would have led to the ^{14}C variations observed in tree-rings. Fig. 3 shows these production rate variations plotted together with inverse sun spot numbers.

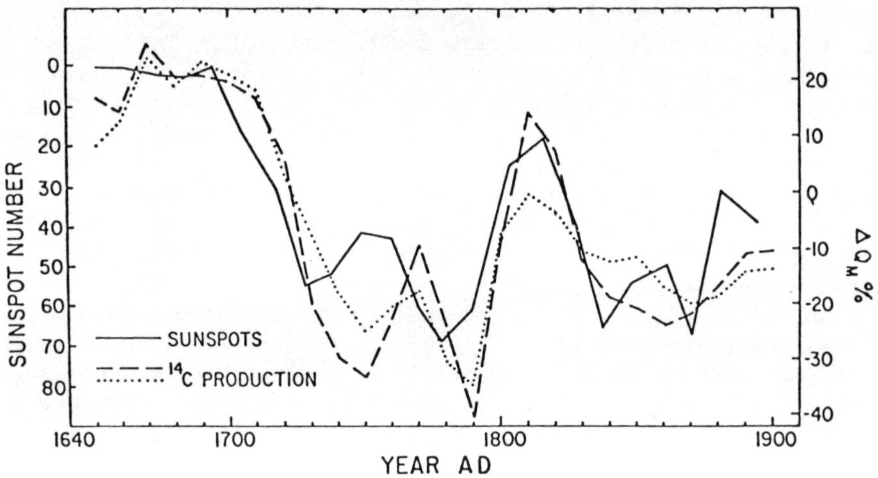

Figure 3: ^{14}C production as calculated from tree-ring ^{14}C concentrations by means of the box-diffusion model. The dashed and the dotted curve correspond to two different assumptions for the biosphere. The good correlation with sunspot numbers (solid line) suggests that indeed solar modulation of the cosmic radiation is the cause for the ^{14}C variations.
From Stuiver and Quay (6).

A good correlation is obtained. Stuiver and Quay compared the magnitude of these geochemically derived production changes with ^{14}C production changes derived from atmospheric neutron flux measurements. A good agreement is obtained for the 20th and 19th century data, but for the Maunder Minimum period, A.D. 1654 to 1741, a greater dependence on sun spot numbers is suggested, which Stuiver attributes to an additional ^{14}C production increase during periods when sun spots were absent. A generally satisfactory agreement is obtained, suggesting that for cyclic perturbations with a characteristic time of 30 years the model seems to give reasonable answers.

Since the production rates cannot, at this time, be estimated independently in a reliable way, the ^{14}C variations do not permit a quantitative model test. The essential point is that the model does not show a tendency to underestimate the attenuation (or dilution) as suggested if there were a large biospheric CO_2 input.

MODEL CALIBRATION AND VALIDATION

The rate of CO_2 uptake by the ocean is determined by two processes:
- gas exchange through the atmosphere - ocean interface,
- vertical mixing of water masses with its total CO_2 (CO_2, HCO_3^- and $CO_3^=$).

The gas exchange rate is determined by different methods.
- The net flux of pre-nuclear ^{14}C through the atmosphere - ocean interface has to be equal to radioactive decay of ^{14}C in the ocean. Using the reservoir values given in Fig. 2 the CO_2 flux is estimated to 19 moles (230 g C) per m^2 of ocean surface per year. This corresponds to a mean residence time of 7.5 years in the atmosphere before a CO_2 molecule is transferred into the mixed layer of the ocean. This residence time corresponds to the inverse of the exchange coefficient k_{am}. The estimated uncertainty of this value is ± 30%.
- The second method is based on the deficiency of radon ^{222}Rn observed in the mixed surface layer of the sea. ^{222}Rn ($T_{1/2}$ = 3.9 days) is produced by radioactive decay of ^{226}Ra dissolved in sea water. Within its mean life (∿ 6 days) some of the radon produced escapes to the atmosphere. The fraction escaping depends on the gas transfer rate between sea and air which can, therefore, be determined by measuring the radon deficiency in surface water. Translated to CO_2 an exchange flux of 16 mols per m^2 of ocean surface and year is estimated (19), which is in satisfactory agreement with the result obtained by the ^{14}C method.

The rate of vertical mixing is determined by the measurement of tracers with different characteristics. The ratio of the average pre-nuclear ^{14}C concentration for the whole ocean (84%) to that of the mixed layer (95%) corresponds to an eddy-diffusion coefficient of 4000 m^2y^{-1} (2). Further useful tracers are ^{14}C and ^3H produced by nuclear weapon tests. Injection of ^3H into the ocean surface occurred mainly in the northern hemisphere in the period 1962-1964. The atmospheric ^{14}C level showed in 1963/64 an increase by 100% and 60% in the northern and southern hemisphere respectively; at present, the ^{14}C excess in both hemispheres is about 30%.

Broecker et al. (20) used GEOSECS ^{14}C and ^{3}H data in the Atlantic and Pacific Oceans, measured by Oestlund and Stuiver, to calculate apparent eddy diffusion coefficients describing the uptake of bomb-produced ^{14}C and ^{3}H. Their best estimate is $K = 7000\ m^2/yr$, significantly higher than the value of $4000\ m^2/yr$ derived from the pre-industrial ^{14}C distribution. An eddy diffusivity of $7000\ m^2/yr$ corresponds, together with the standard values for the other parameters, to an ocean dilution factor $D^{ocean} = 0.72$, compared to a standard value of 0.60. These different results for K show that ocean mixing is not ideally represented by vertical eddy diffusion with a constant diffusivity; the diffusivity should perhaps be assumed as variable with depth, and besides diffusion, advection must play an essential role. The characteristic scales for the steady-state ^{14}C distribution (∿ 1000 yr) and bomb-produced isotopes (∿ 10 yr) are rather different, so that the two kinds of data sets test different scales of oceanic circulation. In view of this, the disagreement between the two values for K is not very large. A specific feature which may be responsible for the relatively high apparent diffusivity obtained for bomb products is the formation of intermediate and deep water. In restricted areas of the cold ocean surface, vertical exchange with intermediate and deep water is much faster than for the rest of the ocean. One-dimensional models cannot take into account this feature explicitly; this may lead to an underestimate of the total CO_2 uptake if a model is calibrated by pre-bomb ^{14}C only.

From all these model checks it is concluded that the response of the CO_2 system to disturbances is reasonably well understood, at least concerning the role of the oceans. We will now turn back to the problem of the CO_2 balance but firstly summarize the knowledge on the ocean dilution factor and its possible range.

THE OCEAN CO_2 DILUTION FACTOR; MOST PROBABLE VALUE AND POSSIBLE RANGE

In an earlier paragraph it was stated that with the box-diffusion model an oceanic dilution factor of 0.60 is calculated. Using 50% higher values for the dynamic parameters an upper limit of 0.73 was obtained, i.e. only 22% above the standard value.

As discussed earlier, the factor limiting the uptake by the ocean is the rate of vertical mixing, rather than the gas exchange rate. In order to get an idea of the limitation of uptake we assume that the ocean were mixed infinitely rapid over its entire depth. In this case we get, with $h_{oc} = H_{oc}$

$$e(\mu) = \frac{\xi h_a \; k_{am}}{\xi h_a \; k_{am} + \mu H_{oc}} = 0.357$$

and for the ocean dilution factor:

$$D^{ocean}(1/28yr) = 0.357 \cdot 1 \cdot 6.58 = 2.35.$$

The corresponding airborne fraction for constant biomass would be 0.30 (= 1/3.35). Such a rapid mixing is, of course, in complete disagreement with all the ^{14}C and ^3H ocean data. Therefore, the true ocean dilution factor must be significantly smaller than 2.35. The well-mixed case could, however, be an upper limit for the CO_2 uptake in polar areas where isotopic data indicate much more intensive vertical mixing. We could take this phenomenon into account by assuming that some fraction of the ocean is well mixed, the rest behaves according to the box diffusion model. Taking 20% as the well-mixed fraction, which is a rather high value, we get as an upper limit for the ocean dilution factor

$$D^{ocean}(1/28yr) = 0.8 \cdot 0.60 + 0.2 \cdot 2.35 = 0.95$$

Based on these considerations we think that the most probable value of the oceanic dilution factor is in the range 0.60 to 0.73 (0.67 ± 0.07) and that a value of 1.0 can be considered as an upper limit.

THE EXCESS CO_2 BUDGET

At the present time it is hardly possible to directly determine the magnitude of the present and past net CO_2 inputs from the biosphere; it is not even certain if the biosphere actually is a net source or a net sink of CO_2. The best estimate of the net biospheric contribution and its uncertainty range can probably be obtained from a budget:

Inputs			Uptakes	
fossil +	biospheric =	atmospheric +	oceanic +	biospheric
I_{fo}	I_{bio}	U_a	U_{oc}	U_{bio}

The net biospheric contribution is:

$$I_{bio} - U_{bio} = U_a + U_{oc} - I_{fo}$$

The observed apparent airborne fraction is 0.56, as stated at the beginning; for the oceanic dilution factor we take 0.67. For the net biospheric production we thus get the most probable value

$$I_{bio} - U_{bio} = 0.56\ I_{fo} + 0.56 \cdot 0.67\ I_{fo} - I_{fo} = -0.06\ I_{fo}$$

The biosphere therefore appears to be a sink, taking up several percent of the fossil CO_2 production. Within the uncertainties of the observations and of the model results, the net biospheric flux may well be zero. A maximum value for the net biospheric production is obtained if the maximum ocean uptake is introduced into the balance ($D^{ocean} = 1.0$):

$$I_{bio} - U_{bio} = 0.56\ I_{fo} + 0.56 \cdot 1.0\ I_{fo} - I_{fo} = 0.12\ I_{fo}$$

The resulting maximum net flux from the biosphere is only 12% of the fossil CO_2 production.

In this way, the net biospheric contribution is estimated. According to most ecologists who considered anthropogenic biomass changes, there is a CO_2 release due to deforestation and soil modification. The difference between net flux and anthropogenic release must be taken up by the biosphere (U_b), which could be caused by CO_2 fertilization. A release considerably larger than 10 percent of the fossil input therefore implies a remarkable stimulation of plant productivity by the increased CO_2 concentration.

The whole discussion has been based on an exponential growth with an e-folding time near 28 years. If there had been a biospheric input deviating from this exponential function, the dilution factors would be different; the total CO_2 input would, however, have to have deviated strongly from this exponential growth in order to cause significantly different results.

The oceanic uptake would be larger if the buffer factor were smaller than the value of about 10 derived from chemical equilibrium calculations. These calculations have recently been confirmed experimentally (21, 22). The effective buffer factor would be lowered if dissolution of shallow carbonate sediments would take place. In order to have a significant effect, the rate of this dissolution, which virtually can take place only on a small fraction of the ocean surface, would probably have to be very large.

In summary, we conclude that the understanding of the atmosphere-ocean system is at a stage which permits the estimation of the oceanic CO_2 uptake within reasonable error limits. The corresponding net biospheric CO_2 flux into the atmosphere is probably at most about 10% of the fossil fuel input rate. The dynamic model parameters needed for these estimates are derived from a number of isotope studies - pre-nuclear ^{14}C in the oceans, ^{14}C variations as observed in tree-rings, bomb-produced ^{14}C and tritium, and the $^{222}Rn/^{226}Ra$ disequilibrium in surface waters. It appears improbable that the estimated magnitude of a relevant quantity is seriously in error, or that any major CO_2 sink has been neglected.

THE LAST 30,000 YEARS HISTORY OF THE ATMOSPHERIC CO_2 CONTENT

For an assessment of the CO_2 problem a knowledge of the history of the atmospheric CO_2 content would be of great value. What was the pre-industrial atmospheric CO_2 content? Did it fluctuate or was it rather constant? The answers to such questions would help on the one hand to improve our knowledge on the carbon system and its response to disturbances, and on the other hand provide information regarding the sensitivity of the climate system to atmospheric CO_2 changes.

Probably the only possibility to reconstruct directly the history of the atmospheric CO_2 content (and $^{13}C/^{12}C$ and $^{14}C/C$ ratios) is measurements on natural old ice samples. Ice formed by sintering of dry cold snow contains in its bubbles air with atmospheric composition. Until a few years ago attempts to determine ancient atmospheric CO_2 contents by measuring CO_2 contents of the air included as air bubbles in natural ice seemed to provide unreliable results (23-25). CO_2/N_2 ratios much higher than the atmospheric values were found indicating the presence of additional CO_2 of undefined origin. In the last few years, however, the extraction technique has been further developed and considerably improved by two laboratories (26, 27). The group in Bern obtained CO_2 concentrations for air included in young ice samples of 270 to 370 ppm, with an average of 310 ppm (28). This value is close to the assumed pre-industrial atmospheric CO_2 level. This supports the confidence that on ice samples from very cold accumulation areas, CO_2/N_2 ratios can be measured which do indicate the atmospheric composition at the time of ice formation. The extraction procedure is as follows:

Samples of 300 g of ice are melted in vacuum and the gases produced by the explosion of air bubbles on the melting ice surface are separated. This first extraction fraction is considered to be representative for the composition of the gases in the bubbles. Extraction of the gases is then continued for several hours until CO_2 no longer escapes. Based on the analyses of the two extraction fractions an estimate can then be made of how much CO_2 is contained in the air bubbles and how much in the ice lattice. Analysis of the gas composition (N_2, O_2, Ar, CO_2) is made by gas chromatography.

Our experience to date suggests that current measurement and analysis techniques allow the reliable detection of variations in the atmospheric CO_2 content of 30 percent or more. Our first measurements have been made on samples covering the last 30,000 years, a period of major climatic change which might possibly have led to a change in the atmospheric CO_2 content. Ice cores suited for such a study were available from the successful drilling by U.S. scientists in 1966 to the bedrock of the Greenland ice cap in Northwest Greenland (Station Camp Century) and in 1967/68 to the bedrock of the West Antarctic ice shield at Byrd Station.

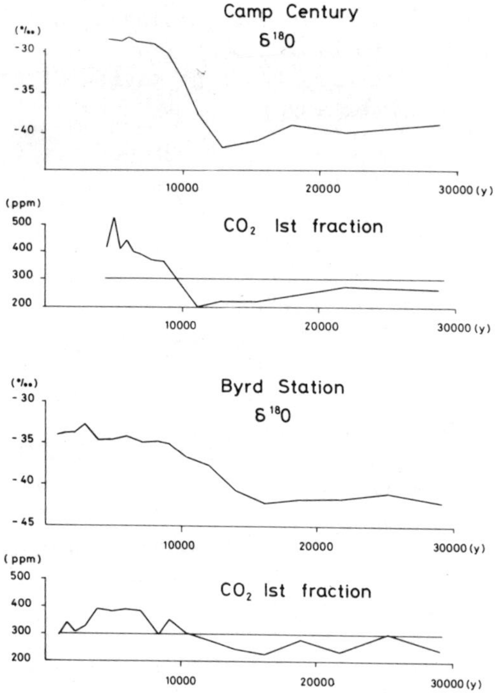

Figure 4: Camp Century, Greenland, and Byrd Station, Antarctica, ice cores: CO_2 contents of the first gas extraction fraction and $\delta^{18}O$ profiles (29,30). The ages are calculated according to refs. (30,31). Note the correlation between the CO_2 content and the $\delta^{18}O$ shifts. From ref. (28).

In Fig. 4, CO_2 contents measured in the first extraction fraction are plotted together with the $\delta^{18}O$ record for the Camp Century and the Byrd Station cores (29, 30). The dating is based on model calculations given in refs. (30, 31). Both records show similar trends: low values during last glaciation, and then, parallel to the $\delta^{18}O$ transition an increase to higher Holocene CO_2 values. For both cores similar minimum values (200 to 230 ppm) are found for the last glaciation. First fraction measurements on young ice (shown here only for the Byrd core) yield values around 300 ppm in both cores. The first fractions differ, however, in the Holocene: for the Camp Century core higher values are obtained.

The question has to be answered whether the general CO_2 trend in the ice core is mainly due to a change in atmospheric CO_2 content or a climatic effect on CO_2 enclosure process. The most probable explanation for the general trend - low values during Glacial and higher values during Postglacial - is a corresponding change in the atmospheric CO_2 content. The difference between the two profiles in the Holocene may partly be due to a contribution of CO_2 trapped in melt layers during the climatic optimum or due to another climatic effect on CO_2 enclosure at Camp Century. The atmospheric CO_2 concentration during the climatic optimum was therefore not necessarily higher than at present, as might be suggested by Fig. 4.

A POSSIBLE EXPLANATION FOR A CO_2 CHANGE IN THE ATMOSPHERE

The partial pressure of CO_2 in the atmosphere depends on total CO_2, alkalinity and temperature of the ocean surface. During the last glaciation the ocean volume was smaller by about 3% and the ocean surface temperature lower by about 2.3°C. For the first effect a CO_2 increase by about 1%, for the second a decrease by about 6.4% can be estimated using chemical equilibria. This results in an overall estimated decrease of the atmospheric CO_2 content during last glaciation by about 5.4%, i.e. these effects cannot account for the large observed difference.

A possible explanation for the change in atmospheric CO_2 content is the fact that atmospheric CO_2 levels strongly depend on the total CO_2 content of the ocean surface. Due to the buffer effect, a change in total CO_2 in the ocean surface leads, for constant alkalinity, to a tenfold relative change in the atmospheric CO_2 content. The total oceanic CO_2 is determined partly by marine biospheric activity, which leads to a depletion at the ocean surface compared to the ocean average. A relatively slight change in biospheric activity could therefore lead to a significant decrease in atmospheric CO_2.

CONCLUSIONS

These data demonstrate the importance of experimental work on natural records of ancient states of the CO_2 system. Though not definitely substantiated, these data imply that

- the atmospheric CO_2 content during the last 30,000 yrs has not necessarily been as constant as hitherto assumed;
- that only slight changes in the total CO_2 content of the ocean surface could lead to considerable changes in the atmospheric CO_2 content;
- that a shift in atmospheric CO_2 content at the end of the last glaciation may have reinforced the climatic change or even induced it.

REFERENCES

1) Bacastow, R.B. & Keeling, C.D. 1973: In: Carbon and the biosphere (G.M. Woodwell & E.V. Pecan, eds.), pp. 86-135, USAEC, Springfield, Va.
2) Oeschger, H., Siegenthaler, U., Schotterer, U. & Gugelmann, A. 1975: Tellus 27, pp 168-192.
3) Bolin, B. 1977: Science 196, pp 613-615.
4) Woodwell, G.M., Whittacker, R.H., Reiners, W.A., Likens, G.E., Delviche, C.C. & Botkin, D.B. 1978: Science 199, pp 141-146.
5) De Jong, A.F.M., Mook, W.G. and Becker, B. 1979: Nature 280, pp 48-49
6) Stuiver, M. and Quay, P. 1980: Science 207, pp 11-19
7) Keeling, C.D., Bacastow, R.B., Bainbridge, A.E., Ekdahl, C.A., Guenther, P.R., Waterman, L.S. and Chin, J.F.S. 1976: Tellus 28, pp 538-551.
8) Keeling, C.D., Adams, J.A., Ekdahl, C.A. and Guenther, P.R., 1976: Tellus 28, pp 552-564.
9) Rotty, R. 1980: Experientia, in press.
10) Bacastow, R.B. 1976: Nature 261, pp 116-118.
11) Siegenthaler, U. & Oeschger, H. 1978: Science 199, pp 388-395.
12) Loomis, R.S. 1979: In: Carbon Dioxide Effects, W.P. Elliott and L.Machta, eds. (U.S. Department of Energy), pp 51-62.
13) Keeling, C.D. 1980: Environment International 2, in press.
14) Broecker, W.S., Takahashi, T., Simpson, H.J. and Peng, T.-H. 1979: Science 206, pp 409-418.
15) Tans, P.P., de Jong, A.F.M. and Mook, W.G. 1979: Nature 280 pp 826-828.
16) Siegenthaler, U., Heimann, M. and Oeschger, H. 1980: Proc. 10th Internat. Radiocarbon Conference, Radiocarbon, in press.

17) Suess, H.E. 1980: Proc. 10th Internat. Radiocarbon Conference, Radiocarbon, in press.
18) Neftel, A. 1980: Manuscript in preparation.
19) Peng, T.-H., Broecker, W.S., Mathieu, G.G. & Li, Y.-H. 1979: J. Geophys. Res. 84, pp 2471-2486.
20) Broecker, W.S., Peng, T.-H. & Engh,R. 1980: Proc. 10th Internat. Radiocarbon Conference, Radiocarbon, in press.
21) Keeling, C.D. & Guenther, P.R. 1980: To be published in SCOPE report on the modelling of the global carbon cycle.
22) Takahashi, T., Broecker, W.S., Werner, S.R. & Bainbridge, A.E. 1980: To appear in: Isotope Marine Geomechistry, K. Sarukashi (ed).
23) Scholander, P.F., Hemmingsen, E.A., Coachman, L.K. & Nutt, D.C. 1961: J. Glaciology 3, pp 813-822.
24) Matsuo, S. and Miyake, Y. 1966: J. Geophys. Res. 71, pp 5235-5241.
25) Raynaud, D. & Delmas, R. 1977: Proc. of the Grenoble Symp.1975, Isotopes and Impurities in Snow and Ice. IAHS-AISH publication no 118, pp 377-381.
26) Berner, W., Oeschger, H., & Stauffer, B. 1980: Proc. 10th Internat. Radiocarbon Conference, Radiocarbon, in press.
27) Delmas, R.J., Ascencio, J.-M. & Legrand, M. 1980: Manuscript in preparation.
28) Berner, W., Stauffer, B. & Oeschger, H. 1978: Nature 275, pp 53-55.
29) Dansgaard, W., Johnsen, S.J., Clausen, H.B. & Langway Jr. C.C. 1971. In: Turekian, K.K, ed. The late cenozoic glacial ages, New Haven and London, Yale University Press, pp 37-56.
30) Johnsen, S.J., Dansgaard, W., Clausen, H.B. & Langway Jr., C.C. 1972. Nature 235, no 5339, pp 429-434.
31) Hammer, C.U., Clausen, H.B., Dansgaard, W., Gundestrup, N., Johnsen, S.J. & Reeh, N. 1978: J. Glaciology 20, no 82, pp 3-26.

THE OCEANS AND BIOSPHERE AS FUTURE SINKS FOR FOSSIL FUEL CARBON DIOXIDE

Charles D. Keeling

Scripps Institution of Oceanography
University of California at San Diego
La Jolla, CA 92093, USA

ABSTRACT. The prediction of future concentrations of atmospheric CO_2 is made difficult by the uncertainty in how rapidly and to what extent fossil fuels will be utilized. But if we focus on the consumption of fossil fuels over the entire period of use we find that almost any growth in consumption above the present, even 1.5 percent per year, will lead eventually to a large increase in atmospheric CO_2. If a major fraction of the physically minable coal is used as fuel, a four fold increase in atmospheric CO_2 concentration is likely to be reached within three centuries. The oceans during the entire fossil fuel era remain a major sink for industrial CO_2, but near the time of maximum fuel consumption, when the need to remove atmospheric CO_2 is greatest, the oceans will be a less effective sink than at present, because of increased acidity of the water attending CO_2 uptake. Because of the large amounts of carbon involved, it appears unlikely that absorption of CO_2 by land plants and soils will significantly reduce CO_2 levels beyond that produced by oceanic uptake.

INTRODUCTION

The world oceans at the present time constitute the principal geochemical reservoir removing fossil fuel produced CO_2 from the air. Modeling calculations, such as those discussed by Oeschger et al. in this volume (1), indicate that the oceanic uptake of industrial CO_2 from fossil fuel combustion is about equal to the amount accumulating in the atmosphere. Considerable uncertainty attends estimates of the carbon flows between the atmosphere and land biosphere (living biomass, surface detrital carbon, and

W. Bach, J. Pankrath, and J. Williams (eds.), Interactions of Energy and Climate, 129–147.

soil carbon) but as Oeschger et al. indicate (1), the observational data on carbon abundances and isotopic ratios suggest that the oceanic uptake cannot be seriously missestimated. Thus, one is inclined to infer that biosphere is presently maintaining a near balance between man-induced releases of CO_2, such as from timber harvesting and land clearing, and increased uptake of CO_2 as previously disturbed areas recover. Also, photosynthesis is possibly being enhanced in favorable locations by the recent world-wide rise in atmospheric concentration.

This rise has been without interruption over the period since accurate observations began, some 25 years. The rise has probably been occurring since early in the last century (1). If the influence of seasonal effects and an oceanic perturbation related to the southern oscillation (1, 2) are subtracted from the observational record, the adjusted trend in CO_2 concentration is nearly exponentially rising. The e-fold time is calculated to be 23 years, in agreement with that of fossil fuel production between A.D. 1945 and 1977.

The rate of rise influences the extent of oceanic CO_2 uptake (1): a slower rate of rise would lead to a larger fraction of the CO_2 from industry being taken up by the oceans, a higher rate would lead to a smaller fractional uptake. If the industrial production of CO_2 ceased altogether, the oceans would eventually take up nearly 7/8ths of the industrial CO_2 produced. The time required to achieve this fraction is, however, very long, of the order of several thousand years, owing to the slow rate of mixing and turnover of deep ocean water. Still more CO_2 must eventually be taken up by the oceans as silicate rocks and deep sea sedimentary carbonates slowly react with the CO_2 enriched air, ground water, and sea water.

The biospheric response to the exponentially rising concentration of atmospheric CO_2 is not known, but if plants typically react to higher CO_2 levels as do plants grown in glass houses (3), the world-wide average response is probably linear until the ambient CO_2 concentration reaches considerably higher levels than at present, i.e. if an x% rise in CO_2 concentration over the preindustrial level leads to an εx% increase in plant growth, then a 2x% rise in concentration will lead to a 2εx% increase in growth. As in the case of the oceans, we therefore expect that a slower or more rapid rise in atmospheric CO_2 would lead, respectively, to a larger or smaller fraction of the CO_2 from industry taken up by the biosphere. If the industrial production of CO_2 ceased altogether, we would expect the biosphere to continue to be fertilized for some period of time to an extent which, however, is possible to estimate only very crudely from present knowledge.

Should we expect the final partitioning of industrial CO_2 between the atmosphere and the biosphere to be similar to the 7/8ths factor predicted for the oceans? Or should we expect the plants to draw down the CO_2 concentration to some definite level such as the preindustrial concentration, irrespective of the present extent of the biosphere? In either case we would expect much of the adjustment to occur within a relatively short time, because the characteristic time of growth of forests and other plant communities is only of the order of 100 years. This time factor should principally govern the rate of adjustment except, perhaps, for increased storage of carbon in deep soil which might be longer. Any such final state of the biosphere is impossible to estimate, however, owing to the lack of direct evidence that plant communities respond to higher ambient CO_2. Only a few recent models of the biosphere even consider the question of such a final state (4, 5).

Thus, although we can identify quite well the role of the oceans in the present and distant future disposal of industrial CO_2, we have a poor understanding of either the present or distant future behavior of the biosphere.

It is clearly not possible for the oceans and biosphere to remove all of the industrial CO_2 from the air while it is still being produced, and therefore the use of fossil fuel in the near future and over the next several centuries is certain to produce higher atmospheric CO_2 concentrations than the present. Over the next 20 or 30 years it is reasonable to extrapolate from present conditions using models which correctly predict the recent rise in atmospheric CO_2. The fraction of industrial CO_2 remaining airborne is not likely to change greatly over this relatively short time (1). Over longer periods, and assuming continued industrial production of CO_2, extrapolation becomes more uncertain, but we are aided by some simplifications of the problem which arise, especially if the cumulative amount of CO_2 produced becomes much larger than at present. Thus, as discussed below, the problem of at least crudely estimating future CO_2 levels does not appear to be insurmountable, in spite of the shortcoming of present knowledge.

The methodology for predicting the future partitioning of industrial CO_2 between the atmosphere, oceans, and biosphere has been considered in several recent studies (4, 5, 6, 7, 8, among others). It is possible here to give no more than a summary and conclusions. As I will indicate, none of the principal conclusions depend on highly precise data and models. Indeed, it is important that conclusions drawn today, not depend on precise knowledge of the carbon cycle, since even the most up-to-date information falls short of that needed for accurate prediction.

PREDICTING THE FUTURE CONSUMPTION OF FOSSIL FUEL

Forecasting atmospheric CO_2 concentrations over the next several centuries depends on being able to estimate future fossil fuel consumption and the fraction of the fuel-derived CO_2 which will remain airborne. I will first consider the former problem, and then discuss geochemical model calculations of the latter.

There have been numerous attempts to predict future production of industrial CO_2 over the next few decades by extrapolating from present trends (4,6,7,8,9, 10, 11, among others). These exercises, often called scenarios, do not claim to be true predictions since economic, political, and social factors make realistic predicting a difficult, if not impossible task. They do attempt to set limits on what we may expect to happen. Nevertheless, even "high" and "low" scenarios have tended to fluctuate from year to year as though reflecting recent short-term tendencies in the energy scene. Forecasting future world energy demand has become far more difficult since the disruption in oil supplies in 1973 than it was during the previous two decades when steady growth occurred at the rate of about 5% per year (11). The uncertain nature of the immediate future suggests difficulties in projecting present trends even over the next few decades. In looking further than 30 years or so, if we are still interested in predicting what is likely to occur during any particular decade, we must accept large margins of error. Even with the most advanced techniques such as detailed economic models, it is unrealistic to forecast the energy demand of any given decade beyond about A.D. 2020.

An alternative approach to predicting industrial CO_2 production is to give up attempting to predict the time of arrival of a given fossil fuel production level, and to focus instead on exploitation patterns over what I might call the entire "fossil fuel era", i.e. the whole period of fuel use until near exhaustion of fuel resources. Since the amount of CO_2 involved is expected to be of the order of several times the present atmospheric amount, it is immediately seen to be likely, irrespective of the fossil fuel growth scenario involved, that some time during this era rather high atmospheric CO_2 concentration will be reached. What can we say in general about this period of high concentration without focusing on its precise time of arrival?

Even this generalized approach involves uncertain factors involving human actions, because the total amount of fossil fuel to be consumed during the fossil fuel era must be estimated. How great are the fuel resources which can be consumed? Will they be consumed?

The second question depends on scientifically obscure social and political factors, even including whether mankind will deliberately avoid the use of fossil fuel because a peril is perceived. We can hardly attempt to consider the fossil fuel era in these terms without first considering what may happen if fuel resources are used. As to what these resources are, this question can be answered within some kind of limits today. With further geological exploration and consideration of possible fuel recovery techniques, a few more years of study should fairly precisely answer the question of ultimate resources. At least a realistic lower limit should be forthcoming.

For oil and gas, which are expected to last only into the next century (11, 12), economic and technical considerations based on current means of recovery are probably reliable bases for estimating reserves. The data of Rotty & Marland (11) and of Sassin (12) in this volume are in reasonable agreement with other apparently authoritative recent reports. For coal, the occurrence of deposits for many areas is known within fairly narrow limits, but the estimates of "recoverable" coal vary widely. Evidently we cannot be at all sure how much coal can be mined

The estimate of Sassin (12), 1579 TW yr (TW yr = 10^{12} watts for one year), assumes $ 50 per ton as an upper limit on cost of recovery. This suggests that total coal resources scarcely exceed the upper limit of gas and oil resources combined, for which Sassin estimates 1375 TW yr. Perry and Landsberg (13) in a study prepared for the U.S. National Academy of Sciences on the other hand estimate coal resources at about 5600 TW yr, 3.6 times larger than Sassin's, thus making coal by far the most abundant fossil fuel resource. Other studies can be found which support either of these view points.

Sassin evidently considered economic constraints to be far more restrictive than did Perry and Landsberg. How can we resolve this difference in approach? For the purpose of estimating coal resources over the entire fossil fuel era it may be appropriate to set limits, not based on economic constraints such as $ 50 per ton, but proceeding from physical limiations. What will economics such as that of the United States of America do in 100 or 200 years if no extensive substitute for fossil fuel has been found, (or judged acceptible), oil and gas are gone, and the energy producing industries, now running mainly on coal, find that the cost per ton is approaching \$ 50 per ton. Will they phase out production? Surely an enormous effort will first be made to exploit lower grade coals under incentives which can hardly be imagined today.

The ultimate limit for coal exploitation is reached when the energy required for extracting and utilizing a given amount of fuel becomes a substantial fraction of the useful energy produced (14). This limit is probably well above \$ 50 per ton; it leads to estimates of coal resources of the order of 10,000 TW yr (15,16), i.e. about 50% of the coal in minable seams (17). As can be seen, the estimate of Perry and Landsberg is well below this figure and thus takes into account some degree of economic limitation.

Although the exploitation of shale oil is far less certain than coal, a physical limit may also reflect the extent to which this potential fuel will be utilized. Finally, it should be recognized that the total amount of organic carbon in the earth's sedimentary rocks, most of which can be reached by modern drilling techniques, is the order of 10^{21} moles of carbon equivalent to about 10,000,000 TW yr if it could be burned to CO_2. The part which we judge to be exploitable during the fossil fuel era is the order of 1/1000th to 1/10000th of the total. How can we be confident that we have established the upper limit when the rejected inventory is so many times larger, and the time for mankind to find a way to use this carbon is hundreds of years?

On the other hand, the social and political obstacles to exploiting the hardest-to-get physically exploitable coal and shale oil may be so great that no nation or society will organize itself to do so. Economic limits such as \$ 50 per ton are perhaps an indirect means of avoiding too much expectation of human organization and ingenuity.

With these considerations in mind, I have accepted the estimates of Perry and Landsberg as realistic enough for purposes of calculation. These authors predict that the total fossil fuel resource will produce CO_2 equal to 8.2 times the preindustrial amount in the atmosphere assuming an average atmospheric concentration of 290 ppm (parts per million of dry air by volume).

The patterns of exploitation assumed below are based on a modified logistic function (8, 19) which predicts that the relative rate of increase in N_f, the cumulative amount of carbon produced by fossil fuel combustion, follows the differential relationship:

$$\frac{1}{N_f}\frac{dN_f}{dt} = \mu \left\{1 - \left(\frac{N_f}{N_{f_\infty}}\right)^n\right\} \qquad \{1\}$$

where μ is the reciprocal of the e-fold time for exponentially growing production of CO_2,
$N_{f\infty}$ denotes the ultimate amount of carbon produced, and
t represents time.

The term in brackets expresses parametrically the predicted extent to which approach to exhaustion of fuels will impact the overall rate of cumulative increase in CO_2 produced. The CO_2 production rate dN_f/dt, thus simulates the characteristic rise and fall associated with the utilization of a limited non-renewable resource, under a variety of situations. In one set of calculations the adjustable growth cut-off paramter, n, was varied while μ was fixed. The relative rate of increase in N_f was made to be 4.5% for A.D. 1972, thus consistent with the average annual rate of increase in industrial CO_2 production from 1947 through 1972 (19). In the other set of conclusions, μ was varied with n fixed at unity. No attempt was made to distinguish between the separate patterns of use of coal, oil and natural gas on the grounds that the total is the important criterion over the long period, and because the overall exploitation depend mainly on the most abundant fuel, coal.

PREDICTING THE FUTURE RESPONSE OF THE OCEANS TO INDUSTRIAL CO_2

The fraction of industrial CO_2 taken up by the oceans, as already noted, has probably until now remained nearly constant except for short-term variations. The latter relate to air-sea exchange and upwelling, and correlate with a natural meteorological phenomenon, the southern oscillation (2) which suggest that they probably influence only slightly the airborne fraction over the full twenty years of observations. In the future, however, if fossil fuel consumption continues to rise until a major part of the total reserves are used up, the ability of the oceanic surface water to respond to rising atmospheric CO_2 concentration will considerably diminish. This is because of a shift in the chemical equilibrium between dissolved CO_2, HCO_3^-, and $CO^{3=}$ of sea water associated with increasing acidity attending CO_2 uptake as noted by Oeschger et al. in this volume (1).

Following these authors, let us express the time variable rate of uptake of CO_2 by the oceans, F_{am}, as a product, $k_{am}N_a$, where k_{am}, a constant, denotes the coefficient for transfer of CO_2 from the atmosphere to the oceans, and N_a denotes the total amount of carbon in atmospheric CO_2 at any given moment. For ocean water in equilibrium with the atmosphere, the return flux to the atmosphere, F_{ma}(equil.) has the same value, or in terms of h_a, the thickness of a surface ocean layer containing the same amount of carbon as the preindustrial atmosphere:

$$F_{ma}(\text{equil.}) = k_{am}h_aAC_m \tag{2}$$

where C_m denotes the time variable concentration of total dissolved inorganic carbon in the water, and
A is the area of the ocean surface.

During the fossil fuel era equilibrium is not realized. The departures in the fluxes from preindustrial values to a close approximation are expressed by:

$$\Delta F_{am} = k_{am}(N_a - N_{ao}) \tag{3}$$

$$\Delta F_{ma} = \xi k_{am}h_aA(C_m - C_{mo}) \tag{4}$$

where N_{ao} and C_{mo} indicate preindustrial values and ξ refers to the CO_2 evasion, or "buffer" factor which depends on the thermodynamic state of the sea water system.

The higher is ξ, the larger the return flux, F_{ma}, for a given sea water carbon concentration, C_m, and the less is the net uptake, $\Delta F_{am} - \Delta F_{ma}$, for a given increment of industrial CO_2 introduced into the atmosphere. Although ξ is presently about 10 it is predicted to rise to approximately 35. In the absence of other mechanisms of removal of CO_2 from the atmosphere, the yearly airborne fraction of industrial CO_2 will rise from its present value of about 55% to the vicinity of 80%. Even the cumulative fraction is likely to rise to over 70%, i.e. a considerably larger fraction of industrial CO_2 than present will be airborne at the time most critical for the world's environment.

Although this circumstance may have grave environmental consequences, it makes the modeler's task easier. The behavior of the oceans below the surface layer becomes progressively less important to predicting the uptake of industrial CO_2 when approaching the time of large amounts. The magnitude of the evasion factor, ξ, because it has been determined by laboratory experiments, is not in doubt, and therefore unless some unforeseen process of oceanic CO_2 uptake occurs, such as massive dissolution of carbonates in shallow water marine sediments or greatly accelerated sedimentary burial of marine organic carbon, the predictions of the present model cannot be very far off. A major effort is underway to explore all possible previously overlooked pathways by which the oceans might remove additional CO_2 from the air (20), but at present no proposed mechanism appears to be important.

Because we do not know at present whether the biosphere is supplementing the oceanic sink, we cannot predict with any reliability what biospheric response may occur in future years. For the moment let us assume that irrespective of the net transfer of CO_2 between the atmosphere and the biosphere today that this transfer ceases to be important after A.D. 2025. After we have calculated the expected atmospheric CO_2 rise under this assumption we will reexamine the question of a long-term biospheric response.

To predict future atmospheric CO_2 concentration, I will summarize results obtained by the model shown in figure 1 and explained in detail in reference (19). The model significantly

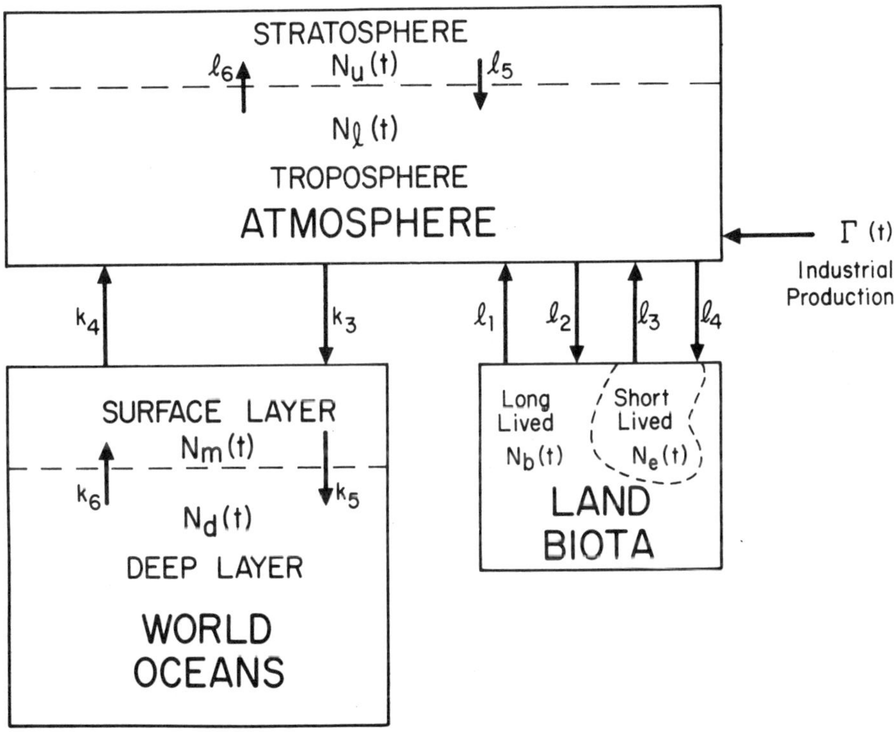

Fig. 1: Six-reservoir model of the carbon cycle. The mass of carbon in each reservoir is represented by the $N_j(t)$, and the transfer coefficients between reservoirs are given by the k_j and l_j. The magnitudes of the initial values of the $N_j(t)$ and of the (in some cases) time-dependent values of the k_j and l_j are listed in reference (19).

differs from the BD, "box diffusion", model of Oeschger et al. (1, 21) only in the treatment of the subsurface ocean, part of which is included in an enlarged surface reservoir and the remainder in a single donor controlled deep reservoir. For an eddy diffusion coefficient of $K = 3000\ m^2 yr^{-1}$ the present model gives a response almost identical to the BD model over the past 25 years (21). Oeschger and coworkers have used a somewhat higher value of K of about $4000\ m^2 yr^{-1}$ in their model calculations, and Broecker et al. in a recent assessment of the BD model

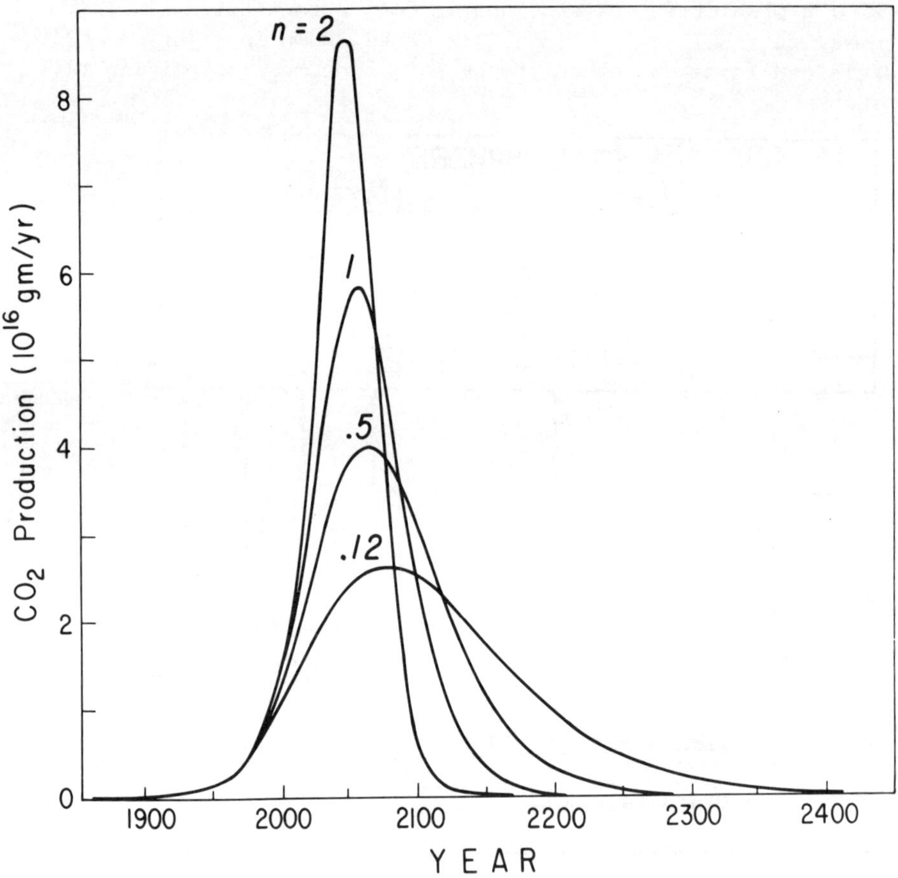

Fig. 2: Industrial CO_2 production in grams of carbon per year for various assumed patterns of fossil fuel consumption. The ultimate production of industrial CO_2 is fixed at 8.2 times the amount of CO_2 in the preindustrial atmosphere. The parameter n in equation {1} is varied between .12 and 2. with exponential factor $\mu = .0453$.

suggest that K = cir. 8000 m^2yr^{-1} is a still more likely value (22). Since the uptake of CO_2 below the surface ocean in the BD model depends very nearly on the square root of K (see ref. (1), equation {5}), the differences in all three cases are not very large, with respect to calculations of oceanic uptake for the recent past, and still less for periods of higher evasion factor, ξ.

For the period A.D. 1860 to 1974 the model uses yearly average estimates of industrial CO_2 taken from historic records of fuel production with estimates of the carbon fraction of each major fuel type. For the period 1800 to 1860 an exponential backward extrapolation was used, and after 1974, equation {1}. For various assumed patterns of future fuel consumption, the CO_2 input patterns are as shown in figures 2 and 3.

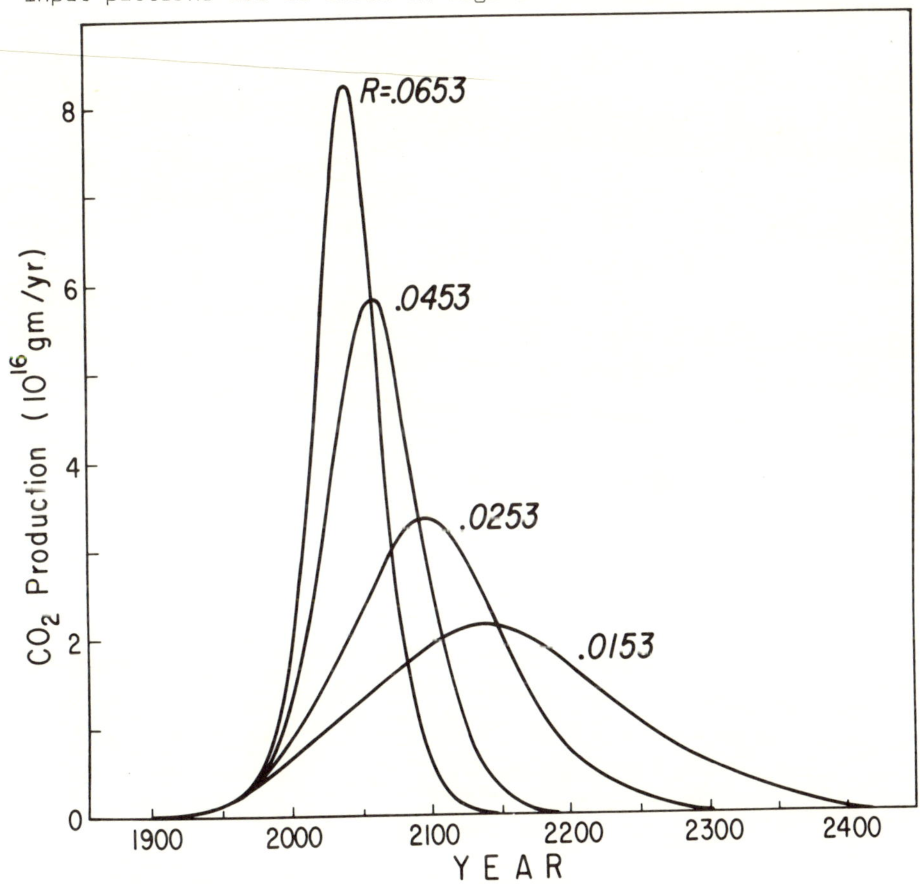

Fig. 3: Additional patterns of industrial CO_2 production similar to figure 2 except that the parameter, μ, is varied from 0.153 to .0653 with n = 1.

The model results, figures 4 and 5, show that the predicted CO_2 concentration for any given year during the next several centuries is sensitive to the assumed rate of fuel use. But when the patterns of CO_2 concentration are viewed over longer time periods, the assumed rate of use is not very important. Even if the growth rate of fuel consumption is abruptly reduced now to 1.5% and is further reduced later on as the total reserves diminish appreciably, still the major trend of CO_2 concentration over the fossil fuel era is nearly the same: an eventual rise to several fold the preindustrial level and thereafter a very slow decline.

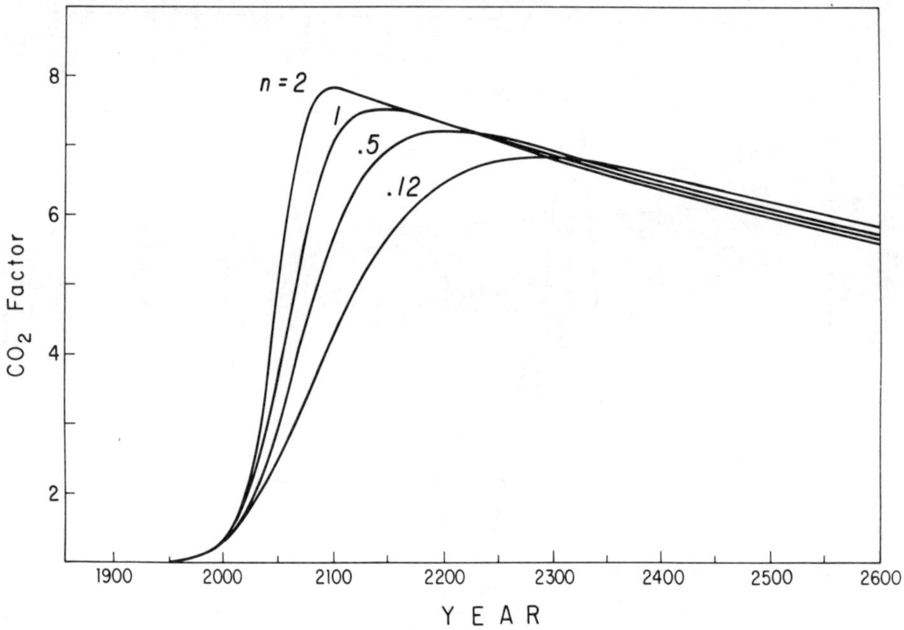

Fig. 4: Predicted increase in atmospheric CO_2 from A.D. 1900 to A.D. 2600. Curves are shown for the four simulated fossil fuel consumption patterns shown in figure 2. The CO_2 factor expresses the amount of CO_2 in the atmosphere as a multiplier of the preindustrial atmospheric CO_2 abundance. The land biota (land biosphere) is assumed to become static after A.D. 2025 by setting the biospheric growth factor, ε, linearly to zero between A.D. 2000 and A.D. 2025.

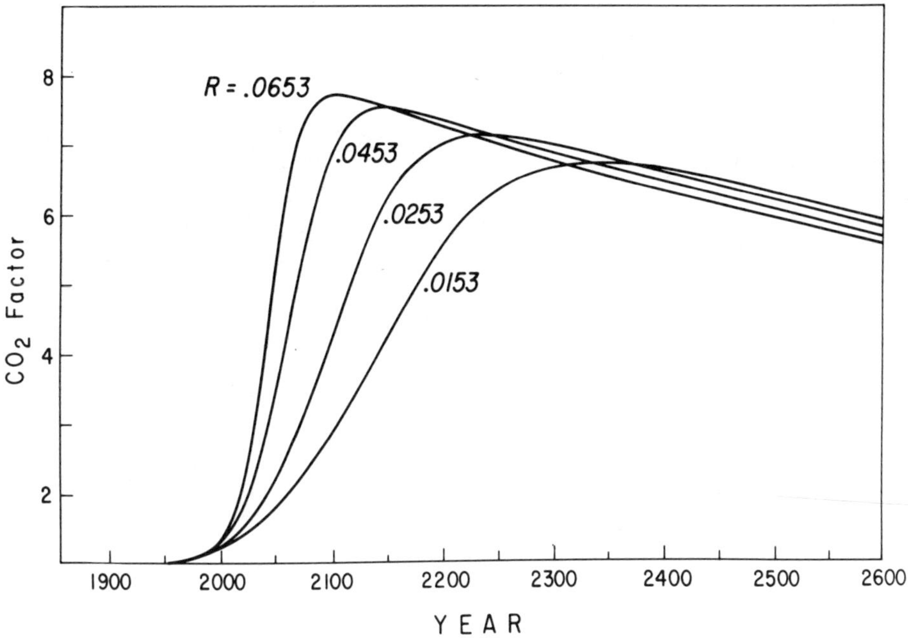

Fig. 5: Additional predictions of atmospheric CO_2 increase similar to figure 4 except that the simulated fossil fuel consumption patterns are as shown in figure 3.

The effect of including the maximum possible dissolution of sedimentary carbonate in the deep oceans, figure 6 and reference (19), is to reduce the CO_2 concentration significantly only after about 1000 years. The BD model predicts a nearly 25% greater oceanic uptake of CO_2 near the time of peak concentration than does the present model, but owing to the strong influence of the evasion factor which has the same thermodynamic form for both models, the airborne fractions differ by only a few percent, see figure 7 and reference (8).

PREDICTING THE FUTURE RESPONSE OF THE BIOSPHERE TO INDUSTRIAL CO_2

With respect to the land biosphere, it is worthwile to consider at least qualitatively the possible uptake of industrial CO_2 by both the biomass of living plants and the mass of carbon in detritus and soils. Releases of carbon from these reservoirs to the atmosphere stemming from future human disturbances will also be considered.

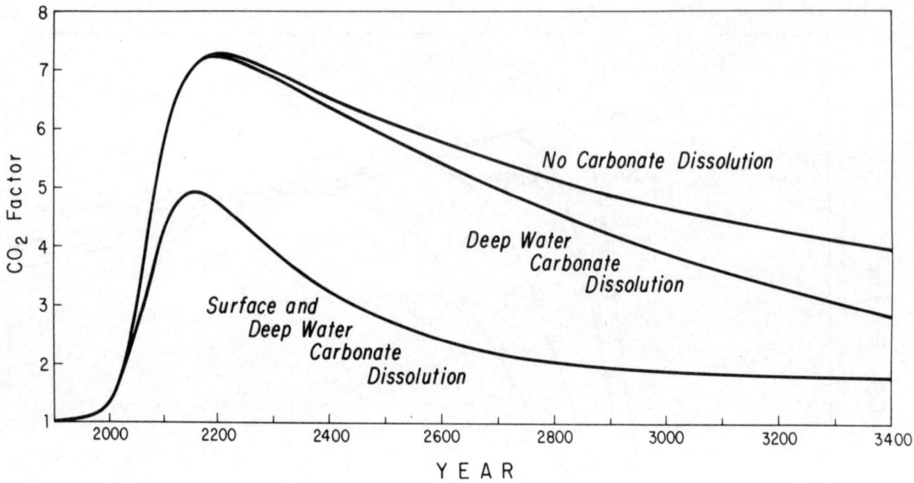

Fig. 6: Predicted increase of atmospheric CO_2 from A.D. 1900 to A.D. 3400 with and without dissolution of carbonate sediments. The pattern of fossil fuel consumption is as shown by the case n = 0.5 of figure 2. The CO_2 factor has the same meaning as for figures 4 and 5. The curves for carbonate dissolution assume thermodynamic equilibrium between sea water and calcium carbonate for the entire sea water reservoir in which dissolution is considered to occur. The case of combined surface and deep water dissolution is shown for illustrative purposes only, since most near surface marine carbonates are too insoluble to dissolve during the fossil fuel era, and the remaining carbonates are believed to be in too low abundance to change substantially the uptake of CO_2 from the case where only deep water dissolution occurs.

Under natural conditions approximately 60 Gt (1 gigaton = 10^{15}g) of carbon per year are transfered between the biosphere and the atmosphere (23). This two-way flux is the order of 10 times the present CO_2 production from fossil fuel of 5 Gt per year (24). Thus, even a few percent change in either the biospheric uptake or release of CO_2 may be important today, because the net effect may be a significant fraction of the CO_2 produced industrially. In the future, however, if the production of industrial CO_2 continues to increase as predicted by any of the curves of figures 2 and 3, we will reach a point after which only very large net changes in the transfer of CO_2 between the atmosphere and biosphere can significantly influence atmospheric CO_2.

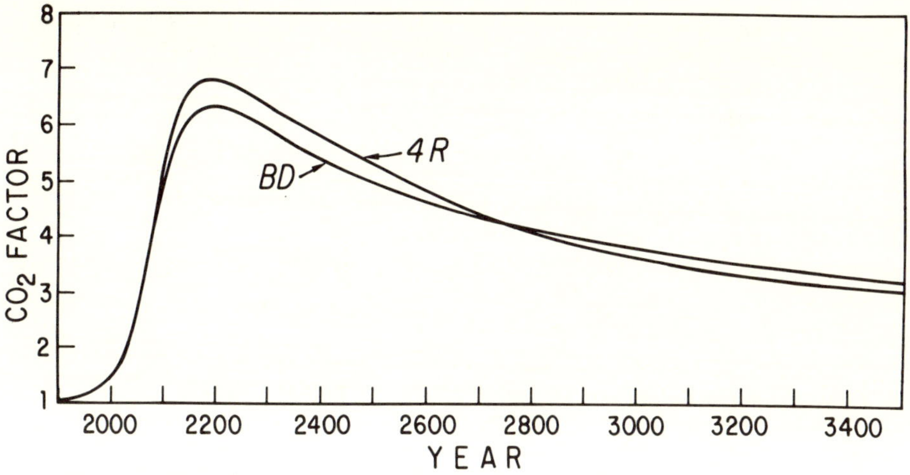

Fig. 7: Increase in atmospheric CO_2 plotted as in figure 6 for a four-reservoir (4R) model and the box-diffusion (BD) model. The former is a simplified version of the six reservoir model shown in figure 1.

With respect to biospheric storage of carbon, a similar declining interest in small changes occurs. The total inventory of carbon in living plants, some 600 Gt (23) would have to increase several fold to provide an effective sink for CO_2 if the atmospheric CO_2 abundance were to reach the order of 3000 Gt of carbon as predicted by figures 4 and 5. Since the upper limit on living biomass increase during the fossil fuel era is almost surely less than double the present biomass, an increasing storage of carbon in living plants is not a likely pathway for significantly reducing atmospheric CO_2 in the future. Since detritus is a considerably smaller reservoir (estimated at about 60 Gt of carbon in ref. 23), this reservoir is still less likely to be important.

The world's soils, including deep organic soils, contain the order of 2000 Gt (23) to 3000 Gt (24) of carbon, and thus are a sufficient reservoir to take up much or all of the otherwise airborne industrial CO_2 without being required to more than double their inventory of carbon. High storage would require, however, a greatly accelerated formation of that fraction of the soil carbon which remains in the soil for long time periods, because that fraction must perforce form only slowly under natural conditions (24).

Thus massive soil storage of industrial CO_2 remains only a distant possibility. Revelle and Munk (4) using a simple donor controlled one reservoir model of the soil, coupled to a fixed size reservoir of living plants, estimated a cumulative uptake of carbon by soil during the 23rd century of 1100 Gt for an industrial CO_2 input of approximately 5000 Gt, i.e. they predicted that the soils might remove about 20% of the industrial CO_2. This uptake lowers the predicted atmospheric CO_2 at the time of peak concentration from about 5.4 times the preindustrial level (in their model) to about 3.7 times.

Focusing again on the fluxes of carbon involved, we perceive the difficulty of removing significant amounts of CO_2 from the air by biospheric uptake, if we compare the rate of production of industrial CO_2 with that part of the biospheric uptake which involves transfer of carbon to long term storage reservoirs. About half of the carbon initially fixed by plants during photosynthesis (net primary production) is returned to the atmosphere within a few years or less by decay of leaves and small roots and woody parts (25). The order of 30 Gt yr^{-1} becomes stored for longer times in stable detritus and soils. The present industrial CO_2 production is about 20% of the long term biospheric flux, and if this production grows as projected in figures 3 and 4, it is likely to approach, perhaps even exceed, the long term biospheric flux near the time of peak industrial CO_2 production. Very considerable increases in the rate of formation of long stored plant and soil carbon must thus take place to affect atmospheric CO_2 concentrations significantly near the time of peak industrial CO_2 production.

The biospheric storage of carbon would be increased substantially if the biosphere could be managed deliberately on a world-wide basis so as to limit atmospheric CO_2. Increasing plant growth and soil carbon storage in areas not needed for human food and industrial fiber production would contribute to reducing atmospheric CO_2 as would the growing plants for fuel to replace fossil fuel. The latter management program, indeed, would merely hasten the time when renewable fuels become attractive because of the exhaustion of fossil fuels.

In regard to managing the biosphere to increase storage, this must, of course, be done not only by growing more plants, but also by preventing the accelerated flux of CO_2 back into the atmosphere after plants die and become part of the detrital carbon pool. Typical present practices of land use do not work in the desired direction (26). The near certainty (11) of a much larger world human population at the time of high industrial CO_2 production is a serious drawback to managing the biosphere to increase storage of carbon. If human population reaches close to the limit of food supply, as seems likely, it is difficult to imagine maintaining biospheric conservation programs extensive enough to take significant amounts of CO_2 out of the atmosphere and hold them in storage for the remainder of the fossil fuel era. But the growing of plants to substitute for fossil fuel perhaps remains an option even in a crowded world, since the cost of producing fossil fuels is likely to be so high in the next century that plant biomass may become an economic competitor world-wide.

In summary, it is not now possible to predict future biospheric responses related to the fossil fuel era. These responses would have to be very large, however, to reduce significantly the amount of CO_2 in the atmosphere beyond that reduction caused by oceanic uptake. Large world population makes unlikely the increased storage of biospheric carbon. The substitution of plant biomass for fossil fuel might prove an effective means of reducing CO_2 concentration or at least reducing the rate of rise.

CONCLUSIONS

If the amount of coal consumed during the fossil fuel era is determined by physical limitations related to the energy required to extract and utilize the coal, then the amount of CO_2 produced is 8 or more times that of the preindustrial atmosphere. This amount is so large that neither the oceans nor the biosphere, during the next several centuries, appear able to remove from the air more than a small fraction of the CO_2 added. The concentration of atmospheric CO_2 will rise to a peak approaching that predicted if all of the CO_2 were remaining airborne. Only if growth in the consumption of fossil fuels is reduced to well below 1.5% per annum is a substantially lower peak concentration likely.

REFERENCES

(1) Oeschger, H., Siegenthaler, U., Heimann, M.: 1980, The Carbon Cycle and its Perturbation by Man, this volume.

(2) Bacastow, R.: 1979, Dip in the Atmospheric CO_2 Level During the Mid-1960's, J. Geophys. Res. 84, pp. 3108-3114.

(3) Witwer, S.H.: 1970, Aspects of CO_2 Enrichment for Crop Production, Trans. Amer. Soc. Agr. Eng. 13, pp. 249-251.

(4) Revelle, R., and Munk, W.: 1977, The Carbon Dioxide Cycle and the Biosphere, Chap. 10 in Energy and Climate, U.S. National Academy of Sciences, Washington, D.C. pp. 140-158.

(5) Heimann, M.: 1978: Ueber ein geophysikalisches Modell des globalen CO_2-Kreislaufs, Lizentiatsarbeit (2. Teil), Universität Bern.

(6) Zimen, K.E., Offermann, P., Hartmann, G.: 1977, Source Functions of CO_2 and Future CO_2 Burden in the Atmosphere, Zeit. Naturforsch. 32a, pp. 1544-1554.

(7) Siegenthaler, U., and Oeschger, H.: 1978, Predicting Future Atmospheric Carbon Dioxide Levels, Science 199, pp. 388-395.

(8) Bacastow, R.B., and Keeling, C.D.: 1979, Models to Predict Future Atmospheric CO_2 Concentrations, In Workshop on the Global Effects of Carbon Dioxide from fossil Fuels, W.P. Elliott and L. Machta, editors, U.S. Dept. of Energy, Washington, D.C., pp. 72-90.

(9) Baes, C.F., Goeller, H.E., Olson, J.S., and Rotty, R.M.: 1977, Carbon Dioxide and Climate: The Uncontrolled Experiment, American Scientist 65, pp. 310-320.

(10) Niehaus, F., Williams, J.: 1979, Studies of Different Energy Strategies in Terms of Their Effects on the Atmospheric CO_2 Concentration, J. Geophysical Res. 84, pp. 3123-3129.

(11) Rotty, R.M., Marland, G.: 1980, Constraints on Fossil Fuel Use, this volume.

(12) Sassin, W.: 1980, High Energy Supply Scenarios, this volume.

(13) Perry, H., Landsberg, H.H.: 1977, Projected World Energy Consumption, Chap. 1 in Energy and Climate, U.S. National Academy of Sciences, Washington D.C., pp. 35-58.

(14) Zimen, K.E.: 1971, Nuclear Energy Reserves and Long-Term Energy Requirements, Angew. Chemie 10, pp. 1-11.

(15) Energy: Global Prospects 1985-2000, Report of the Workshop on Alternative Energy Strategies, C.L. Wilson, editor, McGraw-Hill Book Co. New York.

(16) World Energy Conference: 1978, World Energy Resources 1985-2020, IPC Science and Technology Press, New York.

(17) Hubbert, M.K.: 1969, Energy Resources, Chap. 8 in Resources and Man, National Academy of Sciences, Washington D.C., pp. 157-242.

(18) Garrels, R.M., Lerman, A., and Mackenzie, F.T.: 1976, Controls of Atmospheric O_2 and CO_2: Past, Present, and Future, American Scientist 4, pp. 306-315.

(19) Keeling, C.D., and Bacastow, R.B.: 1977, Impact of Industrial Gases on Climate, Chap. 4 in Energy and Climate, U.S. National Academy of Sciences, Washington D.C., pp. 72-95.

(20) Carbon Dioxide Effects Research and Assessment Program: 1979, Carbon Dioxide Research Progress Report, Fiscal Year 1979, U.S. Depart. of Energy, Washington, D.C.

(21) Oeschger, H., Siegenthaler, U., Schotterer, U., Gugelmann,A.: 1975, A Box Diffusion Model to Study the Carbon Dioxide Exchange in Nature, Tellus 27, pp. 168-192.

(22) Broecker, W.S., Peng, T.H., Engh, R.: 1980, Modeling the Carbon System, Submitted to ^{14}C Conference, Bern and Heidelberg, August 1979.

(23) Ajtay, G.L., Ketner, P., and Duvigneaud, P.: 1979, Terrestrial primary production and phytomass, In SCOPE 13: The Global Carbon Cycle, B. Bolin, E.T. Degens, S. Kemps, and P. Ketner, editors. Internat. Council of Scientific Unions, J. Wiley and Sons, Chichester, pp.

(24) Rotty, R.M.: 1979, Present and Future Production of CO_2 from Fossil Fuels - A Global Appraisal, In Workshop on the Global Effects of Carbon Dioxide from Fossil Fuels, W.P. Elliott and L. Machta, editors, U.S. Dept. of Energy, Washington, D.C., pp. 36-43.

(25) Keeling, C.D.: 1973, The Carbon Dioxide Cycle: Reservoir Models to Depict the Exchange of Atmospheric Carbon Dioxide with the Oceans and Land Plants, in Chemistry of the Lower Atmosphere, S.I. Rasool, editor, Plenum Press, New York. pp. 251-329.

(26) Woddwell, G.M., Whittaker, R.H., Reiners, W.A., Likems, G.E., Delwiche, C.C., Botkin, D.B.: 1978, The Biota and the World Carbon Budget, Science 199, pp. 141-146.

THE ROLE OF THE BIOSPHERE

U. Hampicke

Arbeitsgruppe A.U.G.E., Universität Essen/Gesamthochschule, 4300 Essen, Federal Republic of Germany

ABSTRACT. The land biota is an active compartment in the global carbon cycle, and transfers to and from the atmosphere are to be taken into account in sufficient detail in model calculations. Most probably, net losses of carbon of considerable size from some terrestrial ecosystems to the atmosphere are at present balanced by net gains of other ecosystems. Even for quite different assumptions regarding the behaviour of land ecosystems and the ocean, the future atmospheric CO_2 increase will primarily depend on the quantities of fossil fuels burned.

1. INTRODUCTION

The earth's land vegetation and soils currently comprise 2,000-2,500·10^{15} g of organic carbon (3-3.5 times the amount stored in the atmosphere). Thus, they are an important compartment in the global carbon cycle (Table 1). Their present role with respect to the atmospheric CO_2 budget has been a matter of vivid discussion in recent years (22,23,and following ref.). While geochemical models generally require the carbon pool of the land biota to increase or to be in a steady state at present (3,5,8,28), it has been suggested that, due to man's destructive impact in the tropics, land ecosystems are, on the whole, a net source of CO_2 for the atmosphere (1,6,10,14,37-40). If true, this would require that either sinks unknown to-date are taking up the additional CO_2, or that the ocean is a more effective sink than calculated by geochemical models - something which is at variance with established knowledge on ocean chemistry and vertical mixing.

In this paper, I do not want to repeat all the arguments that have

W. Bach, J. Pankrath, and J. Williams (eds.), Interactions of Energy and Climate, 149–167.

Table 1. Distribution of carbon in various reservoirs in ~1980.

Reservoir	Carbon stored at present, 10^{15} g	Calculated from reference
Land vegetation	~550-650	(2,16)
Atmosphere	710	a)
Dissolved inorganic C in mixed layer of the ocean (0-75 m)	~700	(8)
Dissolved and particulate organic C in the ocean	~1,000	(26)
Terrestrial soil organic matter[b)]	1,500-2,000	(2,10,16,37)
Recoverable fossil fuels	5,000-10,000	c)
Dissolved inorganic C in thermocline and deep sea (75-3,800 m)	~38,000	(8)
Non-concentrated organic C in rocks	$\sim 20 \cdot 10^6$	(17)
Carbonate rocks	$\sim 80 \cdot 10^6$	(17)

a)Corresponding to 335 ppm as the most recent value.b)Including peat.c)Range of available figures, for details see (30).

been raised in recent years. Rather I would like to address myself to three questions: (1) What do we know about the behaviour of land ecosystems in the global carbon cycle and what do we not know? (2) What is most probably the role of land ecosystems at present? (3) To what extent will the future atmospheric CO_2 increase depend on the various roles of land ecosystems?

2. SOME RELEVANT ASPECTS OF LAND ECOSYSTEMS

2.1. Potential variability

Land ecosystems do have the capacity to influence the atmospheric CO_2 budget. If the amount of carbon presently stored in vegetation and soils would change by only one per mil per year, this would result in a net transfer to or from the atmosphere of $2\text{-}2.5 \cdot 10^{15}$ g/yr. of C, which is equivalent to 40-50% of the present annual input due to fossil fuel burning. Man is no doubt able to induce changes of this order of magnitude.

2.2. Historical reduction

As shown in Table 2, man has reduced the amount of organic C stored in land vegetation of the earth by more than 40%, and the amount stored in soil organic matter by 18%. The total carbon stored in land ecosystems is now smaller by 25% than it used to be before man became an ecological factor on earth; the difference between the potential and the present value amounts to about $700 \cdot 10^{15}$ g. In view of the low accuracy of figures of this kind, these figures should be evaluated critically, of course. Remarkably, however,

Table 2. Potential and present organic carbon stored in major land ecosystems of the earth. (for explanations see page after next)

Vegetation Unit	Area, $10^{12} m^2$		Carbon per Area in 1977, $10^3 g/m^2$		Carbon stored, 10^{15} g			
					Living phytomass		Humus	
	Pot.[a)]	1977[b)]	Liv.Ph.	Humus[c)]	Pot.[d)]	1977	Pot.	1977
Tropical zone[e)]								
Moist forest[f)]	24.7	14.8	16.6	13.0	422.7	245.3	320.8	192.0
Dry forest and woodland	5.1	3.5	6.8	7.0	38.1	23.6	35.4	24.2
Grassland and bushland	15.0	19.0	2.2	4.2	36.4	42.0	62.8	79.8
Cultivated land	0.0	10.5	0.3	5.9	0.0	3.4	0.0	62.0
Swamp	1.5	1.5	6.8	30.0	11.1	10.1	44.9	45.0
Total[g)]	46.3	49.3	6.6	8.2	508.3	324.4	463.8	403.2
Temperate zone								
Forest and woodland	20.9	12.0	9.7	10.7	267.9	115.9	255.4	131.6
Wet grassland (meadow)	3.6	5.0	0.9	18.9	3.7	4.7	68.5	94.5
Dry grassland (steppe)	8.6	7.5	0.6	18.9	5.6	4.4	161.9	141.8
Cultivated land	0.0	7.5	0.4	8.0	0.0	3.2	0.0	60.0
Swamp	3.2	0.5	3.4	30.0	11.9	1.7	95.7	15.0
Total[g)]	36.3	32.5	4.0	13.6	289.1	129.9	581.5	442.8
Boreal zone								
Closed forest	11.9	6.5	11.3	20.6	133.9	73.1	245.2	133.9
Forest tundra	2.1	2.5	7.6	20.6	16.1	19.1	43.2	51.5
Tundra[h)]	6.7	8.0	0.7	20.4	4.8	5.7	136.7	163.2
Bog	2.3	1.5	2.3	70.0	5.2	3.4	161.4	105.0
Total[g)]	23.0	18.5	5.5	24.5	160.0	101.4	586.6	453.6
Semidesert	17.5	21.0	0.4	5.8	6.1	7.4	101.6	121.8
Extreme desert	10.1	10.5	0.0	0.2	0.4	0.4	1.7	1.8
Ice covered[g)]	13.9	15.5	0.0	0.0	0.0	0.0	0.0	0.0
Land total[g)j)]	147.1	147.3	3.8	9.7	963.9	563.5 (−41.5%)	1,735.2	1,423.2 (−18.0%)

other authors calculated even higher figures for the cumulative losses (10,29).

To a large extent, deforestation and humus oxidation constitute slow processes lasting over many millenia and thereby hardly influencing the atmospheric CO_2 level. In the temperate zone the depletion of organic carbon probably came to a halt by the middle of this century. In the tropics, however, the causes of depletion still persist: explosive population growth, rural poverty, concentrated land ownership, low agricultural productivity and a merely extractive and export-orientated use of forestry resources. It would be unrealistic to assume that the tropics lost 25% of their organic carbon in earlier times, whereas now, ecological deterioration being greatly intensified, there should be no net losses. If, rather, 50-70% of the cumulative historic losses of the tropics of about $240 \cdot 10^{15}$ g C have been oxidized since 1860 with the annual losses increasing exponentially at 2-3.5% p.a., there should be at present an annual loss of $1.8\text{-}4 \cdot 10^{15}$ g C (see Table 4).

2.3. Observed reduction of tropical forest area

The area covered by closed rain forests in the tropics at present amounts to $9\text{-}11 \cdot 10^{12}$ m^2 (2,9,31,35), as compared with an original $15\text{-}17 \cdot 10^{12}$ m^2 (33, predominantly evergreen forests only). Most probably, a considerable part of this shrinkage occurred during the last 30 years. In Thailand, an evaluation of earlier air photos combined with recent satellite imageries revealed that the area of closed forests in this country decreased by approximately 25% between 1961 and 1975 (20, see also 25). It seems justified to extrapolate this finding to the rest of the densely populated humid tropics (South East Asia, West Africa, and Central and parts of South America), which comprise about half the area of the remaining closed forests (about $7.25 \cdot 10^{12}$ m^2, including moist seasonal forests, excluding mangrove), which yields a cumulative loss of $2.4 \cdot 10^{12}$ m^2 of tropical moist forest area in 14 years. On the assumption that annual clearings have increased at 2% p.a. during the last decades, there results a present loss of about $190 \cdot 10^9$ m^2/yr. of tropical forest area, a figure larger than the most recent FAO estimate of $160 \cdot 10^9$ m^2/yr. (21). Yet this estimate implies that large unpopulated areas such as the Amazon basin still remain unaffected.

2.4. Conclusions from socio-economic evidence

The area needed for subsistence agriculture in the tropics is increasing each year. Given an annual population increase in these regions of 2.4%, and considering that, for lack of fertilizer and other equipment, intensification of agriculture is often insignificant, the annual increase in cultivated land must be about 2%. Given a tropical cultivated area of $6\text{-}10.5 \cdot 10^{12}$ m^2 (lower value from 32, higher value from 2), the annual increase is $120\text{-}210 \cdot 10^9$ m^2.

Except for arid regions with irrigated agriculture, the amount of carbon stored in the vegetation of cultivated fields is smaller than in natural vegetation on the same site. The difference depends mainly on the type of vegetation removed on cultivating an area; it is large in the case of forest clearance, and smaller in grassland vegetation. It is not known with any exactness how much of the new agricultural land is reclaimed from forests, but the percentage is certainly considerable.

Unless ecologically sound practices are applied (which is less and less the case), tropical soils invariably lose soil organic matter when cultivated (27,41, compilation of field data in 16). The fraction lost amounts to 30-70%, often it is about 40%. A certain complication results from the fact that carbon losses from the vegetation as well as from the soils occur with considerable delays. The decomposition of partly burned wood debris such as roots and stumps is very slow.

In addition to the land reclaimed for subsistence agriculture there are large commercial clearings for the extraction of forest resources and for plantations and pasturages, mainly in South East Asia and South America. The areal extent is not precisely known, but it may amount to $30\text{-}50 \cdot 10^9$ m^2 per year. In a more detailed calculation, and considering all these aspects, Hampicke and Bach (16) estimated that $1.3\text{-}2.3 \cdot 10^{15}$ g of carbon are annually released to the atmosphere at present as a consequence of non-commercial and commercial land clearings (Table 4).

2.5. Miscellaneous terrestrial sinks

In view of the heterogeneity of the terrestrial biota there are numerous possibilities for land, wetland and coastal ecosystems operating as sinks for CO_2 on a smaller scale, for natural as well as anthropogenic reasons, or a complex combination of both. Boreal peat-bogs may still be increasing (10), and it is not necessarily true that in the higher latitudes of the northern hemisphere soil organic matter, litter and even phytomass are in a steady state, there may still be some accumulation (15). The earth's rivers annually transport roughly $0.23 \cdot 10^{15}$ g of C of atmospheric origin, needed for the weathering of carbonates (19), and roughly $0.16\text{-}0.2 \cdot 10^{15}$ g of organic C in humic materials (15) to the sea.

Explanation to Table 2: a)after (33) b)after (2), slightly modified. c)after (37), except for swamps and bogs. d)10% added in tropical and temperate zones, accounting for a reduction in the average C content per area within defined vegetation units during history (e.g. substitution of climax forests by younger commercial stands). e)approx. 30°N-30°S. f)Evergreen and moist deciduous forests and mangrove. g)Differences between potential and 1977 areas due to different data sources. h)Probably, different definitions of tundra in data sources. j)excluding inland waters.

It is not known to what extent these quantities are restituted to the atmosphere. There may be an accelerated sedimentation of organic C in estuaries and coastal waters due to the present fertilization of these waters with phosphorus (8). All these potential sinks combined may withdraw several 10^{14} g of C from the atmosphere per year.

2.6. Regrowth of temperate forests

After a long period of woodland degradation, modern high forests have been established in Central Europe during the 19th and early 20th century. In the U.S., destructive exploitation of forests was widespread until about 1920. Since approximately 1940, these forests have been in a state of recovery, thereby accumulating large quantities of carbon from the atmosphere. In 1976, the annual net increment amounted to $11 \cdot 10^9$ cubic feet of merchantable timber, a figure that it is difficult to transform into an ecological unit such as grams of carbon, including non-merchantable material (11). An annual sink intensity of 100-150 g/m^2 of C appears none the less probable. It is not clear to what extent this value can be extrapolated to the Eurasian temperate forests. Except for Scandinavia (18) and possibly some other regions, the process of recovery started earlier in Europe than in the U.S. so that the forests might no longer accumulate annual quantities of this size. The Soviet Union and China pose further problems. While in the former country, due to extremely large distances, forests are either over-used (such as in the European part) or under-used (in Eastern Siberia, 24), extensive reforestation activities have been reported from China (6, after Persson). On the whole, it is rather probable that the world's temperate forests, including forest plantations, are presently withdrawing some $0.5 \cdot 10^{15}$ g of C from the atmosphere as a consequence of better or less intensive management (independently from atmospheric CO_2 increase). A value of even $1 \cdot 10^{15}$ g/yr. cannot be ruled out.

2.7. Stimulation of photosynthesis by increasing CO_2

Although a matter of much theoretical controversy as well as of empirical ignorance, this topic has been treated in considerable detail in geochemical models of the global carbon cycle. The amount of carbon taken up by land ecosystems in order to balance the atmosphere's carbon budget is interpreted in these models as a reaction of the world's plant cover to an enhanced atmospheric CO_2 supply. An appropriate "biota growth factor", relating atmospheric CO_2 increase and photosynthesis acceleration, is either calculated from the empirical data or parameterized (3,28,34).

Laboratory experiments show unambiguously that, given optimal supply with other growth factors (light, temperature, water, nutrients), a doubling in atmospheric CO_2 results in a photosynthesis acceleration in C-3 plants of 40-80% (data compilations in 16,22; bibliography in 36). Similar results are obtained with crop plants in the field, well supplied with fertilizer and water (Thomas, 1965; Moss et al. 1961; in 22). Some experiments with trees in natural ecosystems suggest that even under environmental stress, there can be a strong response to an enhanced atmospheric CO_2 content (13), so that this effect might not be restricted to conditions of optimal factor supply, as it is often assumed (12).

The relevance of these findings to the question of how the entire land vegetation of the earth responds to the present CO_2 increase remains unclear, however (4). Experiments suggest that there may be at least a small effect on a global scale. Since the entire land vegetation is potentially affected, even such a small effect would lead to a net withdrawal of carbon from the atmosphere which might be significant in an attempt to balance the atmosphere's CO_2 budget. Assuming that the response of the entire land vegetation to increasing CO_2 is only one tenth of the average response found in laboratory experiments (about 0.6), an atmospheric increase of 15% would lead to an additional annual net photosynthesis of $0.6 \cdot 0.1 \cdot 0.15 \cdot 60 \cdot 10^{15} = 0.54 \cdot 10^{15}$ g of C (global net primary production assumed to be $60 \cdot 10^{15}$ g/yr., after 2). Though by no means proven it is possible therefore that this effect results in a sink intensity of $0\text{-}1.0 \cdot 10^{15}$ g/yr. of C.

2.8. Ocean uptake

In a simplified manner, the downward penetration of anthropogenic CO_2 into the ocean can be thought of as a diffusive process (7,28, 34, for a lucid presentation of the problem see 8). For an exponential increase of the anthropogenic CO_2 excess in the atmosphere the effective penetration depth below the mixed layer is then $\overline{h}=\sqrt{D \cdot T}$. Table 3 presents a very simplified expression for the airborne fraction $f=E/\overline{N}$ which, using the values for the various parameters as indicated in the same table, amounts to approximately 0.62. This means that the fraction $\overline{N}$ of the cumulative anthropogenic CO_2 release N which is not taken up by other reservoirs (such as land ecosystems) is distributed between atmosphere and ocean like 0.62:0.38.

The crucial parameters here are the eddy diffusion coefficient D and the buffer factor ξ, both of which are not accurately known. However, as can be calculated from expression (6) in Table 3, the airborne fraction is rather insensitive to even remarkable variations in D and ξ. If, for instance, D would be 7,000 m^2/yr. instead of 4,000 m^2/yr, as suggested by some tracer studies, the airborne fraction would be 0.56 rather than 0.62; if the buffer factor would be 7.3 instead of 9.6 at present, the airborne fraction would be 0.54, and both changes combined would reduce the airborne fraction from 0.62 to 0.48. In the latter case, land ecosystems would not be required to effect a net uptake of about $0.45 \cdot 10^{15}$ g/yr. of C in 1976 in order to balance the atmospheric CO_2 budget, as under com-

Table 3. Approximate calculation of the distribution of anthropogenic CO_2 between atmosphere and ocean.

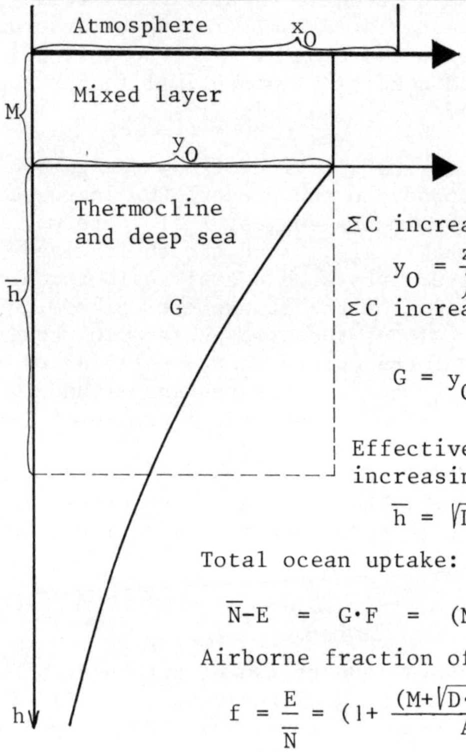

$$x = \Delta p/p = E/A \quad (1)$$

y

ΣC increase per m^2 in surface water:

$$y_0 = \frac{z \cdot \rho}{\xi} \cdot x_0 \quad (2)$$

ΣC increase per m^2 in total water column:

$$G = y_0 \cdot M + \int_0^\infty y\, dh = y_0 \cdot (M+\bar{h}) \quad (3)$$

Effective mixing depth at exponentially increasing E:

$$\bar{h} = \sqrt{D \cdot T} \quad (4)$$

Total ocean uptake:

$$\bar{N}-E = G \cdot F = (M+\sqrt{D \cdot T}) \cdot \frac{z \cdot \rho \cdot F \cdot E}{\xi \cdot A} \quad (5)$$

Airborne fraction of anthropogenic CO_2:

$$f = \frac{E}{\bar{N}} = \left(1 + \frac{(M+\sqrt{D \cdot T}) \cdot z \cdot \rho \cdot F}{A \cdot \xi}\right)^{-1} \quad (6)$$

Symbol	Description	Value
$\Delta p/p$	Atmospheric CO_2 above preindustrial level, %/100	0.14[a)]
z	CO_2 partial pressure increase in mixed layer/ increase in atmosphere (for T=~30yr.)	0.85
ΣC	$CO_2+HCO_3^-+CO_3^{--}$ in ocean water	
ρ	Average ΣC in ocean water	28.5 g/m^3 C
ξ	Buffer factor	9.6
M	Depth of mixed layer	75 m
h	Depth of water column	
$\bar{h}$	Effective penetration depth in thermocline	340 m
D	Vertical eddy diffusion constant	4,000 m^2/yr.
T	Average age of E	30 yr.
N	Cumulative anthropogenic CO_2 release	145 GT
$\bar{N}$	Fraction of N distributed between atmosphere and ocean	130 GT
E	CO_2 Excess in atmosphere	80 GT
$\bar{N}-E$	CO_2 Excess in ocean = total ocean uptake	50 GT
A	Preindustrial atmospheric CO_2 content	615 GT
F	Ocean area	$362 \cdot 10^{12}$ m^2
f	Airborne fraction = $E/\bar{N}$	0.62

a) some values are conventional, and subject to revision.

mon assumptions (D=4,000 m^2/yr, and ξ=9.6). Rather, an additional net release of approximately $0.4 \cdot 10^{15}$ g/yr. of C from biotic sources would be allowed in 1976. This is a very modest quantity, as compared with the quantity tropical land ecosystems are able to release if subject to destructive pressure, modest also in relation to the quantities suggested by some scientists involved in land ecosystems research (39,40). However simplifying, Table 3 shows that modern oceanographic theory has a strong case for rejecting the hypothesis of a large biotic net release of CO_2.

2.9. Unknown sinks

The difficulties involved in the attempt to balance the global carbon cycle encourage speculations that there may be alternative sinks for CO_2 completely uninvestigated to-date. Dissolution of carbonate sediments in shallow waters, leading to a better supply of CO_3^{--}-Ions, is one possibility, an unexpected imbalance in the organic carbon reservoir of the ocean (which comprises some 10^{18} g of C) is another one. However, no compelling hypothesis has been raised to-date which would prove that sinks such as these play a significant role at present.

3. POSSIBILITIES TO BALANCE THE GLOBAL CARBON CYCLE

In principle, three solutions to the puzzle of the global carbon budget are available: (1)Atmosphere-ocean exchange models are correct, and there is no net release of CO_2 to the atmosphere at all except from fossil fuels. (2)Atmosphere-ocean exchange models are grossly mistaken, and there is a significant net release from tropical ecosystems, additional to the one from fossil fuels, the airborne fraction being correspondingly lower. (3)Atmosphere-ocean models are correct, and there is a significant non-fossil CO_2 release in the tropics which is, however, compensated by an equal uptake in other land ecosystems, mainly in non-tropical latitudes. Present knowledge is too insufficient to come to any definite decision as to which of the solutions is the correct one. A critical evaluation of all relevant aspects, however, enables us to judge which is the most probable one. From the above considerations it is clear that solution (3) is most probable. On the one hand, it would be a great surprise if atmosphere-ocean models were completely wrong. On the other hand, the present shrinkage of tropical vegetation and soil organic matter, though not demonstrable in a rigid way, is well established by several independent reasons, as summarized in Table 4. Even if several calculations made in this table should prove wrong in detail, the overall conclusion would still remain valid. Finally, there is no compelling reason opposed to the assumption that non-tropical land ecosystems are a considerable net sink for CO_2, for various reasons. Thus, among all possible solutions, (3) is preferable because it minimizes the amount of disagreement with all kinds of relevant knowledge on the global carbon cycle.

Table 4. Possible net transfer of carbon from tropical land ecosystems to the atmosphere in 1976.

A. EVIDENCE FROM A HISTORIC REDUCTION OF TERRESTRIAL ORGANIC CARBON BY MAN	
Assumptions: (for details of calculations see (16))	
1. Probable total historic losses of C in tropical land ecosystems (Table 2)	$240 \cdot 10^{15}$ g
2. Probable losses since 1860 (50-70% of (1))	$120\text{-}168 \cdot 10^{15}$ g
3. Estimated rate of increase of annual losses	0.02-0.035
4. Resulting annual loss in 1976	$(2.66\text{-}5.98) \cdot 10^{15}$ g
5. Allowance for delayed decomposition (~30% of (4))	$(0.86\text{-}1.98) \cdot 10^{15}$ g
6. Annual transfer to the atmosphere in 1976	$(\sim 1.8\text{-}4.0) \cdot 10^{15}$ g
B. EVIDENCE FROM SOCIO-ECONOMIC DEVELOPMENT	
Assumptions:	
7. Population increase in the tropics in 1976 (rate)	0.024
8. Probable annual increase in cultivated land (rate)	0.02
9. Annual increase in cultivated land in 1976 (2% of $(6\text{-}10.5) \cdot 10^{12}$ m^2	$(120\text{-}210) \cdot 10^{9}$ m^2
10. Average organic C in tropical vegetation (Table 2)	6,600 g/m^2
11. Average phytomass (C) in trop. cropland	300 g/m^2
12. Average loss on cultivation	6,300 g/m^2
13. Average soil organic matter in the tropics (C)	8,200 g/m^2
14. Estimated 40% loss on cultivation	3,300 g/m^2
15. Sum (12) + (14)	9,300 g/m^2
16. Allowance for delayed decomposition	1,700 g/m^2
17. Difference (15) - (16)	7,600 g/m^2
18. Product (9) x (17)	$(0.91\text{-}1.60) \cdot 10^{15}$ g
19. Estimated area of trop. commercial forest clearings, not subsequently used for subsistence cropping	$(30\text{-}50) \cdot 10^{12}$ m^2
20. Average phytomass decomposition in this area, allowing for delayed decay	10,100 g/m^2
21. Average soil organic matter decay in this area	3,600 g/m^2
22. ((20) + (21)) x (19)	$(0.41\text{-}0.68) \cdot 10^{15}$ g
23. (18) + (22), Total annual transfer	$(\sim 1.3\text{-}2.3) \cdot 10^{15}$ g
C. EXTRAPOLATION FROM FOREST AREA DECREASE OBSERVED BY REMOTE SENSING	
Assumptions:	
24. Reduction in area of closed forests (evergreen and moist deciduous) in Thailand, 1961-1975	-25%
25. Area of world's tropical moist forests in populated regions (~50% of total trop. moist forests) in 1976	$7.25 \cdot 10^{12}$ m^2
26. Extrapolated loss in 14 years	$2.40 \cdot 10^{12}$ m^2
27. Loss in 1976 at 2% annual increase	$190 \cdot 10^{9}$ m^2
28. Annual transfer of C in 1976	$1.75 \cdot 10^{15}$ g
29. C-Transfer in 1976, Range of estimates	$(\sim 1.3\text{-}4.0) \cdot 10^{15}$ g
30. Most probable	$(1.5\text{-}2.5) \cdot 10^{15}$ g

4. FUTURE ATMOSPHERIC CO_2 INCREASE

The role of land ecosystems is not only a matter of scientific interest. It may also become relevant to the question of what kind of future atmospheric CO_2 increase has to be expected. If it were true that the airborne fraction is much lower than calculated by current models, the future atmospheric CO_2 burden, as well as the climatic risks associated, could be correspondingly lower - an important aspect in planning long-term energy strategies.

I tentatively investigated this question by means of a very simple model of ocean uptake, the elements of which are described in Table 5. In a first step, the average age of the combined anthropogenic input(minus the uptake by terrestrial sinks) is calculated according to expression (7). The result is inserted in (8), where the airborne fraction is calculated. In using the empirical function (9), as given in (34), the expression (8) is modified into (10), thereby accounting for a variable buffer factor which strongly influences the future atmospheric CO_2 increase. The model can easily be executed on a programmable pocket calculator (TI 59). Its main shortcomings are (1)that the average age of the combined CO_2 emission function is calculated,which is in general not exactly identical with the average age of the anthropogenic CO_2 excess in the atmosphere, (2)that the relation z($\Delta p/p$ in mixed layer:$\Delta p/p$ in the atmosphere) is held constant whereas in fact it is a variable, and (3)that the effective penetration depth is invariably calculated as $\bar{h}=\sqrt{D \cdot T}$ which applies strictly speaking only to an exponential increase. In spite of these deficiencies the model reproduces the predictions of the Box-Diffusion-Model of Oeschger et al.(28) with a maximal error of 5-6 per cent, as proven in test runs using the data in (34). It is important that non-exponential developments are reproduced satisfactorily. I conclude that while the absolute values obtained should be viewed with some scepticism, one should feel justified in using the simple model for comparisons between different scenarios.

Several scenarios are considered which constitute superpositions of a number of simple logistic input and output functions for atmospheric CO_2 (Table 6): a low fossil fuel use function, assuming that only $1{,}000 \cdot 10^{15}$ g C will be used (1), and (2) a high fossil fuel case, assuming that $M=6{,}000 \cdot 10^{15}$ g C. Furthermore, there are two functions for an atmospheric input of CO_2 from tropical vegetation and humus destruction, assuming that 330 and $400 \cdot 10^{15}$ gC respectively have been lost since 1860. In one case (3), the annual input to the atmosphere in 1976 is $1.8 \cdot 10^{15}$ g, and the cumulated input up to that date is $80 \cdot 10^{15}$ g C, while in the other case (4) the annual input in 1976 is $2.95 \cdot 10^{15}$ g, and the cumulated input is $150 \cdot 10^{15}$ g C. It is further assumed (5), that the cumulated C uptake of the vegetation, resulting from atmospheric CO_2 increase, will be $90 \cdot 10^{15}$ g C (about 15% of the C stored in present phytomass), due to a limited

Table 5. Elements of a simple diffusive model to calculate the airborne fraction of anthropogenic CO_2

1. Calculation of average age of anthropogenic CO_2 excess:

$$T_\tau \approx \tau - \frac{\int_0^\tau (N'_t \cdot t)dt}{(N_\tau - N_0)} \quad (7)$$

2. General expression for the airborne fraction (Table 3):

$$E = \frac{\bar{N}_\tau}{\left(1 + \frac{(M+\sqrt{D \cdot T_\tau}) \cdot F \cdot \varsigma \cdot z}{\xi \cdot A}\right)} \quad (8)$$

3. Empirical expression for a variable buffer factor:

$$\xi = X + Y \cdot \beta - Z \cdot \beta^2; \quad \beta = \frac{z \cdot E}{A} \quad (9)$$

4. Inserting (9) in (8), and calculation of anthropogenic CO_2 excess E:

$$E^3 - \left(\frac{Y \cdot A}{Z \cdot z} + \bar{N}_\tau\right) \cdot E^2 - \left(\frac{X \cdot A^2}{Z \cdot z^2} + \frac{(M+\sqrt{D \cdot T_\tau}) \cdot F \cdot \varsigma \cdot A - Y \cdot \bar{N}_\tau \cdot A}{Z \cdot z}\right) \cdot E - \frac{X \cdot A^2 \cdot \bar{N}_\tau}{Z \cdot z^2} = 0 \quad (10)$$

τ present year (e.g. 116 for 1976 and t_0=1860)

X	empirical constants	9	for a buffer factor of 9.6 in 1976 a) (most probable)	7	for a buffer factor of 7.3 in 1976 (hypothetical)
Y		4.9		3.82	
Z		0.1		0.08	

for all other symbols see Table 3. a)after (34).

storage capacity of the vegetation and independently from the exact atmospheric increase. The annual uptake in 1976 is $0.85 \cdot 10^{15}$ g. It is estimated, furthermore, that temperate forests will increase by $30 \cdot 10^{15}$ g C between about 1940 and 2010 (about 30% of their present content), with a peak annual increase of $0.7 \cdot 10^{15}$ g C in 1976 (6). Finally, a number of miscellaneous terrestrial sinks are considered to withdraw C from the air in constant quantities throughout the period concerned, with a lower value of $0.35 \cdot 10^{15}$ g, and a higher value of $0.6 \cdot 10^{15}$ g C (7 and 8).

As shown in Table 7, there are essentially four scenarios: Scenario A/A' implies that there is no biotic influence on the atmospheric CO_2 level except for a small terrestrial sink absorbing $0.35 \cdot 10^{15}$

Table 6. Input and output functions of CO_2 to and from the atmosphere, for various causes.

$$N = \frac{M}{1 - b \cdot e^{-r \cdot t}} \qquad N' = - \frac{M \cdot b \cdot r \cdot e^{-r \cdot t}}{(1 - b \cdot e^{-r \cdot t})^2}$$

		M	b	r	N_{1976}	N'_{1976}
1 Fossil fuel, low	+	1,000	-634.39	0.04033	145	5
2 " , high	+	6,000	-2,434.29	0.03534	145	5
3 Trop. defor., low[a)]	+	330	-97.97	0.02970	80	1.8
4 " , high[b)]	+	400	-64.13	0.03147	150	2.95
5 CO_2-induced growth[c)]		90	-105.65	0.03825	40	0.85
6 Temperate forests[d)]	-	30	-50,326.15	0.09333	15	0.7
7 Miscel. sinks, low	-	e)	-	-	~30	0.35[f)]
8 " , high	-	e)	-	-	~60	0.6[f)]

+ release to the atmosphere, - uptake from the atmosphere. N: cumulative release or uptake, N': annual release or uptake. M: cumulative value for $t \rightarrow \infty$. t_0 = 1860. a) Assuming that 34% of the potential organic C of the tropics are released from approx. 1860-2060, see §2.2-2.4 and Table 2. b) Assuming that 41% are cumulatively released. c) Assuming that the present terrestrial phytomass will increase by 15% due to enhanced atmospheric CO_2. d) Assuming that organic C in temperate forests will increase by 30% from approx. 1940 to 2010. e) Variable, depending on length of period considered. f) Constant annual values throughout the period considered. All values in 10^{15} g C.

g/yr. throughout the period considered. Conventional figures for the eddy diffusion constant (D=4,000 m^2/yr.) and for the buffer factor (ξ=9.6 in 1976) are used. Apart from the fact that the biotic uptake is constant rather than exponentially increasing, this scenario resembles the usual simulations made with the help of computer models. This scenario is therefore called "Reference".

Scenario B/B' represents an extreme case where a biotic source is adding $2.95 \cdot 10^{15}$ g of C to the atmosphere in 1976, additionally to fossil fuel, while there is only a very small biotic sink, as in the previous case. In order to balance the atmosphere's C budget, an eddy diffusion constant of 20,000 m^2/yr. (implying an effective mixing depth of more than 800 m in 1976) is required, which is clearly at variance with oceanographic knowledge. This scenario, therefore, may be termed "Unrealistic".

Scenarios C/C' and D/D' are both characterized by an active biotic carbon pool exchanging large quantities with the atmosphere. In C/C', biotic inputs from tropical vegetation and soil organic matter destruction are balanced by biotic withdrawals resulting from miscellaneous sinks, from regrowth of temperate forests and from CO_2-induced photosynthesis enhancement so that the ocean characteristics (eddy diffusion constant and buffer factor) can be assumed to be the same as in scenario A/A'. In Scenario B/B', a larger biotic input ($2.95 \cdot 10^{15}$ g) is assumed in 1976, requiring an eddy diffusion constant of 7,000 m^2/yr. and a buffer factor of 7.4. Although these values are not very probable they cannot be ruled out, so that D/D' may be termed "Optimistic" (optimistic in the sense of ocean uptake), while C/C' is regarded as the "Most Probable" case of all, according to the statement made in §3 of this paper.

All scenarios are simulated twice: In the first case (A,B,C,D) they are combined with the low fossil fuel use function in Table 6 (LF cases), whereas in the second case (A',B',C',D') a high fossil fuel scenario (HF cases) is assumed. By adjusting the preindustrial atmospheric CO_2 level it is possible in each case to reproduce satisfactorily the observed atmospheric CO_2 increase between 1958 and 1976.

The predictions for the eight cases up to the year 2200 are shown in Figure 1. In the "Reference" scenario, the atmospheric CO_2 level will increase to about 540 ppm (1.85 times the conventional preindustrial level of 290 ppm) in the LF case, and to 2,400 ppm (8.3 times the preindustrial level) in the HF case, with an airborne fraction at the time of the peak concentration of about 0.6 respectively 0.78. As could be expected, the peak concentration in the "Most Probable" scenario differs only slightly from the one in the "Reference" scenario (580 and 2,455 ppm instead of 540 and 2,400 ppm). Due to the fact that the future biotic CO_2 release estimated here slightly exceeds the future biotic uptake, an additional quantity has to be distributed between atmosphere and ocean.

Interestingly, the peak atmospheric concentration in the "Optimistic" scenario falls short of the one in the "Reference" scenario by only 3% in the LF case and by 12% in the HF case (522 and 2,120 ppm). The reason is quite simple, however: In the "Optimistic" scenario, a lower airborne fraction is combined with a higher biotic release. If, in 1976, there is an annual biotic input to the atmosphere from tropical deforestation and humus oxidation of almost $3 \cdot 10^{15}$ g C, as assumed here, a large future cumulative biotic input has normally to be expected. In the LF case, this quantity, estimated here to be $250 \cdot 10^{15}$ g, is large, as compared with the expected fossil fuel release of $855 \cdot 10^{15}$ g. Thus, the effect of a lower airborne fraction is largely set off in the LF scenario by the larger future non-fossil input, so that about the same atmospheric increase has to be expected. There would be no such compensation only if the biotic input could be stopped within a very short time, which is however rather improbable. In a HF scenario, the future biotic input is small as compared with the future fossil input, so that the compensating effect is less pronounced; the difference between "Reference" and "Optimistic" scenario is larger.

Table 7. Scenarios depicting various roles of land ecosystems and the ocean in the global carbon cycle, in combination with fossil fuel burning.

	Scenario[a)]		Functions combined (Table 5)	$A^{b)}$	$\overline{N}$ (1976)	E (1976)	$\overline{N}'$ (1976)	E' (1976)	$f^{c)}$ (1976)	$f'^{c)}$ (1976)	D (m^2/yr)	ζ (1976)
A	"Reference"	LF	1-7	632.8	110.6	73.9	4.65	2.8	.67	.602	4,000	9.6
A'	"	HF	2-7	634.3	110.6	72.4	4.65	2.8	.65	.602	4,000	9.6
B	"Unrealistic"	LF	1+4-7	608.0	260.6	98.7	7.6	2.8	.37	.368	20,000	7.3
B'	"	HF	2+4-7	608.3	260.6	98.4	7.6	2.8	.37	.368	20,000	7.3
C	"Most Probable"	LF	1+3-5-6-8	629.7	119.1	77.0	4.65	2.8	.64	.602	4,000	9.6
C'	"	HF	2+3-5-6-8	631.5	119.1	75.2	4.65	2.8	.63	.602	4,000	9.6
D	"Optimistic"	LF	1+4-5-6-8	614.4	189.1	92.3	5.8	2.8	.48	.483	7,000	7.3
D'	"	HF	2+4-5-6-8	613.8	189.1	92.9	5.8	2.8	.49	.483	7,000	7.3

a)For explanations see text. LF=Low fossil fuel scenario (M=1,000 GT); HF=High fossil fuel scenario (M=6,000 GT).b)Preindustrial level calculated for best fit with measured atmospheric values during 1958-1975.c)$f=E/\overline{N}$; $f'=E'/\overline{N}'$. An atmospheric value of 333.5 ppm in 1976 in assumed. A:Preindustrial atmospheric CO_2 level. $\overline{N}$:Cumulative anthropogenic release to be distributed between atmosphere and ocean. E:Cumulative atmospheric CO_2 excess. $\overline{N}'$:Annual anthropogenic release, to be distributed between atmosphere and ocean. E':Atmospheric CO_2 increase in 1976, as deduced from trend-line. f;f':Cumulative and annual airborne fractions. D:Eddy diffusion constant of the ocean. ζ:Buffer factor. See also Table 6.

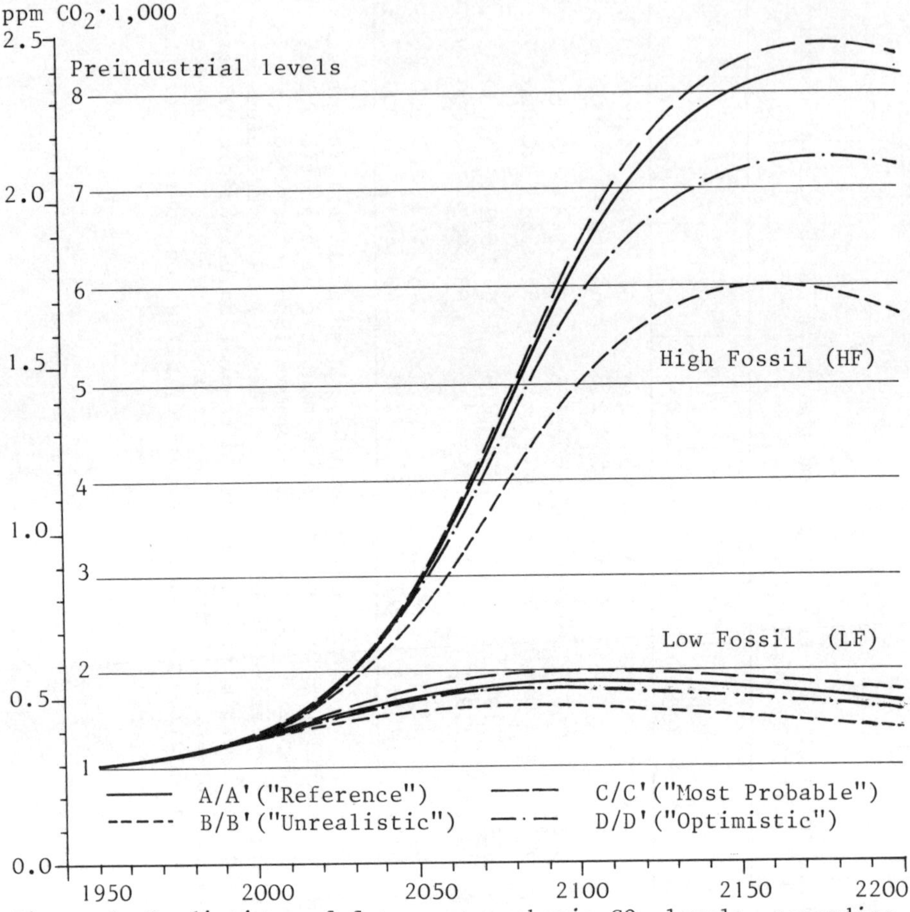

Figure 1. Predictions of future atmospheric CO_2 levels, according to scenarios in Table 7.

Whether or not it would constitute an important difference in terms of the climatic future of the earth, if the atmospheric CO_2 content would rise to only 7.3 times the preindustrial level in the "Optimistic" scenario rather than to 8.3 times, as in the "Reference" scenario, is a question I am unable to answer. However, considering that the CO_2-induced global temperature increase would probably be proportional to the logarithm of the atmospheric CO_2 increase, the climatic consequences would possibly not differ dramatically. Moreover, the lower of both values may be as little permissible in view of mankind's future as the higher one. This being the case, an optimistic but still realistic conception of ocean uptake (involving an eddy diffusion constant of about 7,000 m^2/yr. and a present buffer factor of about 7, or any combination of other causes effecting an acceleration of ocean uptake of this magnitude), combined with a higher biotic net release of CO_2 at present and in the future, would be able to alleviate the CO_2 problem neither in a HF nor in a LF case.

Although of little practical relevance, similar arguments apply even to the "Unrealistic" scenario B/B'. Even if the vertical diffusivity in the sea were three times as large as considered possible in an already optimistic case, the atmospheric CO_2 level would rise to 6 times the preindustrial value in the HF case. In the LF case however, the increase would be considerably reduced, as compared with the "Reference" scenario.

It should be noted that differences between any two scenarios will not be noticeable before the year 2000. The scenarios A',C' and D' will not be discernible by atmospheric measurements before the late 21st century. Among the HF cases, the doubling of atmospheric CO_2 will be delayed by only 10 years in the "Unrealistic" scenario, as compared with the other three scenarios. I propose that calculations such as these be redone using large computer models in order to verify the results. Provided these should prove roughly correct, the following conclusions would be justified:

5. CONCLUSIONS

(1) Any CO_2 quantity released from biotic sources in the future will be insignificant for the future atmospheric CO_2 level if fossil fuels will be consumed in large amounts. Biotic releases will be significant in a low-fossil-fuel scenario, however. If mankind should succeed in limiting the future use of fossil fuels, an additional effort to control the biotic sources would be fruitful. Likewise, biotic sinks such as reforestation will become relevant only in a low-fossil-fuel scenario.

(2) Within the range to be considered possible, any airborne fraction lower than the one assumed at present will be of little relevance for the climatic future of the earth in a high-fossil-fuel scenario, for it will not prevent the atmospheric CO_2 from increasing to presumably intolerable levels. In a low-fossil-fuel scenario a lower airborne fraction will be beneficial only if biotic net inputs of CO_2 to the atmosphere can be reduced effectively within a short period of time.

(3) Although present geochemical models do not consider the role of land ecosystems in sufficient detail, their predictions of the future atmospheric CO_2 increase remain valid at least in a high-fossil-fuel scenario. Considering it most probable that the airborne fraction is not much lower than commonly assumed, that present biotic CO_2 exchanges with the atmosphere are balancing each other, and that there is little hope of reducing future vegetation and soil destruction in the tropics, the atmospheric CO_2 increase will be even slightly higher than hitherto predicted.

References

1 Adams, J.A.S., Mantovani, M.S.M., and Lundell, L.L.:1977, Science 196, pp.54-56.
2 Ajtay, G.L., Ketner, P., and Duvigneaud, P.:1979, in "The Global Carbon Cycle", Bolin, B., Degens, E.T., Kempe, S., and Ketner, P., Eds., SCOPE-Report 13, Chichester New York Brisbane Toronto (John Wiley), pp. 129-181.
3 Bacastow, R., and Keeling, C.D.:1973, in "Carbon and the Biosphere", Woodwell, G.M., and Pecan, E.V., Eds. USAEC CONF 720510, NTIS Springfield, VA, pp. 86-135.
4 Björkman, O.:1979, Research Program Development Paper for ERDA, Stanford, CA, pp. 1-28.
5 Björkström, A.:1979, in "The Global Carbon Cycle", Bolin, B., Degens, E.T., Kempe, S.,and Ketner, P., Eds., SCOPE-Report 13, Chichester New York Brisbane Toronto (John Wiley), pp. 403-457.
6 Bolin, B.:1977, Science 196, pp. 613-615
7 Broecker, W.S., Li, Y.-H., and Peng, T.-H.:1971, in "Impingement of Man on the Oceans", Hood, D.W., Ed., New York (John Wiley), pp. 287-324.
8 Broecker, W.S., Takahashi, T., Simpson, H.J., and Peng, T.-H.: 1979, Science 206, pp. 409-418.
9 Brünig, E.F.:1977, Ambio 6, pp. 187-191.
10 Buringh, P.:1979, Paper presented at the SCOPE Workshop "The Role of Terrestrial Vegetation in the Global Carbon Cycle: Methods of Appraising Changes", Woods Hole, Massachusetts, 6-11 May, 1979, pp. 1-45.
11 Clawson, M.:1979, Science 204, pp. 1168-1174.
12 Goudriaan, J. and Ajtay, G.L.:1979, in "The Global Carbon Cycle", Bolin, B., Degens, E.T., Kempe, S., and Ketner, P., Eds., SCOPE-Report 13, Chichester New York Brisbane Toronto (John Wiley), pp. 237-249.
13 Green, K., and Wright, R.:1977, Ecology 58, pp. 687-692.
14 Hampicke, U.:1979, in "The Global Carbon Cycle", Bolin, B., Degens, E.T., Kempe, S., and Ketner, P., Eds., SCOPE-Report 13, Chichester New York Brisbane Toronto (John Wiley), pp. 219-236.
15 Hampicke, U.:1980, Environment International, to be published.
16 Hampicke, U., and Bach, W.:1979,"Die Rolle Terrestrischer Ökosysteme im Globalen Kohlenstoff-Kreislauf", Bericht im Auftrag des Umweltbundesamtes, Forsch.- u.Entw.-vorh. Nr. 104 02 513, pp. 1-153.
17 Junge, C.:1978, Promet 8, pp. 21-32.
18 Kardell, L.:1978, Ambio 7, pp. 84-92.
19 Kempe, S.:1979, in "The Global Carbon Cycle", Bolin, B., Degens, E.T., Kempe, S., and Ketner, P., Eds., SCOPE-Report 13, Chichester New York Brisbane Toronto (John Wiley), pp. 317-342.
20 Klankamsorn, B.:1976, in "Remote Sensing in Forestry", G. Hildebrandt, Ed., Proceedings of the Symposium held during the XVI. IUFRO World Congress, Oslo, 21-26 June 1976, Freiburg i. Br., pp. 227-233.

21 Köpp, H.:1979, Natur und Landschaft 54, p. 155.
22 Lemon, E.:1977, in "The Fate of Fossil Fuel CO_2 in the Oceans", Andersen, N.R., and Malahoff, A., Eds., New York and London (Plenum Press), pp. 97-130.
23 Loomis, R.S.:1979, in "Carbon Dioxide Effects Research and Assessment Program", Elliot, W.P., and Machta, L., Eds.,US Dept. of Energy, CONF-770385, UC-11, Washington, D.C., pp. 51-62.
24 Maydell, H.J. von:1972, Forstarchiv 43, pp. 245-250.
25 Miller, L.D., Nualchawee, K., and Tom, C.:1978, NASA Tech. Memo. 79545, in "Proceed. of the 12th International Symposium on Remote Sensing of the Environment" Env. Inst. of Michigan, Ann Arbor, pp. 1167-1185.
26 Mopper, K., and Degens, E.T.:1979, in "The Global Carbon Cycle", Bolin, B., Degens, E.T., Kempe, S., and Ketner, P., Eds., SCOPE-Report 13, Chichester New York Brisbane Toronto (John Wiley), pp. 293-316.
27 Nye, P.H., and Greenland, D.J.:1960, "The Soil Under Shifting Cultivation", Comm. Bureau of Soils, Tech.Comm. No. 51, Harpeden, pp. 1-156.
28 Oeschger, H., Siegenthaler, U., Schotterer, U., and Gugelmann, A.:1975, Tellus 27, pp. 168-192.
29 Olson, J.S., Pfuderer, H.A., and Chan, Y.-H.:1978, "Changes in the Global Carbon Cycle and the Biosphere", ORNL/EIS-109, Env. Sci. Div. Publ. No. 1050, Oak Ridge National Laboratory, pp. 1-169.
30 Perry, H., and Landsberg, H.H.:1977, in "Energy and Climate", National Academy of Sciences, Washington, D.C., pp. 35-50.
31 Persson, R.:1974, "World Forest Resources", Royal College of Forestry, Stockholm, pp. 1-264.
32 Revelle, R., and Munk, W.:1977, in "Energy and Climate", National Academy of Sciences, Washington, D.C., pp. 140-158.
33 Rodin, L.E., Bazilevich, N.I., and Rozow, N.N.:1975, in "Productivity of World Ecosystems", National Academy of Sciences, Washington, D.C., pp. 13-26.
34 Siegenthaler, U., and Oeschger, H.:1978, Science 199, pp. 388-395.
35 Sommer, A.:1976, Unasylva 28, pp. 5-24.
36 Strain, B.R., Ed.:1978 "Report of the Workshop on Anticipated Plant Responses to Global Carbon Dioxide Enrichment", Duke University 27706, Durham, NC, pp. 1-23 + 91 pp. bibliography.
37 Schlesinger, W.H.:1979, Paper presented at the SCOPE Workshop "The Role of Terrestrial Vegetation in the Global Carbon Cycle: Methods of Appraising Changes", Woods Hole, Massachusetts, 6-11 May, 1979, pp. 1-31.
38 Wong, C.S.:1978, Science 200, pp. 197-200.
39 Woodwell, G.M., and Houghton, R.A.:1977, in "Global Chemical Cycles and Their Alterations by Man", Stumm, W., Ed., Dahlem Konferenzen, Berlin, pp. 61-72.
40 Woodwell, G.M., Whittaker, R.H., Reiners, W.A., Likens, G.E., Delwiche, C.C.,and Botkin, C.D.:1978, Science 199, pp. 141-146.
41 Young, A.:1976, "Tropical Soils and Soil Survey", Cambridge London New York Melbourne (Cambridge Univ. Press), pp. 1-468.

MODELING THE SURFACE TEMPERATURE CHANGES DUE TO INCREASED ATMOSPHERIC CO_2

W. Lawrence Gates

Department of Atmospheric Sciences
and Climatic Research Institute
Oregon State University
Corvallis, Oregon, USA

ABSTRACT. Although there are considerable differences in atmospheric models' structure and sensitivity, there is a consensus that increasing the concentration of CO_2 in the atmosphere will lead to an overall warming. The more simplified models (which treat the atmosphere as a vertical column in which the temperature is determined by the net absorption of radiation and the redistribution of heat by convection) generally yield a surface temperature rise of several degree C for a doubling of atmospheric CO_2. Other simplified models built around the meridional variation of the surface heat balance (in which the surface radiative, sensible and latent heat fluxes are parameterized in terms of the surface temperature) also yield an average warming of several degree C for a CO_2 doubling. Such heat balance models are particularly sensitive to the feedback between the surface temperature and the albedo of snow- or ice-covered surfaces, and generally show a poleward amplification of the CO_2-induced warming when this feedback is included.

General circulation models or GCMs (in which the three-dimensional distribution of global climate is simulated in response to a wide variety of interacting time-dependent processes) generally confirm the simpler models' warming, and serve to emphasize the climate's sensitivity to the treatment of the oceans. Using idealized geography and a motionless swamp-like ocean of zero heat capacity, GCMs yield a global average warming from about 3°C in the tropics to about 6°C in high latitudes for a doubling of CO_2 under annual average insolation. The importance of the swamp ocean in these calculations is shown by a model with realistic geography and an ocean whose surface temperature is constrained to follow the normal climatological seasonal variation. Such a

W. Bach, J. Pankrath, and J. Williams (eds.), Interactions of Energy and Climate, 169–190.

GCM yields only about 0.2°C overall warming for doubled CO_2, with significantly larger warming over the higher latitude continents in summer. When an atmospheric GCM and a shallow mixed-layer ocean are used with realistic geography and seasonal insolation, a more realistic average warming of about 2°C is inferred for doubled CO_2, with ice albedo-temperature feedback responsible for a several-fold amplification in higher latitudes in winter. This value is regarded as the best estimate of the average surface warming due to doubled CO_2 now available.

To determine the seasonal and regional distribution of a CO_2-induced warming as required for the reliable estimation of energy-related climate impacts, further model simulations are required in which the natural interannual variability of the atmosphere and surface ocean are more adequately represented. These factors, together with the time-dependent response of the surface biomass and of the deeper ocean, will determine the course of climate as a result of future CO_2 increases.

1. INTRODUCTION

As the atmospheric CO_2 concentration continues to rise as a result of the combustion of fossil fuels, the determination of the possible accompanying climatic effects has become a matter of increasing concern. This has motivated the renewed search of past climatic data for analogs of possible future climates, and the increased application of atmospheric models to the estimation of the climatic consequences of increased CO_2. Only through the use of models, however, can we hope to acquire the quantitative information on the seasonal and geographical structure of likely future climates which is necessary for effective use in estimating climatic impacts.

All atmospheric models which have been applied to the estimation of the climatic effects of increased CO_2 include the surface temperature as a dependent variable. This also happens to be the quantity of perhaps the greatest practical importance in connection with energy allocation and use, although some models do not distinguish between the temperature of the air near the surface and the temperature of the earth's surface itself. The types of models used for this purpose constitute a hierarchy, ranging from simplified models which consider only the average heat balance at the earth's surface (and in which all processes are parameterized in terms of the surface temperature) to sophisticated general circulation models (or GCMs). In these latter models the three-dimensional time-dependent behavior of all principal atmospheric variables such as temperature, wind, cloudiness and precipitation are simulated along with the global balances of heat, moisture, momentum and energy.

While the simplified models are relatively easy to solve, their solutions are sensitive to the way in which processes such as radiation, cloudiness and surface effects are represented. Such models also do not provide estimates of the atmosphere's natural variability as a result of synoptic-scale disturbances, and in some instances do not consider even the normal seasonal cycle. The very simplicity of these models, however, facilitates the diagnosis of various physical processes, at least insofar as they are represented in the model.

In GCMs, on the other hand, the global distribution and variation of a wide variety of interacting processes is routinely simulated. This detail and hence presumed higher reliability of GCMs, however, is obtained at a high cost in terms of the computer time necessary to produce a satisfactory solution. Such models' integration proceeds on modern computers at a rate between about 10^2 to 10^3 times nature, while simplified models proceed hundreds or thousands of times faster. The result is that a large number of climate experiments have been carried out with simplified models, whereas relatively few have been made with GCMs.

In most experiments for increased atmospheric CO_2 with either kind of model, only the long-term equilibrium climate is usually sought rather than the climate's evolution. Seasonal effects have frequently been omitted in simplified models, and it is only recently that they have been included in global GCM simulations with raised CO_2. In such experiments attention is now focused on the extraction of the geographical and seasonal distribution of the warming and on the associated changes of surface climate. So far no model has successfully simulated the time-dependent variation of climate in response to the projected future time-dependent rise of atmospheric CO_2, nor has effective modeling of the atmospheric CO_2 interaction with the ocean and biomass yet been achieved.

2. RESULTS FROM SIMPLIFIED CLIMATE MODELS

Although there is some parameterization in all climate models, those in which one or more dimensions of the atmosphere is suppressed through an averaging process are of particular interest in the CO_2-climate problem. Such statistical-dynamical models (or SDMs) include one- and two-dimensional models built around the atmosphere's radiation or energy balance, and two-dimensional zonally-averaged models.

2.1. Radiative-convective models

The simplest climate models which could be used to study the changes of surface temperature in response to increased CO_2 are the one-dimensional radiative balance models, such as that of

Manabe and Möller (1961). In these models the net annual solar radiation absorbed in an atmospheric column is assumed to be balanced by the net outgoing longwave radiation. An increase of CO_2 concentration in the atmosphere increases both the air's emission and absorption of longwave radiation (and also slightly increases its absorption of shortwave radiation) by increasing the opacity in those infrared wavelengths dominated by CO_2 (e.g., between 12-18 μm). This results in an increase of the atmosphere's absorption of the longwave radiation emitted by the surface (and by clouds), and thus in a direct radiative warming of the troposphere. At the same time, however, the increased CO_2 also increases the downward emission of longwave radiation by the troposphere, and this in turn results in a warming at the surface. The net result is a reduction of the longwave radiative flux divergence in the troposphere, as measured, say, by an increase in the net radiative flux at the tropopause or by a net heating of the earth-atmosphere system. If all other atmospheric properties remained the same upon CO_2 doubling, the annual global average of this increase in the net (downward) radiative flux at the tropopause is estimated to be about 4 W m^{-2} (Ramanathan et al., 1979), while that at the surface is estimated to be about 1 W m^{-2} (Newell and Dopplick, 1979; Ramanathan et al., 1979). If the models' atmospheric temperatures are allowed to adjust to these radiative flux changes, as only seems reasonable, then the net increase in the longwave flux at the surface due to doubled CO_2 is increased to about 3-4 W m^{-2}.

The amount by which the surface temperature will rise in response to this radiative flux change is influenced by factors which are not always considered in a simple climate model, and as we shall see, several of them involve feedback processes which effectively increase or decrease the basic radiative warming. If in the simplest case of a globally-averaged radiative balance model the earth is assumed to behave as a blackbody, the surface temperature change in response to an increased surface radiative flux of 4 W m^{-2} accompanying doubled CO_2 is readily calculated to be about 1°C. A useful measure of this effect is the sensitivity or feedback parameter $\lambda = \partial(\text{temperature})/\partial(\text{heat flux})$, which in this case of radiative balance gives $\lambda = 0.25$ deg $W^{-1}m^2$. A similar average surface warming of 0.8°C was found by Rasool and Schneider (1971) for doubled CO_2, using a one-dimensional radiative balance model with fixed lapse rate and fixed cloud heights.

When such purely radiative models are used to calculate the vertical structure of the atmosphere, they generally result in temperatures which are too low due to their neglect of the vertical flux of heat by atmospheric motions (Manabe and Möller, 1961). In an effort to improve the radiative balance models, Manabe and Strickler (1964) introduced the effects of convective heat transport by restoring the vertical lapse rate to a pre-assigned stable value (6.5 deg km^{-1}) whenever it became greater due to radiative heat transfers. This effective parameterization of the vertical heat flux by convective-scale motions is referred to as convective adjustment, and serves to compensate the net radiative heat loss by the atmosphere. Following the work of Manabe and Strickler, a wide variety of such radiative-convective models have been developed, and many of them have been used in the estimation of the warming due to increased CO_2. The results of some of these investigations in terms of the globally-averaged increase of annual surface temperature resulting from a doubling of CO_2 are shown in Fig. 1.

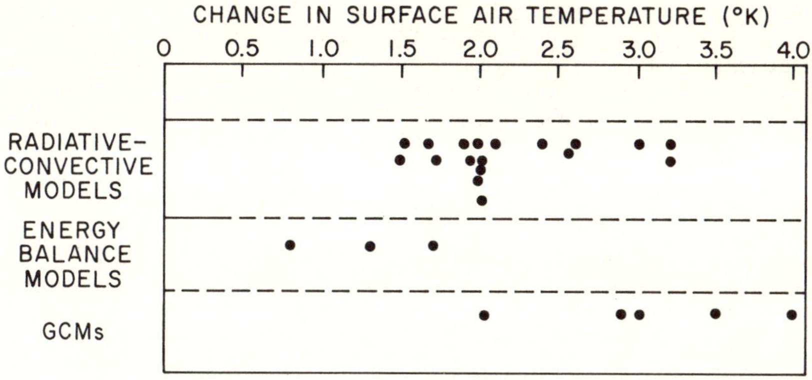

Fig. 1. The change in globally-average annual surface air temperature resulting from a doubling of present atmospheric CO_2, as given by a variety of radiative-convective models, energy balance models, and general circulation models (GCMs).

The first and best-known application of a radiative-convective model for doubled CO_2 was that by Manabe and Wetherald (1967), using the model developed earlier by Manabe and Strickler (1964). With assigned average cloudiness, and by fixing both the altitude of the cloud tops and the relative humidity, they found an average global surface warming of 2.4°C for doubled CO_2. Because of the strong effects of clouds and humidity on the fluxes of both solar and infrared radiation, their model was found to be sensitive to the treatment of these features. For clear skies, for example, and with fixed relative humidity, an average surface warming of 2.9°C (rather than 2.4°C) was found for doubled CO_2, while the corresponding warming with fixed absolute rather than relative humidity was 1.4°C. With an improved version of the treatment of radiative transfer in the model, Manabe (1971) subsequently found the somewhat lower value of 1.9°C for the average surface warming accompanying doubled CO_2, while further improvement by Ramanathan (1975) resulted in a warming of only 1.5°C. These results were summarized by Schneider (1975), whose own radiative-convective model version for doubled CO_2 yielded an average surface warming of 1.4°C, in good agreement with Ramanathan's (1975) results for average cloudiness.

The next major advance in modeling the increased surface temperature to be expected with doubled CO_2 with radiative-convective models was made by Augustsson and Ramanathan (1977), using the improved radiative model of Ramanathan (1976). By including all active CO_2 and H_2O absorption bands, allowing a variable relative humidity, and fixing the cloud top temperature, they showed that an average surface warming from about 1.5°C to about 3°C could result, depending upon the assumed humidity and cloudiness, with an average surface warming for doubled CO_2 of about 2°C under average conditions. Augustsson and Ramanathan were able to show, moreover, that the smaller warming of about 1.5°C found earlier by Ramanathan (1975) and Schneider (1975) was due to these latter authors' neglect of the effect of weak CO_2 absorption bands and their assignment of cloud top altitude rather than cloud top temperature. The lowering of the surface warming by this latter effect was also noted by Wang et al. (1976). Other recent studies with similar radiative-convective models have given results within the range of about 1.5°C to 2.5°C for doubled CO_2 (Rowntree and Walker, 1978; Hummel and Reck, 1979), and are generally consistent with the results of Augustsson and Ramanathan (1977) described above.

In a recent application of a model similar to that of Manabe and Wetherald (1967), Hunt and Wells (1979) have coupled a 300 m oceanic mixed layer to a radiative-convective model in which the lapse rate was not permitted to exceed the moist adiabatic, and with the surface sensible and latent heat fluxes determined by the conventional bulk aerodynamic formulae. They found an average equilibrium surface warming of 1.8°C, while noting that the thermal inertia of the coupled ocean introduced a lag of about 8 years in the response of the surface temperature over the ocean. Were deeper portions of the ocean effectively coupled to the surface, then a longer lag in the surface temperature's approach to equilibrium in response to increased CO_2 could be expected, as has been recently emphasized by an ad hoc study group of the Climate Research Board (National Academy of Sciences, 1979).

In a study combining a radiative-convective model with a surface heat balance model (see below), Ramanathan et al. (1979) have found that the seasonal variations of the CO_2-induced warming have no appreciable effect on the annual average surface temperature rise, although the greatest warming occurs in high summer latitudes due to the feedback effect of melting ice. This study also confirmed the earlier result of Cess (1976) that the changes of albedo and infrared opacity caused by changes in cloudiness effectively compensate each other as far as the surface heat balance is concerned.

Building upon the earlier summary of climate modeling by Schneider and Dickinson (1974), Ramanathan and Coakley (1978) have presented a thorough review of radiative-convective models, including their application to experiments with increased CO_2 as discussed above. They conclude that it is the positive feedback between temperature and water vapor which is primarily responsible for the approximate doubling of the change in surface temperature found in radiative-convective models compared with that which would otherwise occur. In terms of the feedback parameter introduced earlier, this corresponds to $\lambda = 0.5$ deg $W^{-1}m^2$, or twice the value of a purely radiative model's response. With a net change of surface heat flux of 4 W m^{-2} for doubled CO_2, this represents an average surface warming of about 2°C.

2.2 Energy balance models

Instead of calculating the vertical distribution of temperature in a representative atmospheric column as in the radiative-convective models, a widely-used modeling approach is to consider the variation of the surface heat balance itself. In such models the fluxes of sensible and latent heat at the surface are parameterized in terms of the surface temperature, with the surface fluxes of solar and infrared radiation parameterized in terms of the surface temperature and an assumed atmospheric structure. By requiring a balance of surface heating (or by specifying an effective surface heat capacity and calculating the time-dependent temperature response), such heat or energy balance models allow the determination of the surface temperature. We may, in fact, regard such models as a sub-class of the radiative-convective models discussed above, inasmuch as they prescribe the vertical temperature distribution.

The first energy balance model to be applied to the estimation of the surface temperature changes resulting from increased atmospheric CO_2 (aside from the study of Rasool and Schneider (1971) cited earlier) was that of Sellers (1973). Using an improved version of his earlier two-dimensional energy balance model (Sellers, 1969), including the feedback between ice albedo and surface temperature and with fixed cloudiness and relative humidity, he found a global average warming of only 0.1°C for doubled CO_2. Sellers (1974) subsequently showed that this low warming was due to poor approximations in the determination of the infrared emission to space, and upon their improvement found an average surface warming of 1.3°C. This is in fair agreement with the value of 1.7°C found by Temkin and Snell (1976) using a two-dimensional zonally-averaged model with parameterized diffuse thin cloudiness.

Aside from the hybrid model of Ramanathan et al. (1979) discussed above, the most recent applications of energy balance models to the determination of CO_2-induced warming have involved the use of zonally-average versions of atmospheric circulation models along with the requirement of a surface heat balance. With a zonally-averaged two-level quasi-geostrophic atmospheric model coupled to a surface whose heat balance excluded ice-albedo temperature feedback, Ohring and Adler (1978) found an average surface warming of 0.5°C in response to doubled CO_2; upon the inclusion of this feedback, the average surface warming was increased to 0.8°C as a result of greatly increased warming in the higher latitudes. While these results appear low in comparison to those found with other models, they do illustrate the amplification of surface temperature changes by the positive feedback between the extent of ice (or snow) and the surface air temperature. By including this effect in a zonally-averaged nine-level primitive equation model, along with a shallow surface oceanic mixed layer, Potter (1978) found an average surface warming of 2°C for doubled CO_2. Even this result, however, is somewhat less than the average 3.2°C warming found under similar conditions by Ramanathan et al. (1979). Since these latter authors demonstrate that their result is due to ice albedo feedback in higher latitudes, it is probably the more reliable. In terms of the feedback parameter discussed earlier, the warming of Ramanathan et al. (1979) corresponds to a further increase of λ by about 50%, i.e., to about $\lambda = 0.7$ deg $W^{-1}m^2$, which is consistent with the results to be discussed below.

3. RESULTS FROM ATMOSPHERIC GENERAL CIRCULATION MODELS

The most complex models which have been applied to the study of the climatic changes resulting from increased atmospheric CO_2 are the general circulation models (or GCMs). These are models in which the three-dimensional behavior of the global atmosphere is determined by the time-dependent solution of the dynamical equations governing large-scale atmospheric motion, and in which the important radiation- and water vapor-temperature feedbacks are routinely included. Although the processes of convection, friction, cloud formation and other sub-grid scale phenomena are parameterized in terms of the resolved synoptic-scale features in a GCM, the results from such models are generally regarded as more accurate than those from the more simplified models (in which one or more dimension is suppressed and the atmosphere's large-scale motion may be neglected altogether). In GCMs a heat balance condition is usually imposed at the earth's surface when it consists of land, ice or snow, while a variety of conditions may be imposed over an ocean surface. As the results to be discussed below will show, it is this treatment of the ocean's surface temperature which determines a major portion of the atmospheric GCM's response to increased CO_2.

3.1. GCMs with idealized geography and a swamp ocean without seasons

The first and most widely known use of an atmospheric GCM for doubled CO_2 was made by Manabe and Wetherald (1975). Using a nine-level version of the GFDL model with an idealized sectorial distribution of land and ocean approximating that in the Northern Hemisphere, Manabe and Wetherald carried out an integration for doubled CO_2 with fixed clouds, while treating the ocean as a swamp of zero heat capacity. With a swamp-like ocean the sea-surface temperature is determined from the requirement of instantaneous thermal equilibrium with the overlying atmosphere, in the same way that the surface temperature is found in the heat balance models discussed above. Manabe and Wetherald (1975) found an average surface warming of 2.9°C under conditions of annual average insolation (as required when using a swamp ocean), with an increase from values near 2°C in the lower latitudes to about 6°C toward the pole. Since the swamp ocean serves as a ready source of water vapor and may form sea ice as a result of its changes of temperature, it seems likely that the simulated feedbacks of both water vapor and ice albedo are serving to amplify the surface temperature's response to doubled CO_2 in this model. Consistent with this warming, the results of Manabe and Wetherald (1975, 1979) also showed a lowered level of eddy kinetic energy in mid-latitudes, and a lowered mid-latitude meridional heat transport. In the tropics, however, the strength and width of the Hadley circulation was slightly increased with the doubling of CO_2, along with an increase in the intensity of the hydrological cycle by about 7%.

These general results were confirmed in a subsequent study by Manabe and Wetherald (1980) with a similar model, but in which they permitted a simplified form of cloud interaction. The change of surface temperature found with both doubled and quadrupled CO_2 with this model version (in which the calculations were extended to the pole) is shown in Fig. 2. These results also serve to display the approximate proportionality of the temperature change to the logarithm of the CO_2 concentration as suggested by radiative theory. In the experiment with doubled CO_2 Manabe and Wetherald (1980) found an average surface warming of 3.0°C, in close agreement with their previous result.

3.2 GCMs with realistic geography and a climatological ocean with seasons

If the swamp ocean used by Manabe and Wetherald (1975, 1980) is regarded as a limiting case in which there is no heat storage in the ocean (and consequently one in which the sea-surface temperature reacts instantaneously to the surface heat flux), it is of interest to examine the contrasting case of an ocean whose surface temperature is given by monthly climatology. While such an ocean's behavior is not therefore influenced by the model atmosphere, it

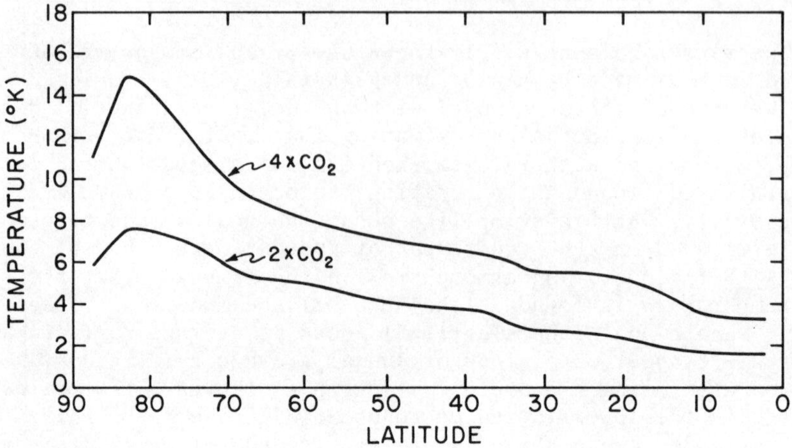

Fig. 2. The change in annual average surface temperature as a function of latitude for both doubled and quadrupled CO_2, as found by Manabe and Wetherald (1980) with an atmospheric GMC and swamp ocean of idealized shape.

freely exchanges heat with the atmosphere without displaying an anomalous change of surface temperature; in this sense it behaves as though it had an infinite heat capacity.

The first uses of an atmospheric GCM with such a climatological ocean in experiments with doubled CO_2 have recently been reported by Mitchell (1979) and Gates and Cook (1979). Using the five-level global GCM of the Meteorological Office (Mitchell) or the two-level OSU GCM (Gates and Cook) with realistic geography and seasonally-varying insolation, each of these studies gave a rise of annual average surface temperature of about 0.2°C with doubled CO_2 in comparison to control runs of the respective models with normal CO_2 levels. Since these GCM experiments are the first to be made for doubled CO_2 with a realistic global distribution of land and ocean together with seasonally-varying insolation, it is of interest to examine their results in some detail. For this purpose we will use the results of Gates and Cook (1979) as illustrative of the analysis which might be undertaken with models containing a more comprehensive treatment of the oceans.

The behavior of the mean tropospheric temperature during the first month of the integrations reported by Gates and Cook (1979) is shown in Fig. 3 for the control (or standard) level of CO_2 (320 ppm) as well as for doubled and quadrupled CO_2. Here it is evident that there is a progressive warming in the cases of increased CO_2 relative to a standard parallel case (with all integra-

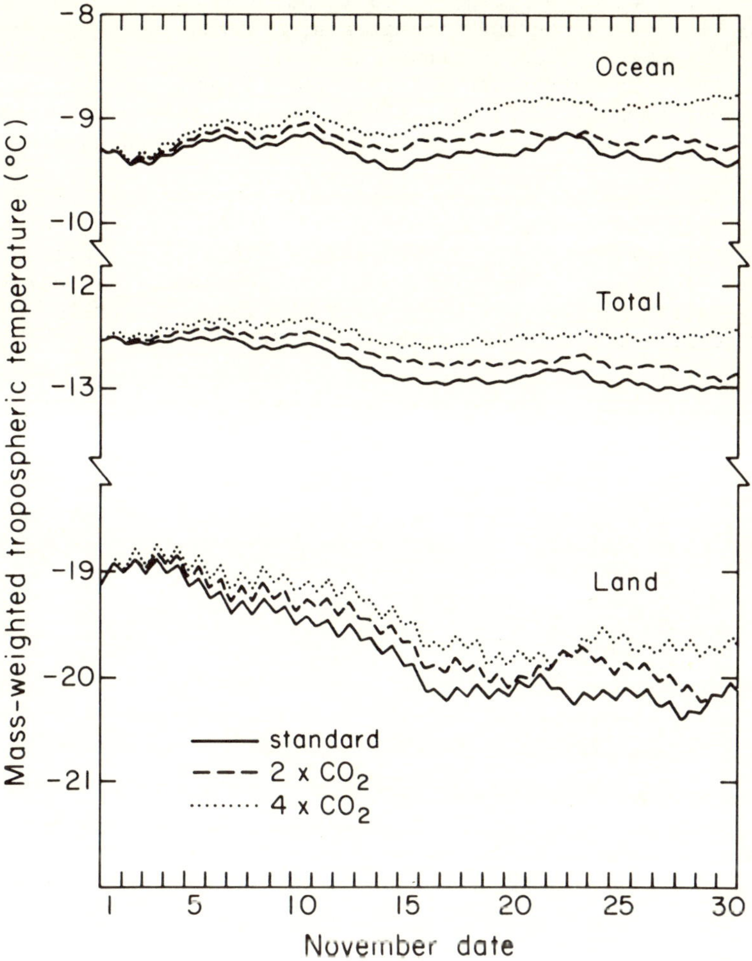

Fig. 3. The variation of the globally-averaged mean tropospheric temperature during the first 30 days (November) of the experiments with doubled and quadrupled CO_2 reported by Gates and Cook (1979). Here the temperature is plotted from the solutions every six hours, and the total refers to the global response over ocean and land surfaces combined.

tions starting from identical initial and boundary conditions except for CO_2 concentration), with the case of quadrupled CO_2 generally warmer than that of doubled CO_2. We may also note that in their earlier portion each integration shows similar synoptic fluctuations, but that after about three weeks' time there is less coherence in their evolution, especially over land areas. After

November (not shown here) there is no coherence in the synoptic evolution of the three integrations, and the cases of increased CO_2 are sometimes colder than the control, especially over land surfaces. These results are therefore a demonstration of the limited synoptic predictability of GCMs, and illustrate the essentially statistical nature of the CO_2-induced warming.

The integrations were continued for eight additional months, with both the insolation and sea-surface temperature following their prescribed normal seasonal variations, i.e., without seasonal anomalies. The behavior of the surface air temperature over this period is shown in Fig. 4 for the cases of doubled and quadrupled CO_2, along with the mean monthly variation of surface air

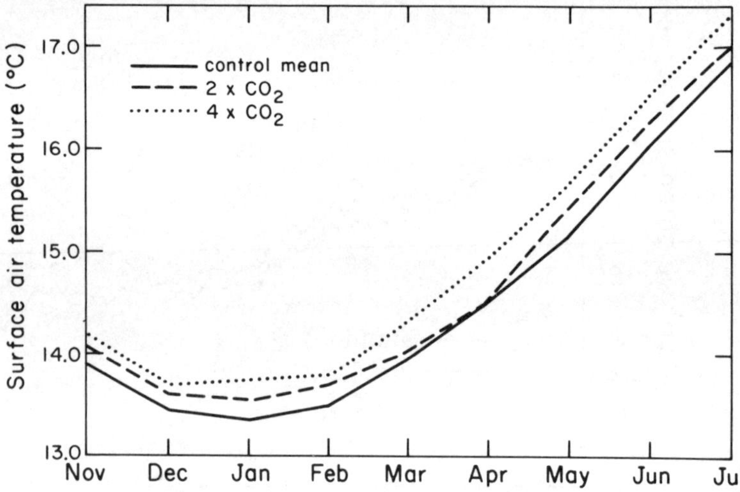

Fig. 4. The variation of the monthly mean of the globally-averaged surface air temperature during the experiments with doubled and quadrupled CO_2 reported by Gates and Cook (1979), along with the average monthly variation found in the control run.

temperature found in an extension of the control integration over three years. Here the average surface air temperature is seen to be systematically higher with doubled and quadrupled CO_2 than in the control, with the CO_2-induced warming reaching a statistically steady value within the first month of the integration.

Selecting January and July as representative of the winter and summer seasons, the latitudinal variation of the change in surface air temperature with doubled and quadrupled CO_2 is shown in Fig. 5. With respect to the average monthly temperature and its rms variability in the 3-year control integration, the

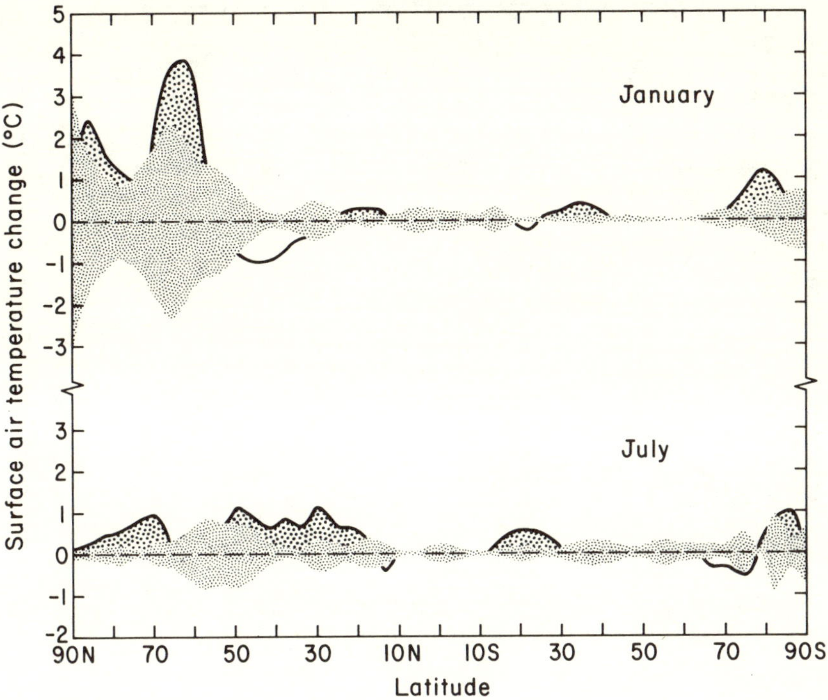

Fig. 5. The latitudinal variation of the change of surface air temperature during January and July in the GCM experiment of Gates and Cook (1979) for doubled CO_2. The lightly-shaded area represents the variability of the monthly average temperature simulated in the appropriate month of the 3-year control, with the control rms plotted symmetrically both above and below the zero line. Only that portion of the CO_2-induced temperature change which exceeds in magnitude the control rms is shown (by the full line), with the cases of warming indicated by stippling.

doubling of CO_2 does not appear to have significantly warmed the surface in January, although in July there is warming of about 1°C in portions of the summer (northern) hemisphere. The corresponding results for quadrupled CO_2 shown in Fig. 6 confirm this systematic summer warming, especially in the Northern Hemisphere. The large changes of surface temperature simulated in the higher latitudes in both Figs. 5 and 6, on the other hand, are apparently not significantly greater than the relatively high levels of natural variability found in these regions in winter as a result of high levels of synoptic activity associated with baroclinic instability.

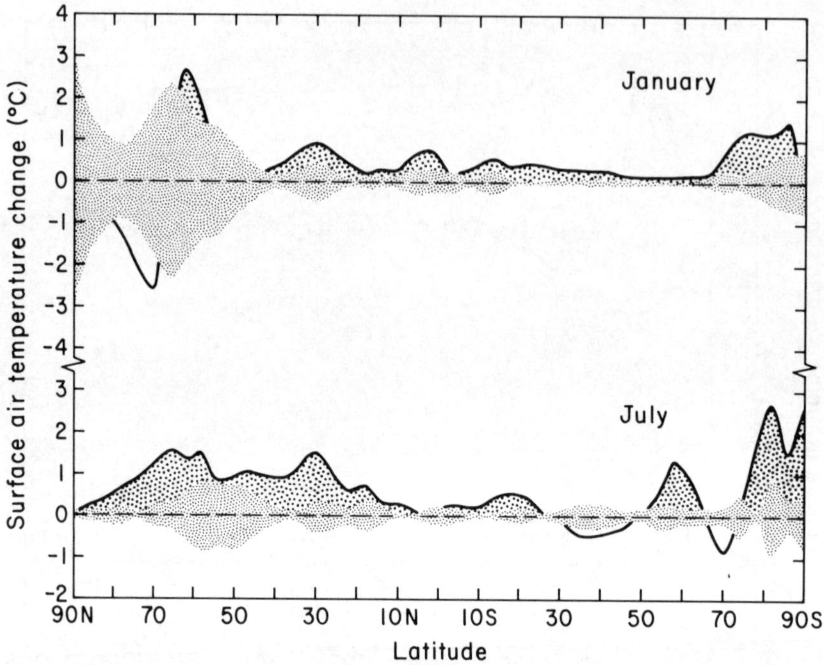

Fig. 6. Same as Fig. 5 except for quadrupled CO_2.

With these zonally-averaged results as a guide, it is of particular interest to examine the corresponding global geographical distribution of the changes of surface air temperature in January and July as shown in Fig. 7 for doubled CO_2. In Fig. 7a for January we note extensive areas in the higher northern latitudes over which the surface air temperature in the raised CO_2 case is either higher or lower than the control January average, with the corresponding results for July shown in Fig. 7b. As might have been anticipated, the largest temperature changes are over the land, with local anomalies as large as 10°C. In view of the high levels of natural (and essentially unpredictable) variability to be expected in association with wintertime synoptic activity in middle and high latitudes, the temperature changes in the higher northern latitudes in January and in the higher southern latitudes in July in Fig 7 are not believed to be statistically significant. During the summer, however, and especially in July over the Northern Hemisphere (Fig. 7b), there are predominantly increases of surface air temperature greater than 2.5°C over large portions of the interiors of North America and Asia.

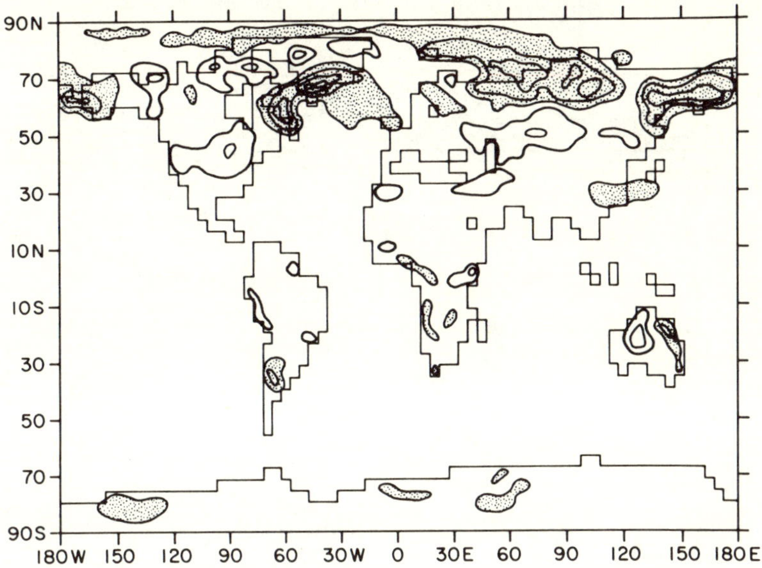

Fig. 7a. The geographical distribution of the average change (with respect to the control mean) of surface air temperature simulated during January in the GCM experiment of Gates and Cook (1979) with doubled CO_2. The isolines are drawn at intervals of 2.5°C, with regions of warming greater than 2.5°C shaded.

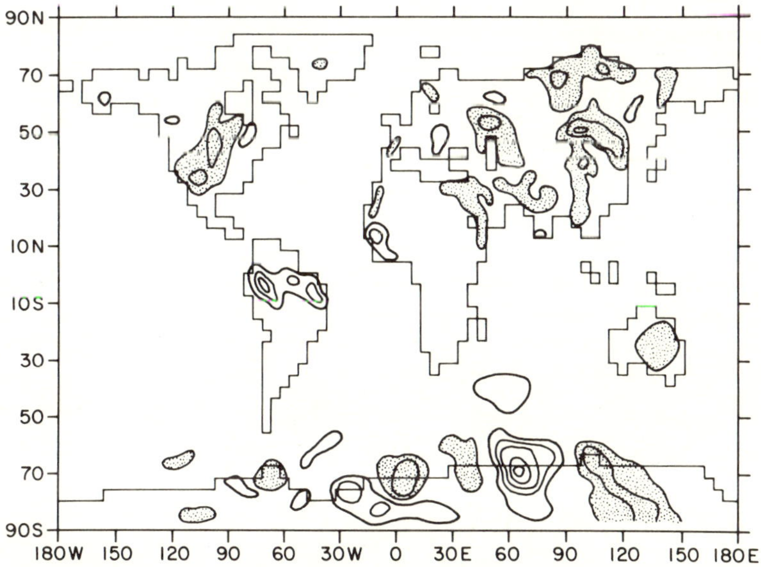

Fig. 7b. Same as Fig. 7a except for July.

These same general features are seen in the seasonal changes of temperature with quadrupled CO_2 shown in Fig. 8. Here the

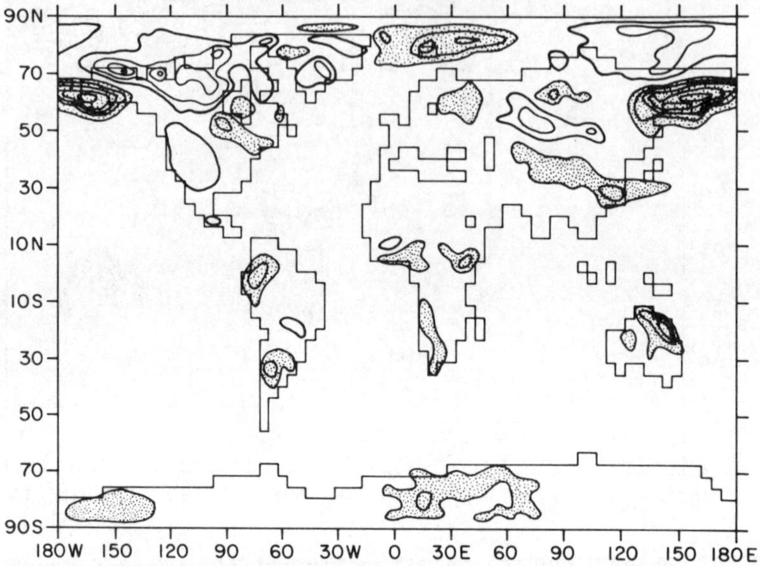

Fig. 8a. Same as Fig. 7a except for quadrupled CO_2.

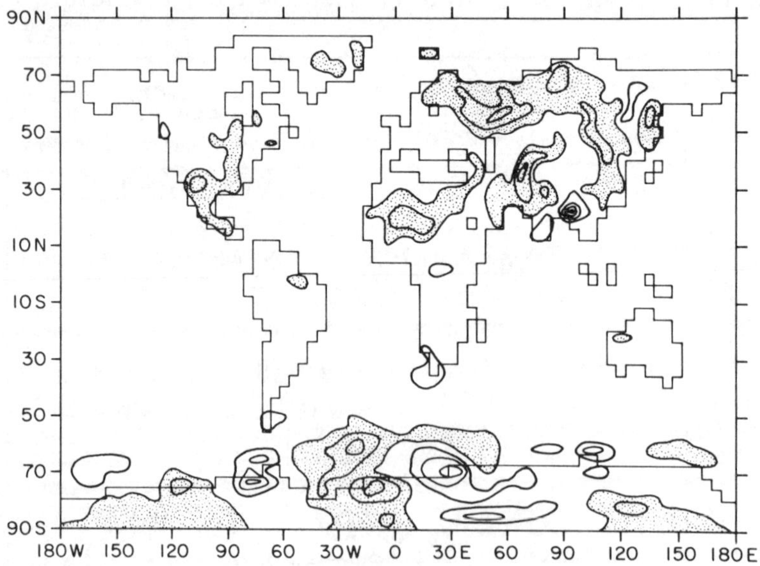

Fig. 8b. Same as Fig. 7b except for quadrupled CO_2 and for July.

coherence of the warming in the Northern Hemisphere in Figs. 7b and 8b (and the relative incoherence of the corresponding changes in Figs. 7a and 8a) suggests that summer warming over the continents is a systematic effect of the increase in CO_2.

Although these GCM experiments have undoubtedly underestimated the final or equilibrium warming which would occur with increased CO_2 because of their virtual exclusion of the significant amplification of surface temperature changes by the feedback with ice albedo (and water vapor via evaporation from a warmer ocean), they do reveal the seasonal and geographical dependence of the background radiative response. As shown in Table 1, both the CO_2-induced changes of net surface heating and surface temperature are markedly different over land and ocean portions of the globe. If

Table 1. Area-averaged Anomalies over Ocean and Land Areas with Doubled and Quadrupled CO_2 as Simulated in the GCM Experiments of Gates and Cook (1979).

Quantity	Month	Doubled CO_2 Anomaly* Ocean	Land	Total	Quadrupled CO_2 Anomaly* Ocean	Land	Total
Net surface heating ($W\ m^{-2}$)	January	2.7	0.3	1.8	9.4	0.7	6.5
	July	5.5	0.9	3.8	9.0	0.9	6.1
Surface air temperature (°C)	January	0.10	0.34	0.18	0.19	0.75	0.38
	July	0.03	0.48	0.18	0.14	1.14	0.49

*Anomaly = CO_2 experiment average - control mean

these data are further partitioned between the hemispheres, a maximum average warming of 0.9°C for doubled CO_2 occurs over the land areas of the Northern Hemisphere in July. These results are generally consistent with those of the radiative-convective models discussed earlier, and correspond to a sensitivity $\lambda = 0.7$ deg $W^{-1}m^2$ in the annual average over land areas for doubled CO_2. (Over oceans a very small value of λ is found as a result of the small changes in surface air temperature permitted by the assigned sea-surface temperature.) In comparison with the results of Manabe and Wetherald (1980) for a swamp ocean (and the results with a mixed-layer ocean described below), the experiments of Mitchell and of Gates and Cook at least serve to show the critical role of the ocean in the atmosphere's response to increased CO_2.

3.3. GCMs with realistic geography and a mixed-layer ocean with seasons

Although experiments for increased CO_2 with fully-coupled atmospheric and oceanic GCMs have not yet been made, a significant advance toward this goal was recently reported by Manabe and Stouffer (1979). Using a spectral version of the GFDL GCM with fixed cloudiness, seasonal insolation, and an underlying static mixed-layer ocean of 68 m depth, they found that when surface thermal equilibrium was reached after approximately 10 years of model simulation, the average surface warming with quadrupled CO_2 was about 4.1°C. The inferred rise of surface air temperature of about 2°C for doubled CO_2 is significantly less than the 3°C warming found by Manabe and Wetherald (1975, 1980). This suggests that the combined effects of continental geography and a seasonally interacting ocean serve to place the GCM's response to doubled CO_2 close to that found with the best of the simplified climate models.

Although Manabe and Stouffer (1979) do not show the global geographical distribution of the CO_2-induced changes of surface air temperature, their zonally-averaged results are shown as a function of season in Fig. 9. Here the large surface warming which

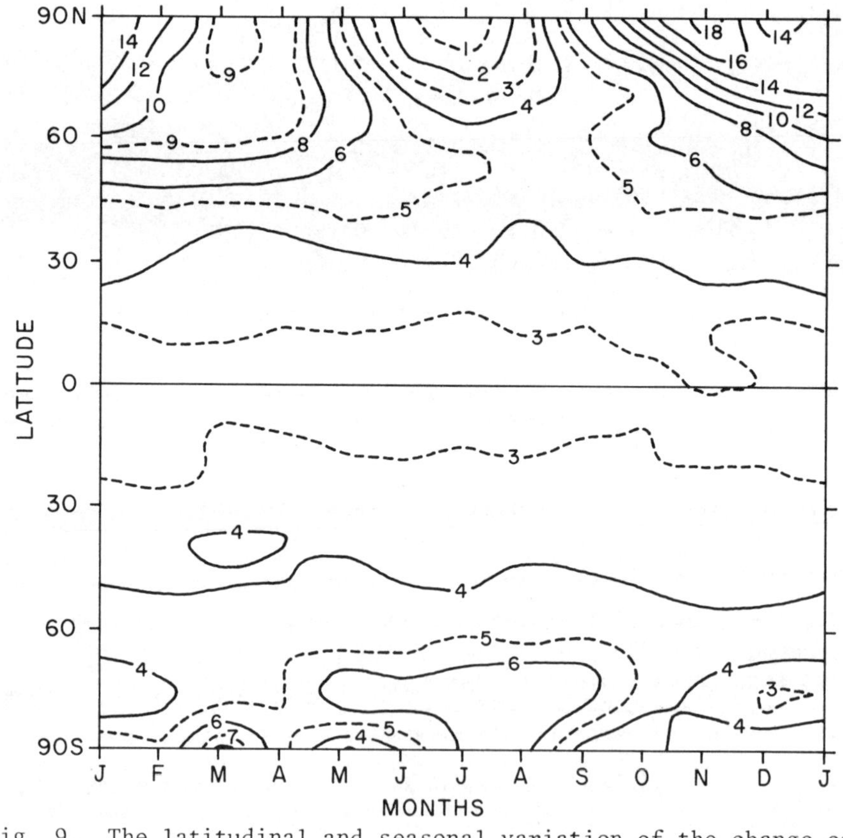

Fig. 9. The latitudinal and seasonal variation of the change of surface air temperature in the experiment of Manabe and Stouffer (1979) for quadrupled CO_2.

occurs in higher and especially northern latitudes during winter is due to the positive feedback between the surface temperature and the albedo changes as a result of the retreat of surface snow- and ice-cover. Pending further calibration of the spectral GCM itself, these results should therefore be representative of those to be found over decadal time scales with a coupled ocean.

4. OUTSTANDING PROBLEMS AND FUTURE RESEARCH

While a variety of models have contributed much to our understanding of the basic physical responses of the atmosphere to rising levels of CO_2, there remain a number of outstanding questions requiring further research. Perhaps the most important of these is the problem of satisfactorily coupling the ocean to the atmosphere over the time scales of decades to centuries which are of most concern in connection with the climatic effects of increased CO_2. As emphasized in recent reviews of the CO_2-climate question (Bach, 1978; WMO, 1979; National Academy of Sciences, 1979), this will require improved knowledge of the effective rates of vertical mixing in the ocean and of the mechanics of air-sea interaction over a wide range of time scales.

It is important to note that a greatly expanded program of numerical experimentation with a variety of models will be required in order to satisfactorily document the characteristic seasonal and regional changes of surface air temperature and associated climatic elements which are the result of increased atmospheric CO_2. Much more research is also needed with climate models which address the time-dependent nature of the atmospheric CO_2 concentration as a function of anthropogenic and biospheric interactions.

REFERENCES

Augustsson, T., and V. Ramanathan, 1977: A radiative-convective model study of the CO_2 climate problem. *J. Atmos. Sci.*, *34*, 448-451.

Bach, W., 1978: The potential consequences of increasing CO_2 levels in the atmosphere. In *Carbon Dioxide, Climate and Society*. IIASA/WMO Workshop, Laxenburg, Austria, 21-24 February, 1978, J. Williams, Ed. Pergamon Press, Oxford, pp. 141-167.

Cess, R. D., 1976: Climate change: an appraisal of atmospheric feedback mechanisms employing zonal climatology. *J. Atmos. Sci.*, *33*, 1831-1843.

Gates, W. L., and K. H. Cook, 1979: Preliminary analysis of experiments on the climatic effects of increased CO_2 with the OSU atmospheric general circulation model. Climatic Research Institute, Oregon State University, Corvallis (unpublished manuscript).

Hummel, J. R., and R. A. Reck, 1979: Carbon dioxide and climate: The effects of cloud specification in radiative-convective models. Research Publication GMR-3151, General Motors Research Laboratories, Warren, MI, 6 pp.

Hunt, B. G., and N. C. Wells, 1979: An assessment of the possible future climatic impact of carbon dioxide increases based on a coupled one-dimensional atmospheric-oceanic model. *J. Geophys. Res.*, *84*, 787-791.

Manabe, S., and F. Möller, 1961: On the radiative equilibrium and heat balance of the atmosphere. *Mon. Wea. Rev.*, *89*, 503-532.

Manabe, S., and R. F. Strickler, 1964: Thermal equilibrium of the atmosphere with a convective adjustment. *J. Atmos. Sci.*, *21*, 361-385.

Manabe, S., and R. T. Wetherald, 1967: Thermal equilibrium of the atmosphere with a given distribution of relative humidity. *J. Atmos. Sci.*, *24*, 241-259.

Manabe, S., 1971: Estimates of future changes of climate due to increase of carbon dioxide concentration in the air. In *Man's Impact on Climate*, W. H. Matthews, W. W. Kellogg and G. D. Robinson, Eds. MIT Press, Cambridge, MA, p. 256.

Manabe, S., and R. T. Wetherald, 1975: The effects of doubling the CO_2 concentration on the climate of a general circulation model. *J. Atmos. Sci.*, *32*, 3-15.

Manabe, S., and R. J. Stouffer, 1979: A CO_2-climate sensitivity study with a mathematical model of the global climate. *Nature*, *28*, 491-493.

Manabe, S., and R. T. Wetherald, 1980: On the distribution of climate change resulting from an increase in CO_2 content of the atmosphere. *J. Atmos. Sci.*, *37*, 99-118.

Mitchell, J. F. B., 1979: Preliminary report on the numerical study of the effect on climate of increasing atmospheric carbon dioxide. Met. 0. 20. Technical Note No. II/137, Meteorological Office, Bracknell, 12 pp.

National Academy of Sciences, 1979: *Carbon Dioxide and Climate: A Scientific Assessment*. Report of an Ad Hoc Study Group on Carbon Dioxide and Climate. Climate Research Board, National Academy of Sciences, Washington, D.C., 22 pp.

Newell, R. E., and T. G. Dopplick, 1979: Questions concerning the possible influence of anthropogenic CO_2 on atmospheric temperature. *J. Appl. Meteor.*, *18*, 822-825.

Ohring, G., and S. Adler, 1978: Some experiments with a zonally-averaged climate model. *J. Atmos. Sci.*, *35*, 186-205.

Potter, G. R., 1978: Preliminary zonal model analysis of the climate change resulting from increased atmospheric carbon dioxide. Preprint UCRL-81575, Lawrence Livermore Laboratory, University of California, Livermore, CA, 30 pp.

Ramanathan, V., 1975: A study of the sensitivity of radiative-convective models. In *Proc. Second Conf. Atmos. Radiation*, American Meteor. Soc., Boston, MA, pp. 124-125.

Ramanathan, V., 1976: Radiative transfer within the earth's troposphere and stratosphere: A simplified radiative-convective model. *J. Atmos. Sci.*, *33*, 1330-1346.

Ramanathan, V., and J. A. Coakley, Jr., 1978: Climate modeling through radiative-convective models. *Rev. Geophys. Space Phys.*, *16*, 465-489.

Ramanathan, V., M. S. Lian and R. D. Cess, 1979: Increased CO_2: Zonal and seasonal estimates of the effect on the radiation energy balance and surface temperature. *J. Geophys. Res.*, *84*, 4949-4958.

Rasool, S. I., and S. H. Schneider, 1971: Atmospheric carbon dioxide and aerosols: Effects of large increases on global climate. *Science*, *173*, 138-141.

Rowntree, P. R., and J. Walker, 1978: The effects of doubling the CO_2 concentration on radiative-convective equilibrium. In *Carbon Dioxide, Climate and Society*. IIASA/WMO Workshop, Laxenburg, Austria, 21-24 February, 1978, J. Williams, Ed. Pergamon Press, Oxford, pp. 181-191.

Schneider, S. H., and R. E. Dickinson, 1974: Climate modeling. *Rev. Geophys. Space Phys.*, *12*, 447-493.

Schneider, S. H., 1975: On the carbon dioxide-climate confusion. *J. Atmos. Sci.*, *32*, 2060-2066.

Sellers, W. D., 1969: A global climatic model based on the energy balance of the earth-atmosphere system. *J. Appl. Meteor.*, *8*, 392-400.

Sellers, W. D., 1973: A new global climate model. *J. Appl. Meteor.*, *12*, 241-254.

Sellers, W. D., 1974: A reassessment of the effect of CO_2 variations on a simple global climate model. *J. Appl. Meteor.*, *13*, 831-833.

Sellers, W. D., 1976: A two-dimensional global climatic model. *Mon. Wea. Rev.*, *104*, 233-248.

Temkin, R. L., and F. M. Snell, 1976: An annual zonally-averaged hemispherical climatic model with diffuse cloudiness feedback. *J. Atmos. Sci.*, *33*, 1671-1685.

Wang, W. C., Y. L. Yung, A. A. Lacis, T. Mo and J. E. Hansen, 1976: Greenhouse effects due to man-made perturbations of trace gases. *Science*, *194*, 685-690.

Wetherald, R. T., and S. Manabe, 1979: Sensitivity studies of climate involving changes in CO_2 concentration. In *Man's Impact on Climate* (Proceedings of an International Conf., Berlin, 14-16 June, 1978), W. Bach, J. Pankrath and W. Kellogg, Eds. (Developments in Atmospheric Science No. 10), Elsevier, Amsterdam, pp. 57-64.

WMO, 1979: Climatic effects of increased carbon dioxide. Report of an ad hoc Study Group to the CAS Working Group on Atmospheric Carbon Dioxide, 30-31 October, 1979, Boulder, CO. World Meteorological Organization, Geneva, 23 pp. (unpublished manuscript).

CONSTRAINTS ON FOSSIL FUEL USE

Ralph M. Rotty and Gregg Marland

Institute for Energy Analysis, Oak Ridge Associated Universities, Oak Ridge, Tennessee 37830, U.S.A.

ABSTRACT. The exponential growth of fossil fuel use over recent decades has resulted in a 4.3 percent annual increase in the carbon dioxide emitted to the atmosphere. The question addressed here is: When (and to what extent) will constraints limit the use of fossil fuels and the subsequent production of CO_2? We discuss three types of possible constraints: resource constraints, fuel demand constraints, and environmental constraints. An analysis of the next 50 years suggests that resource constraints will not provide severe limits. The impacts of atmospheric carbon dioxide will not reach levels that cause mankind to take action soon. Fuel-demand constraints will probably limit the use of fossil fuels to levels that keep the atmospheric carbon dioxide concentration below 450 ppm(v) for the next 50 years. In spite of this conclusion we see a continuing, long term problem and urge that full efforts be made to understand and continually monitor the CO_2 problem and to be alert to any changes that may require action.

INTRODUCTION

For more than a century the world has experienced extraordinary growth — growth in population, growth in economic output, growth in health care and general welfare, and growth in the use of energy. With the exception of times of world conflicts and world depressions, the commercial energy supply fueling this development has grown at a rate of about 5.3 percent per year since 1860 (1). Nearly all of the commercial energy used up to the present has been derived from fossil fuels.

W. Bach, J. Pankrath, and J. Williams (eds.), Interactions of Energy and Climate, 191–212.

A natural question arises: Will the use of fossil fuels continue to increase unabated, or are there constraints on mankind's continued dependence on fossil fuels? Although they are not independent or truly separable there are three possible constraints on fossil fuel use that must be considered: 1) resource constraints – there is a finite amount of fossil fuel within the Earth's crust (even though this limit is not well known now); 2) fuel demand constraints – there is a broad spectrum of economic and social factors that limits the rate of fossil fuel consumption; and 3) environmental constraints – the impact on the environment may be so great as to preclude the production of otherwise desired fuels. Of course all of these issues would become irrevelant if we had a cheaper, more appealing option to fossil fuel burning.

In this paper we explore these three potential constraints on fossil fuel use.

1. RESOURCE CONSTRAINTS

As indicated above, global demand for commercial energy (mostly from fossil fuels) has been growing steadily – at about 5.3 percent per year – since 1860 (except for hiatuses during World Wars I and II and the Great Depression). During this interval the fuel mix has shifted from largely coal initially, to an emphasis on petroleum products, with natural gas contributing a rapidly increasing fraction of the total in the years following World War II. Because of the shift toward fossil fuels with a higher hydrogen to carbon ratio, the energy produced per unit of carbon burned has increased. Thus, the annual rate of energy increase is greater than the 4.3 percent per year (2) increase in the CO_2 emissions from fossil fuels.

When resources are plentiful and easy to recover, steady growth in their use is not unexpected. The issue of interest here is the point at which limits to supply begin to be felt. Many people suspect we have now reached such a point in regard to oil. An examination of recoverable resources of all the world's fossil fuels could indicate when the growth rate will be forced to change.

For ultimately recoverable conventional crude oil, we have relied heavily on the report of the 1977 World Energy Conference (3). (A reading of Parent (4) helps provide perspective on the WEC numbers.) Between September 1976 and April 1977, the WEC conducted a poll of the world's leading petroleum resource experts to benefit from their collective experience and subjective judgments for a consensus estimate. The questionnaire requested an estimate of ultimately recoverable conventional crude oil (including polar and deep offshore regions plus condensate and heavy oils of the type presently under production and classified as movable — and excluding oil already recovered.) The assumption was to be made that recovery would reach 40 percent of oil in place and that the future price would enable the best present and future technologies of exploration, production and recovery to be used. Replies were sought from 42 individual, corporate, and government entities with 28 responding and 33 providing sufficient detail to be employed in a statistical treatment. The responses fell into three discrete categories from which the WEC derived a "pessimistic opinion" (two replies), an "optimistic opinion" (five replies), and a central "majority opinion" (15 replies). The overall mean fell 15 percent higher than the mean of the "majority opinion" and both of these are within the range that those familiar with the issue would consider to be conventional estimates. The mean of all estimates, 291.5 x 10^9 tons of petroleum (standard deviation = 95 x 10^9 — our calculation from their data) is therefore taken as our best estimate of ultimately recoverable conventional oil.

To establish an upper limit for conventional crude oil, we considered taking the maximum of the 22 separate estimates for each of the various global regions and summing these. This would recognize that the experts differed widely on their estimates for the various global subdivisions — estimates varied over a factor of 8 for even relatively well-known areas like North America — but yields an exorbitantly high 832 x 10^9 tons. Instead, we have chosen to use the overall mean plus two standard deviations, 482 x 10^9 tons. Only one of the WEC's 28 respondents suggested the world total might exceed 500 x 10^9 tons.

The 291.5 x 10^9 tons of crude oil is distributed globally as shown in Table 1. Values in Table 1 were taken directly from WEC tables except for the mean value for polar and deep offshore regions which were calculated from their data. Our sum is then 291.5 x 10^9; theirs was 302.4 x 10^9 tons.

For conventional natural gas, the WEC tabulates 2,362 x 10^{18} joules of proved reserves plus 8,147 x 10^{18} joules of remaining undiscovered resources for a total of 10,509 x 10^{18} joules yet to be recovered. (26 x 10^9 cubic meters of gas are required to deliver 10^{18} joules.) Parent (4) presents this plus 12 other estimates of ultimately recoverable natural gas made between 1967 and 1977. The mean of these 13 estimates is 275 x 10^{12} m^3 (10,610 x 10^{18} joules) with a standard deviation of 85 x 10^{12} m^3. One must subtract from this the gas already produced to get a number for remaining recoverable gas. Parent adopted 250-265 x 10^{12} m^3 as his year-end 1977 assessment of remaining recoverable conventional gas. Recognizing that these values are relatively close, we adopt here the WEC number as a best estimate and assume it is suitable for year-end 1977. Our estimate then becomes 273.4 x 10^{12} m^3 with a standard deviation of 85 x 10^{12} m^3. The mean value plus two standard deviations, 443 x 10^{12} m^3 is exceeded by only one of the 13 estimates in Parent, and is adopted as an upper limit value for ultimately recoverable conventional natural gas. The regional distribution of the best estimate value is shown in Table 2.

Table 1. Global Distribution of Crude Oil
(10^9 tons – yet to be recovered)
Already produced 45 x 10^9 tons
(adapted from WEC, 1978)

U.S.S.R., Eastern Europe and China	59.4
U.S.A. and Canada	28.5
Middle East and North Africa	109.1
Africa, South of Sahara	11.3
Western Europe	11.2
Latin America	22.9
Japan, Australia, New Zealand East and South Asia	15.1
Deep Offshore and Polar Areas	34.0
Total	291.5
	(2040 x 10^9 barrels)

Table 2. Global Distribution of Natural Gas
(10^{12} m^3 – yet to be recovered)
Already produced 24.17 x 10^{12} m^3
(from WEC, 1978)

North America	50.74
Western Europe	12.15
Japan, Australia, New Zealand	7.10
U.S.S.R. and Eastern Europe	78.50
China	10.43
OPEC	95.07
Latin America	11.60
Middle East	1.17
North Africa	1.04
Africa, South of Sahara	0.39
East and South Asia	5.21
Total	273.40
	(9,659 x 10^{12} standard cubic feet)

We have also taken our coal resource estimate from the WEC report (3). WEC tabulates tons coal equivalent (tce) where 1 tce = 7 x 10^9 cal = 29.29 x 10^9 joules = 27.78 x 10^6 Btu.

Coal is classified as either hard coal or brown coal, the boundary between them is taken to be at 5,700 kcal/kg = 23.8 x 10^6 joules/kg = 22.62 x 10^6 Btu/metric ton = 20.5 x 10^6 Btu/short ton.

The WEC obtained data from the open literature, from a questionnaire to countries with coal deposits, and from interviews conducted in some of the countries. Results show both "geological resources," "resources that may become of economic value to mankind at some time in the future," and "recoverable reserves," those that are technically and economically recoverable today (WEC, p. 64).

Coal resource estimates have been traditionally less variable than corresponding oil and gas estimates, but the lack of certainty is evident in the observation that the 1977 WEC survey totals more coal than the 1976 WEC summary by 10^{12} tons coal equivalent (10 percent). The 1977 WEC report (3), (p. 64) suggests these numbers still "might represent a lower limit," given the acknowledged lack of exploration in many areas.

Because the coal resources are so large, it is apparent that, insofar as potential impact on atmospheric CO_2 is concerned, the uncertainty in conventional oil and gas resources is a small factor compared to the uncertainty in coal resources and the fraction thereof that can be exploited. In line with our best estimates of gas and oil resources it seems reasonable to accept the WEC estimate of conventional coal resources (with the global distribution shown in Table 3) and to assume further that 50 percent of this, that is, 5,063 x 10^9 tons of coal equivalent is ultimately recoverable. This acknowledges the problem of exploiting deep or thin coal seams, the difficulties with developing seams with small vertical separation, the fundamental inefficiencies of underground mining, etc. On the other hand, it can be easily argued that developments in mining techniques, and in situ extraction strategies, along with economic pressures and new discoveries, could result in eventual exploitation of a considerably larger quantity. As an upper limit for ultimately producible coal, we shall therefore use 90 percent of the currently tabulated in-place resources, i.e., 9,113 x 10^9 tons coal equivalent.

The "unconventional" fossil fuel with the greatest prospect for major development is shale oil. A conventional figure often quoted (ultimately from Culbertson and Pitman (5)), is that the Green River Shales of Colorado, Utah, and Wyoming contain 1,800 x 10^9 barrels of oil (257 x 10^9 tons) at greater than 10 gallons oil per ton[1] of shale (about 3.4 wt. percent). Extended worldwide, the total identified oil resource in shales with over 10 gallons per ton comes to 3,098 x 10^9 barrels (445 x 10^9 tons) of which 678 x 10^9 barrels (97 x 10^9 tons) are at grades better than

25 gallons per ton (8.5 wt. percent) (Culbertson and Pitman (5)). Recognizing that this is "in-place" oil, we include 50 percent of it in our best estimate of recoverable oil, i.e., 225 x 10^9 tons (1550 x 10^9 barrels) of recoverable oil. The WEC appraisal is that known, in-place resources are 400 x 10^9 tons with 5-10 percent exploitable now.

Table 3. Global Distribution of Coal Resources
Geological Resources (in 10^9 tons coal equivalent)
(from WEC, 1978)

U.S.S.R.		4,860
U.S.A. and Canada		2,685
China		1,438
Western Europe		420
Latin America		32
Africa		172
Japan, Australia, New Zealand		272
Eastern Europe		171
Other Asia		76
	Total	10,126

To extend this to a speculation of unappraised or undiscovered shale oil resources, the commonly quoted estimates of world resources are those of Duncan and Swanson (6). Their numbers are based on the statistical distribution of carbon in sedimentary rocks and thus must be regarded as very rough estimates. Categories of oil yield are also rough and merit some explanation. The Duncan and Swanson (6) calculation is based on the mathematical logic; 6.25 percent organic C implies 10 percent organic matter which implies 10 gallons oil yield per ton of shale. The recovery fractions are not out of line for conventional retorts in the western U.S. and the hydrogen retorts suggested for development of eastern U.S. shales (Weil et al. (7)). Duncan and Swanson assume that all shale with 10-25 percent organic matter will yield exactly 10 gallons of oil per ton of shale, and that half of this in-place oil is recoverable. The estimate is clearly speculative and overestimates the potential yield from some deposits while underestimating the yield from others. As an upper estimate for conceivably recoverable shale oil, we shall assume that one-fourth of the oil thus indicated at greater than 10 gallons per ton is exploitable (the remaining three-fourths not exploitable because of thickness, depth, etc.) and that half of the exploitable oil is recoverable (or would be oxidized during recovery). Of the total resource tabulated by Duncan and Swanson (half of the in-place oil) at 342,000 x 10^9 barrels of oil in shales with greater than 10 gallons per ton (17,000 x 10^9 barrels of it at greater than 25 gallons per ton), we suggest that one-fourth (85,500 x 10^9 barrels, or 12,391 x 10^9 tons), might be considered ultimately recoverable.

As evidenced by current projects in Canada and discussions in Venezuela, there are also significant resources of tar sands or very heavy crudes that will be ultimately developed. Major known deposits in Canada, Venezuela, and the Soviet Union dominate the known resource picture. While, as is the case with shale oil, it is impossible to give a truly meaningful resource number, Sande (p. 46-47 in (3)) has provided an order of magnitude total for the world's largest deposits — deposits which are believed to contain the overwhelming bulk of the global resource. Sande's estimate is that these 21 deposits contain about 327 x 10^9 tons of oil, of which 5-10 percent is recoverable by currently available methods and perhaps as much as 30-50 percent might ultimately be recovered. We will use 98 billion tons (30 percent) as a best estimate of the ultimately recoverable resource, and as an upper limit (and to allow for unknown deposits), 261 billion tons (80 percent).

At this point we have not included any unconventional gas in our resource estimate. Although there is some speculation on the amount of methane contained in U.S. coals, tight sands, and geopressured reservoirs, it is not at all clear how much of this might ultimately be recovered or how big the world resource might be. Thus, while unconventional gas is potentially a very important source of CO_2 emissions, we are not able to provide even tentative numbers at this time.

Table 4 gives the amount of carbon in the recoverable fossil fuel reservoir on the basis of our best current estimate and an upper limit speculation. In the case of each of the fossil fuels, the fraction of the total that has already been used is still so small that the physical quantities cannot yet be perceived as presenting a real constraint.

2. FUEL-DEMAND CONSTRAINTS

The second constraint is embedded in the growth of the global social/economic system — in the balance between supply and demand and the choices involved between demand for fuels and other needs. It is, of course, impossible to predict the future. Most analysts who deal in economic or energy futures develop scenarios for which they argue "feasibility." These are "predictions" in a sense, although they are founded on a bed of "plausible" boundary conditions which must be recognized. It is interesting that, in general, the more recent the scenario the lower the energy demand for a given point in the future. To some extent this reflects the very rapid rise in energy prices, and the slower growth

Table 4. Energy and Carbon Contained in Fossil Fuel Reservoirs

	Energy Content of Best Estimates (TW yr.)***	Best Current* Estimates (10^{15} g C)	Upper Limit* Speculations (10^{15} g C)	Historic Consumption up to 1978 (10^{15} g C)
Ultimately recoverable conventional petroleum resources	420	230	380	35
Ultimately recoverable conventional natural gas	331	143	230	15
Ultimately recoverable conventional coal	4,690	3,510	6,315	100
Shale Oil	313	173**	9,530**	
Tar Sands and Heavy Oil	136	75	200	
Total	5,890	4,131	16,655	150

*Not including that produced prior to 1978.

**Not including CO_2 driven off from carbonate rock in retorting.

***Assuming both conventional and unconventional crude oils have a density of 0.9 g/cm^3 and an energy content of 44×10^9 joules per ton.

suggested by economic analyses. Another factor that affects all scenario builders is the "mood of the times" when the scenario is developed. For example, the instability of the Mid East – including the crisis in Iran and the threat of a major conflict erupting over the Soviet excursion into Afganistan – has influenced the attitudes of the public. Although recent events may prove to have minor impact on the energy demand 25, 50, or 75 years from now, they are prominent in the minds of the scenario builders.

Several groups around the world have engaged in the "game" of building energy scenarios. For the United States, the high energy scenarios developed only four or five years ago now seem unreasonably high, and in the international community too, the recent revisions are nearly all downward. It is now generally believed that global energy demand cannot continue to grow at the historic 5.3 percent/year rate and the carbon dioxide produced from fossil fuels cannot grow at the 4.3 percent/year rate of the past. (Projecting at a 4.3 percent/year increase in CO_2 emissions and an airborne fraction of 0.5 produces an atmospheric concentration of 600 ppm(v) about 2030.) In addition to the global energy scenarios developed at the International Institute for Applied Systems Analysis (8), by Perry and Landsberg (9), WAES (10), and others, a group of us at the Institute for Energy Analysis has developed a scenario for energy use and CO_2 production.

As is the case with most scenario building teams, we claim that our scenario **is** plausible. Because the world's severe energy problems are still problems of the future and because long-range planning requires examination of plausible futures, we believe our scenarios serve a useful purpose. The goals for our scenarios are to be internally self-consistent and to have boundary conditions as simple and as transparent as possible. Our scenarios are updated as techniques are refined and our perceptions of the boundary conditions change.

One of the primary factors embedded in nearly all of the energy demand scenarios is world population. Barring a major worldwide military conflict or other social catastrophy, the world population can be projected 50 years or so with much greater certainty than most other factors affecting global energy demand. The assumption of 7.5 to 8 billion people for a global population in AD 2025 is much easier to defend than some of the other assumptions and agrees with projections made by Littman and Keyfitz (11).

Another of our fundamental perceptions is that the economy of the developed world will grow much more slowly between now and AD 2000 than it has over the past 30 years. The reduced capability to assemble capital to increase fuel supply, the push toward conservation, the shift toward industries that are less energy-intensive, and the escalating fuel costs all point toward slowed increase in energy demand. The numerical projections summarized in Table 5 are based on detailed examination of data for many political subdivisions. Information on the U.S.S.R., Eastern Europe, and the People's Republic of China is from Allen (12). Allen estimates the quantities of each fuel consumed annually, but his projections extend only to AD 2000. He has both economic and energy growth rates that evolve during the 1975-2000 period. For the purposes of this paper, we have projected his 1990-2000 average growth rates for an additional 25 years. The data for the OECD countries of Western Europe, North America, Japan, Australia, and New Zealand are from IEA reports for individual countries (13), (14), (15), (16), (17), (18), and a summary of OECD energy requirements now being prepared by Davison (19). The projections for each nation are heavily dependent on population forecasts, with population information enabling the estimation of labor force. Combining the growth in labor force with worker productivity trends yields the projected economic growth. An elasticity of additional energy requirements per unit of gross product has been calculated for the recent past and estimated for the future. It should be emphasized that this procedure gives a changing rate of growth during the period to AD 2000. To extend the projections, we have assumed that the mean 1990-2000 growth rates continue until 2025 (see Table 5).

Similar analyses for the developing nations of the world are just beginning and no substantive results are yet available. Consequently, for this part of the world we have looked at growth rates of the recent past and made rough estimates of growth rates for the future. These are coupled, in part, to the growth of the developed parts of the world and in all cases are considerably lower than those of the recent past. There was a conscious recognition of efforts toward improvement in the per capita energy use in the extremely poor parts of the world. While these increases are not large and are not sufficient to wipe out poverty and suffering in many nations, they are significant and do narrow the gap between the rich and the poor nations. We anticipate that population growth in the developing parts of the world will continue to be much greater than in the developed countries, and the high population growth will contribute to the faster growth in energy requirements.

Per capita energy requirements for 1975 and 2025, consistent with the data of Table 5, are in Table 6. The Mid East, Latin America, and Communist Asia all show 2025 per capita energy levels suggestive of improved living standards for most of their inhabitants. Per capita energy consumption for 2025 in Africa and South Asia suggests the persistence of many pockets of extreme poverty.

TABLE 5. WORLD ENERGY SCENARIO — 2025
(with average growth rates: population, economy, energy)

	1975		2025	50-Year Average Growth Rate		
	Population (millions)	Energy (TW yrs/yr)	Energy (TW yrs/yr)	Population (percent/year)	Economy (percent/year)	Energy (percent/year)
North America	237	2.72	4.74	0.6	1.8	1.05
a. United States	(214)	(2.47)	(3.93)		(1.7)	(0.88)
b. Canada	(23)	(0.25)	(0.81)		(3.3)	(2.68)
Western Europe	305	1.70	2.47	0.2	2.0	0.75
a. F.R. Germany	(62)	(0.33)	(0.50)	(-0.13)	(2.7)	(0.91)
b. United Kingdom	(56)	(0.27)	(0.24)	(-0.07)	(0.15)	(-0.42)
c. France	(53)	(0.22)	(0.38)	(0.3)	(2.9)	(1.07)
d. Italy	(56)	(0.18)	(0.27)	(0.4)	(2.0)	(1.00)
e. Others	(78)	(0.70)				
Eastern Europe & Siberia	359	1.90	6.54	0.7	2.7	2.70
a. U.S.S.R.	(253)	(1.33)	(5.00)	(0.8)	(2.7)	(2.7)
b. Others	(106)	(0.57)	(1.54)	(0.6)	(1.9)	(1.9)
Other Developed Countries	128	0.55	2.02	0.6	3.8	2.60
a. Japan	(112)	(0.43)	(1.60)	(0.6)	(4.0)	(2.63)
b. Australia, New Zealand	(16)	(0.12)	(0.42)	(1.0)	(3.0)	(2.50)
Asia, Planned Economies	1,029	0.63	3.43	1.1	3.5	3.50
a. China	(943)	(0.58)*	(3.14)			
b. Others	(86)	(0.05)	(0.29)			
Latin America	323	0.30	2.22	1.8	3.0	4.00
Mid-East	110	0.11	1.72	2.3	5.0	5.50
Africa	370	0.06	0.94	1.7	4.5	5.50
South Asia	1,170	0.23	2.80	1.5	4.0	5.00
WORLD TOTALS	4,031	8.20	26.88			2.39

*1978 value

Table 6. Rate of Energy Use Per Capita (kW/person)

	1975	2025
North America	11.5	14.6
Western Europe	5.6	7.3
Eastern Europe	4.7	12.8
Other Developed Countries	3.4	11.7
Asia, Planned Economies	0.61	1.9
Latin America	0.93	2.8
Mid East	0.97	5.0
Africa	0.17	1.1
South Asia	0.21	1.1

Given the energy use scenario described above and summarized in Table 5, the conclusion of principal interest here is that about 50 years from now the world energy demand will be three to four times that of the present; not 14 times the present value as would be achieved by continued 5.3 percent per year growth. Fossil fuel use is expected to grow even more slowly than total energy demands.

3. ENVIRONMENTAL CONSTRAINTS

Environmental constraints on fossil fuel use are many and varied. Although petroleum and natural gas are considered relatively benign as compared to coal, there is still widespread concern about oil spills (especially along seacoasts, in scenic areas, and in delicate habitats) and local concern around facilities like refineries and LNG terminals. Concentrations of petroleum-burning vehicles can also create problems (urban smog), as can the encroachment of industry and the influx of large numbers of people in environmentally sensitive areas. Much greater environmental concerns are associated with coal extraction, transportation, and combustion. Most of the severe problems associated with petroleum stem from accidents or defects in the system, while many of the problems with coal are inherent in the system. Nonetheless, there are "technical fixes" (though sometimes expensive) for many of coal's environmental impacts. Reclamation of mined land is effective in many places, particulates emitted to the atmosphere can be effectively controlled at the power plant and to a lesser extent at the mine, and transportation can be made cleaner and safer. Discharge of sulfur oxides and the subsequent formation of acid rain are more serious problems, with international implications, but here too we sense that effective control technologies exist. The essential point is that although both natural and synthetic fuels have a variety of deleterious environmental impacts, and in spite of the fact that they can have important repercussions on both the short time scale and the local geographic scale, these things can be technically restrained within acceptable bounds — although sometimes with sizable costs.

There remains one environmental impact from fossil fuel use for which there is no known reasonable "technological fix." And this impact — the emission of CO_2 during combustion of fossil fuel, and the resulting increase in the concentration of CO_2 in the atmosphere — is on a global scale. The collection of CO_2 from the stacks of large stationary sources is possible (with an energy expenditure of somewhat more than half the amount released in the initial combustion of fuel (20)), but collection from small and mobile sources does not appear feasible, nor does the recovery of CO_2 from the atmosphere or ocean. Thus the climate change that could result from increasing atmospheric CO_2 concentrations could be a real constraint on the use of fossil fuels to meet global energy needs.

Although CO_2 is an essential by-product of all fossil fuel combustion, the problem is closely tied to coal. To put the potential contribution of oil and gas in perspective, consider the resources of fossil fuels. Based on the best estimates of resources, as discussed above, recoverable conventional oil and gas together can yield only 373 x 10^9 tons carbon in the CO_2 produced (see Table 4). If about half of this quantity remains airborne (2), burning all of the remaining oil and gas would result in only a 28 percent increase in the concentration of CO_2 in the atmosphere. The resources of coal, on the other hand, are sufficient to supply the energy requirements (including significant amounts of liquids and gases) in our scenario for 2025, and beyond. Table 4 shows the relative amounts of carbon in the recoverable coal and in oil and gas. In the extreme of development of very dilute and inaccessible sources of reduced carbon, the atmospheric CO_2 concentration could be increased many-fold.

As technology continues to develop, however, it is likely that the fraction of global energy demand that is met with fossil fuels will decrease. Both solar and nuclear technologies will improve and assume larger portions of the global energy supply picture. From the 92 percent level of today the fossil fuel portion could be in the range of 70-80 percent 50 years from now. Note that 30 percent of the 2025 requirement of Table 5 is about 8 TW yr/yr (1 TW = 10^{12} watts), the total global use at present. This is a formidable goal for solar and nuclear technologies!

On the assumption that such a goal can be approached, we have estimated the rate of fossil fuel use in AD 2025 will be about 21 TW yr/yr. Such a fossil fuel rate will release CO_2 to the atmosphere at a rate of 13.6 x 10^9 tons carbon per year (compared to 5.2 x 10^9 tons carbon in 1977), a growth rate in

CO_2 production equivalent to only 2.0 percent/year! Table 7 gives the breakdown for CO_2 emissions in AD 2025 according to this scenario.

Table 7. CO_2 Releases — 2025

	Total Energy Use (TW)	Fossil Fuel Energy Use (TW)	CO_2 Released (10^6 tons)
North America	4.74	3.56	2260
Western Europe	2.47	1.73	1100
U.S.S.R. and Eastern Europe	6.54	5.37	3415
Other Developed Countries	2.02	1.52	964
China and CTP* Asia	3.43	3.30	2205
Latin America	2.22	1.67	1059
Mid East	1.72	1.29	820
Africa	0.94	0.71	450
South Asia	2.80	2.10	1336
World Total	26.88	21.25	13609

*Planned Economies

If one accepts this scenario, the cumulative emission of CO_2 from fossil fuels between 1975 and 2025 is given by

$$N_{1975}^{2025} = 4.83 \times 10^9 \int_0^{50} e^{.02t} \, dt$$

$$= 241.5 \times 10^9 \; [2.718 - 1] = 415 \times 10^9 \text{ tons C},$$

where 4.83×10^9 was the CO_2 emission rate for 1975 and 0.02 is the average annual growth rate from 1975 to 2025. If we assume that 53.5 percent of this remains in the atmosphere (i.e., that net biospheric changes have been negligible and that fossil fuel combustion has been the sole source of anthropogenic CO_2 during the past 20 years (2)), the atmospheric concentration of CO_2 in 2025 will be

$$331 + (0.47)\,(0.535)\,(415) = 435 \text{ ppm(v)},$$

where 331 is the 1975 concentration in ppm(v) and 0.47 converts from billions of tons of carbon to ppm(v) of atmospheric CO_2. If we accept the CO_2 source from tropical forest clearing as suggested by Bolin (21) and assume that this source ceases within the next decade or so and that the other biospheric and oceanic sources and sinks continue to behave as they have for the past two decades, then we conclude that only 43 percent (2) of the cumulative emissions are remaining in the atmosphere and the concentration in 2025 will be

$$331 + (0.47)\ (0.43)\ (415) = 415 \text{ ppm(v)}$$

To illustrate this graphically, Figure 1 is a plot of the annual rate of CO_2 production with data points indicated for the period 1948-1978 and a single point at 2025 plotted in agreement with the scenario presented above. The equation

$$\frac{dN}{dt} = 0.043N \left(1 - \frac{N}{4281}\right)^x ,$$

where N is the amount of carbon (in 10^9 tons) in fossil fuels that have been burned at time t, and 4281 x 10^9 tons is the best estimate of total recoverable carbon (past and future), can be used to provide a family of curves in which different values of x attach a greater or lesser emphasis to the fraction of resources remaining. The shape of the curves in Figure 1 suggests a near exponential growth in fuel use during the early period of small use of resources, and leveling to a maximum rate of production when $N = 4281/(x+1)$. The value of $x = 4$ gives a curve that very nearly passes through our 2025 point of 13.6 x 10^9 tons of carbon per year.

Figure 2 indicates cumulative values for CO_2 released from fossil fuels, starting at the estimated 57.8 x 10^9 tons C released up to 1948. The two central curves in the band representing atmospheric concentration are based on the $x = 4$ curve of cumulated CO_2 production with assumptions that the airborne fraction is 53.5 percent (no net biospheric source) and 43 percent (approximately 20 percent additional source from the biosphere in the 1948-1980 period) respectively. To accommodate, at least partially, the fact that the whole procedure of estimating the future is built on assumptions that may or may not be borne out, we have arbitrarily widened the band representing "projected atmospheric concentration" to include 25 percent greater and 25 percent lesser increases. We call attention to the fact that this gives a maximum concentration in 2025 of 470 ppm(v) and that even our highest estimates do not reach 600 ppm(v) until around 2060.

The global climate change that can result from an atmospheric CO_2 increase is best estimated by atmospheric models. Probably the most frequently quoted model is that of Manabe and Wetherald (22). As pointed out by Flohn (23), other "greenhouse" gases

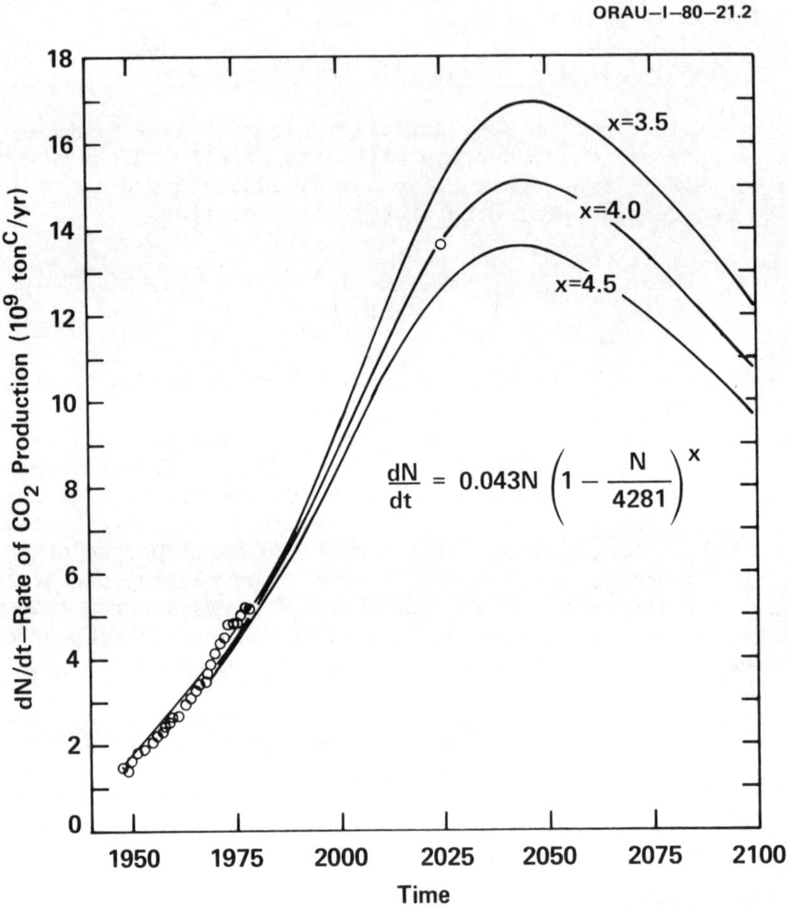

FIGURE 1. Annual rate of CO_2 production from fossil fuels. Circles indicate 1948-1978 actual production and 2025 production for the scenario developed in the text. The equation shown is used to describe future production, with N as the amount of carbon in fossil fuels that have been burned at time t.

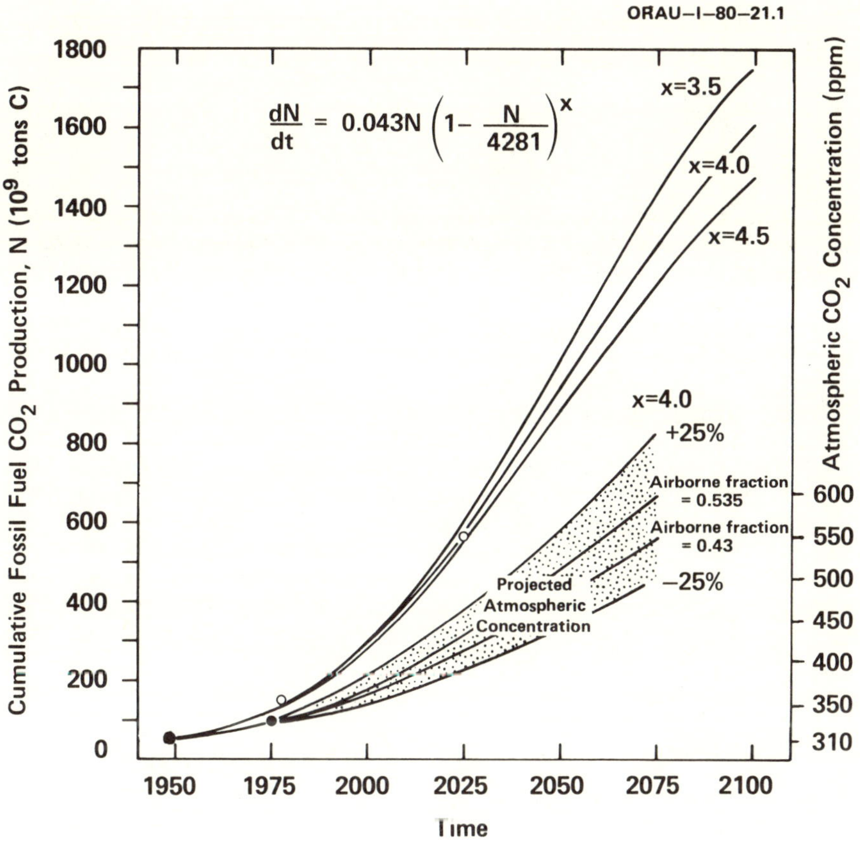

FIGURE 2. Cumulative CO_2 released from fossil fuels and projected atmospheric concentrations of CO_2. The upper curves are cumulative production (58×10^9 tons C in 1948, 150×10^9 tons C in 2025). Values of x = 3.5, 4.0, 4.5 refer to the curves of Figure 1. The lower curves are the atmospheric concentrations of CO_2 resulting when x = 4.0 and airborne fractions are 53.5 percent and 43 percent.

could add to the CO_2 effect and cause larger temperature rises, but we are concerned here only with the fossil fuel impacts and, therefore, have continued to quote Manabe and Wetherald. They indicate an average global warming at the Earth's surface of 2.9°C for a doubling of atmospheric CO_2 concentration. Thus, an AD 2025 concentration of 435 ppm(v) would give a warming (over the 1975 average temperature) of

$$\Delta T \begin{matrix} 2025 \\ \\ 1975 \end{matrix} = 2.9 \left(\frac{\ln \frac{435}{331}}{\ln 2} \right) = 1.14°\text{C}.$$

For a concentration of 415 ppm(v), the mean warming, at equilibrium, would be 0.95°C. If our scenario for 2025 underestimates the cumulative fossil fuel use by as much as 50 percent instead of the 25 percent used in Figure 2, and we use the assumption of 53.5 airborne fraction, by AD 2025 the CO_2 concentration would reach 488 ppm(v) and the mean surface temperature would increase 1.62°C.

Although global warming of about 1°C to 1.5°C over a 50-year time span is enough to cause concern among climatologists, it is probably not enough to cause a revision in policy of fossil fuel use — especially at the beginning of the 50-year period, and when fossil fuels play a major role in the global economy. Until we, in the climatology community, can document with greater confidence what a 1°C average warming within 50 years will cost the global society, the "atmospheric CO_2 problem" probably will <u>not</u> (and probably should not) prove to be a constraint on the use of fossil fuels.

4. CONCLUSION

Of the three possible constraints on the growth of atmospheric CO_2 from fossil fuel use, we believe we have shown that over the next 50 years:

A. There will be no real constraint as a result of a physical shortage of fossil fuel resources. We will likely be forced to use some solid fuel to make more convenient liquids and gases, but the total resources are an order of magnitude greater than what we will need during the next 50 years. The unequal geographic distribution of these resources probably will continue to be a source of international stress.

B. There will be no real constraint, at least during the early portion of the 50-year period, as the result of the "atmospheric CO_2 problem." Until we have a more convincing case regarding climatic changes induced by a CO_2 increase and a much better grasp on what the consequences of a 1°C change in mean surface temperature over a 50-year period will be for the world's social and political structure (particularly the capability of feeding the growing population) the CO_2 threat probably will not motivate a major change in the world energy supply system.

C. The major constraint limiting the use of fossil fuels will be a slowing in growth of fuel demand dictated by social/economic factors. The demand for energy will grow more slowly than in the recent past and the ability to generate capital to explore for and produce fossil fuels will not sustain past rates of growth.

The foregoing should not lull us into a state of inactivity and complacency. This is not our intention, and we strongly caution against it. Even a 1°C average temperature rise within 50 years could produce serious impacts in some areas. The magnitudes of the changes in other meteorological variables that could occur in association with this average warming are largely unknown, especially on a regional basis. It is also true that changes, once induced, are largely irreversible on the time scale of concern to mankind. It is probable that global society could "cope" and adapt to such changes, at least for awhile, but there are no estimates of the cost of adaptation.

Continued reliance on coal, as well as unconventional oil and gas, into the middle and latter parts of the next century can be expected to push the CO_2 concentration in the atmosphere higher (as suggested by Figure 2) and make the climate changes more extreme. How much?, how soon?, and how resilient is society to such changes?, remain questions of fundamental importance.

Our key conclusion is that we believe there is time — disaster is not imminent — to conduct a carefully designed program to understand thoroughly the processes involved and the probable consequences. The need is not for immediate, short-range, and probably ill-conceived crash programs, but rather for long-range, carefully developed programs in the area of alternatives to fossil fuels. As research gives better estimates of the consequences, the scientific community will be in a much stronger position to call for government or intergovernmental action and regulation.

FOOTNOTES

[1]Gallons of oil per ton of shale is a commonly used unit in the shale oil industry. Here the ton is 2000 pounds; approximately 10 percent smaller than the metric ton used elsewhere in this paper.

REFERENCES

1. Rotty, R.M.: 1979(a) *Growth in Global Energy Demand and Contribution of Alternative Supply Systems*, Energy 4, pp. 881-890.

2. Rotty, R.M.: 1979(b) *Uncertainties Associated with Global Effects of Atmospheric Carbon Dioxide*, ORAU/IEA-79-6(O), Institute for Energy Analysis, Oak Ridge Associated Universities, Oak Ridge, Tennessee.

3. World Energy Conference: 1978, *World Energy Resources 1985-2020*, IPC Science and Technology Press, Guildford, U.K.

4. Parent, J.D.: 1979, *A Survey of United States and Total World Production, Proved Reserves, and Remaining Recoverable Resources of Fossil Fuels and Uranium as of December 31, 1977*, Institute of Gas Technology, Chicago, Illinois.

5. Culbertson, W.C. and Pitman, J.K.: 1973, "Oil Shale," in D.A. Brobst and W.P. Pratt, eds., *United States Mineral Resources*, U. S. Geological Survey Professional Paper 820, pp. 497-503.

6. Duncan, D.C., and Swanson, V.E.: 1965, *Organic-Rich Shale of the United States and World Land Areas*, U. S. Geological Survey Circular 523.

7. Weil, S.A., Feldkirchner, H.L., Punwani, D.V., and Janka, J.C.: 1978, *The IGT Hytort Process for Hydrogen Retorting of Devonian Oil Shales*, paper presented at the Chattanooga Shale Conference, Oak Ridge, Tennessee, November 14, Institute of Gas Technology, Chicago, Illinois.

8. Häfele, W.: 1979, *Global Perspectives and Options for Long-Range Energy Strategies*, Energy 5, pp. 745-760.

9. Perry, H., and Landsberg, H.H.: 1977, "Projected World Energy Consumption," in *Energy and Climate*, U.S. National Academy of Sciences, Washington, D.C., pp. 35-50.

10. Workshop on Alternative Energy Strategies: 1977, *Energy, Global Prospects 1985-2000*, McGraw-Hill Book Company, New York, New York.

11. Littman, G., and Keyfitz, N.: 1977, *The Next Hundred Years*, Working Paper Number 101, Center for Population Studies, Harvard University, Cambridge, Massachusetts.

12. Allen, E.L.: 1980, *The People's Republic of China: Economic Growth and Energy Consumption* (Draft), *The Soviet Union: Economic Growth and Energy Consumption* (Draft), *Eastern Europe: Economic Growth and Energy Consumption* (Draft), *Centrally Planned Economies: Projected Energy Supply and Demand to 2000* (Draft), All forthcoming from Institute for Energy Analysis, Oak Ridge Associated Universities, Oak Ridge, Tennessee.

13. Davison, C. and Allen, E.L.: 1979, *Italy: Estimates of Future Energy/GDP Relationships*, ORAU/IEA-79-20(M), Institute for Energy Analysis, Oak Ridge Associated Universities, Oak Ridge, Tennessee.

14. Edmonds, J.A. and Allen, E.L.: 1979, *Japan: Estimates of Future Energy/GNP Relationships in Energy Use*, ORAU/IEA-79-21(M), Institute for Energy Analysis, Oak Ridge Associated Universities, Oak Ridge, Tennessee.

15. Williamson, R.B., and Allen, E.L.: 1979, *Federal Republic of Germany: Estimates of Future Energy/GDP Relationships*, ORAU/IEA-79-22(M), Institute for Energy Analysis, Oak Ridge Associated Universities, Oak Ridge, Tennessee.

16. Allen, E.L.: 1979, *France, Estimates of Future Energy/GDP Relationships*, ORAU/IEA-79-23(M), Institute for Energy Analysis, Oak Ridge Associated Universities, Oak Ridge, Tennessee.

17. Allen, E.L., Edmonds, J.A., and Williamson, R.B.: 1979, *Canada: Estimates of Future Energy/GNP Relationships*, ORAU/IEA-79-24(M), Institute for Energy Analysis, Oak Ridge Associated Universities, Oak Ridge, Tennessee.

18. Gehman, J.C., and Allen, E.L.: 1979, *United Kingdom: Estimates of Future Energy/GDP Relationships*, ORAU/IEA-79-25(M), Institute for Energy Analysis, Oak Ridge Associated Universities, Oak Ridge, Tennessee.

19. Davison, C.: 1980, *OECD, Energy Supply and Demand in 2000* (Draft), forthcoming from Institute for Energy Analysis, Oak Ridge Associated Universities, Oak Ridge, Tennessee.

20. Baes, C.F., Jr., Beall, S.E., Lee, D.W., and Marland, G.: 1980, *Options for the Collection and Disposal of Carbon Dioxide*, this volume.

21. Bolin, B.: 1978, *Changes of Land Biota and Their Importance for the Carbon Cycle*, Science 196, pp. 613-615.

22. Manabe, S., and Wetherald, R.T.: 1975, *The Effects of Doubling the CO_2 Concentration on the Climate of a General Circulation Model*, J. Atmos. Sci., 32, pp. 3-15.

23. Flohn, H., 1979: "Can Climate History Repeat Itself? Possible Climatic Warming and the Case of Paleoclimatic Warm Phases," in *Man's Impact on Climate*, Elsevier Scientific Publishing Company, Amsterdam, The Netherlands, pp. 15-28.

THE U.S. DEPARTMENT OF ENERGY AND THE CARBON DIOXIDE ISSUE

David H. Slade

Manager, Carbon Dioxide and Climate Research Program, Office of Health and Environmental Research, Office of Environment, Department of Energy, Washington, D.C.

ABSTRACT. A comprehensive research program dealing with the environmental and societal impacts of increasing atmospheric concentrations of carbon dioxide is being developed by the U. S. Department of Energy. The program, when fully realized, will be a cooperative effort with other U. S. agencies, such as the National Oceanic and Atmospheric Administration and the National Science Foundation, with the U. S. National Climate Program Office and with a variety of other national and international organizations.

The program has five components:

1. Research on the global carbon cycle.
2. Research on the effects of carbon dioxide on climate.
3. Research on the effects of a climate change and carbon dioxide increase on the terrestrial biosphere, the oceans, and the cryosphere.
4. Research on the impact of a total environmental dislocation on the global social, political and economic order.
5. Research on methods to ameliorate undesirable effects and on options for adaptation.

A sixth program component - a review, integration and evaluation of all available research information, often called an "impact assessment" - will also be conducted.

This paper describes the content, planning and execution of this program.

Congressman George Brown of California has said that,

W. Bach, J. Pankrath, and J. Williams (eds.), Interactions of Energy and Climate, 213–234.

"To address the (carbon dioxide) problem requires mounting an urgent effort in pre-crisis management." Rather than dwelling on difficulties of such an effort, Congressman Brown prefers to stress the opportunities:

1. The carbon dioxide issue offers the opportunity for forging new interdisciplinary and interinstitutional bonds.
2. The carbon dioxide issue could serve as a catalyst for the ultimate development of a global strategic planning capability with applicability to a variety of other long-term problems.
3. An international carbon dioxide program could function as a vehicle for transferring scientific and technological capabilities to less-developed countries.
4. Focusing on the carbon dioxide issue would increase the global awareness of our vulnerability to climate changes or other major environmental dislocations.

The United States Department of Energy, working closely with the recently-created National Climate Program Office, has begun the development of a national CO_2 research program which could accelerate activity in the areas outlined by Mr. Brown. The remainder of this paper will be devoted to a description of the growing U. S. CO_2 research program and to a brief discussion of examples of initiatives other countries might take over and above their current CO_2 research activities.

The Department of Energy's Carbon Dioxide and Climate Research and Assessment Program began in 1976. Its goal is:

> To develop the ability to predict the agricultural, environmental and societal consequences, both national and international, of increasing atmospheric concentrations of carbon dioxide with sufficient confidence to permit informed policy decisions to be made on the use and development of energy and other resources.

To reach this goal, three broad approaches are required:

1. To create a national and international focus for the carbon dioxide issue.
2. To act as a lever to assure adequate funding of the scientific community for the conduct of a coordinated and interdisciplinary national and international research effort.
3. To conduct a review, integration, and evaluation of all relevant information in order to provide estimates of the consequences of the CO_2 issue for use in developing policy alternatives.

The creation of the national focus called for in the first of these approaches was achieved by the National Climate Program Act signed into law by President Carter in late 1978. This Act, in large part the result of the interest and effort of Congressman Brown, created a National Climate Program Office in the United States charged with, among other responsibilities, the role of "...the lead entity responsible for the administration of the (Climate) Program." Working closely with the Office and with other involved U. S. agencies, the Department of Energy is developing a comprehensive plan for carbon dioxide research and assessment within the context of the entire U. S. climate effort.

We have, unfortunately, not been successful in moving toward an international focus for this issue. The excellent international cooperation by scientists in the carbon dioxide field does not have an institutional parallel for coordinated funding, planning and, most importantly, for advocacy among national, multinational and international bodies with an interest or capability in carbon dioxide research and assessment.

The development of our second strategy, obtaining the necessary research funding, constitutes the substance of the Department's activity. We decided at the outset, that a request for the sizeable funds required could only be justified by a plan for a research program based on the authority and wisdom of the world's leading scientists in various aspects of the CO_2 issue. Considering that the goal of the program would, of necessity, address virtually the total global environment and all of society that dwells therein, the planning process has become a continuous function of our Program.

The Program and the planning involved are divided into six interactive categories:

1. Research on the Carbon Cycle
2. Research on CO_2 and Climate
3. Research on the Effects of Climate Change and CO_2 Increase on the Remainder of the Environment
4. Research on Social, Political and Economic Costs and/or Benefits of Global Environmental Change
5. Research on the Amelioration of Global and Sectoral Impacts
6. Review, Integration and Evaluation (Assessment)

These Research Categories are further divided into Research Subcategories and then into Research Issues, the primary building blocks of the program. Table 1 shows the current listing of Categories, Subcategories, and Issues. A brief description and discussion at the Research Category level follows.

TABLE 1

A NATIONAL PROGRAM ON CARBON DIOXIDE, ENVIRONMENT AND SOCIETY

RESEARCH CATEGORY	RESEARCH SUBCATEGORY	RESEARCH ISSUE
I RESEARCH ON THE CARBON CYCLE	A. Net Sources of Atmospheric CO_2	1. Industrial Sources
		2. Terrestrial (Biospheric Sources)
		3. Oceanic and Cryospheric Sources
	B. Atmospheric Storage	1. Observations of Atmospheric CO_2
	C. Sinks for CO_2	1. Transfer Across the Air-Sea Boundary
		2. CO_2 Exchange Between the Upper and Deep Oceans
		3. Carbonate Dissolution and Biological Transport
		4. Biospheric Sinks
	D. Historical Records of CO_2	1. Same
	E. Models of the Carbon Cycle	1. Same
II RESEARCH ON CO_2 AND CLIMATE	A. Modelling the Effects of CO_2 Increase	1. Model Improvement and Modification
		2. Model Verification Studies
		3. Model Sensitivity Studies
		4. CO_2 Perturbation Studies
	B. Reconstruction of Past Climates	1. Reconstruction of Paleoclimates
		2. Reconstruction of Historical Climates
	C. Climate Scenario Development	1. Same
	D. Evidence of Climate Change	1. Same
III RESEARCH ON THE EFFECTS OF CLIMATE CHANGE AND CO_2 INCREASE ON THE REMAINDER OF THE ENVIRONMENT	A. Effects on the Oceans	1. Effects on Oceanic Circulation
		2. Effects on Marine Biota
	B. Effects on the Cryosphere	1. Effects on the West Antartic Ice Sheet
		2. Effects on Arctic Sea Ice
		3. Physical and Hydrological Effects on Permafrost
	C. Effects on the Less-Managed Biosphere	1. Carbon Balance in Northern Eco-systems and Effects of Climate Change
		2. Effects on Grazing Lands, Wildlife and Animal Husbandry
		3. Paleoclimatic Data as a Predictive Tool for Biospheric Effects
		4. Effects on Forest Ecosystems
		5. Effects on Freshwater Ecosystems
		6. Effects on Human and Animal Health
		7. Ecosystem Response to CO_2 Enrichment
	D. Effects on Agriculture and the Managed Biosphere	1. Effects of Increased CO_2 on Photo-synthesis and Agricultural Productivity
		2. Alleviation of Environmental Stress on Renewable Resource Productivity
		3. Effects on Managed Forests and Range Lands
		4. Effects on Crop Use Water Efficiency and Water Requirements
		5. Effects on Plant Pests
		6. Effects on Food Crops in LDCS
		7. Effects on Animal Agriculture and Livestock Production

TABLE 1 (Continued)

A NATIONAL PROGRAM ON CARBON DIOXIDE, ENVIRONMENT AND SOCIETY (Cont'd)

RESEARCH CATEGORY	RESEARCH SUBCATEGORY	RESEARCH ISSUE
IV RESEARCH ON SOCIAL, POLITICAL AND ECONOMIC COSTS AND/OR BENEFITS OF GLOBAL ENVIRONMENTAL CHANGE	A. Social and Institutional Responses	1. Cultural Determinants of Perceptions of Climate Change and Social System Reactions
		2. Climate Change, International Law and Institutions
		3. Factors Affecting Vulnerability of Society to Environmental Change
		4. Political and Institutional Decision-Making in Response to Climate Change
		5. Perceptions of and Reactions to Environmental Change in Relation to Decision-Making
		6. Historical Dimensions of Effects of Climate on Human Societies
	B. Economic and Geopolitical Consequences	1. Economic Policy Evaluations of CO_2 Abatement
		2. Assessment and Resolution of Climate-Induced Strains on the International Economic System
		3. Mitigating Losses from Disruptions Caused by Environmental Change
		4. Planning for Environmental Change: Scenario Construction and Evaluation
V RESEARCH ON THE AMELIORATION OF GLOBAL AND SECTORAL IMPACTS	A. Preventative Measures	1. Same
	B. Curative Measures	1. Same
	C. Moderating Measures	1. Same
	D. Adaptive Measures	1. Same
VI REVIEW, INTEGRATION AND EVALUATION (ASSESSMENT)	A. The 1984 Review, Integration & Evaluation (Assessment) Report	1. Same
	B. CO_2 Information and Data System	1. Same

RESEARCH ON THE GLOBAL CARBON CYCLE

Description

This research area includes the roles of the atmosphere, terrestrial biosphere, and the oceans in distributing and sequestering CO_2 released from fossil fuel combustion, forest clearing, and other land surface disturbance. The development of scenarios for future fossil fuel releases of CO_2 is also included.

Discussion

The primary reason for studying the mechanisms by which CO_2 injected into the atmosphere is distributed among the various environmental compartments is one of timing. If all new CO_2 injected into the atmosphere were to remain there, the future rate of increase of CO_2, while uncertain, would be primarily determined by the rate of fossil fuel combustion and by the nature of land use practices. However, the extent to which the oceans and terrestrial biosphere redistribute CO_2 injected into the atmosphere introduces an uncertainty in the rate of atmospheric increase and, therefore, into estimates of the time at which a given atmospheric concentration will be reached. The importance of the timing of the increase lies in the fact that decisions on the strategies that may be necessary to deal with the CO_2 issue are profoundly dependent upon the length of time available for their implementation.

Fossil fuel combustion is currently adding about 6×10^9 tons of carbon annually to the global atmosphere. This rate has been increasing by about 4.3% per year over much of the past century. The quantity of carbon dioxide released by deforestation and cultivation is far less well known. Various estimates hold the terrestrial biosphere to be a major source for the release of carbon dioxide to the atmosphere. Contrary estimates suggest, however, that the biosphere might even be a weak sink. If it is a substantial source, then either the ocean must be a more effective sink than measurements indicate, or there must be another, as yet unknown, sink.

A prime objective for research dealing with the carbon cycle is the resolution of this dilemma. First, it will be necessary to establish the quantity of the CO_2 released from deforestation and soil disturbance. Second, if a substantial amount of CO_2 has been added, it will be necessary to determine its fate.

Not all of the transfer coefficients describing the exchange of CO_2 between major reservoirs are adequately known. In particular, the exchange of CO_2 between surface (0-100m depth) ocean waters to intermediate layers (100 to 1,000m depth), allows for too large an uncertainty in making predictions of future atmospheric CO_2. Finally, the reality of CO_2 fertilization of the world forests can play a major role in the carbon cycle. Would the forest carbon storage grow as it is exposed to higher atmospheric CO_2 concentrations or is it limited by nutrients, water, light or other factors?

At the same time these questions are addressed the search for CO_2 sinks not now considered important must continue; if such sinks exist they will have a bearing on the carbon cycle, regardless of the reality of biospheric sources of CO_2. This requires a major emphasis on ocean transport and ocean sediment chemistry.

A continuing atmospheric CO_2 monitoring program is currently providing an atmospheric history. This information will be assembled into a credible model allowing for both prediction of likely future CO_2 concentrations and the uncertainty attached thereto.

RESEARCH ON THE EFFECTS OF INCREASING CARBON DIOXIDE ON CLIMATE

[Note: The usage of the terms "climate" and "environment" in the following sections differs according to the context in which they are discussed. When used in the sense of the climate system or a climate model, "climate" subsumes all of those environmental features which explicitly interact with atmospheric components of the system or model. These include such characteristics as the heat capacity of the oceans, the frictional properties of the terrestrial surface and the existence of a snow/ice cover. When the impact of climate change on the remainder of the environment is considered, emphasis is placed upon those perturbations in the oceans, on the terrestrial surface or cryosphere which, in themselves, have implications for man's social or economic well-being, regardless of whether or not those changes are interactive with the atmosphere.]

Description

This research area includes studies to determine the effects of increasing CO_2 concentrations on the global atmosphere and its component parts. Included are studies of the global radiative balance, temperature and moisture structure and circulation. It also includes studies of the interaction and feedback between the atmosphere and the oceans, terrestrial land masses, and biota and cryosphere, as well as retrospective evidence of climate and CO_2 relationships. Current atmospheric data will be evaluated to search for evidence of climate change.

Discussion

Present concentrations of CO_2 in the atmosphere, acting in combination with other absorbing gases (e.g., H_2O and O_3), play an important role in maintaining the current climate. The best available, but still significantly simplified, model studies indicate that a doubling of present CO_2 concentrations would lead to an increase in average annual surface temperature of 1.5 to 3°C and as much as 8-10°C in the arctic. Studies of past interactions of CO_2 and climate, while still somewhat speculative, form a further basis for concern that the potential effects may be important.

CO_2 concentrations are, however, only one factor among the many that determine the overall behavior of the climate system. The climate system includes the oceans, the land surface (both ice and snow-covered) and the biota. The uneven distribution of these features around the globe, coupled with differences in land elevation, interact with various driving forces to comprise the climate system. These are also astronomical forcing functions which probably affect climate on longer time scales.

The atmosphere-ocean-land system maintains the earth's climate by transporting heat and water vapor between equatorial and polar latitudes so as to achieve an energy balance. If the temperature gradient between tropics and poles decreases, as current models suggest would happen if atmospheric CO_2 increases substantially, then the current energy and water vapor exchange would also be expected to change and new circulation patterns could lead to substantially different regional climates (i.e., temperature, precipitation, winds, etc.). Due to the complexity of the climate system, these regional changes could be greater than, and in some cases opposite in sign to, changes in the global average.

The causes of the natural variability of climate are little understood, although they have affected man's activities significantly throughout history. Even without man's intervention, these natural variations would continue. The point at which man's impact will be noticeable among the natural fluctuations and cycles and whether the CO_2 effects would add to or counteract the as yet unforeseeable natural changes (or even other man-made factors) are complex questions. At least some of these issues will be addressed by this program.

At present, plans are being formulated for intensified research programs and activities, both nationally and internationally, to improve the understanding of the natural climate system and its variations. Study of the CO_2 climate problem should be (and is) integrated into these emerging international programs.

Clearly, a better understanding of the climate's "unperturbed" state will help build confidence in our predictions of the CO_2 perturbed climate.

The Department of Energy stresses those increments of a total climate research effort which are primarily directed toward the uncertainties of increasing CO_2 on climate.

RESEARCH ON THE EFFECTS OF CLIMATE CHANGE AND CARBON DIOXIDE INCREASE ON THE REMAINDER OF THE ENVIRONMENT

Description

Attention here is directed toward the interactions between a new atmospheric state induced by increasing CO_2 on the one hand and ocean circulation, chemistry and biology, the extent and stability of the marine and terrestrial-based cryosphere, the global and regional agricultural systems and the major components of the natural terrestrial biosphere on the other. Feedbacks and other interaction between the oceans, cryosphere, terrestrial biosphere and the atmospheric system are included here as well as under the climate research area.

Discussion

With few exceptions, the international research program dealing with the CO_2 issue has been devoted to studies of the global carbon cycle and the atmospheric impacts of increased CO_2. This work is characterized by well developed scientific constituencies, active and capable spokesmen, and vigorous research efforts. Comparatively little attention, however, has been paid to research on the environmental effects of an essentially non-reversible climate change accompanied by an increase in global concentrations of carbon dioxide. There are a number of reasons for this. Perhaps foremost is that estimates of future CO_2-induced climate perturbations have been only recently available. This is even more so in the case of regional and seasonal descriptions of perturbations. Thus, there has been insufficient information available for even a preliminary consideration of climate impacts on the environment.

Again, most attention to the impacts of climate change has concerned the effects of climate fluctuation around a long term mean, not with progressive, non-reversible changes. Finally, the lack of development of this aspect of CO_2 research is its unprecedented nature. Other impacts on the global atmosphere, primarily the fluorocarbons and SST emissions, are not associated with a cascade of environmental effects which can potentially affect the physical, chemical and biological functioning of the global environmental system.

The individual elements of the problems to be faced regarding the impact of climate change on the environment are generally in well-established fields of scientific activity: ecology, biology, oceanography, etc. However, an interdisciplinary intellectual community concerned with questions of climate change impacts on the global environment is only now emerging.

For the reasons noted, there will be two primary functions of this portion of the planned research program: the development of researchable issues and the implementation of the research program; the development of a "constituency" or a cadre of discipline-oriented environmental scientists who will apply their knowledge and methodologies in an interdisciplinary and organized manner.

This process has begun formally only this year. Thus, the currently-perceived important research issues in the area of climate change and CO_2 effects on the remainder of the environment as shown in Table 1 can be expected to change.

Should climate shifts occur, all other environmental compartments are likely either to participate in these changes or be affected by them. A variety of environment features is considered explicitly in the current generation of sophisticated "climate" models. Thus, the modeled response of the cryosphere to an increase or a decrease in predicted temperature might be expressed through a changing albedo in arctic areas caused by the melting of the sea ice - a "feedback effect". Many such feedback effects involving the atmosphere with the oceans and land masses could be included in more complete models.

Another class of changes caused by a climate change is not likely to be found in climate models however, since they do not have the potential for geophysical feedback to modify atmosphere processes. This would include changes in oceanic acidity as a result of increased atmospheric carbon dioxide concentrations, changes in the nature and distribution of fisheries due to temperature, stability, current and nutrient variations stemming from carbon dioxide-induced climate change, impacts on the distribution and diversity of plant communities, impacts on the range and number of insects and, of course, impacts on the cultivated biosphere.

A DOE-AAAS Workshop held in Annapolis, Maryland in April 1979 was perhaps the first interdisciplinary attempt to begin planning a research program on disciplinary impacts of climate change and CO_2 increase on the remainder of the environment.

The primary concerns regarding the effects on the oceans were markedly interdisciplinary; various climate elements affect either the circulation or the biogeochemistry of the oceans. The behavior of the cryosphere in the fact of climate change was believed to center on the interaction and feedbacks between the cryosphere and the atmosphere; the dislodgement of a huge ice sheet in the Antarctic could cause a dramatic or even catastrophic rise in sea level.

The question of effects on the unmanaged terrestrial biosphere appeared to revolve around the influence of climate change and CO_2 increase on species and ecological systems - with very little feedback to the climate system. A somewhat different research orientation may be required when addressing the question of effects on the global agricultural system--the managed biosphere of food, fiber, and forests. While the impacts of changes in the climate system on the parameters of agriculture should be matters for research, the workshop participants were primarily concerned with improving the adaptability of the agricultural system to environmental stress.

RESEARCH AND THE SOCIETAL COSTS AND/OR BENEFITS OF GLOBAL ENVIRONMENTAL CHANGE

Description

This research area includes studies of mechanisms to reduce the vulnerability, or increase the adaptability, of social elements to climate change. Investigations will include historical case studies, regional studies and studies of the perception of information. Methodologies for the systematic construction of social and environmental scenarios will be developed. The magnitude and type of current international stresses and their potential exacerbation will be investigated.

Discussion

As already noted, the goal of the DOE program is the prediction of the social and economic impacts of increasing CO_2. We have no illusions that this will be easy to achieve. First, "...it is impossible to forecast with confidence what the world's social and economic picture fifty or a hundred years from now will look like." Second, quantitative estimates of societal effects of increasing CO_2 are the product of a chain of multiplicative uncertainties associated with the cascade of effects set into motion by increasing CO_2.

It is obviously impossible to forecast the future, but we are destined, nevertheless, to make decisions today that have long-term implications. Even if such decisions are beset with uncertainty, it is our grave responsibility to attempt to improve the quality of these decisions and that of the decision-making process. To this end, two approaches will be adopted: We will concentrate on case studies of current, historical or regional situations which will investigate the dynamics of societal reaction to climate variation. Here, we will rely on such disciplines as anthropology, political science, law, history, psychology and geography.

We will also develop a coherent approach to climate scenario development and the use of these scenarios to estimate social consequences. Here, we will focus on societal consequences rather than on an investigation of the scientific reliability and consistency of the climate scenario itself.

RESEARCH ON THE AMELIORATION OF GLOBAL AND SECTORAL CARBON-DIOXIDE-INDUCED IMPACTS

Description

Research in this area consists of studies of strategies and technological initiatives to minimize the rate of CO_2 buildup and to permit ordered and planned societal adaptation to environmental dislocations.

Discussion

This is a relatively undeveloped field both in the DOE program and in CO_2-related activities elsewhere. Certain aspects of it are currently being dealt with in those environmental and societal portions of the program which address mechanisms to increase the adaptability of global institutions to stresses associated with major climatic dislocations.

Some thought has already been given to technological "fixes" but these appear to be grossly uneconomic. Changes in global energy use policy could be an attractive alternative but would be difficult given the need for international cooperation of unprecedented degree. Adaptation of both physical and social systems might be the most attractive option for the United States, if not to other nations including those inthe developing world since such a strategy could imply an energy policy which would permit only the slowest possible growth to inevitable CO_2 levels.

The five areas described above constitute the research component of the National Program on Carbon Dioxide, Environment and Society. A comprehensive research program, in itself, is necessary but not sufficient to meet the need for information upon which energy and other decisions are based. It is important, also, to review, integrate and evaluate the relevant information and provide both feedback to the research program and decision-oriented information to the decision makers. This component of the CO_2 Program is described below.

REVIEW, INTEGRATION AND EVALUATION (ASSESSMENT)

Description

This program category involves determining the environmental, economic, political, and social costs/benefits resulting from increasing levels of carbon dioxide and estimating the confidence levels that can be attached to the cost estimates. It also entails determining the costs of the strategies and technological initiatives that could be taken to mitigate or take advantage of the consequences of CO_2 increases. Finally, it includes an investigation of whether the global research program is properly directed toward answering the key questions regarding the costs/benefits of the carbon dioxide issue.

Discussion

A CO_2 assessment program of the size and scope envisioned here is unprecedented. For this reason, a description of the philosophy and elements of such a program is necessarily tentative.

Consider, for example, the following:

- The CO_2 impact has the potential for causing a significant dislocation in virtually all sectors of the global environment; a feature that has not had to be faced previously even in those global environmental impacts that have been postulated as results of fluorocarbon or SST emissions.

- A virtually total environmental dislocation would have effects, ranging from trivial to traumatic, on various sectors of the entire global social system - a scope of impact that is unprecedented.

- An assessment of the consequences of the CO_2 issue depends upon a chain of multiplicative uncertainties, many elements of which have, even now, not been subjected to coherent intellectual scrutiny.

- Despite considerable discussion, there has been little development of approaches for undertaking a global assessment program; philosophies, methodologies and mechanics are not yet well developed. Only a very few potential centers of excellence have been identified.

One of the most important objectives of this facet of the program will be to foster the development of a multidisciplinary intellectual community dedicated to the specific question of a global assessment of the impacts of the CO_2 issue. Another objective of this portion of the program will be to prepare an authoriatative assessment of the entire CO_2 issue by 1984.

The Department of Energy planning process alluded to earlier is extensive. It begins with a major workshop in one or more of the six areas outlined above. These workshops are interdisciplinary gatherings of specialists in the areas under discussion. The primary purpose of these workshops is to identify the research issues that are not being addressed, or at least not being addressed with sufficient vigor on a world-wide basis. The issues so identified are then assigned to leading scientists in the specific areas and these individuals are asked to prepare a research plan. The individual issue research plans are then assembled, coordinated, and eventually become a part of what we call a "Comprehensive Plan for Carbon Dioxide Research and Assessment."

It is important to recognize that the plan that emerges from this activity is not strictly a Department of Energy plan, or even a U. S. plan. It is our intent to develop a plan that can be subscribed to by any institution having an interest or capability in a particular research area. In our own country, a number of agencies, particularly the National Oceanic and Atmospheric Administration (NOAA) and the National Science Foundation (NSF), are participating in the research activity. We hope other nations and multinational and international organizations will join us in both the planning and the conduct of the research.

Two major workshops have been held thus far. The first, in 1977 in Miami, Florida, addressed problems of the carbon cycle and climate. Its chairman was Dr. Lester Machta of NOAA. The second, held in 1979 in Annapolis, Maryland, dealt with impacts of climate change on the remainder of the environmental system and with social and economic impacts. This workshop was chaired by Dr. Roger Revelle of the University of California at San Diego, and was conducted for us by the American Association for the Advancement of Science. Other more narrowly defined workshops are held as required. For example, we have sponsored a workshop on carbon cycle models and the measurement of the global biomass with the Scientific Committee on Problems of the Environment and Matter, on the role of organic soils and temperate forests in the carbon cycle with the Institute for Ecology, and, yet another, on dendrochronology with the Lamont-Doherty Geological Observatory.

The Department of Energy's CO_2 research program is now primarily concerned with the carbon cycle and climate areas. The results emerging from the Annapolis Workshop will broaden our program considerably in the areas of environmental and societal impacts over the next few years.

A partial listing of our currently-sponsored research programs includes:

Lamont-Doherty Geological Observatory	Investigation of the Surface Water pCO_2 in the Equatorial Pacific Ocean
Lamont-Doherty Geological Observatory	Large-Scale Deliberate Trace Releases in the Ocean Thermocline
Lamont-Doherty Geological Observatory	Natural Catalysis of the CO_2-HCO_3 Reaction in Surface Sea Water
Lawrence Livermore Laboratory	Miniaturized Rapid Response Sensor for Atmospheric Concentrations of Carbon Dioxide
Lawrence Livermore Laboratory	Modeling Climate Effects of Carbon Dioxide Increases
Markley, O. W.	Sociopolitical Impacts of CO_2 Buildup in the Atmosphere Due to Fossil Fuel Combustion
University of Miami	Coordination and Technical Planning for the DOE Carbon Dioxide Effects Research Oceanographic Subprogram
University of Michigan	Successional Forest Stands as Sinks for Atmospheric Carbon Dioxide
The MITRE Corporation	Studies, Analyses and Technical Support to the Carbon Dioxide Research and Assessment Program

National Bureau of Standards	Development of standards for measurement of atmospheric carbon dioxide
National Center for Atmospheric Research	Enhanced Research Program on the Long-Range Climatic Effects of Increasing Atmospheric CO_2
National Oceanic and Atmospheric Administration	Partial Support for the First GARP Global Experiment
*National Research Council	Partial Support of the Climate Research Board
Oak Ridge Associated Universities/Institute for Energy Analysis	Global Effects of Increasing Levels of Carbon Dioxide
Oak Ridge National Laboratory	Global Carbon Cycles and Climatic Risks
Pacific Northwest Laboratory	Atmospheric CO_2 Abundance - An Archival Study of Spectroscopic Data
*Oregon State University	Research on the Dynamics of Climate
San Diego State	Modeling Ecosystem Impacts of Climate Change from Energy Development
Scientific Committee on Problems of the Environment	Support of International Research Activities on the Global Carbon Cycle
*Scripps Institution of Oceanography	A Study of the Abundance of Carbon Dioxide in the Atmosphere and Its Exchange with the Oceans

*Indicates joint funding with another federal agency

Scripps Institution of Oceanography	Carbon dioxide measurement in the equatorial Pacific
Scripps Institution of Oceanography	Nutrient Measurements for NORPAX
Scripps Institution of Oceanography	Technical Support for Collection of C-14 Samples During FGGE
Scripps Institution of Oceanography	Verification of Carbon Dioxide Buffer Factor of Surface Ocean Water by Direct Observation
University of Washington	Bomb Radiocarbon Sections Along the Hawaii-Tahiti Shuttle Track
University of Washington	Geochemical Determination of Biospheric CO_2 Fluxes to the Atmosphere
University of Washington	Historical C-14 Measurements from the Atlantic, Pacific and Indian Oceans
*University of Alaska	Gases in Sea Ice
American Association for the Advancement of Science	AAAS-DOE Workshop on Environmental and Societal Impacts of a Possible CO_2-Induced Climate Change
Aspen Institute for Humanistic Studies	The Consequences of a Hypothetical World Climate Change Produced by Increased Atmospheric Carbon Dioxide
*University of Arizona	Dendrochronology of the Bristlecone Pine

Brown University	Global Temperature Patterns 6000 Years Ago
University of California	Carbon Balance of World Forests and Associated Soils
University of Colorado	An Assessment of the Possible Cryospheric Responses to a CO_2-Induced Global Warming
Commission on Atmospheric Chemical and Global Pollution	Partial Support of the Symposium on the Budget and Cycles of Trace Gases and Aerosols in the Atmosphere
University of East Anglia	On the Geographical Patterns of Climatic Change over Recent Millenia, Particularly Since About 800 AD
Environmental Research Laboratories/NOAA	Expansion of Global Atmospheric Monitoring of Carbon Dioxide
Environmental Research Laboratories/NOAA	Scientific Direction to the Carbon Cycle portion of the DOE CO_2 Program and Support Research
University of Florida/ Cornell University	Models of Carbon Flow in Tropical Ecosystems with Emphasis on Their Role in the Global Carbon Cycle
The Institute of Ecology	Problem Definition and Feasibility Workshops Terrestrial Biosphere Subcomponents on the Global CO_2 Balance
Lamont-Doherty Geological Observatory	Coral Growth Rings and the Temporal History of Nuclear C^{14} and SR^{90}

Lamont-Doherty Geological Observatory	International Meeting on Dendro Isotopic Research
Lamont-Doherty Geological Observatory	Interpretation of Equatorial Hydrographic, Nutrient Element and Radioisotope Data
Lamont-Doherty Geological Observatory	Investigation of the Atmosphere and Surface Ocean CO_2 Partial Pressure in the Indian Ocean
Lamont-Doherty Geological Observatory	Kr-85 Measurement Facility
Woods Hole Oceanographic Institution	Construction of a ^3H- ^3He Facility
University of Miami	Improvements to the ^{14}C Measurement Facility
National Academy of Sciences/National Research Council	Workshop on Flux of Organic Carbon to the Ocean
Underseas Medical Society, Ind.	Determination of the Carbon Dioxide Effects of Mammalian Organisms
American Association for the Advancement of Science	The Development of Options for Research to Determine the Environmental and Societal Consequences of a Possible CO_2-Induced Climate Change
Brookhaven National Laboratory	Atmospheric CO_2 Control

We are in the process of publishing documents describing our plans, workshop results, and research progress. These can be made available to those of you who wish to contact my Office.

Having described our effort thus far, it is necessary to look at the problems that still remain. Those in the environmental research component are generally well known. Some are being vigorously addressed, such as climate modeling. Others have not yet been undertaken; the effect of a changing climate on the oceans, biosphere and cryosphere is an outstanding example.

Another group of problem areas has not yet even been formulated. The question of societal perception and reaction to a climate change and the concomitant environmental dislocation is the subject of considerable intellectual ferment; but clear directions of investigation are just now being developed.

Perhaps the single greatest problem facing us, all of us, is that, if the chain of conjecture regarding the CO_2 issue proves to be correct, we are faced with a global issue of unprecedented gravity and scale which can only be addressed by the world community acting in concert. Fortunately, interest is high in the international scientific community. However, it appears that interest in or, indeed even recognition of, the CO_2 issue is, with a few notable exceptions, minimal at the policy level of most national governments. Three approaches, all of which are mutually supportive, suggest themselves.

First, there are existing international institutions already dedicated to climate research and impacts studies; and while their purview is generally narrower than that required in a carbon dioxide-related program, full use should be made of their capabilities.

Second, there would appear to be an obvious need for nations and international bodies to increase their sensitivity to the CO_2 issue. Certainly, self-interest would suggest that the various individual world communities, faced with a possible but undefined environmental dislocation, would be amenable to considering policies and strategies for dealing with the impact of this issue on their future.

Third, the CO_2 issue requires an acceleration of environmental research on a broad range of relevant questions. Nations possessed of some CO_2-related environmental feature could be expected to concentrate their attention on those features. Just to cite a few examples:

1. The question of biospheric sources and sinks of CO_2 requires quantitative estimates of deforestation, reforestation, and other changes in land use. Deforestation is, of course, well known and almost immediately detected but the matter of reforestation may be almost as important: countries north of the Sahara, e.g., Libya and Algeria, have a major tree planting program; China is said to be planting millions of trees along river banks and between agricultural plots,

etc. Natural changes are also important. Many nations could develop and contribute information in this area.

2. Countries with ecosystems such as Canada's boreal forests and bogs can study their local carbon budget. In addition to Canada, there are Scandinavia, the Soviet Union (boreal forests, bogs, peat accumulation), Brazil, Columbia, Malaysia (tropical rain forests), Australia (subtropical forests, desert area), Ivory Coast (savanna), etc. Information on the typical carbon content in living matter, dead matter, and soils per unit area, response to temperature and precipitation changes in terms of growth rates, soil carbon accumulation, soil carbon loss would be valuable. Such studies would also help a country to understand its own natural resource situation.

3. Almost any nation, regardless of its state of scientific development, could contribute usefully to a program of CO_2 measurement as part of the WMO monitoring program.

4. Countries with oceanographic research vessels could contribute to the U. S. transient tracer work. Perhaps some joint arrangements for support by U. S. laboratories for analysis of samples could be undertaken. At a minimum, any country with an oceanographic vessel is capable of making gas chromatographic measurements for F-11 and F-12 (freons).

5. The systematic reporting of major natural phenomena, such as major fires and volcanic eruptions, which might alter the climate over large areas is important. Such information will, of course, help to interpret the observed climatic records, also a vital aspect of the predictions from CO_2 increases.

6. Countries with major fishing industries would certainly be concerned with questions of climatic impacts. Included here are the USSR, Japan, Iceland, Great Britain, the Scandanavian countries, and a number of the South American coastal nations.

7. The question of the stability of the West Antarctic ice sheet in the face of climate change would be an excellent topic for consideration by those nations with territorial claims in this region: New Zealand, Australia, France, the United Kingdom, Norway, Chile and Argentina.

8. The important question of the behavior of Arctic sea ice is important to the nations rimming the Arctic Ocean. Their study of this would be useful both in a broad international sense as well as in their genuine self interest.

9. The stability, or fragility, of grazing lands faced with climatic change would clearly be addressed by many nations. Progress on such studies, carried out in the framework of existing research institutions, would be followed with great interest by importing nations dependent on meat and dairy products produced on the rangelands.

10. Historic and geologic temperature fluctuations can be derived from local sources such as literature and soil cores, etc. Such past natural variability is important for the assessment of the CO_2 issue. Often this may involve collaboration with and the use of facilities by scientists from other countries.

11. Many institutional systems exist within which these research activities may be carried out. However, there should be a concerted effort to urge those not currently involved in the CO_2 issue to familiarize themselves with what must be done. This could be done by attendance at the many scientific meetings on the CO_2 question. A concerted effort should be made to invite the participation of representatives of all concerned nations to such gatherings.

* * * * * * * * *

I have outlined some of the issues as they are now perceived, the DOE and U. S. research approach, and some examples of opportunities for cooperation. I hope this discussion will be a stimulus to a heightened interest and a spur to the increased cooperation which would appear to be in the international interest.

DISCUSSION

Flohn:
Do you think that the high Holocene CO_2 values at Camp Century and Byrd Station are realistic?

Oeschger:
The Holocene values for Camp Century are higher than those for Byrd Station probably due to the contribution of CO_2 trapped in melt layers during the climatic optimum. The values found for the Holocene from the Byrd core might be close to the atmospheric concentrations, indicating higher CO_2 -concentrations. However, many more measurements are necessary in order to answer this question.

Flohn:
Would you like to comment on Manabe-Stouffer's mixed-layer ocean model, especially on the result of seasonal sea-ice removal after a quadrupling of CO_2?

Gates:
Manabe and Stouffer have estimated an average surface warming of about 2^oC from a CO_2 doubling, using a global circulation model with a 68 m ocean mixed layer. This warming is less than that with a swamp ocean due to the effect of the seasons, but is still dominated by the albedo-temperature feedback of seasonal ice variations in the higher latitude oceans. The model ocean introduces a lag in the atmosphere's temperature response compared to a swamp ocean although it neglects oceanic advecting and upwelling.

Kellogg:
Having worked in the Secretariat of the WMO in Geneva for over a year to design the World Climate Programme, which has now been officially approved, I am acutely aware of the problems encountered in such an international effort. It must be recognised that the international bodies like WMO, UNEP and ICSU do not do the actual work in international programs, since that is carried out in or by the participating countries. Rather, the international bodies make it possible for the programs to be established, to be co-ordinated by international groups of experts, to exchange and archive data, to arrange bilateral agreements between countries and so forth. UNEP (and occasionally WMO) may support specific research or technical co-operation projects with funds, but only as "seed money" for special purposes. Thus, the success of any international

W. Bach, J. Pankrath, and J. Williams (eds.), Interactions of Energy and Climate, 235–241.

initiative, such as the World Climate Programme, depends in the end on the willingness of countries to work together.

Niehaus:

1. If cutting down of forests also had exponential growth, we would not be able to distinguish it from exponential growth of energy consumption.
2. For our present consideration the actual amount of fossil fuel resources is not terribly important as long as we agree that it is at least enough to supply 50 TW for 100 years.
3. Your paper also pointed out how dangerous it is to delay significantly the implementation of new energy technologies like nuclear power.

Keeling:

I agree with these statements.

Lieth:

1. What is your data base for Tables 2 and 4?
2. What is your base for the assumption that 0.2 Gt C are released through rivers to the ocean?
3. How would your calculations change if the numbers for river release were 10 times as high?

Hampicke:

1. The values are calculated from Rodin et al. (1975), Ajtay et al. (1979), and Schlesinger (1979), as indicated in the reference list.
2. To my knowledge, about 0.2 Gt C/year are withdrawn from the atmosphere, used for weathering of terrestrial carbonates and subsequently transported to the sea by rivers. The figure was calculated by Kempe (1979, see reference list), based on previous work by Livingston (1963), Baumgartner and Reichel (1975) and others. Furthermore, 0.16 - 0.2 Gt C/yr. in dissolved allochthonic humic materials are carried by the world's rivers, if we extrapolate from the few values we have, mainly from the Amazon (Williams, 1968) and the Baltic Sea (Sen Gupta, 1973).
3. If there were a continously operating sink of this kind, withdrawing 2 Gt C/yr., the peak atmospheric CO_2 concentration would be reduced by about 8% in the high fossil fuel scenarios, depending on the assumptions regarding present and future biotic releases of CO_2 to the atmosphere. Furthermore, the past atmospheric CO_2 increase (before 1958) must have been different from what is generally assumed, for a very large fraction of the anthropogenic CO_2 release must have been taken up by this sink, rather than by the sea. (I briefly discussed this hypothetical case in Bach et al., Man's Impact on Climate, 1979). The values for river discharge could be wrong by a factor of 2 or 3, due

to insufficient data. I doubt that they can be 10 times as high. I was unable to find detailed data on the organic carbon contained in the sediment load of rivers. However, if the total sediment load were 10 to 20 Gt/yr. (Holeman, 1968; Gregor, 1970; Ocgens et al., 1976), an average organic C content of 10-20% in the sediment load would have to be assumed. This would require some kind of concentration mechanism, for only a few soils on earth contain humus in quantities as high as this. If, for natural reasons, 2 Gt/yr. or organic C have been buried in freshwater and coastal sediments, huge quantities must have accumulated during the Holocene, which should be easy to detect. If, rather, the deposition were a manmade phenomenon, I would hesitate to consider this as a sink for atmospheric CO_2. It would probably result from an accelerated soil erosion, where organic C is merely transported from the land to the sea.

Keeling:
The CO_2 buffer factor is known to be over 9 unless calcium carbonate is dissolving in surface CO_2 water. But if that is occurring so as to reduce the effective factor, to say 7, which you suggested is a low possibility, it can only continue as long as such calcium carbonate is available. This is probably only for a few more years and thus cannot influence predictions for the time of highest CO_2 inputs into the atmosphere in the next century.

Hampicke:
I agree. All I want to point out is that even _if_ the buffer factor were as low as 7 at present, as it is sometimes suggested in the literature, the peak atmospheric CO_2 concentration would be only slightly reduced in both the high-fossil and the low-fossil case of my prediction.

Quirk:
Why does your forecast of CO_2 concentration to the year 2025 vary so little for scenarios A B C & D? The diffusion rates vary by a factor of 5, it would seem there should be a greater difference between the scenarios.

Hampicke:
With atmospheric CO_2 increasing exponentially, a variation in the diffusion rate by a factor of 5 will result in a variation of the ventilated ocean volume by a factor of less than 5, or 2.2. In a high fossil fuel scenario, the percentage of anthropogenic CO_2 taken up by the ocean at peak atmospheric concentration is only slightly more than 20% (the airborne fraction reaching almost 80% due to a high buffer factor). Thus, a doubling of the ventilated ocean volume will roughly take up another 20% of the CO_2 produced by man so that the atmospheric excess will be reduced

from 80% of the cumulative release to about 60%, or by 25%. In a low fossil fuel scenario, the difference should be larger due to a smaller buffer factor. However, a larger diffusivity is coupled in the scenarios with a higher future CO_2 release from biotic sources. Unlike the high fossil fuel scenario, in the low fossil fuel scenario this kind of release is large, compared to the fossil release. Therefore, the effect of a higher diffusivity is set off here by a larger combined CO_2 release, and the atmospheric increase is not reduced substantially.

Waltz:
In view of the fact that GCMs are essentially timeless in their results, can you give us some idea of the prospects for exploring the dynamics of climatic changes which would result from increases in atmospheric CO_2?

Gates:
The GCMs' simulated statistics, including variances and higher-order moments, must first be systematically determined and the levels of natural (model) variability identified as a function of variable, season and location. When this is done for a variety of boundary conditions and alternate physical parameterizations, a statistical search for significant anthropogenic climate changes can be made.

Keeling:
Svante Arrhenius in Lehrbuch der Kosmischen Physik (1903) gives a detailed description of the Greenhouse Effect of increased CO_2 and recapitulates earlier studies. Thus the CO_2 problem has been recognised for a long time. Can you estimate the time delay in the warming of air owing to ocean heat capacity, knowing the strength of the seasonal solar heating variation and the seasonal temperature variation of the ocean water as a function of depth in the water?

Gates:
I thank you for pointing out Arrhenius' early discussion of the greenhouse effect of increased CO_2. The seasonal variation of solar radiation only partly controls the seasonal variation of sea-surface temperature; upwelling and evaporation are also important locally. In bulk, the seasonal amplitude of SST can be used to estimate an effective interacting water depth on the time scale of seasons, but the actual response is distributed unequally in the vertical. The seasonal lags of ocean temperature with depth do give, however, an estimate of vertical thermal conductivity, which should be useful unless deeper water is involved.

Oke:
The very recent (unpublished) work of S. B. Idso makes the point that the research on the thermal effects of doubling atmospheric CO_2 may be in a methodological 'rut' and needs alternative approaches. He gives such an alternative using measurements and derived surface forcing temperatures resulting in warming of about one order of magnitude less than those currently being suggested and given in your first slide. Would you please comment on the existence of such a 'rut' and whether his figures are a surprise to you?

Gates:
Although I have not seen Idso's work, his results are not surprising to me. Our research serves to show the climate's sensitivity to treatments of the ocean and sea ice; it is certainly possible that these could be modeled differently than at present, with different results. In searching observed climatic data for an early CO_2-induced signal, however, I would start with the models' indications.

Waltz:
Although they were few in number, there were some cases in which the temperature curves (with CO_2 increases) transcended the envelope curves, how do you interpret those cases?

Gates:
The CO_2-changed experiments should first be extended for several simulated years, since they (like the control) will exhibit natural interannual variability. The statistical significance of the apparent warming in the mid-latitude summer hemisphere can then be determined adequately; that it occurs there is probably related to the land's location which is consistent with the generally expected result.

Lovins:
One of the troubling features of the IIASA model described this morning is that it seems to assume both that long run energy demand is quite insensitive to price and that long term energy prices will be comparable to our lower-than-present prices (rather than several times higher as implied by the assumed supply system). Did I correctly understand that your and Dr. Allen's 27-TW scenario implicitly makes similar assumptions?

Rotty:
No, we have attempted to consider increasing fuel prices in our economic studies. This is reflected both in Dr. Allen's projection for the controlled economy countries (through consideration of marginal E/GDP ratios applied to reduced economic growth caused

by prices of fuels) and by reduction of growth ratio in the LDC's from those used in the past, which did not consider effects of fuel price changes by 1 to 2% (e.g. from 5% to 3 or 4%). One can certainly debate the relation of fuel cost to inflation, and inflation to fuel costs; we have attempted to give consideration to these matters.

Keeling:
Since it is probably impossible to predict whether fossil fuel production will peak when one-fourth or half of the ultimate resource is used up, we should perhaps not try to decide, but present the predicted consequences of both possibilities. A similar strategy seems to me to be needed in assessing many of the social economic aspects of the CO_2 problem, as these in general are not subject to precise physical modeling such as is possible for, say, the CO_2 buffer factor.

Rotty:
I don't believe one should approach the future fossil fuel production by assuming a fraction at which the production will peak; we haven't done this. The form of the equation for fitting future production is useful mainly for calculating the cumulative production from the present to some future time and it makes little difference what the mathematical expression is if it fits the data (and projected data) - at least up to the maximum production point. The assessment of many of the social/economic aspects is fraught with many uncertainties. One only argues "plausibility". For planning one must consider such things in spite of the uncertainties involved. Unfortunately the precision in the social/economic areas does not allow modeling in the same manner as, say, the buffer factor.

Laurmann:
In order to get at the reliability of your estimates of projected fossil fuel use, I would like you to address the question of stability of the socio-economic constraint. What are the chances that a change of magnitude comparable to that occurring eight years ago might happen again?

Rotty:
Such changes in the social/economic constraint can be expected to occur within the next fifty years. Such "plausible" scenarios as developed in our paper are based on trends and changes that can be, to some extent, anticipated on the basis of population, geographical, and resource statistics. We believe they are helpful in planning policy and planning long range research priorities. They do not attempt to include the effects of discontinuities, i.e. either through cartel economic restraints

or major wars. However, the scenarios considered should have some resilience to enable the social/economic system to recover from such discontinuities.

PART III

CLIMATIC IMPACTS OF OTHER GASES AND AEROSOLS

THE EFFECTS OF CHLOROFLUOROMETHANES ON CLIMATE

D. H. Ehhalt

Institut für Chemie III: Atmosphärische Chemie,
Kernforschungsanlage, 5170 Jülich, FRG.

ABSTRACT. The chlorofluoromethanes, $CFCl_3$, CF_2Cl_2 are chemically quite inert. Their atmospheric destruction is slow and takes place essentially in the stratosphere. With present release rates the tropospheric CFM concentration is expected to increase until a steady state of a 2.1 ppb is reached. Their climatic impact proceeds through a variety of pathways which makes an evaluation more complex and more uncertain than for other trace gases. The CFM have strong infrared absorption bands between 9 to 12 µm, wavelengths at which the atmosphere is relatively transparent. Thus their direct contribution to the greenhouse effect will be relatively large. One-dimensional radiative convective models predict an eventual increase in the surface temperature of about 0.4° C.

Other climatic influences are caused by the CFM induced depletion of the stratospheric ozone layer. Calculations with a one dimensional chemical model predict that stratospheric O_3 concentrations will be depleted by an average of 15 % - 20 %, if the present rates of release of CFM are continued. The depletion of stratospheric O_3 is expected to have the following consequences for the radiation budget and the surface temperature of the earth: (1) A smaller fraction of the solar visible and UV radiation is absorbed in the stratosphere and the stratosphere cools. (2) More solar short wave radiation reaches the earth's surface; the surface temperature increases. (3) Because of the decrease in stratospheric O_3 concentration and temperature, less infrared radiation is emitted to the troposphere; the surface temperature decreases. Effects (2) and (3) nearly cancel leaving a small net cooling of the earth's surface. (4) The change in stratospheric temperature changes the stratospheric

W. Bach, J. Pankrath, and J. Williams (eds.), Interactions of Energy and Climate, 243–255.

density structure and thus influences the stratospheric circulation.

The overall effect, direct and O_3 mediated, on the surface temperature is a small warming of 0.15 K with an uncertainty range of 0-0.6 K.

INTRODUCTION

The possible impact of chlorofluoromethanes (CFM) on the environment was recognized in 1974 when Molina and Rowland pointed out that chlorine atoms released by the CFM might deplete the stratospheric ozone layer. Their argument ran as follows: The chlorofluoromethanes, mainly $CFCl_3$ and CF_2Cl_2 do not occur naturally. They are completely man-made. The CFM are methane-like molecules whose H-atoms have been replaced by Cl- and F-atoms. Due to this structure the CFM have low toxicity, and high chemical stability as well as a number of other desirable physical chemical features. They are, therefore, widely used commercially for example as propellants of aerosol spray cans and as refrigerants in all kinds of cooling units. With that usage most of the CFM produced are eventually released to the atmosphere at a rate approaching 1 million tons per year. Because of their stability the CFM are virtually not destroyed in the troposphere. Only after they have reached stratospheric altitudes above 25 km, the CFM begin to be exposed to UV radiation from the sun with wavelengths shorter than 220 nm, and begin to be photolyzed. By that photolysis chlorine atoms are liberated which destroy O_3 in a catalytic cycle. The resulting depletion of the stratospheric O_3 layer could be considerable.

This fact caused much concern. Atmospheric O_3 is the major absorber of solar UV radiation. A depletion of stratospheric O_3 would increase the incidence of biologically damaging UVB radiation (290 - 310 nm) at the earth's surface at a ratio of about 2 % increase in UVB for a 1 % depletion in stratospheric O_3. Climatic considerations entered somewhat later. First discussed were possible climatic effects due to the changes in the stratospheric O_3 profile. But as was pointed out by Ramanathan, 1975, the CFM have also a direct influence on the infrared (IR) radiation budget, which appears to have a bigger effect on the tropospheric temperature change than the stratospheric O_3 depletion. This discussion will concentrate on the combined climatic effects resulting from the CFM; the biological effects caused by increased UVB will be left aside. To remain concise I will also restrict the discussion to one release scenario of the CFM and I will present conclusions derived from 1-D chemical and radiative convective models only.

I will however point out some of the uncertainties introduced by the simplified calculations. Generally speaking the same reservations have to be made as in the modelling of the climatic impact of CO_2 (cf. Gates, 1980).

The release scenario chosen is shown in Figure 1. Up to 1978 figures on the $CFCl_3$ and CF_2Cl_2 release rate have been published by the industry. They are accurate to $\pm$ 5 %. Until 1974 the release rates increased more or less exponentially. The decrease since 1974 reflects the economic recession, the attempts of government regulation of production and release and possibly also a negative attitude among the consumers. Taking note of this - at least temporary - decrease the future release rates were assumed to remain constant at the 1977 values (NAS 1979a). It is of course quite possible that future production and release of CFM will accelerate again.

In any case continued release leads to further accumulation of the CFM in the atmosphere. The increase in the average tropospheric CFM mixing ratio corresponding to the chosen release scenario is also shown in Figure 1. The concentration increase is controlled by the constant release rate and the increasing stratospheric destruction. The destruction rates are essentially determined by the time it takes to transport the CFM to the altitudes where they are photolyzed. The resulting atmospheric lifetimes are 50 years for $CFCl_3$ and 80 years for CF_2Cl_2. The latter is somewhat larger because CF_2Cl_2 absorbs at shorter wavelength and has, therefore, to be transported to higher altitudes for efficient removal. As a consequence the steady state concentration of CF_2Cl_2 is enhanced and approached more slowly than that of $CFCl_3$. The steady state mixing ratios are 0.7 ppb for $CFCl_3$ and 1.5 ppb for CF_2Cl_2. The present mixing ratios are 0.14 and 0.26 ppb respectively. Consequently we expect a future increase in CFM concentration by about a factor of 5. This increase is of course uncertain - not only because of the projected release rates, but also because of the errors in the estimated lifetimes which alone amount to about 20 %.

Finally, the lowest panel shows the resulting mean stratospheric O_3 depletion in % of the present column density as predicted by a 1-D chemical model of stratospheric chemistry. The perturbed steady state O_3 column density is about 20 % lower than the present one which averages about 0.3 cm STP O_3. The new state is approached with a characteristic time which is essentially a weighted mean of the $CFCl_3$ and CF_2Cl_2 lifetimes, i.e. about 65 years. Due to uncertainties in a number of model parameters, mostly in chemical kinetics, the predicted ozone depletion is uncertain with a relative error, $\sigma/\Delta\ O_3 \simeq 0.40$, as indicated by the hatched area (cf. NAS 1979a, Ehhalt et al., 1979).

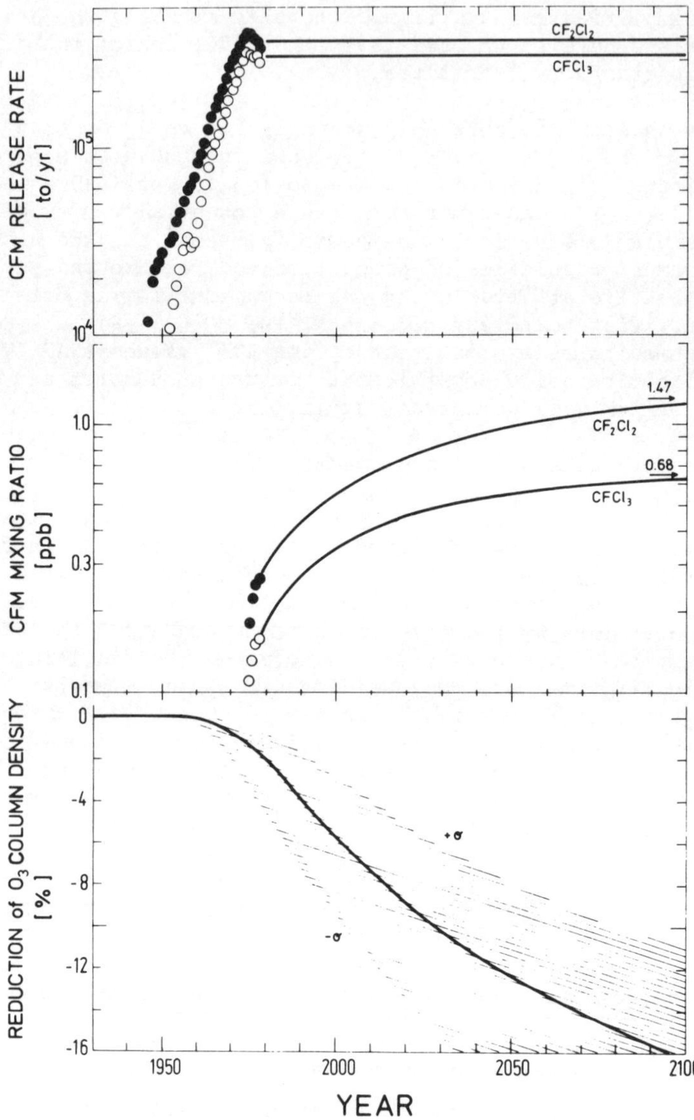

Figure 1: Variation with time of the release rates, and mean tropospheric mixing ratios of $CFCl_3$ and CF_2Cl_2. The lowest panel gives the resulting reduction of the O_3 column density in the stratosphere. The past release rates (data points) were taken from UAP Bulletin 79-5, the past mixing ratios from Singh et al. (1979). The O_3 depletion is taken from Ehhalt (1978a).

Since both the CFM and O_3 absorb in the IR and O_3 also in the visible and UV wavelength range, the predicted changes in the future concentrations of these gases will cause perturbations in the global radiation balance. This in turn will influence the temperatures at the earth's surface as well as within the atmosphere itself.

Direct effect of IR absorption by CFM

Surface warming: As for CO_2 the main radiative effect of the CFM proceeds through absorption of the upward directed thermal radiation from the earth's surface. Since the IR emission by the CFM takes place at the lower atmospheric temperatures and in all directions, the net infrared radiation flux emitted to space is reduced. This additional trapping of surface radiation provides a contribution to the so-called greenhouse effect and tends to increase the surface temperature. $CFCl_3$ and CF_2Cl_2 have strong absorption bands in the spectral range between 8.5 and 12 μm positioned away from the bands of the more abundant IR absorbers CO_2 and O_3. These spectral band positions make the CFM very efficient IR absorbers, and the added greenhouse effect is relatively large despite the small increase in absolute concentration: The increase in CFM to a total steady state concentration of 2.1 ppb is estimated to cause an increase in surface temperature of $\Delta T_S = 0.4$ K, as calculated from a 1-D radiative convective model (cf. Ramanathan, 1975; NAS 1979(b)). As long as the CFM blanket remains optically thin, the surface temperature change increases linearly with the atmospheric CFM concentration: A doubling of the future CFM concentration will lead to a doubling in the surface temperature increase.

Tropopause warming: The presence of CFM modifies also the temperature profile within the atmosphere. As shown in Figure 2 for CF_2Cl_2 the present vertical CFM distribution is virtually uniform in the troposphere. In the stratosphere the CFM mixing ratio decreases rapidly with altitude. Such a distribution is also expected for the future CFM profiles. The expected change in the local atmospheric heating rate can be estimated from Figure 3 which is taken from Dickinson et al. (1978) and has been calculated for the tropical atmosphere. The heating rate increases in the upper troposphere and lower stratosphere due to the presence of CFM. However because of the rapidly decreasing CFM concentration in the stratosphere the effect is larger in the upper troposphere. Indeed the major effect of the changed heating rate profile is an increase in the temperature of the tropical tropopause by about 2.5 K for the expected increase in total CFM concentration to 2.1 ppb.

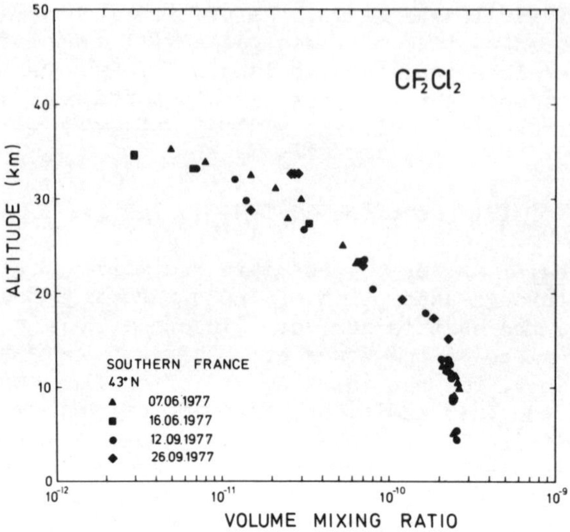

Figure 2: Vertical profile of the CF_2Cl_2 mixing ratio over Southern France (Ehhalt, 1978b).

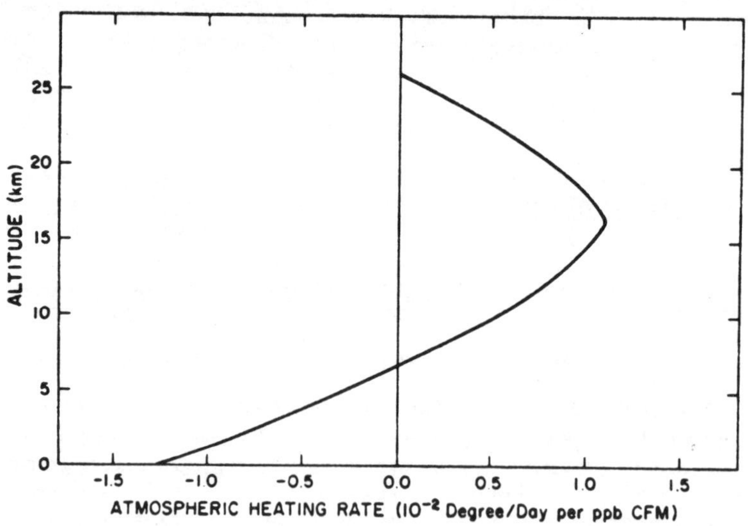

Figure 3: Vertical profile of the change in atmospheric heating rate per ppb CFM for the tropical atmosphere (Dickinson et al., 1978).

This effect is potentially important because the cold equatorial tropopause temperature is believed to be the limiting factor for the transport of tropospheric water vapor into the stratosphere. A 2.5 K warming of the tropical tropopause could increase the stratospheric water vapor concentration by as much as 60 % with appreciable effects on the ozone chemistry and the radiation budget of the stratosphere.

In summary the salient points of the direct IR effect on the temperature from the CFM are:

1.) A $CFCl_3$ increase to a new steady state concentration of 0.7 ppb.
2.) A CF_2Cl_2 increase to a new steady state concentration of 1.5 ppb.
3.) An increase of the global surface temperature of 0.4 K.
4.) An increase in the tropical tropopause temperature of 2.5 K.

All these conclusions are based on a scenario of continued CFM release at the 1977 rate. Other scenarios will lead to different predictions. Finally it is worth remembering that like CO_2 the CFM reside mainly in the troposphere. To a certain approximation their direct influence on surface temperature distribution should be similar to that of CO_2: In particular, a two- or three-dimensional climate model which predicts a latitudinal increase of the surface temperature change for CO_2 should predict the same for the CFM (cf. Dickinson and Chervin, 1979).

Climatic effects of stratospheric O_3 depletion

In addition to the direct climatic effects of the CFM there are climatic effects caused by the stratospheric O_3 depletion. The stratospheric chlorine chemistry involved in the depletion is rather complex and I will only outline its basic features. There are two catalytic cycles directly involved in the destruction of O_3:

$$Cl + O_3 \rightarrow ClO + O_2 \quad (1)$$
$$ClO + O \rightarrow Cl + O_2 \quad (2)$$
$$\text{net} \quad O + O_3 \rightarrow O_2 + O_2$$

and

$$ClO + HO_2 \rightarrow HOCl + O_2 \quad (3)$$
$$HOCl + h\nu \rightarrow Cl + OH \quad (4)$$
$$Cl + O_3 \rightarrow ClO + O_2 \quad (1)$$
$$OH + O_3 \rightarrow HO_2 + O_2 \quad (5)$$
$$\text{net} \quad O_3 + O_3 \rightarrow O_2 + O_2 + O_2$$

The first cycle consisting of reactions (1) and (2) is the one originally proposed by Stolarski and Cicerone (1974) and is the one responsible for most of the stratospheric O_3 destruction. Its completion requires O-atoms which are rare in the lower stratosphere. At these altitudes the second cycle consisting of reactions (3), (4), (1), (5) eventually takes over because it does not require O-atoms. Inclusion of the second cycle is quite recent - in fact the pertinent rate constants were published only last year.

This raises an interesting point: Since the first recognition of the Cl perturbation of stratospheric O_3, additional reaction pathways of importance were discovered at a rate of one in every two years. As a consequence the complexity of the chemical scheme to be considered has increased, and despite the fact that today's knowledge is much greater than five years ago, the presently estimated uncertainty of the predicted O_3 depletion appears not to be any better. The presently estimated uncertainty (± 40 %, see Figure 1), however, should be more realistic.

There are, of course, other reactions besides the destruction reactions (1) to (5) which control the Cl-atom concentration in the stratosphere. Figure 4 gives an overview over the major pathways of stratospheric Cl chemistry as we understand it today. The concentration values given in Figure 4 show that much of the stratospheric chlorine is stored in the form of HCl, $ClONO_2$, and HOCl, compounds which do not react with O_3. Hence a large fraction of stratospheric chlorine is inactivated with respect to O_3 destruction. Nevertheless the destruction rate of O_3 due to chlorine remains considerable.

Despite the simplification, Figure 4 already contains 16 trace species and 14 reactions. The whole chemical system is so complex that a prediction of the future O_3 depletion can only be done by computer simulation. Such calculations show that the O_3 reduction is not uniform with altitude. Figure 5 shows the steady state depletion calculated from a 1-D model (NAS, 1979a, p. 183). Clearly the reduction is largest - close to 50 % - at 40 km altitude and decreases to 10 % by about 20 km altitude. Incidentally the calculation of this particular O_3 reduction profile does not include the catalytic cycle (3) - (6). Its incorporation would increase the O_3 reduction in the lower stratosphere to about 20 %. Integrated over all altitudes one obtains an overall depletion of the O_3 column density of about 20 %. In absolute numbers: The present column density averaging about 0.3 cm STP O_3 will be reduced to 0.25 cm STP O_3. Whereas most models agree fairly well in the calculated reduction of total column density, the agreement is poor for the vertical profile of O_3 reduction.

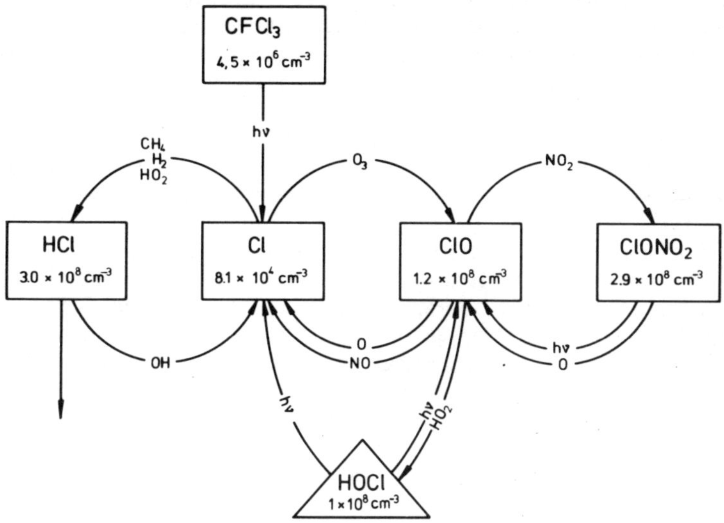

Figure 4: Schematic flow diagram of the major reaction paths of stratospheric chlorine. The concentrations shown have been calculated for 30 km altitude by a 1-D model (cf. Ehhalt, 1978a). All trace species, shown except, HOCl have been identified in the stratosphere.

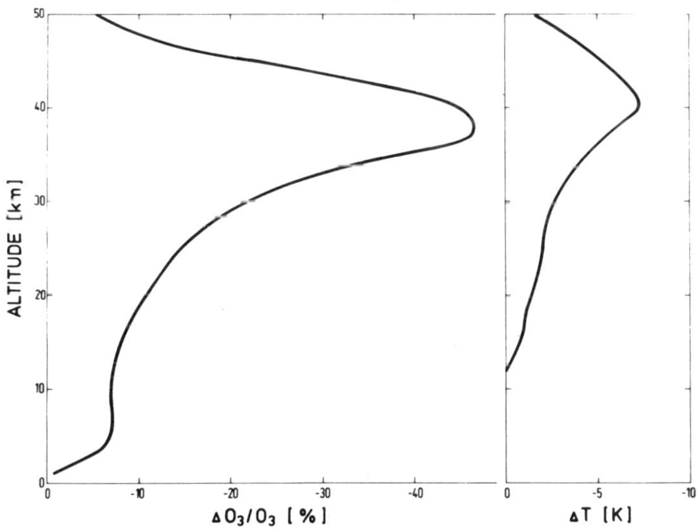

Figure 5: Vertical profile of the relative O_3 depletion induced by the CFM and the resulting change in stratospheric temperature (NAS report, 1979a).

The total O_3 reduction as well as the structure in its vertical distribution have consequences for the stratospheric and tropospheric radiation budget and thus also for the respective temperatures. As mentioned above, O_3 is the major atmospheric absorber of solar radiation in the near UV. It also absorbs in the visible spectral range. The absorbed radiation energy is dissipated locally as thermal energy. Absorption of solar radiation and IR by O_3 is the major radiative energy input in the stratosphere. Thus a decrease in O_3 causes a decrease in stratospheric temperature. The expected temperature decrease calculated from a 1-D radiative convective model is shown in the right hand side of Figure 5. The temperature profile follows the trend of the O_3 depletion profile, and gives a considerable temperature decrease in the stratosphere with a maximum of about -7 K at 40 km. Such relatively drastic temperature changes will also modify the stratospheric density field and wind field. So the stratosphere is quite strongly affected by the expected O_3 decrease. It appears however that little of the dynamic change couples into the troposphere as has been shown by the general circulation calculations of Mahlmann et al., 1978.

The effects on the tropospheric temperature from the stratospheric O_3 depletion are mainly caused by changes in the radiation budget:

a) The solar UV and visible radiation which - because of the reduced O_3 - is no longer absorbed in the stratosphere penetrates into the troposphere. Thus more radiation energy, about 0.3 %, is absorbed mostly at the earth's surface. This effect tends to increase the surface temperature. The effect depends mainly on the reduction of the total O_3 column density.

b) Because of the stratospheric cooling but also because of the decrease in O_3 concentration less IR radiation is emitted from the stratosphere into the troposphere. This effect cools the troposphere, mainly its upper part. The extent of this cooling is most sensitive to the O_3 reduction in the lower stratosphere, that is, it depends strongly on the vertical distribution of the stratospheric O_3 reduction.

With the present O_3 reduction and its vertical distribution the decrease in the stratospheric IR radiation to the troposphere is the bigger factor. The net result of the O_3 decrease is a decrease of the earth's surface temperature. The presently estimated reduction in O_3 column density of about 20 % leads to a global decrease in surface temperature of about -0.25 K. This value was estimated using a O_3 reduction uniform with altitude (NAS, 1977, p. 37). Again the temperature decrease scales linearly with the O_3 decrease. A doubling of the O_3 reduction would lead to twice as large a decrease in surface temperature. It should be noted that the change in surface temperature caused by the stratospheric O_3 depletion is obtained as the difference of two large effects of which the IR cooling is rather uncertain. Correspondingly the surface temperature effect from O_3 depletion is very uncertain, and even may change sign (NAS, 1977).

In summary, the salient features of the climatic changes derived by 1-D models from the CFM induced stratospheric O_3 depletion are:

1.) A strong temperature perturbation everywhere in the stratosphere with a maximum around 40 km altitude and about -2 K at the tropopause.

2.) A weak and rather uncertain decrease of the global surface temperature of -0.25 K.

The total effect on the surface temperature of the earth is obtained by adding the direct contribution and that from the O_3 depletion. The result is a net warming of 0.15 K. The errors of this estimate are large. The NAS report (1979, (b)) estimates an error range of 0.0 to 0.6 K.

CONCLUSION

In conclusion I would like to add a few words of caution. The results presented were derived from 1-D studies. Already 2-D chemical model calculations exist which show that the stratospheric O_3 depletion caused by CFM depends on latitude: The decrease in column density is a factor of two bigger at high latitudes than in the tropics. This will cause a similar change with latitude in the O_3 controlled part of the radiation budget. Secondly there are other anthropogenic perturbations of stratospheric O_3 for example through N_2O or CO_2. The net change of O_3 from all perturbations cannot be obtained from the sum of the single perturbations mainly because of the interference between the N_2O and CFM perturbation. To obtain a realistic O_3 change the various perturbations have to be treated simultaneously. This has not been done mainly because of the lack of a realistic scenario for the global N_2O increase (cf. Isaksen, 1980). Until then the estimated climatic changes caused by anthropogenic O_3 perturbations will have even larger errors than given above.

REFERENCES

Dickinson, R.E., Liu, S.C., and Donahue, T.M.: Effect of Chlorofluoromethane Infrared Radiation on Zonal Atmospheric Temperatures, J. Atmos. Sci. 35, pp. 2142-2152, 1978.

Dickinson, R.E., and Chervin, R.M.: Sensitivity of a General Circulation Model to Changes in Infrared Cooling due to Chlorofluoromethanes with and without Prescribed Zonal Ocean Surface Temperature Change, J. Atmos. Sci. 36, pp. 2304-2319, 1979.

Ehhalt, D.H.: Chlorfluormethane und ihr Einfluß auf die stratosphärische Ozonschicht, Berichte der Kernforschungsanlage Jülich, Jül-1545, 1978a.

Ehhalt, D.H.: Cryogenic Sampling Measurements of Stratospheric Gases, Proc. WMO Symp. on the Geophysical Aspects and Consequences of Changes in the Composition of the Stratosphere, Toronto, 26-30 June 1978b, WMO-NOSU pp. 71-78.

Ehhalt, D.H., Chang, J.S., and Butler, D.M.: The Probability Distribution of the Predicted CFM-Induced Ozone Depletion, J. Geophys. Res. 84, pp. 7889-7894, 1979.

Gates, W.L.: Climatic Effects of Increasing Atmospheric CO_2 Levels, This volume

Isaksen, I.S.A.: The Impact of Nitrogen Fertilization, This Volume

Mahlmann, J.D., Sinclair, R.W., and Schwarzkopf, M.D.: Simulated Response of the Atmospheric Circulation to a Large Ozone Reduction, Proc. WMO Symp. on the Geophysical Aspects and Consequences of Changes in the Composition of the Stratosphere, Toronto, 26-30 June 1978b, WMO-NOSU, pp. 219-220.

Molina, M.J., and Rowland, F.S.: Stratospheric Sink for Chlorofluoromethanes: Chlorine Atom Catalyzed Destruction of Ozone Nature 249, pp. 810-812, 1974.

National Academy of Sciences, Response to the Ozone Protection Sections of the Clean Air Act Amendments of 1977: An Interim Report, NAS Washington DC, 1977.

National Academy of Sciences, Stratospheric Ozone Depletion by Halocarbons: Chemistry and Transport, NAS Washington DC, 1979a.

National Academy of Sciences, Protection against Depletion of Stratospheric Ozone by Chlorofluorocarbons, NAS Washington DC, 1979b.

Ramanathan, V.: Greenhouse Effect due to Chlorofluorocarbons: Climatic Implications, Science 190, pp. 50-51, 1975.

Singh, H.B., Salas, L.J., Shigeishi, H., Smith, A.J., Scribner, E., and Cavanagh, L.A.: Atmospheric Distributions, Sources, and Sinks of Selected Halocarbons, Hydrocarbons, SF_6 and N_2O, Stanford Rès. Inst. Report Proj. 4487, 1979

Stolarski, R.S., and Cicerone, R.J.: Stratospheric Chlorine: A Possible Sink for Ozone, Can. J. Chem. 52, pp. 1610-1615, 1974.

Upper Atmospheric Programs Bulletin, UAP 79-5, pp. 2-6, 1979.

THE IMPACT OF NITROGEN FERTILIZATION

I.S.A. Isaksen

Institute of Geophysics, University of Oslo
P.O. Box 1022, Blindern, Oslo 3, Norway.

ABSTRACT. A main source of anthropogenic nitrous oxide in the future will be the increasing use of industrially produced nitrogen fertilizers. Biological processes in the soil convert fixed nitrogen to N_2O which is emitted to the atmosphere. Large uncertainties are connected with these production processes, and reliable estimates of their efficiencies are difficult to make. Atmospheric observations of N_2O show small temporal and latitudinal variations in the troposphere, which indicate a rather slow production. It is therefore not likely that any noticeable effect on climate or stratospheric chemistry will occur before late in the next century.

1. INTRODUCTION

Nitrous oxide (N_2O) may have a direct effect on the earth's climate through the absorption of long wave radiation in the 7.8 μm and 17.0 μm bands. Increased N_2O levels will increase the greenhouse effect and raise surface temperatures on the earth. It may also have an indirect effect on the climate through its influence on the stratospheric gaseous distribution. A key question in this respect is how it influences the ozone distribution. Ozone shields the earth's surface from lethal UV solar radiation ($\lambda<320$ nm), and affects the temperature distribution in the stratosphere through absorption of short and long wave radiation.

Although N_2O has been known to be present in the earth's atmosphere for more than 40 years (1), it was not before 1970 that its sources, sinks and distribution in the atmosphere became interesting. Crutzen (2) showed that it is the main source of nitro-

W. Bach, J. Pankrath, and J. Williams (eds.), Interactions of Energy and Climate, 257–268.

gen oxides in the stratosphere. This discovery stimulated a lively debate on the strength of its sources and sinks, and its effect on the stratospheric ozone. Most of the discussion has been about the possible role of human activities on the future levels of N_2O. A main difficulty in assessing the effect of N_2O arises from the fact that N_2O is produced by biological processes in the soil, and in the oceans. These processes are poorly known.

The human influence may either be indirectly through the use of industrially produced fertilizers or directly through the release of N_2O through combustion processes. An accurate evaluation of the importance of these processes is only possible if the natural production and loss processes are reasonably well known. Observations of the distribution and temporal variations seem at present to provide us with the best information of the N_2O turnover in the atmosphere.

2. THE ATMOSPHERIC N_2O DISTRIBUTION

Observations point to a rather uniform distribution of N_2O in the troposphere. The mixing ratios do, however, vary depending on the observational techniques used (3). While measurements using thermal conductivity detectors or the ratio of N_2O to CO_2 indicate values less than 300 ppbv, measurements using infrared absorption spectra and gas chromatographs with electron capture or helium ionization detectors point to levels above 300 ppbv. Pierotti and Rasmussen (3) have performed an extensive series of observations between 80^oN-90^oS over a 2 year period. They obtained an average mixing ratio of approximately 330 ppbv. The standard deviation was less than a few %. Similar values were obtained by other groups. Goldan et al. (4) measured average values of 326 ppbv at several locations in the troposphere during a 6 months period servational period with a standard deviation of 5.3 ppbv. Rasmussen and Pierotti (5) have conducted an intercalibration program for N_2O between different laboratories. All sets of data, except one, showed a rather consistent N_2O distribution in the troposphere. The average mixing ratio was 323.5 ppbv, with a standard deviation of 8.7 ppbv in the Northern Hemisphere. The conclusion was that the average N_2O distribution in the troposphere is in the range 315-330 ppbv. Observations in the upper troposphere from balloons and aircraft show similar consistency. Tyson et al. 6) find no significant hemispheric differences from aircraft for heights about 10 km. The average values were approximately 310 ppbv. Fabian et al. (7) observed 330 ppbv of N_2O in the upper troposphere over southern France (44^oN).

The use of observed temporal variations of N_2O to obtain residence time and thereby identify the strength of sources and sinks in the troposphere has revealed large discrepancies between the results obtained by different groups. Some of the first systematic observations of N_2O variations were started in 1966, and were performed by two different groups. In the USA Goody (8) measured N_2O over Massachusetts, while Junge's group (9) measured N_2O over Germany. Although they measured quite different absolute values, probably a result of the different methods used, both series of measurements showed a pronounced increase in N_2O levels (17%), starting at almost the same time. The measured increase took place during late 1967 and early 1968. The fact that this was obtained by different groups strongly supported the existence of large tropospheric sources and sinks. Junge (10) has given an expression where the atmospheric turnover time T is inversely proportional to the mean standard deviation σ_d: $\sigma_d T=0.14$, with σ given in per cent and T in year. Based on the early N_2O measurements, he obtained atmospheric turnover times of a few years, and concluded it was less than 10 years. The more recent series of observations (3), (4), show significantly smaller variations, and hence a considerably longer residence time for N_2O. The extensive observations made by Pierotti and Rasmussen (3) and by Goldan et al. (4) both show standard deviations of the order of 1%. Other observations made over the Pacific Ocean even showed standard deviations as small as 0.5%. All of these series indicate that the residence time is well above 10 years. As pointed out by Goldan et al. (4), it is worth noting that with the relatively small variations found in the N_2O levels, an unknown but possibly significant part of the variations is due to measurement errors, and does not reflect a variation in the N_2O levels. The residence times obtained must, therefore, be considered as lower limits only.

An interesting experiment of the N_2O variability was performed on the Princeton 3-D general circulation model (10). It was shown that with a uniform source in the troposphere, sufficient to balance the known photochemical sink in the stratosphere, changes in the wind pattern during the different seasons were sufficiently strong to result in standard deviations in the range 0.1-0.8%. These values are quite similar to what is obtained in some of the observations referred to above. In such cases the N_2O fluctuations may reflect changes in the transport processes rather than in the chemical activity in the troposphere. The use of Junge's formulation to obtain tropospheric turnover times will, therefore, not be valid. The small variations obtained in the recent observations point to a rather small biological production of N_2O.

3. ESTIMATES OF SOURCES AND SINKS

An intensive research activity has been carried out over the last 10 years to determine the sources and sinks of N_2O. However, the uncertainties are still large, and we do not know how the N_2O production is distributed over the earth's surface. The only sink which is reasonably well known is photochemical decomposition in the stratosphere (above ~25 km):

(1) $N_2O+h\nu \rightarrow N_2+O$
(2) $N_2O+O^* \rightarrow N_2+O_2$
(3) $N_2O+O^* \rightarrow 2\ NO$

$h\nu$ represents a solar photon, and O^* is an energetic oxygen atom (^1D state). Reaction (1) is the dominant loss reaction. Reaction (3) is of particular interest for stratospheric photochemistry as it provides the main production of nitrogen oxides. The strength of these sink processes depends on the rate at which the N_2O gas is supplied to the stratosphere. Based on a large series of stratospheric observations at different latitudes, Schmeltekopf et al. (1) obtain a flux into the stratosphere of $\sim 15\times10^{12}$g(N) per year. As the total atmospheric mass of N_2O is approximately 1500 Tg(N), the stratospheric loss corresponds to a residence time of 100 years.

Recent observations of the diurnal variation of N_2O over local areas, and of the distribution in the oceans indicate that the production takes place both through nitrification and denitrification processes. It is also reasonable to believe that oceans and land areas act as source and sink regions. Estimated production rates of N_2O in the soil by denitrification are usually based on the amount of nitrogen which is formed by fixation each year. Table 1 gives the most likely values for nitrogen fixation by different processes at present:

Table 1. The amount of nitrogen fixed each year by natural processes and by man's acitivity (references (12), (13), (14), and (15)).

Fixed in the soil:	180	Tg(N) per year
Lightening:	6	"
Combustion (1974):	20	"
Burning/Forest fires:	10	"
Fertilizers (1974):	40	"
	256	Tg(N) per year

The natural system is assumed to be in equilibrium. It is therefore believed that the amount of nitrogen (in the form of N_2 and N_2O) which returns to the atmosphere must balance the amounts fixed each year. Production of N_2O will therefore be determined by the ratio N_2O/N_2 in the denitrification process. The ratio varies strongly with soil conditions, and is difficult to determine. Values in the range 0.06-0.60 have been used as global averages for the N_2O/N_2 ratio in investigations of atmospheric N_2O production (12)-(15).

The observations of local variations over agricultural areas are not conclusive with regard to N_2O production or uptake. Brice et al. (16) observed marked diurnal variations over Great Britain with minimum concentrations in the early morning hours, indicating a surface sink. Matthias et al. (17) on the other hand, observed diurnal variations with a maximum in the morning. Cicerone et al. (18) have made similar measurements close to the ground in Michigan, USA (~0.5 m) and found only marginal variations over the day. In most cases the soil seemed to act as a source. It is not possible from these measurements to derive the magnitude of the N_2O sources and sinks.

Based on the estimates of nitrogen fixation, the following production rates of N_2O from land areas are given by different authors:

Table 2. Estimated N_2O release rates from land areas by denitrification.

Liu et al. (12)	14 - 50 Tg(N)	per year
Mc Elroy et al. (13)	50	"
Crutzen and Ehhalt (14)	7 - 50	"
Hahn and Junge (15)	6 - 65	"

The importance of the oceans for the atmospheric N_2O budget is equally difficult to assess. Hahn (19) and Hahn and Junge (15) found the oceans to be a large source area for atmospheric N_2O. This has not been confirmed by other investigators. The objection to their investigation has been that the results were based largely on a few observations from the North Atlantic, and that an extrapolation to global average values, strongly overestimated the production. McElroy et al. (13) found the oceans to be a large sink for atmospheric N_2O. Recently Cohen and Gordon (20) have estimated the production from a large number of N_2O profiles in the oceans from widely different locations. Their conclusion differs from that of McElroy et al. (17) in that they find nitrification, and not denitrification to be the process taking place in most of the ocean areas. They further find the oceans as a whole, to be a source region for atmospheric N_2O. Based on the rate of mineralization in oceans and on a ratio of N_2O/NO_3^- formation (0.1-0.4%) they find rather small N_2O production rates. Estimates of oceanic production or loss by the three groups discussed above are given in table 3. It should be clear that our knowledge of the natural production of N_2O is strongly limited. The most recent observations and estimates point to a small tropospheric production and destruction. This is in agreement with an atmospheric turnover time of the order of 20-100 years.

Table 3. Estimated sink and source strengths of oceanic N_2O production.

	sink	source	Tg(N) per year
McElroy et al. (13)	∿40		
Hahn and Junge (15)		16-160	"
Cohen and Gordon (20)		4-10	"

4. ANTHROPOGENIC SOURCES OF N_2O

The atmospheric concentrations of N_2O may increase due to man's activity in two ways: 1) indirectly through the fixation of nitrogen by industrial processes (see Table 1), or 2) directly through the release of N_2O to the atmosphere from certain combustion processes. In the former case the same uncertainties will apply to N_2O yield from fixed nitrogen as was discussed in the previous chapter. In addition, as the man-made releases are continuously increasing, the turnover time of fixed nitrogen in the soil is crucial for the N_2O production. In the following some of the estimates of this man-made production of N_2O will be discussed.

It is obvious from the values given in Table 1 that the man-made contribution to nitrogen fixation is significant at present. As a result of increased demand for food in the future, it has been assumed that the use of fertilizers may increase by as much as 6% per year well into the next century. This requires a production of ∿200 Tg(N) per year around the turn of the century. Similarly, combustion processes may produce as much as 50 Tg (N) per year. In the latter case nitrogen oxides (NO, NO_2) are emitted to the atmosphere, and rapidly oxidized in the atmosphere to nitric acid (usually within a few days). Nitric acid is taken up in the soil where it is converted to nitrate. It should be pointed out that the predicted increase in man-made nitrogen fixation toward the beginning of the next century, probably is strongly overestimated, as it at present seems likely that neither the production of fertilizers nor the release of nitrogen oxides will increase at such large rates as previously assumed.

Estimates of the turnover time of fixed nitrogen in the soil differ greatly. The discrepancies are first of all connected to the estimates of the fraction of the fixed nitrogen which returns to the atmosphere within a relatively short time (a few tens of years at most), and of the fraction, which is transported to non-agricultural systems like sediments, where it may be stored for thousands of years. Estimates given by the different groups give a turnover time which varies from a few years (15). to as much as several hundred or thousand years (12). In Table 4 some estimates of future increases in atmospheric N_2O as a result of the use of industrially produced fertilizers are shown. The large difference obtained for the beginning of the next century reflects the large differences adopted for turnover times in the soil.

Table 4. Estimated future increases in atmospheric N_2O levels (in %) due to the use of industrially produced fertilizers. The use of fertilizers is assumed to increase by 6% per year up to the year 2000, whereafter it will remain constant.

Year	2025	2050	2100
Liu et al. (12)	4	20	
McElroy et al. (13)			>100
Crutzen and Ehhalt (14)	15		60
Hahn and Junge (15)	50-70	90-140	100-160

Direct release of N_2O from combustion processes has been measured to take place both when coal and fuel oil are burnt (21), (22). Burning of coal seems to give the largest contribution to the direct release of N_2O. Estimates of the global release rate are in the range 1.6-2.1 Tg (N) per year. In view of the expected increase in the use of coal in the future, direct release of N_2O to the atmosphere from combustion processes (estimated to increase by 3.5% per year) may become an important anthropogenic N_2O source.

Long term observation series, which may reveal an anthropogenic influence on the atmospheric N_2O levels, are not at hand yet. A comparison between series of measurements is difficult to make, as the uncertainties with regard to the absolute N_2O level are large. Pierotti and Rasmussen (3) report two series of observations ten years apart made by the same group at the same location and time of the year, which showed only a small increase (<2%). If this is taken as an indication of the anthropogenic influence on N_2O, it has indeed been very small so far. It points to a rather small contribution from the use of fertilizers.

5. DIRECT TEMPERATURE EFFECTS FROM INCREASED N_2O

N_2O absorbs infrared radiation mainly in two regions of importance for atmospheric temperatures: ν_1=7.8 μm and ν_2=17 μm. In both regions the H_2O absorption is significant, although a noticeable fraction of the radiation escapes the earth's atmosphere. Estimates of the N_2O greenhouse effect have been performed by Yung et al. (23) and by Donner and Ramanathan (24). In both cases a one dimensional radiative convective model was used. A doubling of atmospheric N_2O levels (280 ppbv→600 ppbv) results in a global surface temperature increase of approximately 0.3-0.5 K. A further doubling in N_2O levels will raise surface temperatures by almost 1K as compared with the unperturbed N_2O levels.

In order to assess the importance of anthropogenic increases in N_2O for the earth's climate, it is necessary to know the time it takes for N_2O to be raised to levels which cause a significant temperature increase. If we assume a temperature increase of 0.5 K to be significant, N_2O levels have to be raised to approximately 600 ppbv. From the foregoing discussion, it is clear that estimates of N_2O increases are connected with large uncertainties, although it is most likely that man's influence on N_2O in the coming decades will be small. It is therefore felt that a doubling in N_2O levels from industrially fixed nitrogen will not take place before the end of the next century. Estimates of the future increase of CO_2 give a doubling in atmospheric levels around the year 2050. The corresponding estimates of the increase in average surface temperatures are of the order of 3 K (25). The N_2O effect on the average surface temperature, which is comparable to the effect of CFM absorption of infrared radiation, is therefore small compared to CO_2 surface heating. Nevertheless, it may contribute to the global heating in the future.

6. EFFECTS ON THE STRATOSPHERIC OZONE CHEMISTRY

N_2O is linked to the atmospheric gas chemistry through reaction (3) whereby nitrogen oxides are produced. Schmeltekopf et al. (11) estimated that approximately 10% of the N_2O loss in the stratosphere (∿1.6 Tg (N)) per year proceeds through this reaction. NO_x accounts for most of the ozone loss through the catalytic cycle:

$$NO+O_3 \rightarrow NO_2+O_2 \qquad (4)$$
$$NO_2+O \rightarrow NO+O_2 \qquad (5)$$

$$\text{net effect: } O_3+O \rightarrow 2O_2$$

Increased NO_x levels due to increases in N_2O, will therefore reduce ozone through the above reactions. However, recent investigations have shown that the stratospheric NO_x chemistry is highly complex, partly due to its strong coupling to other chemical cycles. One example is the efficient coupling with the chlorine chemistry:

$$ClO+NO_2 \overset{M}{\rightarrow} ClONO_2 \qquad (6)$$

As ClO participates in catalytic ozone destruction in a way similar to nitrogen oxides, an increase in NO_x reduces the ozone destruction by chlorine. This is of particular importance between 25 and 35 km in the stratosphere where a major fraction of chlorine goes into $ClONO_2$. An important consequence of this with regard to future ozone depletion, is that the effect of increased N_2O depends on the future development in stratospheric chlorine levels. This is illustrated in Table 5 where estimated changes in total ozone are given as a result of doubled N_2O concentrations in the atmosphere, at different chlorine content.

Table 5. Percentage change in total ozone as a result of doubled N_2O, when the chlorine content in the stratosphere is kept constant at different levels. Numbers in paranthesis give the approximate years when the chlorine is calculated to reach the given levels under the assumption of constant CFM release (on the 1977 level).

Chlorine (ppbv)	1.3 (1980)	2.6 (2000)	4.2 (2040)
$\Delta\Sigma O_3/\Sigma O_3$	2.6	4.0	6.0

If the predicted N_2O increases take place, it seems likely that the ozone depletion by CFMs which is estimated to be

7% in the year 2000, and 13% in the year 2040, is modified noticeably. However, one should be cautious about drawing too strong conclusions as an increase in N_2O due to man-made sources seems to be rather slow. It is worth noting that increases in the N_2O levels have a drastic effect on the ozone distribution with height. Ozone densities are strongly reduced above 30 km, where they are also reduced by chlorine reactions, while they increase below 30 km. The result is a marked change in the stratospheric temperature profile. This will most likely lead to changes in the stratospheric circulation patterns, and in the radiative balance of the atmosphere. How important it is for the surface temperature is difficult to assess. Ozone absorbs in the infrared region at 9.6 μm. Future reductions in stratospheric ozone are estimated to give a surface cooling of a few tenths of one degree due to a reduced greenhouse effect. On the other hand, more shortwave radiation is able to penetrate to the troposphere. The effect is to raise surface temperatures with approximately the same value. The net effect of stratospheric ozone changes is therefore calculated to be small in the troposphere (27). The effects may, however, be strongly enhanced during certain situations (e.g. sudden warmings).

REFERENCES

(1) Adel, A.: 1939, Astrophys. J. 90, pp. 627-628.

(2) Crutzen, P. J.: 1970, Quart. J. Roy. Meteorol. Soc. 96, pp. 320-325.

(3) Pierotti, D., and Rasmussen, R. A.: 1977, J. Geophys. Res. 37, pp. 5823-5832.

(4) Goldan, P. D., Bush, Y. A., Fehsenfeld, F. C., Albritton, D. L., Crutzen, P. J., Schmeltekopf, and Ferguson, E. E.: 1978, J. Geophys. Res. 83, pp. 935-940.

(5) Pierotti, D., and Rasmussen, R. A.: 1978, Geophys. Res. Lett. 5, pp. 353-355.

(6) Tyson, B. J., Vedder, J. F., Arvesen, J. C., and Brewer, R. B.: 1978, Geophys. Res. Lett. 5, pp. 369-372.

(7) Fabian, P., Borchers, R., Weiler, K. H., Smidt, U., Volz, A., Ehhalt, D. H., Seiler, W., and Müller, F.: 1979, J. Geophys. Res. 84, pp. 3149-3154.

(8) Goody, R. M.: 1969, Planet. Space Sci. 17, pp. 1319-1320.

(9) Junge, C. E.: 1974, Tellus 26, pp. 477-488.

(10) Levy II, H., Mahlman, J. D., and Moxim, W. J.: 1979, Geophys. Res. Lett. 6, pp. 155-158.

(11) Schmeltekopf, A. L., Albritton, D. L., Crutzen, P. J., Goldan, P. D., Harrop, W. J., Henderson, W. R., McAfee, J. R., McFarland, M., Schiff, H. I., Thompson, T. L., Hofmann, D. J., and Kjome, N. T.: 1977, J. Atmos. Sci. 34, pp. 729-736.

(12) Liu, S. C., Cicerone, R. J., Donahue, T. M., and Chameides, W. L.: 1977, Tellus 29, pp. 251-263.

(13) McElroy, M. B., Elkins, J. W., Wofsy, S. C., Yung, Y. L.: 1976, Rev. Geophys. Space Phys. 14, pp. 143-150.

(14) Crutzen, P. J., and Ehhalt, D. H.: 1977, Ambio 6, pp. 112-117.

(15) Hahan, J., and Junge, C. E.: 1977, Z. Naturforsch. 32a, pp. 190-214.

(16) Brice, K., Eggleton, A., and Penkett, S.: 1977, Nature 268, pp. 127-129.

(17) Matthias, A. D., Blackmer, A. M., and Bremner, J. M.: 1979, Geophys. Res. Lett. 6, pp. 441-443.

(18) Cicerone, R. J., Shetter, J. D., Stedman, D. H., Kelly, T. J., and Liu, S. C.: 1978, J. Geophys. Res. 83, pp. 3042-3050.

(19) Hahn, J.: 1974, Tellus XXVI, pp. 160-168.

(20) Cohen, Y., and Gordon, L. I.: 1979, J. Geophys. Res. 84, pp. 347-353.

(21) Pierotti, D., and Rasmussen, R. A.: 1976, Geophys. Res. Lett. 3, pp. 265-267.

(22) Weiss, R. F., and Craig, H.: 1976, Geophys. Res. Lett. 3, pp. 751-753.

(23) Yung, Y. L., Wang, W. C., and Lacis, A. A.: 1976, Geophys. Res. Lett. 3, pp. 619-621.

(24) Donner, L., and Ramanathan, V.: 1980, J. Atm. Sci. 37, pp. 119-124.

(25) Manabe, S., and Wetherald, R. T.: 1980, J. Atm. Sci. 37, pp. 99-118.

(26) Isaksen, I. S. A., Hesstvedt, E., and Stordal, F.: 1980, Nature 283, pp. 189-191.

(27) Ramanathan, V., and Dickinson, R. E.: 1979, J. Atmos. Sci. 36, pp. 1084-1104.

CLIMATIC EFFECTS OF ANTHROPOGENIC TRACE GASES

V. Ramanathan

National Center for Atmospheric Research*
P. O. Box 3000, Boulder, Colorado 80307

ABSTRACT. Human activities inject several radiatively and photochemically active trace gases into the atmosphere. Of these trace gases, CO_2 is considered to be the most important one, because of its climatic effects. However, recent photochemical and climate modeling results indicate that anthropogenic sources of trace gases other than CO_2 (which include CH_4, N_2O, CFMs and O_3) may contribute as much as 40% of the surface warming to the combined surface warming effects of CO_2 and these other gases. Roughly 15% of the total warming is due to the projected increase in tropospheric O_3 caused by fossil fuel sources of CO, NO and CH_4. These findings may be subject to a large range of uncertainty resulting from: photochemical modeling uncertainties; possible large errors in estimates for the natural sources and sinks for these trace species; and, finally, a big source of uncertainty is the scenarios for future increase in emission rates for these trace species. Hence, the problem, because of its potential importance, should be examined in more detail.

1. INTRODUCTION

Radiatively and chemically active trace gases are being injected into the atmosphere by human activities at ever increasing rates. Our understanding of the extent to which such human activities can alter the chemical and radiative energy balance of the earth-atmosphere system is constantly being revised and modified. This makes it hazardous to summarize the climatic

*The National Center for Atmospheric Research is sponsored by the National Science Foundation.

W. Bach, J. Pankrath, and J. Williams (eds.), Interactions of Energy and Climate, 269–280.

impacts of human activities, since the summary may become outdated before it appears in print. Hence, the present summary should be considered as a snap-shot of the scope of the problem as it is understood currently.

A particularly striking and important example of the changes that are taking place in our understanding of the problem concerns the possible effects of fossil fuel combustion on tropospheric O_3. Recent research (1) suggests that an appreciable fraction of observed O_3 is produced within the troposphere by photochemical oxidation of CO, CH_4 and other hydrocarbons in the presence of nitric oxide, NO. These species have both biogenic and anthropogenic origin and, in addition, are synthesized within the troposphere by chemical processes. The principal anthropogenic source is combustion of fossil fuels and, on a global basis, the combustion source for each of CO, CH_4 and NO is roughly 25% (or larger) of the natural source (2)-(6). Current photochemical modeling studies estimate that if the present growth rate of emission of these gases is continued into the next century, the tropospheric O_3 concentration may double its present value (7)-(9).

This finding has important climatic implications since, as first estimated by Fishman *et al.* (8) and later by Hameed *et al.* (9), a doubling of tropospheric O_3 can warm the climate by about 0.9 K. As pointed out by Ramanathan and Dickinson (10), although the troposphere contains only 10% of total O_3, due to pressure broadening of the 9.6 μm band of O_3, the IR opacity of tropospheric O_3 is nearly the same as that of stratospheric O_3. Consequently, a uniform percentage increase of O_3 in the troposphere has about the same or more heating effect on the surface-troposphere system as the same percentage increase of stratospheric O_3.

This relatively new aspect of the problem concerning impacts of human activity on climate has not been considered in the earlier estimates (11),(12). In this review paper we will adopt Flohn's (12) approach and compare the model-predicted climatic effects of alterations in trace gases due to human activities with those due to an increase in atmospheric CO_2.

2. NATURE OF THE PROBLEM: INTERACTIONS BETWEEN CLIMATE AND PHOTOCHEMISTRY

The human activities that have been identified, up to date, to have potential climate effects are those arising from combustion of fossil fuels, the use of fertilizers, and emission of refrigerants and propellants. These anthropogenic sources, since they inject radiatively and chemically active trace gases, perturb the photochemical and radiative energy balance of the earth-atmosphere system. Furthermore, the two processes, viz., photochemistry and radiative energy transfer, are strongly coupled and hence realistic estimates of climatic effects of human activities must be made with models that account for the coupling between climate and photochemistry. The coupled nature of the problem is illustrated in Figure 1.

The injection of trace gases, CO_2, CH_4, N_2O and CFMs which have strong IR bands can directly warm the surface by enhancing the atmospheric greenhouse effect; while CO, CH_4 and NO injection can warm the climate indirectly by enhancing the production of tropospheric O_3. For example, the oxidation of CO in the presence of NO leads to O_3 production through the catalytic reaction set (13):

R1) $CO + OH + M \rightarrow CO_2 + H + M$
R2) $H + O_2 + M \rightarrow HO_2 + M$
R3) $HO_2 + NO \rightarrow NO_2 + OH$
R4) $NO_2 + h\nu \rightarrow NO + O$
R5) $O + O_2 + M \rightarrow O_3 + M$

where the net effect is

R6) $CO + 2O_2 \rightarrow CO_2 + O_3$.

The oxidation of CH_4 proceeds similarly, initiated by the reaction with the hydroxyl radical

R7) $CH_4 + OH \rightarrow CH_3 + H_2O$

resulting in a net effect of

R8) $CH_4 + 4O_2 \rightarrow H_2O + H_2 + CO + 2O_3$.

Since R1 and R7 are the reactions which initiate the oxidation process, the chemical effects of fossil combustion sources are determined by their influence on the highly reactive OH which plays a significant role as a cleansing agent in the tropospheric carbon, nitrogen and sulfur cycles (13). However, reactions R1 and R7 seem to be a major sink for tropospheric OH, which has led to the conclusion (14) (15) that increased CO emissions will deplete OH through R1 accompanied by a consequent slowing down of CH_4 oxidation in R7 with a net result that increased CO emission will lead to increased CH_4 concentrations. This increase in CH_4 and tropospheric O_3 will enhance the previously mentioned direct warming of trace gases.

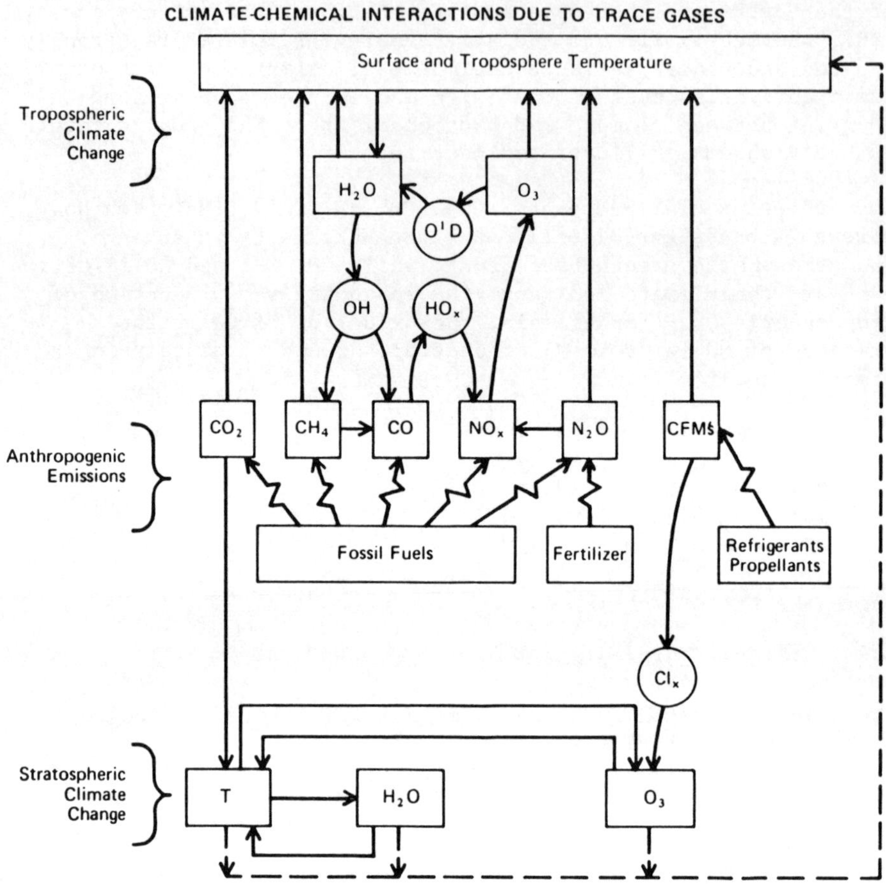

Figure 1

The problem, however, is more complicated because the concentration of OH is strongly coupled to O_3 concentrations and climate (9). The production of OH results from O_3 photolysis

$$\text{R9)} \quad O_3 + h\nu \rightarrow O_2 + O^1(D)$$
$$\text{R10)} \quad O^1(D) + H_2O \rightarrow OH + OH \quad .$$

Consequently OH depletion by R1 and R7 is partly compensated by R10 since increased tropospheric O_3 (due to CO and CH_4 emissions) implies more O^1D through R9. Furthermore, the surface and tropospheric warming caused by trace gases would lead to more H_2O within the atmosphere (16). Hameed et al. (9) estimate that the negative feedback effect of this process reduces the increase in CH_4 by about 30% and the surface warming by about 10%.

As indicated in Fig. 1, stratospheric-tropospheric interactions have also to be taken into account because of the following effects of the trace gases: (i) the Clx compounds from CFMs cause a substantial reduction in stratospheric O_3 which can alter the tropospheric climate; (ii) enhancement in methane oxidation by R8 can increase stratospheric H_2O. A perturbation in stratospheric H_2O can also arise from changes in tropopause temperature since it is believed tropical tropopause temperature controls stratospheric H_2O through the cold-trap mechanism; (iii) stratospheric O_3 decrease can increase tropospheric OH through R9 and R10 since the solar radiation penetrating to the troposphere is modulated by stratospheric O_3 (15).

None of the published studies on anthropogenic effects have accounted for all of the climate-photochemical coupling discussed above and illustrated in Figure 1. Logan et al. (7), have examined in detail the mutual interactions between various chemical processes by performing photochemical calculations of simultaneous injections of CO, NO_x, N_2O and CFMs. Hameed et al. (8) have performed coupled climate-photochemical model calculations for emissions of CH_4, CO and NO_x. Coupled model calculations including simultaneous emissions of all the trace gases in Fig. 1 are currently being undertaken (17). For the present, we will adopt photochemical model results for the change in trace constituents and adopt the model described in (10) and estimate the perturbations in radiative energy budget and surface temperatures. In particular, we feel it is important to consider the changes in various trace species simultaneously, since the IR bands of CFMs, N_2O, CH_4, O_3 and CO_2 overlap with those of H_2O and with each other in some cases, which introduces nonlinearities. Furthermore, the O_3 effect on climate depends very strongly on the vertical distribution of O_3 perturbation and hence it is important to include simultaneously the effects of stratospheric O_3 decrease and tropospheric O_3 increase.

3. FUTURE CHANGE SCENARIO: TRACE GASES

The assumed scenario for potential alterations in trace gases from the various human activities illustrated in Fig. 1 is shown in Table 1. The values are compiled from a variety of sources including observations, carbon cycle models, photochemical models and future energy consumption models, in the following manner:

(i) $\leq$ 1940: CO_2 and N_2O are observed values. CH_4 mixing ratio is reduced by 0.1 ppm from its observed value in 1975 based on photochemical modeling estimates (7),(14) that combustion-related emission of CO since pre-industrial era has resulted in a 5-10% increase of CH_4 from its pre-industrial value.

(ii) 1975: CO_2, CH_4 and N_2O are observed values. Tropospheric O_3 is increased by 10% from its pre-industrial value based on Fishman *et al.*'s (8) suggestions that the roughly 20% larger O_3 in NH troposphere than in SH troposphere is due to combustion sources of CO, NO_x and hydrocarbons. The CFM concentrations are observed values (18).

Table 1. Scenario for global trace gas alterations.

	Mixing Ratios by Volume					% O_3 Change[3]		
				CFMs (ppb)				
Year	CO_2 (ppm)	CH_4 (ppm)	N_2O (ppm)	F11	F12	TR*	ST*	Column
≤1940[1]	300	1.4	0.3	0	0	0	0	0
1975	330	1.5	0.3	0.1	0.17	10	0	1
2025	500	2.1	0.45	0.5	0.8	50	**	-6.4
>2050[2]	600	3.3	0.6	0.7	1.5	100	**	-6.5

[1]Refers to period prior to 1940.

[2]Refers to steady state conditions which is assumed to occur after 2050.

[3]The percentage change in O_3 refers to the change from its pre-industrial value.

*TR: troposphere, alteration is assumed to be uniform between ground and 12 km;

*ST: stratosphere, alteration is a function of altitude.

**See text for vertical distribution of O_3 change.

(iii) 2025: (a) CO_2: The assumed buildup of CO_2 to 500 ppm agrees within 10% of the projections given in (19), (20), and (21). These references give various scenarios; the scenario chosen here makes the following two basic assumptions: the ratio of fossil to nonfossil (solar, nuclear, etc.) source of energy consumption remains roughly the same as that of present; and the emission of CO_2 increases by a factor of about 4 from its 1975 value, corresponding to a growth rate of energy use by roughly 2.5%.

(b) CH_4 and tropospheric O_3: The assumed increase in CH_4 and O_3 from 1975 to 2025 is taken from photochemical model results (9) for a factor of 4 increase in emission rates of CO, CH_4 and NO_x from their respective 1975 values. This is the same factor used for the enhancement of CO_2 emission rates and the implicit assumption made in using the same factor as for CO_2 is that the ratio of CO, CH_4 and NO_x to CO_2 emission remains constant in time. This is an arguable assumption because roughly 66% of the global CO emission comes from transport (22) and, hence, for the CO/CO_2 emission ratio to remain constant the following two additional assumptions have to be made: The growth rate in the consumption of fossil fuels by transport is the same as that for the total consumption of fossil fuels; and future emission control devices will be ineffective in controlling CO (and also NO_x) emission rates. In this context, it should be noted that tropospheric O_3 is much more sensitive to NO_x emissions than to CO emissions. The former assumption may not be unreasonable since a study on global energy prospects for the period up to year 2000 (23) projects that, during this period, consumption of fossil fuels for transportation will have the same growth rate as that for total energy consumption. Another important limitation of this scenario is that it ignores effects of deforestation and biomass burning which could significantly impact CO and CH_4 increase. Deforestation and biomass burning can lead to smaller CO build up. But, if the deforested regions are replaced by industries, then the increase in NO_x emissions may lead to O_3 increases that are as yet poorly described by current models (7),(8),(9).

(c) CFMs and stratospheric O_3: CFM concentrations are taken from (18), which considers a case for which the 1977 emission rates are continued indefinitely into the future. The stratospheric O_3 reduction is inferred from (7), which estimates the steady state reduction in O_3. For the year 2025, we scaled the O_3 change from the steady state O_3 change based on the concentrations of CFMs in 2025 to that in steady state. It should be noted that, if total ozone depletion is as large as that predicted by models, worldwide use of CFMs will stop much before this period, due to international regulations.

(iv) $\geq$ 2050: This last case reflects our assumed steady state conditions and is chosen rather arbitrarily by assuming CO_2 emissions will be controlled when the concentration in the atmosphere has doubled from its pre-industrial value. The change in CH_4 and tropospheric O_3 are taken from Hameed et al. (9) by adopting their case in which CO, CH_4 and NO_x emission rates are increased by a factor of 8 from their values in 1975. Steady state values of CFMs and O_3 are taken respectively from (18) and (9). N_2O concentrations are taken from the lowest of the values predicted by (7). It should be cautioned that the time scales for achieving the steady state are vastly different for the various species, ranging from 2050 to 2150. This scenario, which may be unrealistic, is considered mainly to examine the sensitivity of our results to extreme scenarios.

4. CLIMATIC EFFECTS

Table 2 shows the climatic effects of trace gas perturbations described in Table 1. The effects are shown for two basic climatic variables, namely, the radiative heating of the earth-troposphere system (Q) and surface temperature (T_s). The global mean radiative heating is computed from the model described in (10), with the inclusion of CH_4 and N_2O IR bands as described in Donner and Ramanathan (24). The change in surface temperature is computed by letting

$$\Delta T_s = \lambda \Delta Q \quad , \tag{1}$$

where the parameter λ is the climatic feedback parameter and its value ranges from about 0.5 $K/(W\ m^{-2})$ for a radiative-convective model (with fixed cloud altitude) (25) to about 0.75 $K/(W\ m^{-2})$ for general circulation models (26). For the results shown in Table 2, we adopted the lower value of 0.5 $K/(W\ m^{-2})$ and we also repeated the calculations using a radiative-convective model and obtained ΔT_s values within 5-10% of those obtained from Eq. 1. We adopted the simpler procedure given by Eq. 1 to compute ΔT_s because λ is strongly model dependent and presenting ΔQ and ΔT_s results separately should facilitate intercomparisons between various model results.

The primary conclusion from Table 2 is that the trace gases other than CO_2 contribute roughly 40% of the total warming and the CO_2 increase contributes about 60%. Tropospheric O_3 increase contributes roughly 15% to the total warming. Following Flohn's (12) approach, Table 2 also shows the effective CO_2 fractional increase, f^c_{eff}, to simulate the combined effect of all trace gases, including CO_2 increase. The actual fractional CO_2

Table 2. Change in global mean radiative heating of earth-troposphere system (Q) and surface temperature (T_s).

	1975-2025				<1940->2050			
	f^c	ΔQ	ΔT_s	f^c_{eff}	f^c	ΔQ	ΔT_s	f^c_{eff}
Combined effect of all trace gases in Table 1	1.52	3.9	1.9	1.9	2	7.3	3.6	3.4
CO_2 alone	1.52	2.4	1.2		2	4.1	2	

$f^c = \frac{CO_2 \text{ in perturbed atmosphere}}{CO_2 \text{ in reference atmosphere}}$; The reference atmosphere is 1975 for the 1975-2025 column and it is "pre-industrial" atmosphere for the <1940->2050 column.

f^c_{eff}= is the effective value of f^c, which will yield the same ΔQ or ΔT_s as that of the combined effect of all trace gases.

increase, f^c, and f^c_{eff} can be related as follows. From Ramanathan *et al.* (26)

$$\Delta Q \propto \ell n(f^c) \quad , \tag{2}$$

where f^c is the fractional increase in CO_2; for example, $f^c = 2$ for a doubling of CO_2. From Eq. 2, we can derive

$$f^c_{eff} = [f^c]^n \quad ,$$

where

$$n = \frac{\Delta Q \text{ (all trace gases)}}{\Delta Q \text{ } (CO_2 \text{ only})} \quad .$$

The change in the atmospheric temperature distribution is shown in Fig. 2.

5. CONCLUSIONS

Recent photochemical and climate model results indicate that alterations of trace gases, CO, NO_x, CH_4, CFMs, N_2O and O_3, by human activities can be potentially important in determining future climatic changes. From available estimates for the future increase in the concentrations of these gases, it is indicated that these gases may contribute as much as 40% to

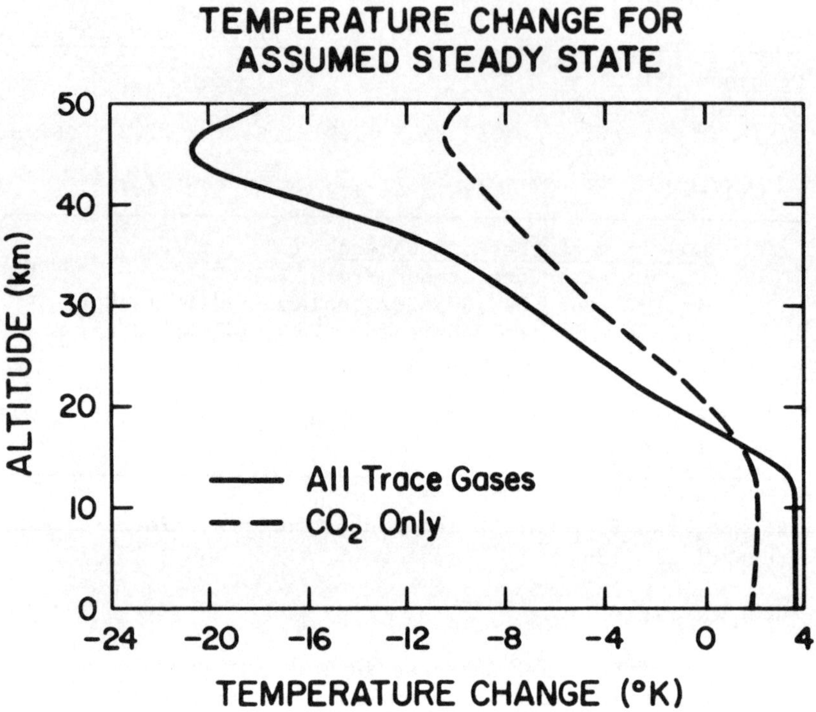

Figure 2

the combined warming of CO_2 increase plus these other trace gases. While this result indicates the potential importance of the problem, it could be subject to a large range of uncertainty because only about 25% of the warming effect of these trace gases is directly due to anthropogenic emission of radiatively active gases and the remaining 75% is a result of anthropogenic emission of photochemically active species which, through a complex series of chemical reactions, lead to an increase in radiatively active trace gases. Most of this photochemically induced increase in radiatively active gases occurs within the troposphere and the field of tropospheric chemistry is a relatively new field. Furthermore, the natural tropospheric sources and sinks for several of the gases, CH_4, CO and NO, are not known, to within perhaps a factor of 2 or more, and any likely significant changes in our current understanding of these sources and sinks will likewise impact the projected increase in these trace gases. For example, it is suspected that natural hydrocarbon emissions from vegetation and their conversion to CO is a major source for CO. Hence deforestation and biomass burning would decrease the CO growth assumed in existing model studies (9). Another large source of uncertainty is introduced by the assumption that the ratio of fossil source CO, CH_4 and NO_x to CO_2 remain constant in time. Nevertheless, the potential importance of the problem suggests a careful and detailed examination of the trace gas problem, particularly those involving increase in CH_4 and tropospheric O_3.

ACKNOWLEDGMENTS

The author is indebted to R. Chatfield and P. J. Crutzen for educating him on new developments in tropospheric photochemistry. He thanks E. Friedman for discussions concerning future energy policy and H. Howard for preparing the manuscript.

REFERENCES

(1) Fishman, J., and Crutzen, P.J.: 1978, Nature 274, pp. 855-858.

(2) Crutzen, P.J.: 1974, Tellus 26, pp. 47-56.

(3) Crutzen, P.J.: 1979, Ann. Rev. Earth Planet. Sci. 7, pp. 443-472.

(4) Ehhalt, D.H., and Schmidt, U.: 1978, Pageoph. 116, pp. 452-464.

(5) Seiler, W.: 1974, Tellus 26, pp. 116-135.

(6) Wofsy, S.C.: 1976, Ann. Rev. Earth Planet. Sci. 4, pp. 441-469.

(7) Logan, J.A., Prather, M.J., Wofsy, S.C., and McElroy, M.B.: 1978, Trans. Roy. Soc 290, pp. 187-234.

(8) Fishman, J., Ramanathan, V., Crutzen, P.J., and Liu, S.C.: 1979, Nature 282, pp. 818-820.

(9) Hameed, S., Cess, R.D., and Hogan, J.S.: 1980, submitted to J. Geophys. Res.

(10) Ramanathan, V., and Dickinson, R.E.: 1979, J. Atmos. Sci. 36, pp. 1084-1104.

(11) Wang, W.C., Yung, Y.L., Lacis, A.A., Mo, T., and Hansen, J.E.: 1976, Science 194, pp. 685-690.

(12) Flohn, H.: 1978, *Carbon Dioxide, Climate and Society*, Ed. J. Williams, Pergammon Press, pp. 227-238.

(13) Crutzen, P.J.: 1973, Pageoph. 106-108, pp. 1385-1399.

(14) Sze, N.D.: 1977, Science 195, 673-675.

(15) Chameides, W.L., Liu, S.C., and Cicerone, R.J.: 1977, J. Geophys. Res. 82, pp. 1795-1798.

(16) Manabe, S., and Wetherald, R.T.: 1975, J. Atmos. Sci. 32, pp. 3-15.

(17) Crutzen, P.J., and Ramanathan, V.: 1980, in progress.

(18) NAS Report: 1979.

(19) Williams, J.: 1978, *Carbon Dioxide, Climate and Society*, Ed. J. Williams, Pergammon Press, pp. 239-248.

(20) Rotty, R.M.: 1978, ibid., pp. 263-274.

(21) Keeling, C.D., and Bacastow, R.B.: 1977, *Energy and Climate*, National Academy of Sciences, pp. 72-95.

(22) Jaffe, L.S.: 1975, *The Changing Global Environment*, Ed. S. Fred Singer, D. Reidel Publishing CO., pp.

(23) Wilson, C.L.: 1977, *Energy: Global Prospects 1985-2000*. Report of the Workshop on Alternate Energy Strategies, McGraw-Hill Book Co.

(24) Donner, L., and Ramanathan, V.: 1980, J. Atmos. Sci. 37, pp. 119-124.

(25) Ramanathan, V., and Coakley, J.A.: 1978, Rev. Geophys. Space Phys. 16, pp. 465-489.

(26) Ramanathan, V.: 1977, J. Atmos. Sci. 34, 1885-1897.

AEROSOLS AND CLIMATE

William W. Kellogg

National Center for Atmospheric Research* and Aspen Institute for Humanistic Studies, Boulder, Colorado, U.S.A.

ABSTRACT. The atmospheric burden of particles, or aerosols, has been measurably increased by human activities, especially in industrialized regions and those where slash-burn agricultural practices are followed. Some of these aerosols are directly produced when fossil fuels or other materials are burned (soot, smoke, fly ash); others are the result of photochemical reactions involving organic molecules, oxides of nitrogen, and sunlight (smog); and a third source is the oxidation of sulfur dioxide, produced when sulfur-bearing fuel is burned, to sulfuric acid thereby forming sulfate particles of droplets.

In all cases, the resulting aerosols scatter and absorb both solar and infrared radiation, and therefore they influence the atmospheric heat balance. The question is the way in which they influence it, and the geographical extent of this influence. While the optical properties of natural and manmade aerosols and their distributions are obviously variable and uncertain, we probably know enough about the subject to venture some generalizations, as follows:

- Anthropogenic aerosols are typically washed out or rained out of the atmosphere after only a few days, so they are more of a regional than a global phenomenon; and futhermore most of them remain in the lower troposphere.

- These aerosols all absorb sunlight, and current estimates of their absorption-to-backscattering ratio indicate

*The National Center for Atmospheric Research is sponsored by the National Science Foundation.

W. Bach, J. Pankrath, and J. Williams (eds.), Interactions of Energy and Climate, 281–296.

that they will generally lower the net albedo of the surface-plus-lower-atmosphere when over land, increase the net albedo over the oceans.

- Since most of them are to be found over land, the overall effect is probably a regional warming.

- Their effect on atmospheric stability is such as to reduce the probability of convective precipitation in certain areas.

- The direction of their influence on precipitation processes due to their nucleating properties is still uncertain.

1. INTRODUCTION

There is no feature of mankind's extensive use of energy that is more obvious to the eye than the brownish palls of smoke and smog that lie over our major cities. Other vast palls of smoke lie over countrysides where slash-and-burn agricultural practices are followed. The fact that these clouds of particles floating in the air (or "aerosols") are so visible has raised the question as to what their effects on our climate might be--their effects on the esthetics of our environment are all too obvious.

This matter has attracted further attention as we have become increasingly aware of the fact that the climate of our cities and the regions downwind from them have been changing (1), (2). Can this change be due to the aerosols, perhaps in combination with the effects of waste heat and the peculiar surface characteristics of our urban centers? And can the climatic effects of the aerosols extend to entire regions, or even influence the climate of the globe?

Indeed, there was a widely quoted theory to the effect that "the Human Volcano," or the total emission of aerosol pollutants around the world, together with sporadic volcanic activity, accounted for the fact that northern hemisphere mean surface temperatures declined after about 1945. It was even stated that anthropogenic aerosols could cause such a pronounced cooling effect that, for a few decades at least, it could counteract the "greenhouse effect" warming of anthropogenic carbon dioxide (3). There is now little doubt that this conclusion is almost the opposite of the true situation.

In the following we will examine the factors that bear on this problem, which are (4), (5), (6):

- Regional trends in lower atmosphere aerosol concentrations.
- Geographical distribution of industrial aerosol pollutants.
- Optical properties of such aerosols and their consequent influence on atmospheric heat balance.
- Implications of all these factors for regional and global climate.

Unfortunately, there is no definitive information on any of these factors, but probably we know enough to be able to say what the direction of their effect will be, whether an influence toward heating or cooling of the lower atmosphere. This information is needed in order to specify whether anthropogenic aerosols will tend to reinforce or counteract the future warming effect of anthropogenic carbon dioxide. The WMO Commission for Atmospheric Sciences (CAS), the IAMAP Radiation Commission, and the GARP/JOC Board for the Climate Dynamics Sub-Programme have all recognized this important need and have recommended a coordinated effort to study the radiative effects of aerosols (7), (8).

It is important to emphasize that we are dealing here with particles emitted at or near the surface, even 300 m stacks, and that they tend to remain in the lower few kilometers of the atmosphere until being removed. This fact can be verified by looking out of an aircraft window almost anywhere over the United States or Europe, especially on a clear day in summer. These low-level aerosols are very distinct from the layers of stratospheric aerosols that spread worldwide following major volcanic eruptions. Such stratospheric aerosols probably do have a direct influence on the temperature of the troposphere (9), but that is another matter with which we will not be concerned here.

It has long been recognized that some kinds of particles act as effective condensation and freezing nuclei, and thus can change the cloud-physical processes that lead to rain and snowfall (10). However, there are so many variables in the theory of precipitation that we cannot even decide on the sign of the influences of aerosols as a whole, much less on its magnitude. In any case, it will not be dealt with further here.

2. REGIONAL TRENDS IN AEROSOLS

There have been a number of stations scattered throughout the world that have made measurements from which recent aerosol trends can be determined, and a few of these have records

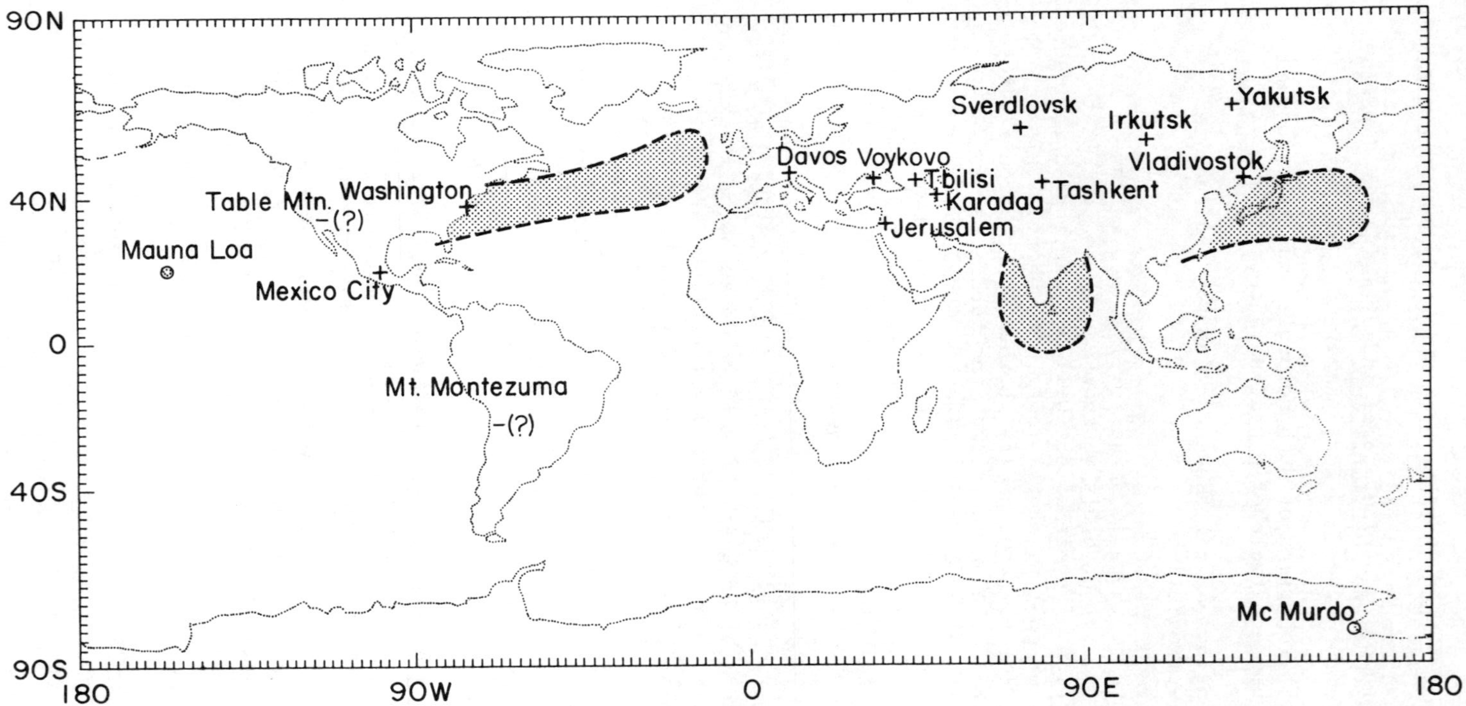

Figure 1. Map of stations measuring aerosol concentration trends over the past several decades. (+) indicates a general rise; shaded areas over ocean also experienced a rise.

extending back into the past century. Figure 1 shows where some of these stations are, and it can be seen that the great majority have seen upward trends in the past few decades, but definitely not all of them. These determinations over land were mostly based on turbidity observations, turbidity being a measure of the reduction in direct-beam radiation from the sun due to aerosol scattering and absorption (11), (12), (13). The determinations over the ocean were based on intermittent electrical conductivity measurements from research vessels over a long period of years, as summarized by Cobb (14).

The record, as can be seen, is far from complete. Other evidence, based on observations of visibility and with high-volume filter samplers at a large number of U.S. urban and non-urban air sampling stations, seems to indicate that, while the trend in the non-urban areas has shown some increase from small particles (15), (16) [presumably mostly secondary sulfate particles (17)], it has actually gone down in the majority of the large cities as a result of air pollution control measures (15). Nevertheless, from the large-scale climatic point of view it is probably the non-urban or widespread trend that is most important.

Although the evidence is somewhat conflicting, it is most likely that the ever-increasing use of coal and fuel oil containing sulfur will cause a continued rise in the sulfate aerosol burden in and downwind from industrialized areas, and this will continue until more widespread use of stack gas "scrubbers" or very low-sulfur fuel is mandated.

Another source of aerosols is from cultivated or overgrazed areas where the soil particles are raised by the wind, and also smoke from slash-and-burn agriculture. There is a widespread belief that all of these sources are on the increase, but facts are hard to find.

3. EXTENT OF INDUSTRIAL AEROSOLS

A distinction must be made between the regional effects of industrial aerosols and their influence on global heat balance. The latter will depend on the fraction of the earth that they cover.

Since there are not enough monitoring stations in the world to permit a global map of industrial aerosol distribution (see Figure 1), a simple exercise, which has already been reported elsewhere (4), (5), (6) was employed to deduce this distribution theoretically. The result is shown in Figure 2.

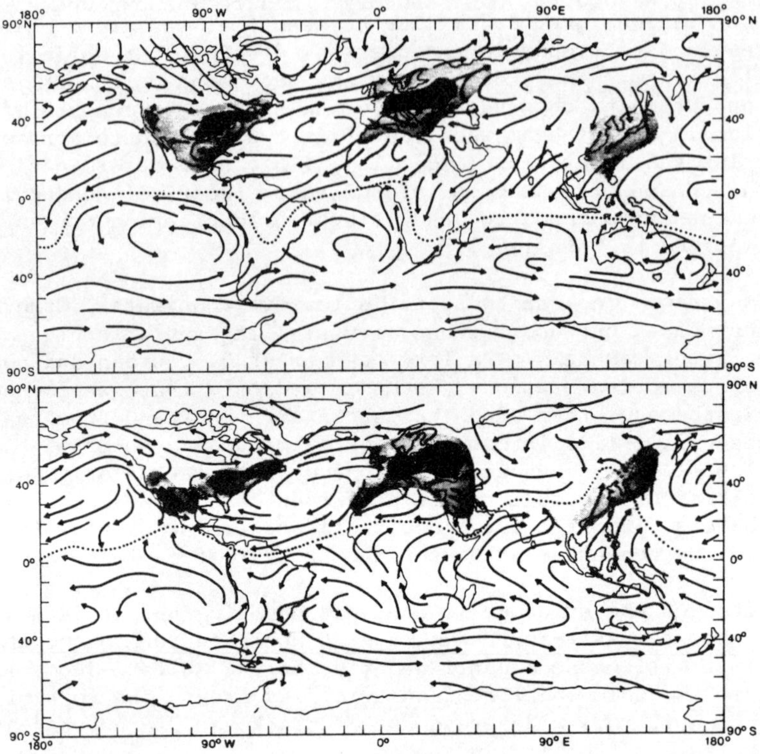

Figure 2. Map showing approximate extent of industrial aerosols, or "gross national pollution" (GNP). See text for explanation of how it was prepared. Top map is for January, bottom map for July.

The assumptions made in creating these rough maps were that aerosols in the lower troposphere, where they are subject to wash-out and rain-out, have a mean lifetime of about five days, that they are carried with the mean surface wind (which is usually not much different from that in the lower few kilometers), and that their source strength in each country is proportional to that country's national product.

The message in Figure 2 is fairly clear. It can be seen that industrial aerosols only cover a small fraction of the earth, and are therefore unlikely to have much effect on the global heat balance. Furthermore, the greater part of these aerosols, estimated to be about 80 percent, exists over the land area in and immediately downwind of their sources. This last point will be important when their effect on the radiation balance is considered.

4. OPTICAL CHARACTERISTICS OF AEROSOLS AND THEIR EFFECTS ON RADIATION BALANCE

Since, as pointed out earlier, anthropogenic aerosols released at or near the surface will remain in the lower few kilometers of the atmosphere, any effect they may have will be confined to that part of the "climate system". The most rational way to assess their net effect on the radiation balance of the surface-plus-lower-atmosphere is to ask how they influence the short-wave (solar radiation) albedo and the infrared emission to space. In this way one can determine whether more or less heat is being supplied to the lower atmosphere when aerosols are present. It is definitely not correct, in this context, to consider only the modified radiation balance at the surface.

The mathematical treatment of aerosol scattering and absorption has been fairly well worked out and described elsewhere (18), (19), (20), (21). It will be sufficient here to briefly review the physics involved and to apply the theory with some of the available data on aerosol optical properties.

A part of the scattered incoming solar radiation is directed upwards and therefore lost. However, the radiation absorbed by the particles heats the particles and thereby the atmosphere in which they are embedded, and one must take account of both the absorption of the incoming solar radiation and that which is reflected from the surface. The theory shows that the net effect of the aerosols on the "effective albedo" of the surface-plus-lower-atmosphere depends onthe absorption-to-backscatter ratio, a/b, and on the surface albedo, α. In this case the backscattered fraction, b or $\beta\omega$, must properly be considered as that which is scattered in the upward direction, and this in turn depends on the solar zenith angle. (ω is here the average single-scattering albedo of the particles.)

The results of the theory are summarized in Figure 3, which is the familiar diagram of Mitchell (22), subsequently refined by more detailed calculations by Coakley and Chylek (18) and others. Furthermore, theoretical work in which the phase-dependence of scattering by typical aerosols was taken into account suggests that the effect of changing solar zenith angle during the day at mid-latitudes may be approximated by choosing an average value for the factor β (in the denominator of the fraction a/b) equal to that for a zenith angle of about 65° (20).

The domain in Figure 3 above the solid line defines conditions when adding aerosols will decrease the effective albedo, and therefore result in more solar radiation being available to the system than would be the case without the aerosols. In short, this is the condition under which aerosols will cause a warming.

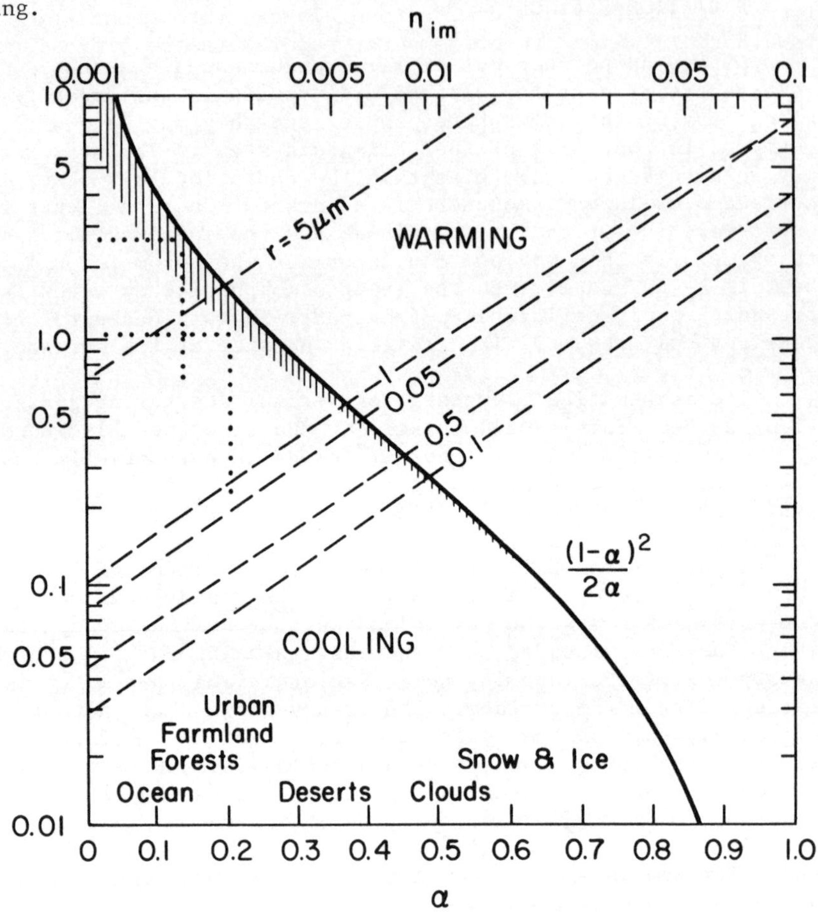

Figure 3. The relationships between absorption-to-backscattering ratio (a/b) and the critical surface albedo (α) (lower scale), and between a/b and the imaginary index of refraction (n_{im}) for a given radius of particle (upper scale). Calculations are for real part of index of refraction equal to 1.5, and for a solar zenith angle of 65° (an average value taking the globe as a whole). Shaded area indicates locus of some critical values when infrared radiation was taken into account along with solar radiation.

While the solid line represents the boundary between warming and cooling when shortwave solar radiation alone is considered, the effect of infrared radiation will be to move the boundary downwards as suggested by the hatched area. This is due to the fact that aerosols absorb and emit infrared radiation. The introduction of a low-level aerosol layer, throughout which there is a normal lapse rate of temperature, lowers the effective temperature of the surface-plus-lower-atmosphere as seen from space. There is thus less radiative loss to space and a consequent warming of the system. (The combined effects of short-wave and infrared radiation in an aerosol layer affect the lapse rate, or stability, of the lower atmosphere as well, and that will be discussed below.)

This infrared influence can be calculated for any specific situation, given the infrared optical properties of the aerosols, the temperature distribution, the water vapor distribution, and so forth. However, it is generally not as large an effect as that due to solar radiation, and the shaded area below the curve in Figure 3 is intended to suggest the loci of a few typical cases that have been considered by Coakley (private communication).

We must now find where the optical properties of anthropogenic (mostly industrial) aerosols fall on this diagram, defined in terms of their values of a/b. In Figure 3 we have replotted the theoretical calculations of Chylek et al. (20) that relate a/b (for a solar zenith angle of 65°) to particle radius and imaginary index of refraction, and then overlaid it on the same diagram as that relating the critical values of a/b to surface albedo. This makes it possible to have all the necessary relationships in one set of plots. Notice that for a given value of the imaginary index of refraction very large and very small particles have larger values of a/b than the size ranges between 0.1 and 0.5 μm, which is the common size range found under haze and smog conditions.

In recent years a number of measurements have been made of the size distributions and optical properties of various kinds of aerosols, and a summary of some of the better observations is given in Table 1. Other summaries of optical properties have been made by Eiden and Eschelbach (23), and Spänkuch (24). Other specific sets of measurements of optical properties are referenced in the table (25), (26), (27), (28), (29), (30), (31), (32). Except for the first two samples, which refer to soil and sand, these observations probably refer to industrial aerosols of uncertain age, and these are generally sulfate aerosols with inclusions of soot, the latter accounting for their absorption properties (27), (32). Where size distributions or mean radius were not provided we have assumed a radius of 0.5 μm as being typical of this kind of industrial aerosol.

It can be seen that about three-fourths of the values of a/b in Table 1 fall in the range of 2 to 4. (The average a/b is between 2.4 and 3.2). These aerosols warm the lower atmosphere

Table 1. Summary of Observations of Optical Properties of Aerosols

Type of Aerosol	Place Sampled	Mean Radius (μm)	Imag. Index of Refraction	Absorption/ Backscatter	Reference
Texas soil	Big Springs	5	0.005	3	(28)
Sahara dust	Atlantic	4	0.008	4	(29)
Mostly aged	Tyson '75	0.5*		1.58	(27)
industrial	Tyson '73	"		2.01	"
"	Milford	"		1.95	"
"	St. Louis U.	"		4.76	"
"	Washington U.	"		2.63	"
"	Henderson	"		5.20	"
"	Trout Farm	"		4.64	"
Rural (?)	Hall Mt.	"		0.83	"
Industrial	N.Y.C.	0.5*	0.04	1.8	(30)
Industrial	Mainz	0.5*	0.013-0.061	0.6-2.5	(25)
Industrial	Denver	0.5*		2.5	(31)
Industrial	Aspendale	0.5*	0.05 -0.083	2-3.5	(26)
Industrial	Deuselbach	0.2-0.9	0.01 -0.1	0.8-4	(32)
"	Deuselbach	2-8	0.001-0.01	0.8-7	"

* Assumed radius Note: "Industrial" includes automotive emissions.

when the surface albedo is above about 0.15, as can be seen by referring to Figure 3. When infrared warming is taken into account, the critical albedo is even lower, perhaps about 0.1, and that is less than the albedo of most typical land surfaces. However, it is still slightly larger than the average albedo of the ocean, which depends on zenith angle but is on the average about 0.08.

One further point needs to be made regarding the effects of shortwave absorbing aerosols. The above discussion has concerned itself with the effects of aerosols on the radiation balance of the lower atmosphere when there are no clouds. When clouds form in air containing such aerosols, their effect is theoretically to lower the albedo of the clouds, except for quite thin clouds (33), (34), and this has been verified observationally (35). It would be difficult to quantify this effect due, among other things, to the extreme diversity of cloud forms in the real atmosphere, but it is safe to say that the direction of the influence of aerosols in a cloudy atmosphere is to cause more solar radiation to be absorbed by the clouds and, therefore, to produce a warming of the atmosphere (34). Their effect in a cloudy atmosphere is therefore everywhere in the same direction as their effect in clear air over land areas.

5. EFFECTS OF LOW-LEVEL AEROSOLS ON VERTICAL STABILITY OF THE ATMOSPHERE

There have been a number of elaborate theoretical studies of the effects of industrial-type aerosols on the stability of the lower atmosphere (19), (21), (36), (37), and the detailed conclusions will not be discussed here. The main effects can, however, be quite easily understood by referring to the schematic diagram in Figure 4.

During the daytime the solar radiation dominates in determining the heat balance, and we can refer to the left-hand side of the figure. Since more solar radiation is absorbed at the top of the aerosol layer and less reaches the ground, the result will be an increase in the lower layer stability during the day. This will tend to inhibit afternoon convective rainfall, an important climatic influence pointed out some time ago by Bryson and Baerreis (38).

At night the infrared radiation will produce the opposite tendency, and lead to a slight destabilization of the lower layers due to cooling at the top of the aerosol layer. This is usually not a significant effect in terms of climate at the ground, though under special conditions it can help to set off nocturnal thunderstorms.

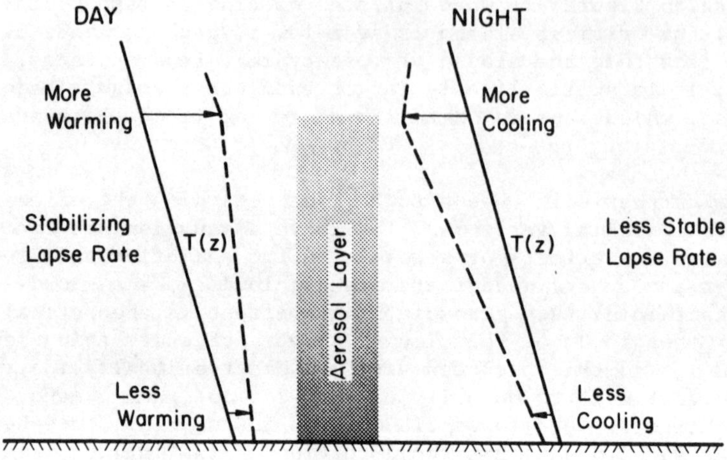

Figure 4. Schematic representation of the modification of lapse rate due to the radiative effects of a low-lying aerosol layer. During the day effects of solar radiation dominate, while at night infrared radiation alone plays a role.

6. DISCUSSION AND CONCLUSIONS

An attempt has been made here to assemble what appears to be a "best guess" about the overall effects of industrial or agricultural (slash-and-burn) aerosols on the radiation balance of the lower atmosphere. The situation is complicated by the variability of virtually all the factors involved and the lack of good information on many of them, but it is nevertheless possible to draw some conclusions from what we already know.

Anthropogenic aerosols, like natural aerosols in the lower atmosphere, have an average lifetime of the order of 5 days, so they can ordinarily only drift 1000 to 2000 km before being removed. They are thus mostly found in and downwind of their source regions and over land areas, and on a global basis can influence only a relatively small fraction of the earth.

An analysis of their effect on the albedo of the surface-plus-lower-atmosphere, the "effective albedo," indicates that over land with its moderately high surface albedo (compared to the oceans) they will cause the effective albedo to be less than the surface albedo, and therefore the effect of the aerosols is to warm the lower atmosphere. This conclusion is based on theory, together with observations of surface albedo and those of aerosol optical properties that have been published.

Over the oceans, on the other hand, in clear air their effect will probably be to raise the effective albedo, and therefore they may cause a cooling here. Recall, however, that for the most part industrial and agricultural aerosols are to be found over land; and furthermore the additional warming due to the influence of aerosols on infrared radiation transfer could partly counteract any cooling action due to shortwave effects alone.

The above statements refer to clear air, without clouds. When there are clouds in the aerosol-laden air, the effect of such aerosols is generally to lower the cloud albedo, thereby warming the atmosphere.

These considerations all lead to the conclusion that anthropogenic aerosols:

(a) can hardly have a significant effect on the overall radiation balance of the earth as a whole;

(b) probably cause a warming of the lower atmosphere over land in or near the source region, and therefore can influence climate regionally; and

(c) when over the ocean they may cause a small cooling in cloud-free areas, but in cloudy areas they also cause a warming over the ocean.

It is recognized that the influence of aerosols on the stability of the lower atmosphere in or near the source region could be even more important to regional climate than their effect on overall heat balance. In this paper the effects of aerosols on precipitation through their influence on cloud physics processes is not taken into account.

REFERENCES

(1) Dettwiller, J., and S.A. Changnon, Jr.: 1976, J. Appl. Meteorol. 15, pp. 517-519.

(2) Bach, W.: 1979, Chapt. 17 in Energy and the Environment: Interactions (Vol. 1), ed. by L. Theodor, A.J. Buonicore, and E.J. Rolinski, CRC Press, West Palm Beach, Fla.

(3) Bryson, R.A. and G.J. Dittberner: 1976, J. Atmos. Sci., 33, pp. 2094-2106.

(4) Kellogg, W.W., Coakley, J.A. and Grams, G.W.: 1975, Proc. WMO/IAMAP Symp. on Long-Term Climatic Fluctuations (Norwich, U.K.) WMO Document 421, Geneva (pp. 323-330).

(5) Kellogg, W.W.: 1977, Effects of Human Activities on Global Climate, World Meteorological Organization, Tech. Note No. 156 (WMO No. 486), Geneva, Switzerland.

(6) Kellogg, W.W., 1979: Proc. WMO Tech. Conf. on Regional and Global Observations of Atmospheric Pollution Relative to Climate (Boulder, USA), World Meteorological Organization, Geneva.

(7) CAS: 1978, Abridged Final Report of the Seventh Session (in Manila, Philippines), WMO, Geneva (Recommendation 3 and Annex D).

(8) GARP/JOC: 1978, Report of the Third Meeting of the JOC Board for the Climate Dynamics Sub-Programme (in Kiel, F.R.G.), ICSU/WMO, Geneva; see also: Report of JOC Working Group on Aerosols and Climate (in Boulder, USA), ICSU/WMO, Geneva.

(9) Schneider, S.H., and C. Mass: 1975, Science, 190, pp. 741-746.

(10) Hobbs, P.V., H. Harrison, and E. Robinson: 1974, Science, 183, pp. 909-915.

(11) Machta, L. and Telegadas, K.: 1974, Chapt. 19 in Weather and Climate Modification, ed. by W.N. Hess, John Wiley and Sons, New York, (pp. 687-726).

(12) Pivovarova, Z.S.: 1970, Proc. WMO/IUGG Symposium (Bergen, August 1968), WMO Tech. Note 104, Geneva, pp. 181-185.

(13) Ellsaesser, H.W.: 1975, Chapt. in The Changing Global Environment, ed. by S.F. Singer, D. Reidel, Dordrecht and Boston, pp. 235-272.

(14) Cobb, W.E.: 1973, J. Atmos. Sci., 30, pp. 101-106.

(15) NOAA: 1975, Geophysical Monitoring for Climatic Change, No. 3, Summary Report - 1974, ed. by J.M. Miller, Environ. Res. Labs., Nat'l. Oceanic and Atmosph. Admin., Boulder, Colorado.

(16) Husar, R.B., Patterson, D.E., Holloaway, J.M., Wilson, W.E. and Ellestad, T.G.: 1979, Proc., Fourth Symp. on Turbulence, Diffusion and Air Pollution (January 1979; Reno, Nev.), publ. by Amer. Meteorol. Soc., Boston.

(17) Weiss, R.E., Waggoner, A.P., Charlson, R.J. and Ahlquist, N.C.: 1977, Science, 195, pp. 979-981.

(18) Coakley, J.A., Jr., and Chylek, P.: 1975, J. Atmos. Sci., 32, pp. 409-418.

(19) Viskanta, R., Bergstrom, R.W. and Johnson, R.O.: 1977, J. Atmos. Sci., 34, pp. 1091-1103.

(20) Chylek, P., Grams, G.W., Smith, G.A. and Russell, P.B.: 1975, J. Appl. Meteorol., 14, pp. 380-387; see also W.J. Wiscombe and G.W. Grams: 1976, J. Atmos. Sci., 33, pp. 2440-2451.

(21) Atwater, M.A.: 1975, J. Appl. Meteorol., 14, pp. 1061-1071.

(22) Mitchell, J.M., Jr.: 1971, J. Appl. Meteorol., 10, pp. 703-714.

(23) Eiden, R., and G. Eschelbach: 1973, Zeitschrift fur Geophysik, 39, pp. 189-228.

(24) Spänkuch, D.: 1978, Zeit. fur Meteorologie, 28, pp. 119-207.

(25) Fischer, K.: 1970, Beitr. Phys. Atm., 43, pp. 244-254.

(26) Fujimoto, F.: 1974, Geophys. Mag., 37, pp. 1-48.

(27) Weiss, R., R.J. Charlson, A.P. Waggoner, M.B. Baker, D. Covert, D. Thorsell, and S. Yuen: 1977, Proc. IAMAP Symposium on Radiation in the Atmosphere (Garmisch-Partenkirchen, FRG, August 1976) Ed. by H.-J. Bolle, Science Press, Princeton, pp. 469-471.

(28) Grams, G.W., I.H. Blifford, D.A. Gillette, and P.B. Russel: 1974, J. Appl. Meteorol., 13, pp. 459-470.

(29) Carlson, T.N., and R.S. Caverly: 1977, J. Geophys. Res., 82, pp. 3141-3152.

(30) Lin, C.-I., Baker, M. and Charlson, R.J.: 1973, Appl. Optics, 12, pp. 1356-1363.

(31) Deluisi, J.J., P.J. Furukawa, D.A. Gillette, B.G. Schuster, R.J. Charlson, W.M. Porch, R.W. Fegley, B.M. Herman, J.T. Twitty, and J.A. Weinman: 1976, J. Appl. Meteorol., 15, pp. 441-454.

(32) Dlugi, R.: 1979, Proc. WMO Tech. Conf. on Regional and Global Observations of Atmospheric Pollution Relative to Climate (Boulder, USA), World Meteorological Organization, Geneva.

(33) Ackerman, T., and Baker, M.B.: 1977, J. Appl. Meteorol., 16, pp. 63-69.

(34) Grassl, H.: 1979, Chapt. in Man's Impact on Climate, ed. by W. Bach, J. Pankrath, and W.W. Kellogg, Elsevier Sci. Publ. Co., Amsterdam (pp. 229-241).

(35) Liou, K.N.: 1976, J. Atmos. Sci., 33, 798-805.

(36) Wang, W.-C., and G.A. Domoto: 1974, J. Appl. Meteorol., 13, pp. 521-534.

(37) Ackerman, T.P.: 1977, J. Atmos. Sci., 34, pp. 531-547.

(38) Bryson, R.A., and D.A. Baerreis: 1967, Bull. Amer. Meteorol. Soc., 48, pp. 136-142.

DISCUSSION

Wallén:

Is it possible, with our present knowledge of industrial production and its changes over the last five generations, to assume a more detailed scenario for the future release of CFMs than you presented, where a constant release after 1980 is assumed? This is important from the point of view of the future management decisions by politicians.

Ehhalt:

The future release scenario will depend on political and socio-economic decisions, which themselves are still in the future and possibly not independent of the predictions discussed today. It is therefore very difficult to make a long range forecast. The best procedure seems to be to make predictions based on the range of plausible release scenarios, and make this information available to the decision makers. In the short range we might very well have a continued increase in CFM consumption due to shift in usage.

Laurmann:

I would like to ask whether there might be a remedial measure available - perhaps by the introduction of chemical agents in the stratosphere - for alleviating the problem were it to become obviously dangerous say fifty or a hundred years in the future.

Ehhalt:

It seems very dangerous to counteract one perturbation by another perturbation since at the present level of understanding not all side effects may be identifiable. In any case the cost of the injection of possible millions of tons of corrective chemicals would be enormous.

Because of the strong coupling of NOX and ClOX chemistry in the stratosphere the various man-made perturbations on O_3 (CFM, N_2O) interact and do not superimpose linearly.

Furthermore, their combined effect will change the vertical profile of O_3 perturbation considerably, probably leading to an increase of O_3 concentration in the lower stratosphere. This has major consequences for the calculated change in the stratospheric temperature profiles and for the surface temperature, which would increase rather than decrease as for the O_3 change resulting from CFM alone.

Breuer:

Would it not be worthwhile collecting old refrigerators instead

W. Bach, J. Pankrath, and J. Williams (eds.), Interactions of Energy and Climate, 297–303.

of discarding them, to avoid CFMs going into the atmosphere and to make a law to this effect?

Ehhalt:
It certainly could be done, but at a cost to society. The question is, will society want to pay this additional cost?

Oeschger:
The expected CO_2 increase in the next century should lead to a decrease of temperature in the stratosphere, affecting chemical reactions. Of what size would the effect be, for example, on future ozone concentrations?

Ehhalt:
There will be an effect. Because of the temperature decrease in the stratosphere, some of the natural O_3 destruction reactions, which are temperature dependent, will slow down. Consequently the stratospheric O_3 concentration will increase. A doubling of CO_2 is expected to lead to a 3 - 5 % increase in O_3 column density.

Siegenthaler:
Can the ocean act as a net sink for anthropogenic N_2O?

Isaksen:
Recent observations of N_2O in the surface waters of the oceans (Cohen and Gordon, JGR 1979) indicate, that most of the oceans are sources of atmospheric N_2O. Only limited areas act as sinks. The net effect is estimated to be a small release of N_2O to the atmosphere (< 10 Tg(N)/yr).
Any moderate changes in atmospheric N_2O due to anthropogenic sources (± 20%) will not affect this conclusion.

Wallén:
In view of the difficulties in understanding sources and sinks of N_2O in the atmosphere, do you favor an increased effort to monitor N_2O and in that case do you have any suggestions for how it should be done?

Isaksen:
In view of the large difficulties in identifying the tropospheric sources and sinks of N_2O and the possible anthropogenic influences, an intensified effort should be made to monitor N_2O in the future. In particular, to reduce uncertainties, intercomparisons of the results obtained by different methods should be given priority.

Ramanathan:

Is the release of N_2O by combustion due to direct emission of N_2O, or does the release depend on denitrification/ nitrification processes?

Isaksen:

Measurements of N_2O in the mixture released from combustion of coal and gas show that it is released directly into the atmosphere. It is therefore not subject to uncertainties connected with the production of N_2O in the soil from fixed nitrogen.

Quirk:

The increase in N_2O is a very sensitive function of fertilizer use. You estimate that fertilizer use will increase by 6% per year. With the increased costs of fossil fuels, the price of fertilizers has been largely increased. It is therefore unlikely that fertilizer use will continue to increase so quickly. Clearly future increases of fertilizer use will be very difficult to estimate.

Bartholomäi:

For me the question is not whether N_2O released from natural fertilizers is greater than that from chemical fertilizers. I think the question in connection with the release of substances to the atmosphere is whether there is a closed cycle or not. That is, whether there is as much uptake from the atmosphere as is released to it on the average during a relatively short time. For instance, the biological C-cycle might be regarded as closed in this connection, but if we consider burning of fossil fuels that have been stored for millions of years in the earth, the C-cycle can't be regarded as closed. It's the same, perhaps, if we take natural fertilizer or mineral or chemical fertilizer.

Hampicke:

Will the quantity of N_2O released from agricultural ecosystems depend on the amount of chemical fertilizers used, or, rather, on the total N involved in plant nutrition, including N from biological fixation and organic fertilizer?

Isaksen:

The NASA Report (1979) points to the importance of non-linear amplification factors for N_2O release from the soil when large quantities of fertilizers are used. This could largely increase the anthropogenic influence on atmospheric N_2O. The mechanism is poorly understood, and should be investigated in the future.

I have made 2-D model estimates of the ozone generation in the troposphere due to man's activities. The results show that there are large latitudinal and seasonal variations, above all due to the uneven distribution of nitrogen oxides. Ozone is most effectively formed at mid - latitudes in the Northern Hemisphere during summer. The present increase is calculated to be 20-30% in the troposphere, at these latitudes. At other latitudes less ozone is formed. Ozone may even be reduced due to effective loss reaction when NOx concentrations are low. Furthermore, there is a marked variation with height in ozone increase, the largest increase taking place in the lower troposphere. In your calculation you used a constant increase with height. Is this of importance for the heating rates you get?

Ramanathan:

Surface-troposphere radiactive heating due to tropospheric O_3 increase is strongly dependent on the vertical distribution of O_3 change. With respect to latitudinal variations in O_3 change, the hemispheric mean radiative heating may not depend very strongly on latitudinal O_3 variations. However, if O_3 increase is primarily concentrated in the Northern Hemisphere, the hemispheric difference in O_3 radiative heating rates may be important in determining the climatic effects. My calculations assume uniform O_3 in the troposphere based on photochemical model calculations of Logan et al.

Schikarksi:

1. What is the effect of hydrocarbons from fossil fuel burning? Is it only small compared to methane?
2. In the energy scenarios we see the strong promotion of biogas. Since biogas is more or less composed of CH_4 or C_mH_n, this would mean a shift from less hydrocarbon emissions to more CO_2 emissions. Did you include this in your future projections?

Ramanathan:

1. This question has not been examined yet because of our poor understanding of atmospheric hydrocarbon photochemistry. It is too early to guess whether the effect is small or big compared to CH_4.
2. No.

Wallèn:

One important problem with regional differences of effects of aerosols over the globe is that such differences might cause changes in the global circulation. Do you think that warming over the continents by low level aerosol could cause such changes in the general circulation?

Kellogg:
In principle, any change in the heat balance of the continental areas will affect the general circulation. However, the most suitable tools that we have now for determining such an effect are the general circulation models (GCMs), and experiments with GCMs to determine the circulation changes resulting from even rather massive additions of heat (thermal pollution) over industrialized regions have failed to reveal a significant general circulation change. By the same token, one would not expect a significant change from aerosol heating.

Gates:
The atmospheric general circulation, at least as simulated by GCMs, is not easily disturbed by large local heat sources or localized modifications of the surface. Such factors primarily modify only the local climate through convective- and meso-scale circulations, while the large-scale circulation remains relatively intact.

Siegenthaler:
Can the effect of aerosols on albedo be observed by satellite measurements?

Kellogg:
No. While stratospheric aerosols have been (and are being) measured by satellite techniques on board Nimbus-7 and Tiros-N, there is no systematic measurement of tropospheric aerosols from satellites. Occasional outbreaks of Saharan dust can be seen moving across the tropical Atlantic, and large forest fire plumes can be observed by satellites, but so far there is no way to monitor the background aerosol level of the lower atmosphere.

Schikarski:
What is the order of magnitude of warming due to anthropogenic aerosols compared with that of CO_2.

Kellogg:
While the exact warming effect of aerosols is not known, it is probably safe to say that on a regional scale it is usually more important now than the effect of CO_2, whereas on a global scale it is less. Efforts to control air pollution in the industrialized countries will tend to keep aerosols from increasing in concentration as fast as CO_2, so the latter will probably be relatively much more important in the future.

Schnatz:

1. As far as I understand your balance diagramm, only cloud free effects of aerosol particles are taken into account. On the other hand, it is known that the radiative properties of aerosols within clouds differ from those in cloud free areas.
2. How is the dependence between particle radius and the imaginary part of the refractive index derived?

Kellogg:

Although I did not stress this point in my talk, in my paper I pointed out that both theory and observation indicate a reduction in the albedo of all but very thin clouds, when aerosols are incorporated in the clouds. Hence, in cloudy areas they usually cause a warming, even over the ocean.
2. The relationships shown between particle radius, imaginary part of the index of refraction, and absorption backscatter ratio are based on Mie theory calculations published by Chylek, Grams, Smith and Russell, J. Appl. Meteorol., 14, 380-387, 1975.

Flohn:

A large man-made contribution to aerosol loading is triggered by destroying the vegetation (especially by desertification and by conversion of natural vegetation into arable land). Very heavy dust loads in the lower troposphere are observed in loess areas (Pakistan, Central Asia). What is the effect of loess particles (mostly calcareous) with respect to your Fig. 3 ?

Kellogg:

I am grateful to Prof. Flohn for reminding us of the unquestionable importance of the influence of human activities, such as soil tillage and overgrazing, on windblown dust and loess.
The optical properties of Saharan dust are included in my table, and their a/b is high enough to cause a warming when over land. I do not have similar data for the optical properties of loess particles, unfortunately.

Quirk:

Over parts of Africa, the Midwestern US and tropical oceans the rainfall maximum is at night. In these regions precipitation may be increased by increased aerosols.

Kellogg:

That is a possibility, since aerosol layers do tend to decrease the static stability of the atmosphere at night due to infrared radiation (and cooling) at their tops.

Roberts:
Are the effects of high clouds produced by jet aircraft contrails (and their frequently-observed growth with time) significant in the tropospheric radiation balance? There is some evidence of increases in such clouds after the mid-1950's, when jet aircraft usage greatly increased.

Kellogg:
Indeed there is some evidence for an increase of cirrus cloudiness along some airways with jet traffic, but I doubt that such increases are very general. Even if cirrus did increase (or decrease) the effect would probably be small as far as the overall heat balance of the troposphere is concerned, since the decrease of solar radiation available due to scattering and albedo increase is very nearly counterbalanced by the decrease in outgoing infrared radiation due to the cold temperature of the cirrus layer. Thus, the influence of jets on the heat balance of the atmosphere is probably minimal.

PART IV

CLIMATIC IMPACTS OF LAND USE CHANGES

IMPACTS DUE TO DEFORESTATION

A. Baumgartner and M. Kirchner

Lehrstuhl für Bioklimatologie und Angewandte Meteorologie, Universität München, Amalienstraße 52, 8000 München 40, Bundesrepublik Deutschland.

ABSTRACT. Forests are very active parts of the earth-atmosphere boundary layer due to their stand characteristics, growth potential and high carbon storage or exchange. High radiation absorptivity and low infrared emission are causes of high net radiation income, which give rise to intensive water consumption. Runoff density and erosion are damped. Great aerodynamical roughness of forest crowns accelerates the turbulent exchange of mass and momentum.

Deforestation changes the functions of forest at the site and in the environment. Runoff and evaporation deplete soil moisture faster, temperature increase accelerates soil biomass metabolism. While timber harvesting in regularly managed forest withdraws carbon fixed from atmospheric CO_2, deforestation of virgin forest, mainly by fires, releases carbon dioxide to the atmosphere probably at the same rate as the CO_2 fluxes due to burning fossil fuels (coal, oil or gas).

1. INTRODUCTION

In terms of its influence on climate, deforestation is one of the most effective changes in land use. The removal of forest stands may be temporary or permanent. Depending on the scale of deforestation, the impacts will be observed mainly at the site of the stand, or in the local, regional or global environment.

W. Bach, J. Pankrath, and J. Williams (eds.), Interactions of Energy and Climate, 305–316.

2. FORESTS AS PART OF THE EARTH-ATMOSPHERE INTERFACE

For the valuation of forests as part of the earth-atmosphere interface a comparison with other earth surfaces is given in Table 1.

Table 1: Earth surfaces and their annual biomass production (1,18,14)

	Area			Production of dry matter (DM)	
	$x10^6 km^2$	%		$x10^9 t/a$	$x10^3 t/km^2 a$
Globe	510	100		155	0,31
Sea	361	71		55	0,15
Land	149	29		100	0,67
			x)		
Forest	50	10	33	65	1,30
Grassland	24	5	16	15	0,67
Cultivated	14	3	9	9	0,65
Woodland	7	1	5	4	0,57
Wildland	30	6	21	7	0,25
Desert	9	1	6	.	.
Polar	15	3	10	.	.

x) related to land surface only

From the total area of the globe about 10% are covered by forests and 1% are in the form of open woodlands as savannah or chaparral and shrubs; but one third of the land surface is forested. One fifth of the total land area is classified as wildland. Tundra, steppe, bogs and water surfaces on the continents are included within this term. Grassland follows with 16%, cultivated land for agriculture with 9%, dry deserts with 6% and polar land regions with 10%.

Important information on the role of forest is given in the two columns at the right side of Table 1: From a global total of 155 billion tons of dry plant matter per year, about 65 billion tons (42%) are produced by forests. The production exceeds that of the sea and amounts to 2/3 of that from all land vegetation. The high areal efficiency amounts to 1300 tons of dry matter per square kilometre and year. Neither cultivated land nor grasslands have such a high productivity. Therefore deforestation results in a reduction of biomass production and in a diminution of the potential to renew the energy resources on earth. The fixed energy in wood is 17 to 21 kJ per gramme and 1.2×10^{18} kJ for the globe. The use of tree plantations for fuel production is in progress, however, in competition with oil, gas or coal, it is not yet economical. The efficiency of storing solar radiation is about

0,3% as an average for the globe, but high-yield-forest stands may fix up to 3% of the solar energy in wood.

Another fact, and related with the production, is the volume of stored carbon or carbon dioxide or plant matter. For the conversion it can be assumed that 1 g dry plant matter (DM) corresponds to 0,46 g carbon (C) and to 1,7 g carbon dioxide (CO_2). The global carbon dioxide storage in the living forest biomass above and below ground is on the order of 3 x 10^{12} tons, the storage of dead biomass or detritus within and above the soil may be 7 x 10^{12} tons. That means that total CO_2-storage by forest amounts to 10 x 10^{12} tons and is more than four times the total CO_2 amount in the atmosphere, which is about 2,5 x 10^{12} tons. Forest is the main organic system in the biosphere for the exchange of carbon dioxide and oxygen. Forestry therefore has a high potential to influence the atmospheric carbon dioxide by deforestation or reforestation.

3. CLIMATIC PROPERTIES OF FORESTS

The functions of forests as a boundary at the earth surface are due to specific characteristics. To make the manifold impacts of deforestation more clear, these characteristics will be briefly described. For further information see (3, 5, 6, 7).

3.1 Radiation and thermal characteristics

For the radiation balance, the volume of wood or the mass of organic matter on the order of 300 to 1000 tons per hectare, which represents a homogenous sheet of only 2-5 cm wood above ground, are not as important as the vertical distribution of stems, twigs, and the assimilation organs. Leaves or needle surfaces cover the soil surface by overlapping 10 to 25 times the area of the projected canopy. Penetrating radiation will be absorbed or scattered and only a small fraction of incoming radiation will reach the ground. There are three remarkable consequences, namely short-wave radiation absorption, infrared emission and net radiation income are high. In Table 2 rough estimations for these elements for different earth surfaces are given.

90 percent of solar radiation is absorbed by conifers and 85 percent by deciduous trees, however grassland absorbs only 75 and cropland in summer only 65 to 70 percent. Small increase above air temperature corresponds with the low infrared radiation emission from the forest canopy.

In summer conditions a forest canopy is a relatively cool surface. The cooling effect is not only due to the stand structure peculiarities, but also to intensive energy use for evapotranspiration, as will be explained later. Grassland and cultivated land or bare soils overheat more. The net radiation income of a forest may be double that of bare soils. As far as energetics are concerned, forests are the most intensive acting element of the land cover.

Table 2: Climatic and hydrologic properties of forest in relation to other earth surfaces.

	Forest	Grass	Crops	Bare	
Absorption (A_s)	85-90	75	70	65	%
Overheating (ΔT)	2-4	4-6	6-10	10-20	K
Net Radiation (Q)	60-65	45-55	45-55	35-40	W/m^2
Potential Evapotranspiration (Ep)	850	550-750	550-750	400-500	mm/a
Evaporation-Ratio (E/P)	70	65	40-50	30	%
Runoff-Ratio (D/P)	30	35	50-60	70	%
Roughness Parameter (z_o)	100-300	5	30	0.1-1	cm

3.2 Hydrological peculiarities

On account of the high net radiation income, forests are able to evaporate more water than other vegetative soil covers. By transforming the net radiation values with the help of the vaporization enthalpy ($r \sim 240$ J/g) potential evapotranspiration may be calculated for the case that radiation would be used up only for evaporation. The water consumption of forest under European conditions could be as much as 850 mm, whereas those of bare soils under the given energy exchange could be only 400-500 mm per year. Another measure of water use is the evaporation ratio (E/P), where evaporation E is related to precipitation P. The ratio decreases with the plant mass on earth. The runoff ratio (D/P), where D stands for discharge, is directly linked with E/P in the hydrologic balance.

$$1 = E/P + D/P$$

This states that the higher the evaporation is, the smaller must be the discharge of water from land by runoff. The inter-relationship between water consumption and runoff is expressed in the dependence of runoff on the amount of forest coverage (f) in river catchments. Deforestation increases runoff and reforestation decreases it. An example has been given for the Ukraine (16):

f	0	20	40	60	100	%
D/P	42	33	26	24	18	%

It is well known that in addition to the runoff quantity forests are most beneficial for land conservation by avoiding or reducing erosion due to reduced runoff density. Deforestation often is disastrous on inclined land with unstable soils as runoff velocity and erosive forces increase with the disappearance of protective tree stands.

3.3 Energy exchanges

As has been demonstrated already, changes in forest cover are accompanied by impacts on radiation balance, thermal behaviour, and on the hydrological cycle. The energy balance is given as

$$Q = S + E_e + H$$

where Q = net radiation, S = soil heat flux, E_e = latent heat flux and H = sensible heat flux in air. The hydrological balance has the form

$$P = E + D \pm R$$

where P = precipitation, E = evaporation, D = discharge and R = reserves as soil or ground water. The two balances are linked by $E_e = rE$. The influence of changes in forest coverage (f) on energy and water balance and the influence of deforestation on climate can be shown very well by the following climatonomy equation (13):

$$(1 - D/P)\ (1 - H/E_e) = Q/r{\cdot}P.$$

The hydrological term D/P is combined with the Bowen ratio H/L and the Budyko-ratio Q/r•P. Deforestation means that the runoff rate D/P and the return flow of energy to the atmosphere by sensible heat increase, but the flow of latent heat $E_e = r{\cdot}E$ must decrease. The energy and water fluxes at the earth's surface are very sensitively correlated with changes in forest coverage. For classification of the influences, the Bowen-ratio and the Budyko-ratio are useful dimensionless numbers. To illustrate the influence of forest cover on the energy transfer at the earth/atmosphere interface the results of a calculation of latent heat released, based on lysimeter studies at Eberswalde, are given:

bare soil	lawn	pines	wet soil	
17	28	35	41	W/m^2 .

Corresponding to the differences in evapotranspiration there are remarkable differences in latent heat flow over the various surfaces. In nature the differences should be smaller, but of the same tendency.

The climatic impacts are difficult to analyse due to the complexity of actions and feedbacks. In a study on the effects of deforestation in the Amazon Basin of Brazil, a system analysis has been used to detect and define the interactions between the energy and hydrological cycles (15). Feedbacks accelerate or suppress the effects. Figure 1 gives an illustration.

The analysis starts with an increase in surface temperature after forest destruction. Increases of heat convection and vapor convection result in more clouds and a higher precipitation rate. Precipitation reduces surface temperature and clouds reduce global radiation, but increase infrared radiation. The higher precipitation rate induces, by increased evaporation, the internal water cycle locally and regionally and increases runoff. The higher soil temperature accelerates the biomass metabolism and mineralisation. Leaching of nutrients degrades the growth potential of the soils.

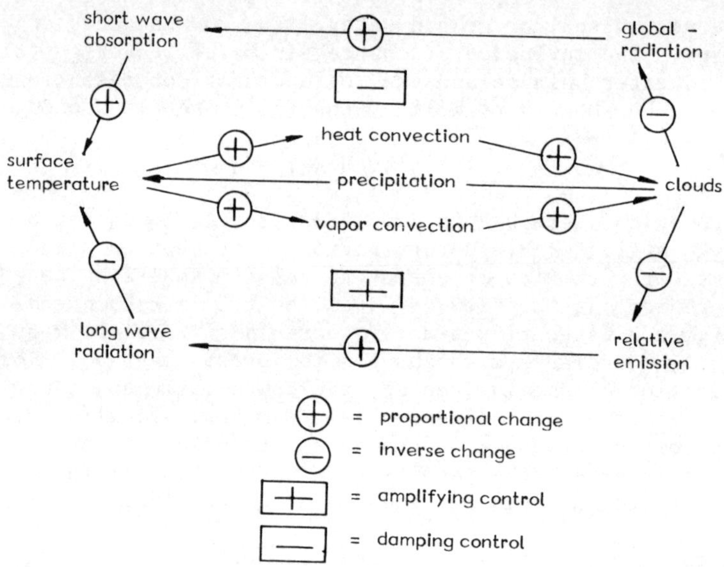

Figure 1: Schematic showing the impacts of deforestation on energy and water balance

3.4 Dynamic properties

The forest crown is one of the roughest earth-atmosphere interfaces. Trees brake the movement of air. The roughness parameter z_o is an expression of the dynamic effectiveness of the earth surface. A comparison of the values is given in the following series.

	Water	Sand	Lawn	Crops	Forest	City	
z_o	0.01	0.2	1-5	10	300	400	cm.

The roughness of forest stands is comparable to the effect of the house or roof irregularities of cities on the movement of air. The surface drag, the energy dissipation and the turbulent mass exchange in air increase proportionally to the roughness. Above forest canopies small vertical gradients of air properties, for instance of air temperature or humidity, are observed. The coverage of forest on continents has a considerable influence also on the angle of deviation of geostrophic wind from the isobar, and thus on the development of the pressure field and on the circulation of atmosphere; quantification of this is given in the last section.

4. CARBON DIOXIDE EXCHANGE

The question of how forestry operations influence the carbon and carbon dioxide transfer between earth and atmosphere has gained considerable interest. In the center of the discussion are the impacts of deforestation in the tropics. We have to distinguish between the organised operations in the managed forests and the disorganised destruction of stands.

In the productive, managed forests only those parts of stands which are at the optimum of biomass increase will be cut. The transport of the timber out of the forest is a withdrawal of carbon from the integrated ecosystem forest-atmosphere. How much and at which time the stored carbon will be released again to the atmosphere depends on the later use of the timber. During the regeneration phase CO_2-assimilation is reduced over years and carbon dioxide from soil biomass is released. To quantify the impacts a carbon dioxide balance for a managed spruce forest at the IBP-Station München-Ebersberger Forst is given in Table 3.

Table 3: Carbon dioxide balance for an 80 year old spruce forest near München (9)

Crown, inflow (F_a)	46 t CO_2/ha, a
Crown, outflow (F_r)	19 t CO_2/ha, a
Crown, net inflow	26 t CO_2/ha, a
Soil, outflow (F_s)	18 t CO_2/ha, a
Net assimilation 1972	8 t CO_2/ha, a

The forest is about to be harvested for timber. It has recently stored about 300 t carbon per hectare and fixes 8 tons of carbon dioxide per hectare and year.

It can be assumed that a considerable part of the timber will be conserved somewhere. Concerning the CO_2-exchange, the managed forest is in an unbalanced state, as the successions to equilibrium between assimilation and respiration are interrupted and therefore this forest type is classified as a sink of CO_2. A calculation, using FAO-Statistics, results in a global total of the CO_2-sink of 1.5×10^9 t per year. This is only about 2 percent of the annual fixation rate of CO_2 by assimilation, which is for the globe on the order of 80 to 100 $\times 10^9$ tons of carbon dioxide.

The destructive and disorganised operations in virgin forests of the tropics have opposite results. By destruction with fire, for example in shifting cultivation, not only the stored carbon in timber, but also the exposure of the unstable soils releases the carbon. High temperatures accelerate the metabolism. For the evaluations a comparison of the forest areas in the different climatic zones is given in Table 4.

Table 4: Productive forests in different climatic zones (8).

	Area		Production above ground	
	10^6 ha	%	$\times 10^9$ tDM/a	%
Tropical	2150	53	29.4	75
Temperate, warm	360	9	5.0	13
Temperate, cool	260	6	2.8	7
Boreal	1300	32	1.8	5

Tropical forests occupy 53% of global forest area and produce 75% of the forest plant mass. Of the 44 million km^2 of non arid tropical land 17 are tropical rain forests, 7.5 tropical seasonal forests, 4,5 tropical woodland or shrubs and 1.5×10^6 km^2 are savannah.

The total carbon in the non arid tropical vegetation amounts to approximately 1×10^{12} t carbon; it corresponds to a potential release of 3.8×10^{12} tons of carbon dioxide. The tropical forests are being destroyed at an annual rate of 0.3×10^6 km^2. The quantity of burnt or decomposed biomass is very uncertain and has been the target of recent research (2, 10). Available calculations of the annual carbon dioxide release are between 2×10^6 t CO/a and 11×10^9 t CO/a. A part of it will be fixed again by the production of the substitute vegetation, which, however, has a lower productivit

Also of considerable uncertainty is the release of CO_2 from soil biomass and detritus. About 1 to 3×10^{12} tons of carbon are stored within or above the soil. This is a potential of 4 to 11×10^{12} tons of carbon dioxide. The annual loss seems to be on the order of

0,1 to 1,5 tons of carbon, this corresponds to a carbon dioxide source of 0,4 to 5 x 10^9 tons.

When evaluating global carbon dioxide balances one cannot neglect the exchanges by forests. The integrated effect of forestry operations may be a source of CO_2 up to 15 x 10^9 tons of carbon dioxide. This is about the magnitude of the carbon dioxide release by burning of fossil fuels (oil, coal, or gas). Therefore experimental research to measure the carbon dioxide exchanges at the forested earth surfaces is urgent.

5. SIMULATION OF A DEFORESTED WORLD

By similation one is able to calculate the effects of total deforestation of the globe. Based on the world maps of albedo, roughness parameter and of the angle between the geostrophic wind and the isobars as given by (4) the effects of global deforestation have been calculated (11).

According to these measurements the difference between the surface albedo of wood (∿ 15%) and green land (∿ 20%) without snow coverage is approximately 5% (with snow coverage: 8-10%) (4, 12). Figure 2 shows the global distribution of the increase of surface albedo in July in the case of a conversion of all forests into

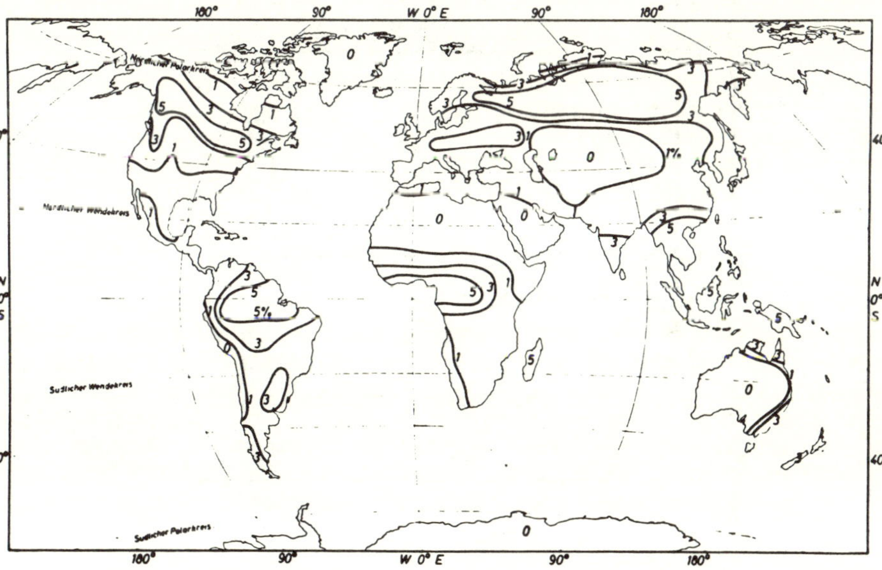

Figure 2: Net effect of deforestation on albedo

grassland. The effects on albedo are largest in the tropics and also in the northern hemisphere. The changes amount to approximately 6% for the winter and 4% annually, within the corresponding latitudinal zones (Fig. 3).

The global average surface albedo would shift from 16.7% to 17.4%. That means that the globe would be brighter and would initially absorb less solar radiation.

The roughness of the earth should decrease significantly in the case of total deforestation. The global average of z_o computed linearly decreases from 15 to 3 cm. Despite the relatively high values of 100-300 cm in the case of forest and 5 cm for grass, the global means are low because the oceans which are very smooth have a larger area than the land. The zonal distribution in Figure 4 shows that the roughness should decrease mainly in the northern hemisphere. The global decrease of the deviation angle between the geostrophic wind and the isobars is roughly 1 degree.

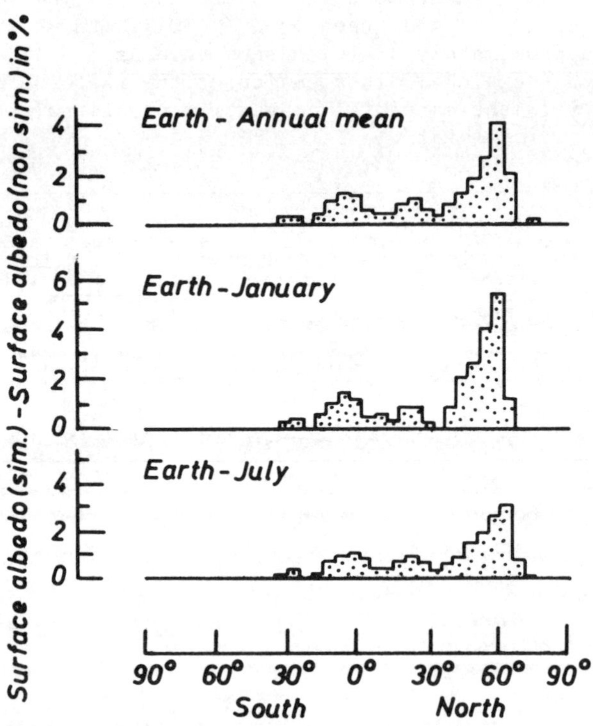

Figure 3: Meridional distribution of measured and simulated albedo changes caused by deforestation

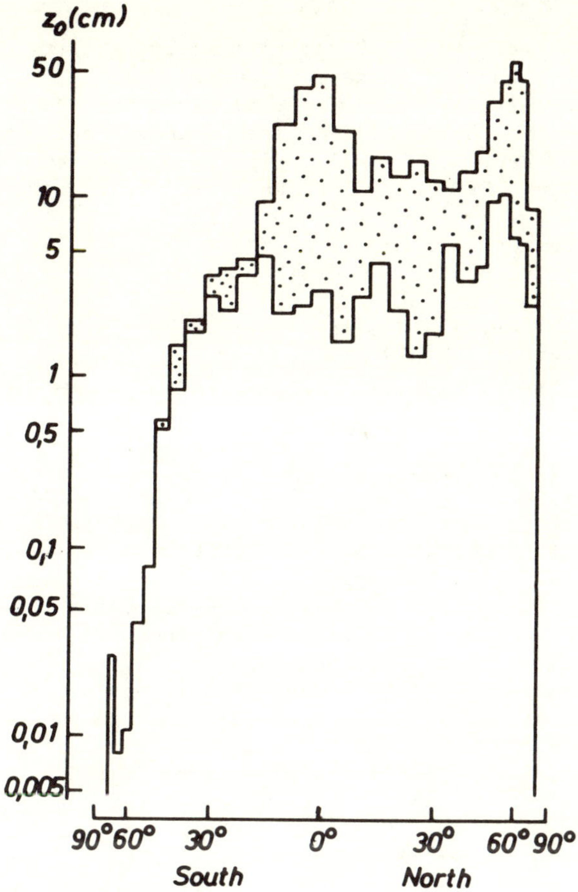

Figure 4: Meridional distribution of measured and simulated roughness parameter changes caused by deforestation

The greatest impact, however, should be on the carbon dioxide cycle. A remarkable rise of atmospheric CO_2 content should follow, caused by the reduction of assimilation and by the release of carbon from biomass and within the soil. Finally it may be noticed that the atmospheric CO_2 stock would be depleted totally by forest production within 26 years, if no CO_2 return from plant metabolism existed.

REFERENCES

(1) Baumgartner, A. (1979): Climate Variability and Forestry. WMO, World Climate Conf. Geneva.

(2) Baumgartner, A. (1978): Forest and the Global Carbon Dioxide Balance. WMO Int. Sympos. Forest Meteor. Ottawa.

(3) Baumgartner, A. (1971): Einfluß energetischer Faktoren auf Klima, Produktion und Wasserumsatz in bewaldeten Einzugsgebieten. IUFRO Congr. Gainsville, pp. 75-90.

(4) Baumgartner, A., Mayer, H., Metz, W. (1976): Globale Verteilung der "Oberflächenalbedo". Meteorol. Rdsch. 29, pp. 38-43.

(5) Baumgartner, A. (1978): Verdunstung im Walde. DVWK, Nr. 41, Koblenz, pp. 39-53.

(6) Baumgartner, A. (1970): Evaporation in Forests. FAO Sympos. Moskau/Rome.

(7) Baumgartner, A. (1977): Klimatische Funktionen der Wälder. Berichte Landw. 55, pp. 708-717.

(8) Brünig, E. F. et al. (1975): Stratification of the moist forest land for land use planning. Mitteilungen d. Bundesforschungsanstalt Forst- und Holzwirtschaft 109, 1-57, Hamburg.

(9) Hager, H. (1975): Kohlendioxyd-Konzentrationen, -Flüsse und -Bilanzen in einem Fichtenhochwald. Wiss. Mitt. Meteor. Inst. München, Nr. 26.

(10) Hampicke, U. (1979): Man's impact on the earth's vegetation cover and its effect on carbon cycle and climate, pp. 139-160. In: Man's Impact on Climate, Elsevier.

(11) Kirchner, M. (1977): Anthropogene Einflüsse auf die Oberflächenalbedo und die Parameter des Austausches an der Grenze Erde/Atmosphäre. Wiss. Mitt. Meteor. Inst. München, Nr. 31.

(12) Kondratyev, K. Ya., (1972): Radiation processes in the atmosphere. WMO - No. 309.

(13) Lettau, H. H. (1969): Evapotranspiration climatonomy I. A new approach to the numerical prediction of monthly evapotranspiration, runoff and soil moisture storage. Monthly Weather Rev. I, 97, pp. 691-699, II, 101, pp. 636-649.

(14) Lieth, W. (1972): Über die Primärproduktion der Pflanzendecke. Angew. Botanik, pp. 1-37.

(15) Molion, L. C. B. (1975): A climatonomy study of the energy and moisture fluxes of the Amazon basin with consideratic of deforestation effects. Thes. Wisc. Univ., Met. Dpt.

(16) Moltschanov, A. A. (1966): Pecularities of hydrology of catchment basins and the determination of the optimum land use. Madrid pap. CFM/GS G. T. IX (6-29).

CLIMATE CHANGE DUE TO ANTHROPOGENIC SURFACE ALBEDO MODIFICATION

Gerald L. Potter, Hugh W. Ellsaesser, Michael C. MacCracken, James S. Ellis and Frederick M. Luther

Lawrence Livermore Laboratory, University of California, Livermore, CA 94550, USA

ABSTRACT. Numerical experiments that dealt with the possible climatic impact of man-induced surface albedo change have primarily focused on individual regions (e.g. increased desert albedo, tropical deforestation, urbanization). Sagan et al. (1) suggest that the combination of anthropogenic albedo changes may have contributed to global climate changes in the past and may be continuing today. Using a statistical dynamic climate model with more realistic surface albedo changes than used in previous experiments, we have conducted a numerical experiment combining desertification of the Sahara and deforestation of the tropical rain forest. Over an area of 9×10^6 km^2 at 20^oN the desert albedo was increased from 0.16 to 0.35 and over 7×10^6 km^2 at the equator and 10^oS the rain forest albedo was increased from 0.07 to 0.16. While the most significant direct climatic responses were observed in the modified zones, high northern latitudes exhibited the greatest cooling through activation of the ice-albedo feedback process. In contrast to Sagan et al. (1), this experiment suggests that anthropogenic modification of surface albedo over the past few thousand years has had an impact on global climate which is likely quite small and probably undetectable.

INTRODUCTION

Modification of the earth's surface albedo through desertification and deforestation has received renewed attention in the past year (1,2). Sagan et al. (1) suggest that the natural climate fluctuation of the past several thousand years may have been affected by large scale modifications of the earth's surface. Earlier numerical model experiments

W. Bach, J. Pankrath, and J. Williams (eds.), Interactions of Energy and Climate, 317–326.

by Charney et al. (3), Ellsaesser et al. (4), and Potter et al. (5) suggest that the main effect of man's activities would be a general increase in surface albedo resulting in lower surface temperatures. These numerical experiments, however, were designed to incorporate only specific scenarios, i.e., tropical deforestation or desertification. The combined effects were not addressed until Sagan et al. (1) suggested the possible long term global ramifications of such massive surface alterations.

This suggestion appears sufficiently serious to warrant investigation in a more sophisticated model. We have available the Lawrence Livermore Laboratory (LLL) statistical dynamic climate model (SDM). Since it has two-dimensional dynamics coupled with radiative and convective transport of heat and moisture, it provides a more realistic appraisal of the possible climatic effects of a massive cumulative alteration of surface albedo than simple energy balance models. As a first step in testing the hypothesis of Sagan et al. (1), we used the LLL SDM to calculate the combined effect of desertification of the northern hemisphere subtropics and deforestation of the tropical rain forest. Other surface modifications suggested by Sagan et al. (1) were not included since they involved relatively small surface areas and since there is less agreement as to the character of their potential impacts on climate.

THE MODEL

The LLL SDM model was conceptually designed for climate studies (6). The model is similar in many ways to general circulation models and uses the basic (primitive) conservation equations to calculate zonal mean fields of pressure, temperature, winds and humidity. Vertically, the model is divided into nine pressure layers and horizontally into a 10° latitude grid. The surface at each latitude is divided proportionately into land (of differing types and elevations) and ocean (which may be partially covered by ice). By using this fractional assignment of various surface types with corresponding physical characteristics (e.g. albedo, heat capacity, bulk transfer coefficients), the model sensitivity to modifications of these parameters can be tested.

Surface fluxes of sensible and latent heat are calculated using the vertical gradients of water vapor and temperature, the surface wind velocity and low-level convective activity. Clouds are randomly correlated vertically and are a prescribed latitude and height dependent function of relative humidity that was calibrated according to data of London (7). Additional details of model structure can be found in MacCracken and Luther (8), Luther and MacCracken (9), and of model performance in Potter et al. (10).

MODEL SENSITIVITY

To relate the degree of response of our model to an imposed perturbation to that of other models, it is useful to compare their responses to a standard perturbation. A good measure of model sensitivity is the response to an increase in solar constant, an experiment that has now been performed by most climate models. The usual unit of comparison is the sensitivity parameter β (equal to 100 times the temperature change resulting from a 1% increase in solar flux).

The planetary energy balance is given by:

$$F = S_o/4\,(1 - \alpha_p)$$

where S_o is the solar constant (1352 Wm^{-2}), α_p is the planetary albedo, and F is the net outgoing longwave flux at the top of the atmosphere. From this relation, Cess (11) has shown that with a solar constant change β can be expressed as follows:

$$\beta = S_o \frac{dT_s}{dS} = \frac{F}{dF/dT_s + (S_o/4)(d\,\alpha_p/dT_s)} .$$

Current experiments with SDM give a global sensitivity factor β = 70K. This is less than half the value quoted for simple energy balance models (11,12) and the GFDL sector model (13). At least some of this reduced sensitivity results from use of a global model due to the greatly reduced sensitivity of the southern hemisphere which has significantly less ice-albedo and land surface albedo feedback than the northern hemisphere.

EXPERIMENTAL DESIGN

The LLL SDM presently approximates the annual average climate by use of a fixed solar zenith angle for each latitude calculated to supply the appropriate annual average solar flux. By eliminating the diurnal and seasonal fluctuations, the model approaches a pseudo-equilibrium state after approximately 200 days of model integration time. This reflects the adjustment time implied by the heat capacities of the atmosphere, land, snow, ice and arbitrarily thin oceanic mixed layer. For this experiment, two model equilibrium states were achieved: one being a control case with standard surface conditions corresponding to a pristine environment; the second, a perturbed case corresponding to the scenario suggested by Sagan et al. (1) in terms of tropical deforestation and desert denudation (see Table 1). The surface bulk transfer coefficients were left unchanged as were the thermal and hydrologic characteristics. Thus, the only surface parameter to be modified for those areas designated as perturbed was the surface albedo which after modification was held constant. This represented a reduction in the degrees of freedom of response both for the SDM and compared to the real world.

TABLE 1. Model albedo changes from Sagan et al. (1).

Process	Land type change	Area km^2	Centered on latitude bands	α_s, surface albedo change
Desertification	savanna to desert	$9x10^6$	20^oN	0.16→0.35
Tropical deforestation	forest to field savanna	$7x10^6$	10^oS and EQ	0.07→0.16

RESULTS

The zonal model response was generally consistent with previous surface albedo modification sensitivity studies (4,10). Figure 1 shows the latitude-height temperature change (perturbed minus control) for the two cases. Generally, the northern hemisphere troposphere cooled with a maximum change near the surface at 70^oN. The southern hemisphere, however, warmed slightly. This response apparently resulted

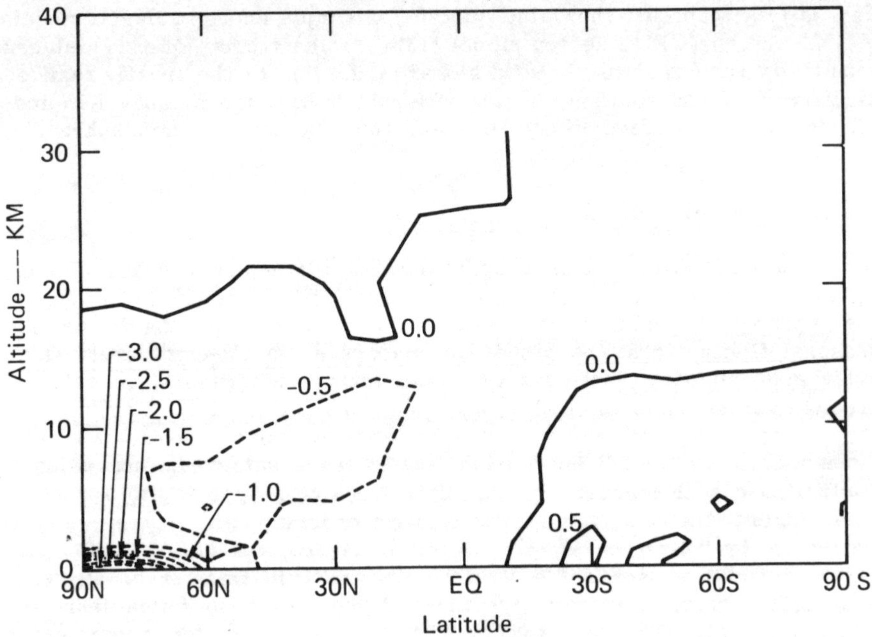

FIGURE 1. Latitude-height temperature difference (perturbed minus control). Contour interval = 0.5K.

from various feedbacks of the model's coupled non-linear processes, although our interpretation of the results is still preliminary.

There appear to be two important factors in the chain of events leading from surface modification to cooling in the northern hemisphere. The first is the direct reduction in insolation in the zones of increased surface albedo (see Table 1). The change at 20°N was sufficiently strong to strengthen the temperature gradient between the Sahara and the equator which, as in our earlier Sahara experiment (4), in turn caused the Hadley circulation to strengthen (in at least the northern descending branch) and to shift southward. This in turn weakened the hydrological activity at 20°N with less water vapor and associated convection and latent heat release, accompanied by maximum cooling in the upper troposphere, lowering of the tropopause and warming of the lower stratosphere. With less solar energy absorbed in the subtropics there was reduced poleward transport of sensible and latent heat and a general cooling of the higher latitudes. The second factor leading to cooling in the northern hemisphere was the ice-albedo-feedback process. This was manifested primarily at 70°N by increased sea-ice coverage resulting from the general cooling caused by the higher subtropical albedo.

Response in the southern hemisphere presents a somewhat more curious picture. A small fraction of the 10°S latitude band was also considered to have an albedo increase from deforestation. Figure 2 shows the zonal average conversion from potential to kinetic energy (Wm^{-2}) due to vertical motion. The heated air in the low level equatorial Hadley Cell rises causing conversion of potential to kinetic energy. In the higher latitudes the sinking motion converts kinetic back to potential energy. This conversion term thus gives an indication of the intensity of the Hadley Cell. In the perturbed case, the Hadley Cell shifted southward with the ascending branch weakening slightly and the southern descending branch weakening more. The southward shift brought increased ascent to the southern subtropics. This and the weakened descending branch permitted increased convective activity, thereby producing clouds with more vertical structure and less horizontal extent. The total cloud coverage thus decreased at 30°S, allowing increased insolation. This increase, along with the albedo increase at the equator and 10°S, weakened the subtropical temperature gradient and presumably further weakened the southern hemisphere Hadley Cell. However, other than the additional solar flux at 30°S, there appears little reason for the slight warming of the southern hemisphere.

The combined cloud and surface albedo effect of the numerical experiment are demonstrated by examining the difference in planetary albedo (Fig. 3). The changes at 20°S, 30°S, and 30°N are related to changes in cloud amount. At 10°S, the equator and 20°N, the increases in planetary albedo are a response to the forced increase in surface albedo. At 70°N, the increase is a response to an increasing

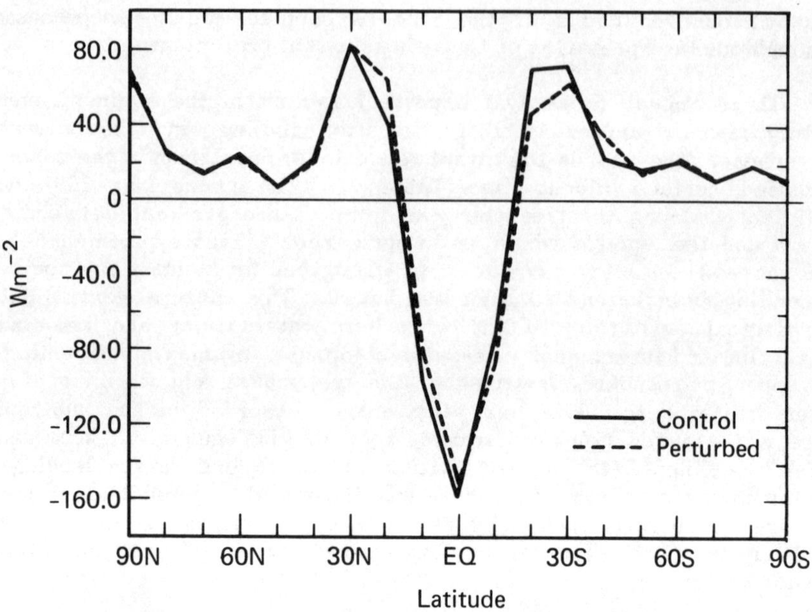

FIGURE 2. Kinetic to potential energy conversion through vertical motion.

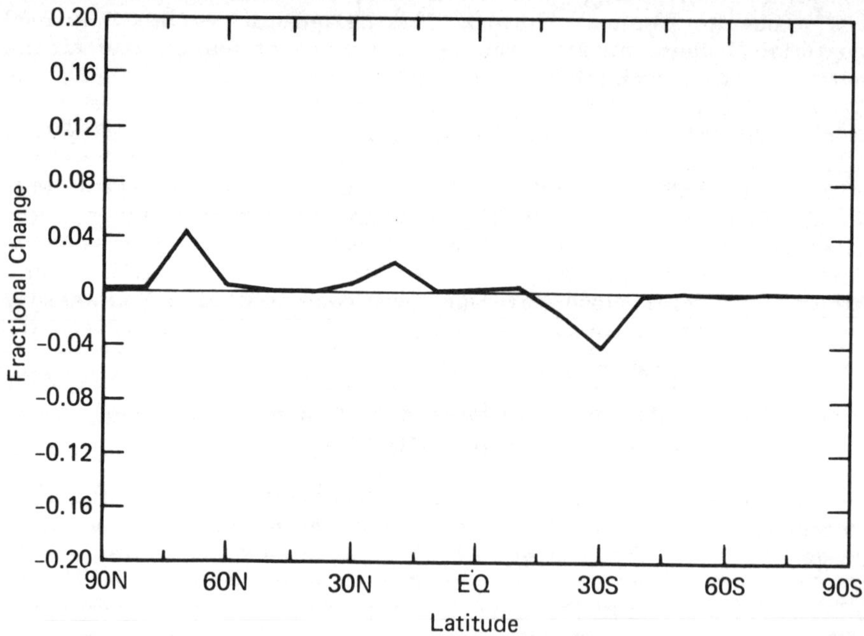

FIGURE 3. Difference in planetary albedo (perturbed minus control).

sea-ice fraction resulting from the overall cooling of the northern hemisphere.

To more fully assess the relative importance of the various model responses to the imposed increase in surface albedo, we can examine the contribution to the change in radiation at the top of the atmosphere as given by

$$\delta R = \delta S - \delta F \quad ,$$

where δS and δF are the changes in the net solar flux and the outgoing longwave flux at the top of the atmosphere, respectively. To first order, the planetary changes may be expressed in terms of separate contributions from each of the various primary mechanisms:

$$\delta S = \delta_r S + \delta_{A_c} S + \delta_{\alpha_s} S$$

$$\delta F = \delta_T F + \delta_r F + \delta_{A_c} F \quad ,$$

where the subscripts r, T, A_c and α_s identify components due to precipitable water vapor, temperature, cloud amount, and surface albedo, respectively. These individual contributions have been calculated by computing the model planetary energy balance keeping all of the input data fields fixed (control case) except for the perturbed data field of interest (e.g. temperature, water vapor, clouds or surface albedo). By changing only one field and recalculating the planetary energy balance, the contribution of this field can be isolated. Table 2 shows the change in solar, longwave and net radiation due to the separate changes in various contributing factors, i.e., T, r, A_c, and α_s.

The two major components in changing the planetary energy balance were found to be clouds and surface albedo. The mean global cloud fraction decreased 0.011, resulting in an increase in longwave loss to space and an increase in solar radiation absorbed by the earth-atmosphere system. The surface albedo increased as a result of both the imposed albedo modification and the ice-albedo feedback, resulting in a decrease in the solar radiation absorbed by the earth-atmosphere system. The mean surface temperature decrease reduced the longwave loss to space because of the cooler radiating temperature, a necessary consequence of the reduced absorption of solar radiation.

Surprisingly, water vapor was the least important component in the longwave response, although it did produce a reduction in longwave loss to space. This small influence may be because water vapor experienced both vertical and horizontal redistributions in conjunction with the hemispheric asymmetries of the total model response. Thus the change in global mean precipitable water vapor was so small (-1.6%) that the radiative response was nearly insignificant.

TABLE 2a. Global mean change in solar, longwave, and net radiation (S, F, and S-F) in Wm^{-2} of the control model due to the separate changes in various contributing factors, i.e., T, r, A_c, and α_s.

	T	r	A_c	α_s
S	—	0.01	1.25	-0.61
F	-0.53	-0.07	.31	—
(S-F)	0.53	0.08	0.94	-0.61

TABLE 2b. Global change in various parameters (perturbed minus control).

Parameter (ψ)	Change	$\frac{\Delta\psi}{\psi}$ x 100%
Global surface temperature (T)	-0.2K	
Mean precipitable water vapor (r)	-0.04 gm/cm^2	(-1.6%)
Global cloud cover fraction (A_c)	-0.011	(-1.9%)
Surface albedo (α_s)	+0.018	(+8.8%)

CONCLUSION

The sensitivity of the LLL SDM to the combined effect of deforestation in the tropics and desertification in the subtropics reveals several interesting responses. The Hadley Cell initially shifted southward and the northern descending branch intensified. Conversely, the southern descending branch decreased in intensity, thereby reducing subsidence. This reduction in stability in the southern hemisphere produced clouds with more vertical and less horizontal extent, allowing more solar radiation to reach the earth's surface and increasing temperatures at 30°S. The northern hemisphere temperature decreased because of the reduced solar radiation reaching the surface coupled with additional cooling due to ice-albedo feedback.

The LLL SDM mean surface temperature cooling of $\sim$0.2K was substantially less than the 1K cooling suggested by Sagan et al. (1) for the possible effect of anthropogenic surface modifications over the past several thousand years. Our northern hemisphere surface temperatures, however, cooled 0.6K. From these results it is apparent that at least in terms of our zonal model, the global temperature response is significantly less and the northern hemisphere temperature response is somewhat less than that of the simple energy balance model cited by Sagan et al. (1). These model results of the possible effects of deforestation and desert expansion suggest that as non-linear feedbacks (such as ice-albedo feedback, large scale dynamics and changes in cloud amount and height) are incorporated into climate models, their sensitivity may decrease because of compensating feedback responses.

ACKNOWLEDGMENTS

The authors would like to thank Dr. W. J. Quirk for his suggestions and comments and Floy Weichel for typing the manuscript.

This work was performed under the auspices of the U. S. Department of Energy by the Lawrence Livermore Laboratory under contract No. W-7405-Eng-48.

REFERENCES

1. Sagan, C., Toon, O. B., Pollack, J. B.: 1979, "Anthropogenic Albedo Changes and the Earth's Climate," Science 206(4425), pp. 1363-1367.

2. Lettau, H., Lettau, K., and Molion, L. C. B.: 1979, "Amazonia's Hydrologic Cycle and the Role of Atmospheric Recycling in Assessing Deforestation Effects," Mon. Wea. Rev. 107(3), pp. 227-238.

3. Charney, J. G., Stone, P. H., and Quirk, W. J.: 1975, "Drought in the Sahara: A Biogeophysical Feedback Mechanism," Science 187, pp. 434-435.

4. Ellsaesser, H. W., MacCracken, M. C., Potter, G. L., and Luther, F. M.: 1976, "An Additional Model Test of Positive Feedback From High Desert Albedo," Quart. J. Roy. Met. Soc. 102(433), pp. 655-666.

5. Potter, G. L., Ellsaesser, H. W., MacCracken, M. C., and Luther, F. M.: 1975, "Possible Climatic Impact of Tropical Deforestation," Nature 258(5537), pp. 697-698.

6. MacCracken, M. C.: 1968, "Ice Age Theory by Computer Simulation," Ph.D. Dissertation, University of California, Davis, CA, 193 pp.

7. London, J.: 1957, "A Study of the Atmospheric Heat Balance. Final Report," Contract AF19(122)-165, Research Division, College of Engineering, New York University, 99 pp.

8. Luther, F. M. and MacCracken, M. C.: 1974, "Initial Validation Studies for ZAM2 Radiation and Large Scale Eddy Transport Mechanisms," in Proceedings of the Third Conference on CIAP, Report No. DOT-TSC-OST-74-15, Department of Transportation, Washington, D. C., pp. 437-449.

9. MacCracken, M. C. and Luther, F. M.: 1974, "Climate Studies Using a Zonal Atmospheric Model," in Proceedings of the International Conference on Structure, Composition and General Circulation of the Upper and Lower Atmosphere and Possible Anthropogenic Perturbations, Melbourne, Australia, January 14-25, 1974, W. L. Godson (Ed.), V. II, 1107-1128, Atmospheric Environment Service, Ontario, Canada, NTIS Report No. UCRL-74887.

10. Potter, G. L., Ellsaesser, H. W., MacCracken, M. C., and Luther, F. M.: 1979, "Performance of the Lawrence Livermore Laboratory Zonal Atmospheric Model," in Report of the JOC Study Conference on Climate Models: Performance, Intercomparison and Sensitivity Studies, WMO Report, Geneva.

11. Cess, R. D.: 1976, "Climate Change: An Appraisal of Atmospheric Feedback Mechanisms Employing Zonal Climatology," J. Atmos. Sci. 33, pp. 1831-1843.

12. Budyko, M. I.: 1974, "Climate and Life," Academic Press, 508 pp.

13. Ramanathan, V.: 1977, "Interactions Between Ice-Albedo, Lapse-Rate and Cloud-Top Feedbacks: An Analysis of the Nonlinear Response of a GCM Climate Model," J. Atmos. Sci. 34, pp. 1885-1897.

FLW/1093/114

IMPACTS OF CHANGES DUE TO LARGE-SCALE DEPLOYMENT OF ENERGY CONVERSION SYSTEMS

Jill Williams

Reinhold-Schneider-Str. 6, 7500 Karlsruhe 51,
F.R. Germany

ABSTRACT. The large scale use of fossil fuels, nuclear energy and solar energy could have climatic impacts. In the case of fossil fuels the major impact of present concern is that due to fossil fuel CO_2 added to the atmosphere, although other gases and particles of climatic importance are also released. Moreover, land use changes due to mining of fossil fuels could also influence climate. The impact of large scale use of nuclear energy is considered to be due to waste heat release. As far as solar energy conversion systems are concerned, the technologies are in an earlier stage of development and it is clear that the deployment on a scale large enough to influence climate will in any case occur in the distant future. In addition the detailed physical characteristics of the systems are not well known. It is possible, however, to discuss the general types of impacts which could arise. An impact on climate could arise due to changes in the surface energy balance, in the surface roughness and in the surface hydrological characteristics. These changes would in reality be extremely complex and at the present it is only possible to make simple assessments of climatic impacts due to individual changes. This paper will consider the potential impacts of the land-use changes due to the deployment of solar thermal electric, photovoltaic and biomass systems. Since the latter two systems are, however, more likely to be deployed in a decentralised fashion with a variety of conversion systems it is felt that they are unlikely to have more than a very localised impact. On the other hand, model results suggest that large scale deployment of solar thermal electric conversion could lead to changes in the cloudiness and precipitation distribution on the scale of the power plant.

W. Bach, J. Pankrath, and J. Williams (eds.), Interactions of Energy and Climate, 327–337.

1. INTRODUCTION

There are basically three energy supply sources which can be used to satisfy future energy demand: fossil fuels, nuclear and solar energy. The large-scale use of any of these will lead to climatic changes on the local, regional or, possibly, global scale. At present most concern centres on the possibility of global climatic changes due to the increasing large-scale use of fossil fuels, which releases carbon dioxide into the atmosphere. Large scale use of nuclear energy could have a climatic impact due to the release of waste heat. Secondary climatic effects of energy systems could also arise through large scale land use changes due to mining activities etc. As far as solar energy is concerned, the conversion technologies are generally in an earlier stage of development and it is clear that deployment on a scale large enough to influence climate will, in any case, occur in the distant future. In addition, the detailed physical characteristics of most of the systems are not well known. It is possible, however, to discuss the general types of impacts which could arise, often without reference to specific systems. A characteristic of solar energy conversion systems is that there is a large variety of system types, each of which is particularly suited to certain locations and the decentralised, small-scale use is unlikely to influence more than the microclimate. Of the many renewable energy sources which are likely to be used in the future, this paper will consider the potential climate impact of three systems: solar thermal electric, photovoltaic and biomass conversion.

2. PHOTOVOLTAIC CONVERSION

Photovoltaic power systems consist of arrays of small individual generating units, the photovoltaic cells, in which semiconductor materials are used to convert the energy in light to electricity. The large interest in photovoltaic electricity generators is largely because they can be made in a wide range of sizes, from small units with a few Watts output to the GW designs of solar power satellites. A second advantage is that photovoltaic units can convert diffuse as well as direct solar radiation, so that they are attractive for the higher, more cloudy latitudes.

With regard to terrestrial applications, results of some studies suggest that photovoltaic systems will be especially applicable in small decentralised units (1, 2). On the other hand it has also been pointed out that larger photovoltaic plants could be connected with a utility grid without additional dedicated storage and that photovoltaic power plants will be very much utility specific and depend on the characteristics of the utility and the geographic location (3).

One important point with regard to photovoltaic systems is that there are no apparent economies of scale (4). Actually, smaller systems have a number of advantages: they can be built more rapidly (reducing interest paid during construction and planning lead times); they can be more effectively tailored to individual load requirements; the requirement for land is reduced if systems are located on roofs; and location near the point of use of the energy reduces transmission losses and equipment needs.

At the present, therefore, the use of photovoltaic conversion in a decentralised fashion appears to be the preferred option. In this case, local/regional impacts on weather and climate would presumably not differ in magnitude from the impacts of urban areas. Isolated applications of decentralised photovoltaic systems cannot be expected to impact the local weather, a single 10 kW array on a rooftop would only exert a micrometeorological influence on the conditions in a narrow layer over the roof. On a larger scale, if photovoltaic systems were used on a high proportion of rooftops, over parking areas etc., in an urban area, then their impact would essentially be an enhancement of the urban heat island effect.

A rough estimate of the impact of a photovoltaic collector system can be made as follows. One can assume that due to the use of non-reflecting surfaces, the cells absorb 95 % of the radiation falling on them and also that the cells cover most of the flat plate (again 95 %). Therefore a very high percentage of the insolation on the flat plates is absorbed. However, cell efficiency is low, only 10 % or so of the energy is converted to electricity and the rest of the energy must be released into the atmosphere as longwave radiation, sensible and latent heat. With cells of only 5 - 8 % efficiency, more energy must be released. Over natural surfaces the amount of energy which is absorbed and subsequently released to the atmosphere varies between 70 - 75 % for desert areas, 75 - 80 % for dried grass, 85 - 93 % for coniferous forests, 90 - 93 % for water (5). Thus if the photovoltaic system were installed in an area where the surface was naturally dark (e.g., forest, wet fields), there would be no significant change in the amount of energy input to the atmosphere. At the other extreme, if cells with a low efficiency were used and the natural area had a high reflectivity (e.g. desert, dry grass), then the photovoltaic system would act as a heat source. However, for the EPRI (6) 200 MW flat plate reference system this heat source would have an area on the order of only 3 km^2. For this size of perturbation the percentage change in the amount of energy input would probably not be enough to cause significant local weather or climate impacts. It should be noted that the photovoltaic systems would lead to an increased proportion of sensible heat input to the atmosphere and a decreased proportion of latent heat input.

The presently conceived photovoltaic conversion systems and the scale of their deployment, therefore, appear to have only a potential impact on micrometeorological conditions, on the scale of the collectors themselves. A significant impact on local/regional climate could only be foreseen in the case of massive deployment of the systems, say in desert areas, or installation of plants with thousands of MWe capacity. Since, however, there appear to be no economies of scale and since there are advantages in smaller sized systems, this is unlikely. Also massive deployment seems unlikely since photovoltaic systems without their own storage facilities could be the most attractive option for electricity utilities (3) in which case the photovoltaic systems within any utility could only contribute part of the electricity supply and other systems with storage would additionally be required.

3. BIOMASS CONVERSION

Biological sources of energy fall into two broad categories: waste from non-energy processes (e.g., food and paper production) and crops grown explicitly for their energy value (7). As fas as wastes are concerned, agricultural residues represent the largest potential source. It is generally felt that the second category of biological energy sources - energy crops - will be limited to marginal lands since population pressure is already forcing food producers onto lands ill-suited to conventional agriculture. However, considerable potential energy cropland does exist. Factors such as availability of water and nutrients will also limit energy cropping, and it has been argued that the cost of transporting bulk biomass should limit energy "farms" to a few thousand hectares or less.

Several technologies are used to transform the biomass. Although 1/3 - 2/3 of the energy is lost in conversion, the converted fuels can be used more efficiently. The technologies include direct combustion, anaerobic digestion, pyrolysis, hydrolysis, hydrogasification and hydrogenation. Anaerobic digestion has received much attention. Biogas generators anaerobically convert cow dung, human excreta and inedible agricultural wastes into a mixture largely of methane (60 - 70 %) and carbon dioxide. The residue of biogas production is a rich fertilizer containing all the original nutrients of the biomass. China has reportedly built more than 4 million biogas plants in the last 3 years and other biogas programmes are underway in India, Indonesia Korea, Taiwan and to a smaller extent in other countries.

Biological sources of energy have many advantages. They can produce solid, liquid and gaseous fuels. Conversion is possible on a local, decentralised scale using generally relatively simple technology and larger scale applications are also possible. The use of biomass in developing countries is recommendable, partly because it already supplies most of the energy used in these countries and also because of the above advantages. In addition, in developed countries substantial sources exist already in wastes and could be increased by energy cropping. Used properly, with due attention to such factors as maintenance of soil nutrient levels, biological energy sources could play an important role in future energy supplies. In addition, possible future non-conventional uses of photosynthesis for such processes as hydrogen production could be introduced (8).

Due to the heterogeneity of biological energy sources, no simple evaluation of the impacts of biomass conversion on climate can be given. In the study for a possible "Solar Sweden", in which it is proposed that half of the energy required in 2015 would be from biomass, energy plantations of 3 million hectares would be required. However, the forestry industry presently uses 23 million hectares (9) and thus the impact of future energy conversion would be presumably smaller than that of present forests. Likewise, a very high proportion of the firewood cut today is burned for cooking food and warming houses. The biomass conversion schemes to use firewood in the future generally are based on technology to use firewood far more efficiently than it is used today, thus providing more energy from less resources and reducing possible climatic impacts.

In view of the potential role of wood in biomass conversion systems, the climate-related properties of forests are of significance. These properties are also considered in the paper by Baumgartner at this conference. From the point of view of energetics, forests are the most active elements of the land cover (10). The energy exchange is not concentrated in a thin layer, because radiation is intercepted and transferred within the stand. Consequently forests do not overheat greatly relative to the ambient temperature and the longwave radiation emission is therefore low. But trees have a low reflectivity for short wave radiation ($\sim$10 % for coniferous and $\sim$15 % for deciduous trees) and the net radiation over forests is high. Baumgartner (10) gives the net radiation for forests as 67 Jm^{-2}, for crops as 53 Jm^{-2}, for grass as 47 Jm^{-2} and for bare soil as 33 Jm^{-2}. The high energy income is used by the forests primarily for evapotranspiration. The potential evaporation from forests (850 mm a^{-1}) is twice that of bare soil (10). Forests consume more water than other vegetation and transfer it by evapotranspiration back into the atmosphere. The increased water vapor in the air results in generally higher precipitation over forest areas. But despite their use of water, forests also act as an important control of surface water. It has been shown for the southern USSR that in totally forested areas only 18 % of the precipitation appears as river discharge, whereas in open land 42 % appears as discharge (10).

The apportionment of net radiation into sensible and latent heat is usually discussed in terms of the Bowen ratio (i.e., the ratio of sensible heat transfer to latent heat transfer), which gives a good measure of the impact of forests on the energy balance. Above irrigated rice fields, swamps and tropical rain forests, the Bowen ratio is about 0.1, whereas under arid conditions the Bowen ratio is well above 1, with values of about 10 in desert regions (11).

As Baumgartner has pointed out, forests also have a dynamic interaction with the atmosphere. The top of the forest is one of the roughest earth-atmosphere interfaces, with the crown braking the movement of the air and absorbing momentum. This effect can be expressed by the roughness parameter, z_0, values of which for various surfaces, are given in Table 1.

Table 1: Roughness parameter, z_0, for various surfaces
Source: Baumgartner (10)

	water	sand	short grass	crops	forest	city
z_0 (cm.)	0.1	.2	1 - 5	10	300	400

The impact of the increased "roughness" of forest compared with open land is illustrated by the influence on wind velocity profiles exerted by forests. Measurements for European forests show that 30 metres of penetration reduces wind velocities 60 - 80 % and the reduction is to 7 % with penetration of 120 m (12). Baumgartner (10) emphasizes that forests influence, thereby, the surface drag, energy dissipation and turbulent exchange of the air, which all depend on the roughness: thus any large scale change in forest cover could have an effect on the development of the pressure field and the circulation of the atmosphere. It is further pointed out that the increased roughness in forests produces high turbulent exchange and mixing rates, which are very active in cleansing polluted air near cities or industries.

A further factor which might be taken into account with regard to biomass conversion is the possible need for irrigation in the case of energy plantation schemes, especially since such schemes are likely to be relegated to marginal agricultural lands, since food production will require the better agricultural conditions. Munn and Machta (5) point out that irrigation projects can be of local climatic significance but the effects are sometimes difficult to detect on a larger scale. They suggest that at a height of 10 m above a large agricultural area under irrigation or at a downwind distance of 1 km from a reservoir the climatic effects may be imperceptible. However, it is possible that such changes in surface conditions can trigger cloud or shower development. One study has found higher rainfall in the vicinity of irrigated regions than in central non-irrigated regions during June, July and August in Kansas, Nebraska and most of Texas, whereas in April, May and September when irrigation was not used, there was no anomaly (5). The impact of irrigation is, of course, most likely to be noticeable in arid regions, where the increased atmospheric humidity may not, however, lead to increased precipitation because of the prevailing large-scale subsidence.

4. SOLAR THERMAL ELECTRICITY CONVERSION

Systems using thermodynamic cycles (turbines) for converting solar energy into electricity have also received attention recently and systems with capacities ranging from 1 kWe to 5 MWe exist (13). One system is the central receiver "power tower" design. The system consists of a central tower with a receiver on the top and several thousand steered mirrors (heliostats) on the surrounding ground, which reflect direct solar radiation onto the receiver. The largest installation at present of this type (10 MWe) is scheduled to be operating in California in 1981. In a location favoured by a high proportion of direct radiation (70 - 80 % of global radiation), a central receiver plant employing a conventional steam turbine may be operated with an overall efficiency of 15 - 18 %.

One example of a system is the design of a 100 MWe solar thermal electric conversion (STEC) power plant by Boeing Engineering and Construction (14). Certain characteristics are noteworthy, especially with regard to their potential impact on climate. The central receiver is 260 m high and is surrounded by heliostats over an area of 1.3 km^2. The heliostats (total of 15,400) cover about 39% of the total ground surface and each collector has an area of about 32 m^2.

The above type of system could be expected to lead to the following major changes in the characteristics of the earth's surface, which influence the local weather and climate:
- changes in the surface energy balance
- changes in surface roughness, since heliostats are several meters high
- changes in surface hydrological characteristics, for example, if a formerly vegetated area is paved.

The changes in surface energy balance would in reality be extremely complex. For instance, the changes in surface thermal characteristics because of paving, the changes of energy fluxes from the surface because of roughness changes and the changes in the amount of sensible and latent heat flux from the surface are only a few of the microclimatological impacts which should be considered.

A simplified estimate (15) suggests that in the presence of the STEC plant the total of the energy flux from the surface into the atmosphere (70 % of the incident solar radiation in the area) does not differ from that without the STEC plant. However, the distribution is changed, the longwave radiation from the surface is reduced from 35 % (no STEC) to 10 % (STEC, winter) or 25 % (STEC, summer). Likewise the sum of sensible plus latent heat fluxes is reduced. These reductions are compensated by a release of waste heat from cooling towers.

The impact of such energy balance changes has been investigated using a two-dimensional mesoscale model of the atmosphere (16). For this study the plant characteristics were based on the central receiver STEC plant system of Martin Marietta. The atmospheric effects of a hypothetical STEC facility located in Southern Spain were investigated. To examine the effects of large-scale installations, an area of about 1000 km^2 was assumed; such a plant could generate about 30 GWe and might be considered as a solar power park. The horizontal dimension of the STEC plant was assumed to be 32 km (oriented in the direction of the prevailing wind), with heliostats covering 8 km (ground cover ratio 25 %). The total horizontal dimension considered was 400 km so that meteorological conditions upstream and downstream of the STEC plant could be included. It was assumed that the STEC system uses wet cooling towers for dissipating waste heat into the atmosphere and this heat is released at heights of 100 m or more in the form of latent (80 %) and sensible heat (20 %).

The model was run for typical conditions of summer and winter in Spain, as derived from meteorological observations. Examination of the model relative humidity field showed that in the presence of the STEC plant the atmosphere was relatively more moist than in the control case, with cloud development (where cloud is defined as an area with relative humidity of more than 100 %). In the control case, clouds are sporadic and cover relatively small areas of domain, whereas they are quite pronounced and persistent in the "with STEC" case. These results are supported by similar considerations of the temperature and vertical velocity fields.

The model was also integrated with initial conditions corresponding to winter meteorological conditions in Southern Spain.

In the winter simulation, no clouds formed anywhere in the model domain throughout the simulation in both the case without STEC and that with STEC. It is suggested that this is due to the strong prevailing winds assumed for the winter simulation.

Bhumralkar et al. (16) emphasize that the model results should be treated as preliminary. They suggest the need to investigate these impacts given a wide range of meteorological conditions and with variations in the design, areal extent, capacity, and location of the STEC system. Bhumralkar (personal communication, 1979) has recently used the same model to study the impact of a 10 MW STEC plant being built in Barstow, California, and preliminary results indicate that the location plus the characteristics of the STEC facility are again sufficient to produce local meteorological effects.

Other studies confirm that changes in the surface energy balance can influence the local atmospheric conditions. These comments are basically applicable to all energy conversion systems which cause similar changes in the surface energy balance. An early suggestion that weather can be modified by changing the reflectivity of the surface was made by Black (17), who suggested that coating the ground with black asphalt could increase soil temperatures as much as 10° C, whereas if the freshly applied asphalt were used to bind a white material such as lime, subsurface temperatures could be reduced as much as 13° C. Black (17) pointed out that surface heat sources play a significant role in localising cumulonimbus clouds and thundershowers, when atmospheric conditions are suitable. Therefore, albedo changes were suggested as possibly effective in modifying weather.

Charney et al. (18) have carried out a number of model simulations to investigate the impacts of albedo changes on the atmospheric circulation. As part of this study, the authors evaluate the least dimension of an area for which albedo and ground moisture changes can be expected to influence convective rainfall. It is suggested that a plausible rule of thumb is that observable effects can be expected when the characteristic time for a change in the moist static energy to penetrate to cloud base is smaller than the time required for new properties to be advected into the region. This suggests a minimum dimension of 40 - 80 km.

A local atmospheric circulation model was developed by Berkovsky (19) and has been used to show that the vertical circulation and therefore cloudiness and precipitation respond to surface albedo changes in desert areas. In particular the model showed that a lowering of the surface albedo in a desert region could lead to increased vertical velocity and possibly increased rainfall.

The impacts of the other potential changes due to STEC systems, i.e., those in roughness and wetness of the surface, have not been investigated in the same detail as the energy balance changes. Some implications of roughness changes and of wetness changes have been described above with reference to biomass systems.

5. CONCLUSIONS

This paper has considered the potential climatic impact of three renewable energy conversion systems. In the case of photovoltaic and biomass conversion, it appears that decentralised small scale applications will be preferred and a wide variety of applications suited to different locations can be expected. For example, biomass conversion could take the form of biogas production in some villages, conversion of agricultural wastes in other areas, energy "cropping" in others and so on. This heterogeneity and localised deployment mean that climatic effects, if they occur at all, would be restricted to the micro or local climate scale. Such effects would mainly arise due to changes in surface reflectivity, roughness or wetness. Solar thermal electric conversion, on the other hand, is a more centralised option and model results suggest that large scale deployment of these systems could lead to cloudiness and precipitation changes on the scale of STEC facilities.

It is clear that the large-scale use of any energy supply source, fossil fuels, nuclear or solar, could cause climatic changes. Since it is not possible at present to predict the nature and extent of these changes, it should be noted that decentralised, heterogeneous small scale energy conversion systems are most likely to have comparatively negligible climatic effects.

REFERENCES

(1) Witwer, J.G., et al.: 1978, "A comparative evaluation of solar alternatives: Implications for Federal RD&D", SRI International, Menlo Park, U.S.A.

(2) Zahn, P.: 1978, "Conceptual design of solar thermal power plants (100W-10MW)", Conference Proceedings (C16), UK-ISES, London.

(3) EPRI: 1978, EPRI ER-685-SY, Electric Power Research Institute, Palo Alto, U.S.A.

(4) Kelly, H.: 1978, Science, 199, pp. 634-643

(5) Munn, R.E., and Machta, L.: 1979, WMO Publication No. 537, World Meteorological Organization, Geneva.

(6) EPRI: 1978, EPRI ER-589-SR, Electric Power Research Institute, Palo Alto, U.S.A.

(7) Hayes, D.: 1977, The Ecologist, 7, pp. 349-356.

(8) Hall, D.O.: 1979, Solar Energy, 22, pp. 307-328.

(9) Johansson, T.B., and Steen P.: 1978, "Solar Sweden", Secretariat for Future Studies, Sweden.

(10) Baumgartner, A.: 1979, WMO Publication No. 537, World Meteorological Organization, Geneva.

(11) SMIC: 1971, "Inadvertent Climate Modification", MIT Press, Cambridge, U.S.A.

(12) Barry, R.G., and Chorley, R.J.: 1968, "Atmosphere, Weather and Climate", Methuen, London, U.K.

(13) Jäger, F., and Jochem, E.: 1979, "Technological forecasting and solar energy", Seminar on Cooperative Technological Forecasting: Solar Energy, UN ECE, Geneva.

(14) EPRI: 1976, EPRI ER-403-SY, Electric Power Research Institute, Palo Alto, U.S.A.

(15) Williams, J., and Krömer, G.: 1979, SR-79-2B, International Institute for Applied Systems Analysis, Laxenburg, Austria.

(16) Bhumralkar, C.M., Williams, J., and Slemmons, A.: 1979, Solar Energy, 23, pp. 393-403

(17) Black, J.F.: 1963, Science, 139, pp. 226-227.

(18) Charney, J.G., Quirk, W.J., Chow, S.H., and Kornfield, J.: 1979, J. Atmos. Sd., 34, pp. 1366-1385

(19) Berkovsky, L.: 1976, J. Appl. Meteor., 15, pp. 1139-1144.

CLIMATIC IMPACTS OF URBANIZATION

T. R. Oke

Department of Geography, The University of British Columbia, Vancouver, B.C., Canada

ABSTRACT. The land-use changes accompanying urbanization have an impact upon virtually all climatic elements as a result of perturbations to the pre-urban cycling of energy and mass by the city-atmosphere system. The paper illustrates how and to what extent these changes occur based on the results of observational studies. By considering the effects of the city upon the processes governing airflow, the water, radiation and energy balances, cloud development and precipitation some understanding is gained for the observed alteration to the pre-urban fields of wind, humidity, air temperature, cloud and rainfall. Direct impacts of land-use change are limited to the micro- and local climatic scales but those related to gaseous, aerosol and heat emissions contribute on all scales up to and including the global.

1. INTRODUCTION

Urbanization, the process of converting an often natural landscape into a human settlement, is by definition accompanied by drastic and almost irreversible land-use change. The replacement of natural surface materials by those used in construction and the release of waste products of urban metabolism create a totally new set of surface aerodynamic, radiative, thermal and moisture properties and air quality. In turn these alterations perturb the pre-urban balances of energy, mass and momentum and lead to modification of the state of all atmospheric parameters (the ensemble average of which represents the climate).

It is the purpose of this paper to review the nature of these climatic impacts. To cover such a wide spectrum of

W. Bach, J. Pankrath, and J. Williams (eds.), Interactions of Energy and Climate, 339–356.

features it will be necessary to condense much of the evidence and those wishing to search more deeply are advised to consult the more extensive reviews available elsewhere[1-7]. In particular the impacts due to urban emissions of gases, aerosols and heat will be compacted because they are fully covered elsewhere in this volume.

Initially we will restrict consideration to impacts on scales up to that of the city itself, and as Oke [8] has noted it is important to distinguish between the micro-scale of the urban canopy layer, UCL (beneath roof-level) and the local scale of the urban boundary layer, UBL (extending from roof-level to the height at which urban influences are absent). Since here we are more concerned with the areally-integrated impact of the city on climate primary attention will be directed to UBL processes and effects but it is often necessary to refer to conditions in the UCL to understand the larger scale responses. Later, in Section 7 we will consider possible urban impacts on regional, synoptic and global climates.

Impacts are usually assessed in a comparative methodological framework. In urban meteorology this often means using estimators such as urban-rural or upwind-downwind differences. Lowry [9] has shown that such surrogate measures of urban impacts are open to criticism and that the only true measure is one involving the difference between urban and pre-urban observations stratified by synoptic weather type. Since such data are only rarely available readers should be cautioned that assessments of urban effects cannot be very precise.

2. IMPACTS OF AIRFLOW CHANGE

Winds in the planetary boundary layer are governed by a balance of the horizontal pressure gradient force, the surface frictional stress and the rotation of the Earth (via the Coriolis force). Urbanization alters the surface and atmospheric characteristics sufficiently to modify the existing balance. In particular the greater aerodynamic roughness of cities increases the frictional drag, and the existence of an air temperature anomaly (known as the urban heat island, Section 5) imposes a three-dimensional deformation upon the pressure field. These changes (and others) combine to cause modification of the horizontal and vertical wind velocity (speed and direction) for the mean and turbulent components of the flow. When the imposed airflow is strong the roughness effects dominate, but when it is weak or absent both the roughness and the pressure effects are involved. In both cases the city is associated with an increase of thermal and/or mechanical turbulence of the order of 30 - 50% [10,11].

With strong winds the impact of the city is fairly clear, at least in principle. Observations are rather scanty but from those available and analogues drawn from wind tunnel and controlled roughness-change experiments the following picture emerges. With medium to strong flow (> 5 m s^{-1}) across the much rougher city, the increased surface drag results in deceleration[12-14]. This in turn produces low-level convergence and uplift of the air mass and a 'plume' of ascending air which can extend to distances as great as 30 km downwind[13]. Conversely, the edges of the 'plume' show descending air appearing rather like roll vortices shed from the outskirts of the city. On some occasions the removal of the city's barrier effect in its lee can produce sharp downward motion giving rise to transport of greater momentum toward the surface and a local maximum in wind speed near the surface. The urban uplift extends through most of the mixed layer and is sufficient to create a 'doming' of the UBL of the order of 0.25 km [15]. The enhanced mechanical turbulence produced by the roughness change also modifies the form of the wind speed profile in the UBL so that its vertical gradient is diminished. Finally, since wind speeds are reduced over the city the Coriolis deflecting force is also decreased. Thus if the heat island effect is small (as it is with strong winds) the pressure gradient force may be assumed constant and therefore the flow will turn towards lower pressure (i.e., cyclonically [13,16]). Downwind of the city, where acceleration takes place over the smoother countryside, the wind direction recovers (turns anti-cyclonically).

The aerodynamic roughness length of suburban terrain is typically 0.5 - 1.5 m and that of densely built-up city centres 1.5 - 4 m[11]. The strength of the roughness effects outlined above are of course dependent upon the magnitude of urban-rural roughness differences. Since the suburban and urban values are considerably larger than those for open countryside and agricultural crops (about 0.01 - 0.2 m) the effects produced by cities surrounded by such land-uses may be expected to be marked. On the other hand those in the midst of jungle or forested areas may show almost no effect.

Conditions in the UCL are very complicated due to the great variety of shapes, sizes and orientations of the buildings. In general street-level winds are decreased in comparison to those in the countryside but increases may occur locally due to 'jetting' down canyons aligned with the wind and in the vicinity of tall buildings which deflect momentum from above[17].

With light or calm winds and the presence of a well-developed heat island the impact of the city upon airflow becomes more complex and difficult to observe. Below some threshold speed, typically about 3.5 - 5.0 m s^{-1}[12,14] synoptic winds may actually be accelerated by the presence of the city. The pressure gradient associated with the heat island (especially on the upwind half of the city) adds vectorially to the velocity and the enhanced turbulent exchange of momentum in the UBL 'folds' energy down into the near surface flow. Reversing the argument used earlier, the increased speeds lead to anticyclonic turning from the upwind edge of the city to the centre and a tendency towards cyclonic re-curvature from the centre downstream because the heat island pressure field is opposing the synoptic flow.

When synoptic winds are absent the heat island pressure field induces a low-level convergent flow from all directions analogous to a sea breeze. This urban breeze system typically has an inflow velocity of about 2 m s^{-1}, an updraft strength of about 0.3 m s^{-1} over the city centre, and extends to a depth of the order of 1 km where divergence results in counter flows aloft [18,19]. This circulation can be present by day or night and it has been shown that for a given heat island intensity the breeze system is more easily established by day because uplift is facilitated by the greater daytime instability of the UBL[20].

3. IMPACTS OF WATER BALANCE CHANGE

The characteristic cycling of water through most natural systems is considerably disrupted by urbanization. The water balance of an urban building-air volume (illustrated in Fig. 1) is given by:

$$p + I + F = E + \Delta r + \Delta S + \Delta A \quad (1)$$

where p - precipitation, I - water supply piped in from rivers or reservoirs, F - water released to the air in combustion, E - evapotranspiration, Δr - net run-off, ΔS - moisture storage and ΔA - net moisture advection. Simple consideration of the input terms (l.h.s. of (1)) suggests that urban areas are in receipt of more water than their environs (unless extensive rural irrigation is practised). Precipitation may be slightly increased (Section 6) and I and F are totally new sources for the city due to anthropogenic activities. On the other hand (and ignoring ΔA) we may expect that the partial 'waterproofing' of the urban surface by largely impervious construction materials will cause a depletion of E and ΔS. Thus we may conclude, in agreement with most hydrologic studies [22-24], that urbanization of a watershed leads to a marked increase of run-off. However, the questions of more concern regarding the impact of the city on the atmosphere are what is the net alteration in the loss of water to the air from natural and anthropogenic sources (i.e., E + F), and what are the resulting effects on urban humidity and fog?

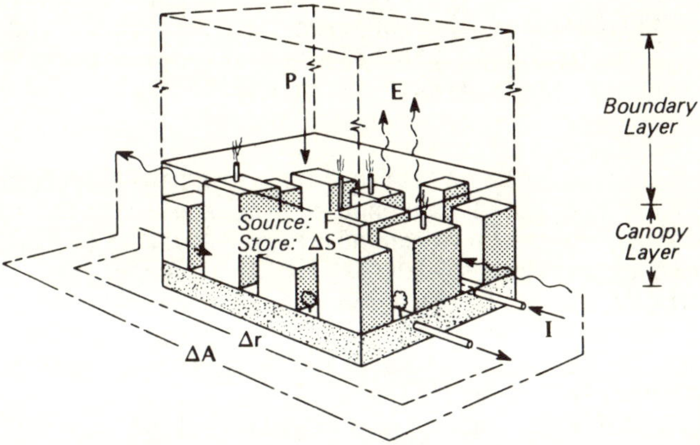

Fig. 1 Water balance of an urban building-air volume (after Oke[21]).

There is little information regarding the strength of urban moisture sources but a reasonable concensus would seem to be that by day in summer the anthropogenic contribution is one to two orders of magnitude less than that by evapotranspiration and can be neglected for most purposes. On the other hand the two sources may be of almost equal strength at night and in the vicinity of industrial areas and anthropogenic releases can become dominant in winter at high latitudes [27,28]. Clearly for most cities evapotranspiration is the primary mode of water input to the atmosphere on most occasions so assessment of a city's impact upon humidity centres on urban-rural comparisons of this term. Somewhat surprisingly, considering the replacement of transpiring surfaces by largely impervious ones, urban evapotranspiration is thought to be only slightly decreased (Section 5)[26,29,30].

The generally accepted impact of urban areas on the humidity field can be tentatively explained in terms of the relative strengths of these urban moisture sources and their comparability to rural evaporation/condensation. In summer the daytime moisture content of the UCL and UBL is less than in the surrounding countryside because E is decreased in the city, F is very small and the greater turbulence and mixed layer depth over the city reduces its concentration. At night in the summer, especially in calm clear weather, the UCL and UBL may be slightly moister with a humidity 'island'(31-33) and 'plume'(25,34). This might be due to the fact that F is present in the city but also because E tends to continue through the night with relatively little depletion of atmospheric moisture by dewfall. In winter, especially in high latitude cities UCL and UBL humidities may be higher than those in the countryside both by day and at night(32,33). Since E is probably negligible in both environments this situation is likely to be due to F which peaks in this season.

There is no simple relation between urbanization and fog except perhaps for the occurrence of ice fog in high latitude settlements due to sublimation of water vapour from combustion at air temperatures below -30°C(27). In other cities the required mix of water vapour, condensation nuclei and cooling may or may not make the urban area a preferred location for fog. At night the city air usually contains more moisture and nuclei but the heat island warmth may be sufficient to offset the formation of fog. As a weak generalization one might regard the city to register more haze, smog and light fog but fewer dense (large droplet) fog occurrences.

4. IMPACTS OF RADIATION BALANCE CHANGE

The city-atmosphere system has profound effects on exchanges of short- and long-wave radiation due both to the presence of radiatively-active pollutants in the air and to the modification of the surface radiative properties. Fig. 2 may help to visualize the radiative fluxes and processes involved.

For the sake of convenience consider that the incoming short-wave radiation at the top of the urban polluted layer (Flux 1 in Fig. 2) is the same as that at the surface of the surrounding countryside. On its way to the surface a significant proportion of the radiation is either scattered or absorbed by the urban gases and aerosols. Some is scattered backwards and is lost to the city-atmosphere system (Flux 2) and the remainder is scattered forward increasing the diffuse-beam portion of the primary incoming short-wave radiation transmitted to the surface (Flux 3). Following reflection from the urban surface the outgoing short-wave radiation (Flux 4) is again subject to pollution scattering and absorption before exiting from the polluted layer (Flux 6). Since Flux 4 is very much smaller than Flux 1 the multiple back-scattered radiation which forms a secondary input to the surface (Flux 5) is not large.

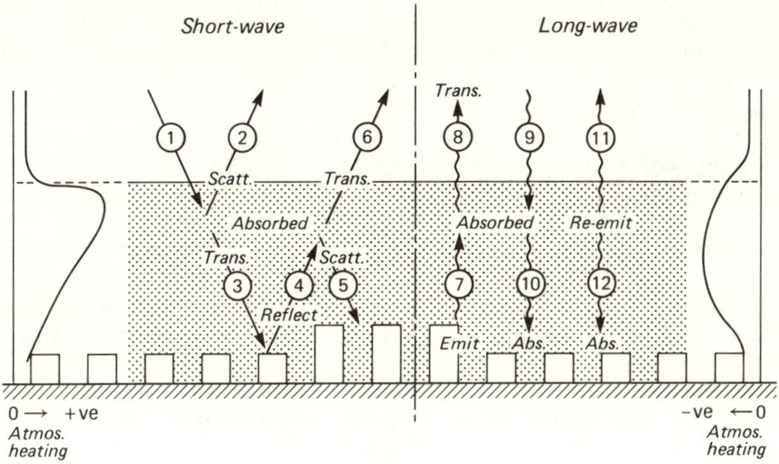

Fig. 2 Radiative fluxes and processes involved in a city-atmosphere system and the suggested vertical profiles of short-wave radiative heating (at the left) and long-wave radiative cooling (at the right) due to the urban aerosol layer (shaded). The numbered fluxes are referred to in the text. Figure modified after Atwater[35].

The overall attenuation of incoming short-wave radiation (i.e., the difference between Fluxes 1 and (3 + 5)) has been the subject of many observational studies [36-40] and to a lesser extent so has its height variation [41,42] and spectral dependence [38,40,43,44]. The accepted concensus until fairly recently was that the polluted atmospheres of large cities were capable of reducing annual receipts of solar radiation by as much as 15 - 20%[4,5]. The more modern view is that this figure should be lower, say 2 - 10% [7,45]. If this signals real temporal changes they may be due to an alteration of the nature of the pollutants (e.g., from large particulates towards small particulates and gases) or to the impact of emission control programmes. What is not yet available is a clear idea of the relative roles of scattering and absorption, so that although the form of atmospheric heating due to short-wave absorption is probably as suggested by profile in Fig. 2, its magnitude is not known with any generality.

The urban albedo is usually reduced by urbanization. This is probably due to the relatively low albedo of roofing materials and to the 'trapping' effect of multiple reflection between the building elements of the UCL[46]. Typical albedo values for mid-latitude cities in summer are 0.12 - 0.17[5,7]. The urban-rural difference depends of course on the nature of the surrounding terrain - with forest environs the difference will be small or absent. For most mid-latitude cities the difference is typically less than 0.10, but in winter with the countryside being more uniformly covered in snow the difference may be much larger[47].

Given a 2 - 10% decrease in incoming short-wave radiation and a decrease of 0.05 - 0.10 in surface albedo it seems that the urban-rural differences of net (absorbed) short-wave radiation will be small because the changes in the component fluxes are off-setting.

The long-wave exchange is perturbed not only by the presence of extra gases and aerosols in the atmosphere and the changed surface emissivity of the urban area but also by the fact that the city air and surface are normally warmer (the heat island). Despite the probability that the urban emissivity is slightly less it appears that the greater surface temperature results in an increased emission of long-wave radiation from the city surface (Flux 7 in Fig. 2) at least at night. A large portion of this is absorbed by the polluted layer and some is re-radiated back to the surface along with that emitted by the warm heat island air (Flux 12) and that portion of the normal incoming atmospheric radiation (Flux 9) that is transmitted to the surface (Flux 10). The combined input (10 + 12) is observed to be larger in urban areas by a small amount at night[48,49] and persists in the area underlying the urban 'plume' in the downwind region. During the daytime it seems possible that the increase is quite marked especially in the morning and this may be due to emission from solar heated pollutants[37]. The net effect of long-wave radiation upon atmospheric temperatures is not entirely clear. Aircraft observations [44] suggest that daytime convergence can produce warming on occasion, whereas most radiation models[35,50] show cooling. Nocturnal observations in the lower UBL almost always show divergence but the UCL exchanges are more complex[51,52].

In terms of the overall impact of urbanization on the surface net long-wave radiation balance the general opinion is that the urban area experiences a slightly greater deficit at night but by day there is little net effect[5]. However, it should be noted that such conclusions are not firm and require further support.

Similar caution is required in dealing with the net all-wave radiation. The few studies available seem to support the suggestion that although all of the component short- and long-wave fluxes are affected by urbanization the net impact is remarkably small[5,7,39,53].

5. IMPACTS OF ENERGY BALANCE CHANGE

The energy balance of an urban building-air volume (Fig. 3) is given by:

$$Q^* + Q_F = Q_E + Q_H + \Delta Q_S + \Delta Q_A \qquad (2)$$

where Q^* - net all-wave radiative flux density, Q_F - anthropogenic heat released by combustion, Q_E and Q_H - latent and sensible turbulent heat flux densities respectively, ΔQ_S and ΔQ_A - net heat storage and net advection in the volume respectively.

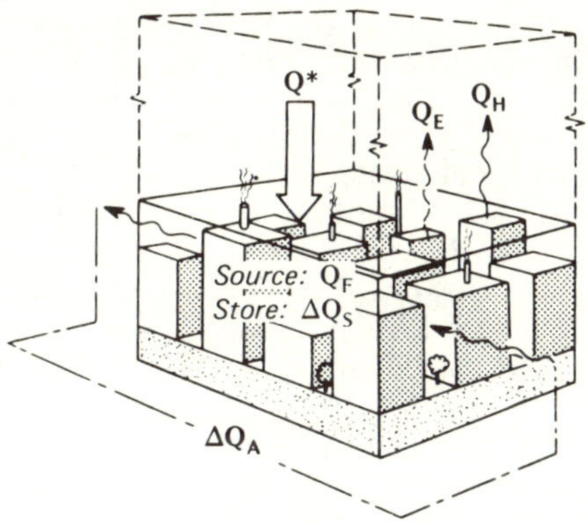

Fig. 3 Energy balance of an urban building-air volume (after Oke[21]).

Comparing urban and pre-urban energy balances for an extensive site where advection can be neglected, and noting from Section 4 that radiation budget differences are likely to be small, we can see from (1) that any differences must stem from the addition of anthropogenic heat (Q_F) and/or from a redistribution of energy between the turbulent (Q_E,Q_H) and storage (ΔQ_S) terms.

The impact of anthropogenic heat emissions is dealt with thoroughly elsewhere in this volume. Suffice it to say that on a city-wide basis this heat source is always present but can vary in impact from negligible (e.g., in most cities in the summer it is about one order of magnitude less than the available daytime net radiation, say Q_F typically $\sim$ 20 W m^{-2} (21,54,55) to dominant (e.g., especially at night in cities with cold winters (21,54,55)). In a concentrated built-up area the average winter value can approach 200 W m^{-2}(21), as in Manhattan, but a value only one tenth as large can have a large impact in a high latitude settlement such as Fairbanks(55,56) where the winter radiation budget can be negligible or negative for considerable periods. Having noted these possibilities, here we will only deal with what at present is the more normal situation for most urban areas - namely Q_F is present but a relatively small component of the energy balance - and consider the possibilities regarding changes in the relative magnitudes of Q_E, Q_H and ΔQ_S due to urbanization.

Results from the few studies where the urban energy balance has been measured show remarkably little disruption considering the impressive land-use changes. As mentioned in Section 3, the latent heat flux associated with evapotranspiration appears to be somewhat reduced in the city(26,29,30), but the change is less than had been anticipated. Average Bowen ratios (Q_H/Q_E) for suburban terrain in summer are typically in the range 0.5 to 1.0(26,29) which is only slightly higher than values quoted for forest, agricultural and grassland areas(55,57). The reasons why the 'waterproofing' effects of urbanization are not more effective may include the increased soil moisture availability by urban irrigation and the creation of micro-'oasis' effects(57). The net result of reduced latent heat transfer is to channel more available energy into sensible heat; either to storage changes in the urban fabric or to the air. Estimation and observation confirms that ΔQ_S and Q_H are somewhat increased in the city but, as constrained by (1), the changes are not large (58-60). For example, the ratio $\Delta Q_S/Q^*$ is typically 0.10- 0.20 for rural terrain and about 0.2 - 0.3 for suburban and urban areas(21,60).

The primary impact of the altered urban energy balance is the production of a positive thermal anomaly - the heat island. However, the exact relationship between the <u>surface</u> energy balance changes (outlined above) and the characteristics of the <u>atmospheric</u> heat island have not been demonstrated. It would appear that in most mid-latitude cities the differential warmth of the city is best displayed at night and is produced by the reduced ability of UCL air to cool. The export of UCL air into the UBL (plus the action of UBL processes) also gives rise to a heat island effect above and downwind of the city. (For a listing of UCL and UBL processes involved see Oke(21)).

Although in annual statistics a large city may only be 1 - 2°C warmer than its surroundings on individual nights with particularly favourable meteorological conditions this difference in the UCL may be as great as 12°C[61] (Fairbanks even has a report of 14°C although there is some difficulty in assigning the non-urban background value[56]). The phenomenon is best displayed in fine weather and has a diurnal variation with a minimum magnitude by day and a maximum a few hours after sunset. Heat islands are found in all settlements and for a given set of weather conditions their intensity is related to some measure of city size[61]. By day the warming influence of a large city may extend to a height of 0.6 - 1.5 km, by night it contracts to a depth of only 0.1 - 0.3 km. The elevated heat 'plume' of the city extends many kilometres downwind before merging with the ambient air (e.g., the St. Louis daytime thermal and moisture plume extends to a maximum distance of about 40 km[62]).

6. IMPACTS OF CLOUD AND PRECIPITATION CHANGE

The impact of cities upon precipitation is a topic of some controversy, concerning both the proof that effects exist and the nature of the processes involved. The present concensus (based on a decade of active research in this field especially in Project METROMEX in St. Louis[63-65]) is that there is reasonable statistical and other support for the hypothesis of precipitation enhancement by large cities but that proof including knowledge of the governing physical mechanisms remains to be demonstrated.

Considering that in addition to water the prime requirements for cloud and precipitation are a supply of condensation nuclei, a means of cooling (usually via uplift) and updrafts sufficient to hold droplets aloft, it is easy to see that the urban atmosphere appears to be a favourable environment. Pollutants provide the nuclei and the heat island/roughness effects on airflow (Section 2) provide the uplift and updrafts.

Braham[64] provides an outline of the impact of urbanization on cloud and haze (Fig. 4). Naturally-occurring Aitken condensation nuclei (ACN) and the larger cloud condensation nuclei (CCN) in the background atmosphere are greatly added to by urban emissions. They are further supplemented by gas-to-particle conversions especially from pollutant gases such as SO_2 and NO_x. In the example shown in Fig. 4 a portion of the urban SO_2 emissions are transformed in the presence of light, water and catalytic surfaces to give small sulphuric acid or sulphate particles (ACN), which can grow by further adsorption and oxidation of SO_2 to CCN sizes. These are then ingested into urban clouds resulting in much higher drop concentrations than non-urban clouds. As a result urban clouds are characterized by a very large number of small droplets. This is not conducive to rain drop growth and the repeated cycling of ACN, SO_2 and CCN through short-lived clouds produces a haze of particles and solution droplets near the cloud base. These may be mixed through much of the depth of the UBL and the downwind PBL to give an extensive and persistent haze layer during stagnant weather.

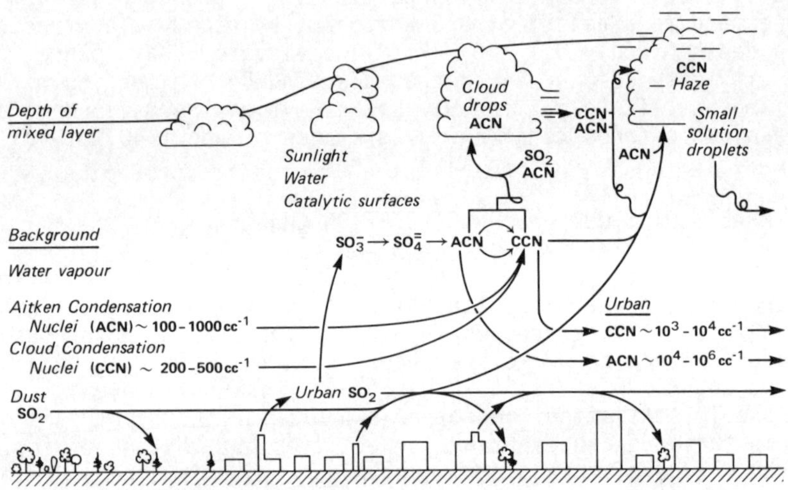

Fig. 4 Schematic outline of processes responsible for the production of urban nuclei and aerosols (after Braham[64]).

This reasonably well understood situation presents what at first sight appears to be a negative urban microphysical influence on rain drop growth and therefore precipitation[64,66]. At least in St. Louis the dominant mechanism for rain drop growth is in-cloud collision/coalescence (most clouds are above 0°C when precipitation starts[67]; so the Bergeron ice crystal mechanism is not applicable). Such a process requires some large droplets to 'sweep up' the small ones. The situation is at least partially rebalanced, however, with the discovery that certain industrial sources produce ultra-giant nuclei and thus provide an appropriate distribution of droplets for droplet growth[64,66,68].

Even where the appropriate seeding of nuclei from urban-industrial sources is present it would seem that more than microphysical modification of clouds will be necessary for significant precipitation enhancement. Thus thermodynamic and mechanical mechanisms are envisaged to be important by most investigators who have suggested hypotheses for urban precipitation modification (64,69-72). The uplift associated with the heat island and mass convergence could well be sufficient to permit urban clouds to penetrate stable layers in the mid-troposphere and to reach glaciation levels, or to produce greater instability in general, or to produce confluence zones of preferred activity. It has also been suggested(70) that the ingestion of warmer, drier urban air into thunderstorms may weaken the cell updraft and accelerate the release of suspended precipitation.

The statistical studies of observed precipitation around cities are always open to criticism concerning the way in which extraneous controls on distributions are handled, e.g., effects of topography and climatic fluctuations. Similar studies of the other climatological variables are also suspect of course (9) but precipitation is particularly vulnerable because of the natural variability of the field and problems of sampling and measurement. The general view now appears to be, given the weight of evidence from many cities, that enhancement probably does occur downwind of urban areas. Most agree that convective rather than frontal events are most susceptible to modification, and severe weather phenomena (thunderstorms, hail) are particularly likely to be affected. The typical size of the increase in precipitation is about 5 to 30% on an annual basis and the maximum effect is usually located a few tens of kilometres downwind of the city. In regions where summer convective rainfall is involved increases of 100% in certain phenomena have been reported(73) and may lead to hydrologic changes (increased streamflow, groundwater contamination) and economic repercussions (decreased crop yields and crop damage)(74).

7. IMPACTS ON LARGER SCALE CLIMATES

It has been estimated(75,76) that the total area of the Earth now converted to urban land-uses is about 1×10^6 km^2 (0.2% total Earth area) and that the present rate of change is about 2×10^4 km^2 y^{-1} (4×10^{-3}% y^{-1}). Therefore if the surface cover changes which accompany urbanization are to have any significant impact on large scale climates they must be drastic. In the light of the review given here such changes do not appear to occur. For example the total cumulative effects of urbanization on the global albedo are estimated to be about 2.5×10^{-5}(76) and are therefore almost negligible. When one also notes the rather small observed modifications to the surface radiation and energy budgets it appears there is little reason to suspect any important impacts on the global climate.

This view, that surface cover changes by cities do not produce large-scale impacts, should not be construed to suggest that cities do not affect climates on scales beyond the local, even up to the global. First, much of the Man-made emission of pollutants (gaseous, aerosol and thermal) have their origin in cities and there is abundant evidence in this volume and elsewhere[4,54,55,77] to illustrate their impacts on all climatic scales. Second, a number of the surface cover effects extend downwind of the city as constituents of its 'plume'. In addition to pollutants these include enhanced turbulence and vertical velocity, altered air temperature and humidity, and increased haziness, cloud and precipitation. Indications of these modifications may persist to as far as 100 km downwind. Third, there is evidence to suggest that cities may modify the path, speed and other characteristics of meso-scale and synoptic systems passing across or near them. Sea breeze and synoptic fronts have been shown to be affected by the frictional and heat island characteristics of large cities[78-80]. Fourth, the clustering together of conurbations into megalopolitan regions brings with it the probability that the effects of one city will overlap those of one or more others thus producing cumulative impacts. There is even the suggestion that a line of cities might be capable of altering the general circulation of the atmosphere[6].

REFERENCES

(1) Kratzer, A.: 1956, Das Stadtklima, Braunschweig, 90, 184 p.

(2) Peterson, J.T.: 1969, The Climate of Cities: A Survey of Recent Literature, NAPCA Pub. No. AP-59, Washington, 48 p.

(3) Landsberg, H.E.: 1956, in Man's Role in Changing the Face of the Earth, W.L. Thomas, ed., Univ. Chicago Press, Chicago, 584-606.

(4) Landsberg, H.E.: 1970, Science, 170, pp. 1265-1274.

(5) Oke, T.R.: 1974, Review of Urban Climatology, 1968-1973, WMO Tech. Note 134, WMO, Geneva, 132 p.

(6) Lowry, W.P.: 1978, Proc. Hung-Amer. Bilateral Seminar Geog. Charact. Urban Develop., Budapest, pp. 231-252.

(7) Oke, T.R.: 1979, Review of Urban Climatology, 1973-1976, WMO Tech. Note 169, WMO, Geneva, 115 p.

(8) Oke, T.R.: 1976, Atmosphere, 14, pp. 268-277.

(9) Lowry, W.P.: 1977, J. Appl. Meteorol., 16, pp. 129-135.

(10) Bowne, N.E. and J.T. Ball: 1970, J. Appl. Meteorol., 9, pp. 862-873.

(11) Counihan, J.: 1973, Atmos. Environ., 7, pp. 673-689.

(12) Chandler, T.J.: 1965, The Climate of London, Hutchinson, London, 292 p.

(13) Angell, J.K., W.H. Hoecker, Dickson and D.H. Pack 1973: J. Appl. Meteorol., 12, pp. 924-936.

(14) Bornstein, R.D. and D.S. Johnson 1977: Atmos. Environ., 11, pp. 597-604.

(15) Auer, A.H., Jr.: 1975, Proc. 2nd US Nat. Conf. Wind Engin. Res., WERC/NSF, pp. I-20-1 to I-20-3.

(16) Lee, D.O.: 1977, Weather, 32, pp. 162-170.

(17) Penwarden, A.D. and A.F.E. Wise: 1975, Wind Environment around Buildings, BRE Report, HMSO, London.

(18) Dannevik, W.P., S. Frisella, J.H. Hwang and R.B. Husar: 1974, Preprint 5th Conf. Weath. Forecast. Analysis, AMS, Boston.

(19) Baxter, R.A.: 1977, Unpub. manuscript, San José State Univ., 37 p.

(20) Shreffler, J.H.: 1978, Boundary-Layer Meteorol., 15, 229-242.

(21) Oke, T.R.: 1978a, Boundary Layer Climates, Methuen, London, 372 p.

(22) Leopold, L.B.: 1968, US Geol. Survey Circ. 554, Washington, D.C.

(23) Moore, W.L. and C.W. Morgan: 1969, Effects of Watershed Changes on Streamflow, Univ. Texas Press, Austin.

(24) Schaake, J.C. Jr.: 1972, In Urbanization and Environment, Detwyler, T.R. and M.G. Marcus (eds.), Duxbury Press, Belmont, Calif., pp. 97-134.

(25) Sisterson, D.L. and R.A. Dirks: 1977, Proc. Conf. Metro. Phys. Environ., USDA Forest Service Tech. Report NE-25, USDA Upper Darby, Pa., pp. 26-35.
(26) Kalanda, B.D., T.R. Oke and D. Spittlehouse: 1980, J. Appl. Meteorol., (in press).
(27) Hage, K.D.: 1972, Boundary-Layer Meteorol., 2, pp. 334-347.
(28) Bornstein, R.D. and Y-T. Tam: 1977, Proc. Conf. Metro. Phys. Environ., USDA Forest Service Tech. Report NE-25, USDA Upper Darby, Pa., pp. 36-52.
(29) Oke, T.R.: 1978b, Proc. WMO Symp. Boundary Layer Phys. Appl. Specific Problems Air Poll., WMO No. 510, WMO, Geneva, pp. 63-69.
(30) Coppin, P.A.: 1979, Ph.D. Thesis, Flinders Univ., Adelaide.
(31) Chandler, T.J.: 1967, Bull. Amer. Meteorol. Soc., 48, pp. 394-399.
(32) Ackerman, B.: 1971, Preprints Conf. Air Poll. Meteorol., AMS, Boston, pp. 154-158.
(33) Hage, K.D.: 1975, J. Appl. Meteorol., 14, pp. 1277-1283.
(34) Dirks, R.A.: 1974, J. Geophys. Res. 79, pp. 3473-3475.
(35) Atwater, M.A.: 1971, J. Appl. Meteorol., 10, pp. 205-214.
(36) East, C.: 1968, Cahiers Géog. Québec, 12, pp. 81-89.
(37) Rouse, W.R., D. Noad and J. McCutcheon: 1973, J. Appl. Meteorol., 12, pp. 798-807.
(38) Sekihara, K.: 1973, J. Meteorol. Soc. Japan, 51, pp. 119-132.
(39) Probáld, F.: 1975, Idöjárás, 79, pp. 69-76.
(40) Peterson, J.T. and E.C. Flowers: 1974, Preprints Symp. Atmos. Diffusion Air Poll., AMS, Boston, pp. 129-132.
(41) McCormick, R.A. and D.M. Baulch: 1962, J. Air Poll. Control Assoc., 12, pp. 492.
(42) Bach, W.: 1971, Geog. Review, 61, p. 573-594.
(43) Randerson, D.: 1970, J. Air Poll. Control Assoc., 20, p. 546.
(44) Berlyand, M. Ye., K. Ya. Kondrat'yev, O.B. Vasil'yev, I.V. Vasil'chenko, V.F. Zhvalev, I.M. Krazhevskiy and L.S. Ivlev: 1974, Meteorol.: Gidrol., No. 1, pp. 14-23.
(45) Peterson, J.T., E.C. Flowers and J.H. Rudisill: 1978, J. Appl. Meteorol., 17, pp. 1595-1609.
(46) Craig, C.D. and W.P. Lowry: 1972, Preprints Conf. Urban Environ. and 2nd Cont. Biometeorol., AMS, Boston, pp. 159-164.
(47) Kung, E.C., R.A. Bryson and D.H. Lenschow: 1964, Month. Weath. Review, 92, pp. 543-564.
(48) Oke, T.R. and R.F. Fuggle,: 1972, Boundary-Layer Meteorol. 2, pp. 290-308.
(49) Aida, M. and M. Yagi: 1979, Boundary-Layer Meteorol., 16, pp. 453-465.
(50) Bergstrom, R.W. Jr. and R. Viskanta: 1973, J. Appl. Meteorol., 12, pp. 901-912.

(51) Fuggle, R.F. and T.R. Oke: 1976, Boundary-Layer Meteorol., 10, pp. 113-120.
(52) Nunez, M. and T.R. Oke: 1976, Boundary-Layer Meteorol. 10, pp. 121-135.
(53) White, J.M., F.D. Eaton and A.H. Auer, Jr.: 1978, J. Appl. Meteorol., 17, pp. 593-599.
(54) Bach, W.: 1976, Reviews Geophys. Space Phys., 14, pp. 429-474.
(55) SMIC: 1971, Inadvertent Climate Modification, MIT Press, 308 p.
(56) Bowling, S.A. and C.S. Benson: 1978, USEPA, EPA-600/4-78-027 Research Triangle Pk., N. Carolina, 150 p.
(57) Oke, T.R.: 1979, Boundary-Layer Meteorol., 17, pp. 167-173.
(58) Yap, D. and T.R. Oke: 1974, J. Appl. Meteorol., 13, pp. 880-890.
(59) Ching, J.K.S., J.F. Clarke and J.M. Godowitch: 1978, Proc. WMO Symp. Boundary Layer Phys. Appl. Specific Problems Air Poll., WMO No. 510, WMO, Geneva, pp. 71-78.
(60) Oke, T.R., B. Kalanda and D. Steyn: 1980, Urban Ecology, (submitted).
(61) Oke, T.R.: 1973, Atmos. Environ., 7, pp. 769-779.
(62) Shea, D.M. and A.H. Auer, Jr.: 1978, J. Appl. Meteorol., 17, pp. 689-698.
(63) Principal Investigators Project METROMEX:, Bull. Amer. Meteorol. Soc., 57, pp. 304-308.
(64) Braham, R.R., Jr.: 1977, Proc. Conf. Metro. Phys. Environ., USDA Forest Service Tech. Report NE-25, USDA, Upper Darby, Pa., pp. 3-17.
(65) Special Issue: 1978: J. Appl. Meteorol., 17, pp. 565-715.
(66) Anon: 1979, Bull. Amer. Meteorol. Soc., 60, pp. 1099.
(67) Braham, R.R., Jr. and M.J. Dungey: 1978, J. Appl. Meteorol., 17, pp. 644-654.
(68) Johnson, D.B.: 1976, Science, 194, pp. 941-942.
(69) Changnon, S.A., Jr., R.G. Semonin and F.A. Huff: 1976, J. Appl. Meteorol., 15, pp. 544-560.
(70) Boatman, J.F. and A.H. Auer, Jr., Proc. 4th Conf. Weath. Modif., AMS, Boston, pp. 366-373.
(71) Atkinson, B.W.: 1975, Occas. Paper No. 3, Queen Mary Coll., Univ. London, 27 p.
(72) Harnack, R.P. and H.E. Landsberg: 1975, J. Appl. Meteorol., 14, pp. 1050-1060.
(73) Changnon, S.A., Jr.: 1978, J. Appl. Meteorol., 17, pp. 578-586.
(74) Changnon, S.A., Jr.: 1973, J. Irrig. Drain. Div., ASCE 99 IRI, pp. 27-41.
(75) Wong, C.S.: 1978, Science, 200, pp. 197.
(76) Sagan, C., O.B. Toon and J.B. Pollack: 1979, Science, 206, pp. 1363-1368.

(77) Bach, W., J. Pankrath and W. Kellogg (eds.): 1979, Man's Impact on Climate, Elsevier, Amsterdam, 327 p.
(78) Belger, W.: 1940, Thesis, Univ. Berlin, 69 p.
(79) Loose, T. and R.D. Bornstein: 1977, Month. Weath. Rev., 105, pp. 563-571.
(80) Bornstein, R.D., P.H. Fontana and W.T. Thompson: 1978, WMO Symp. Boundary Layer Phys. Appl. Specific Problems Air Poll., WMO No. 510, WMO, Geneva, pp. 79-87.

DISCUSSION

Schiefer:
What is the size of the today's reforestation rate compared with the deforestation rate worldwide, and especially in tropical areas? I was told by Professor Hövermann from the Geographisches Institut in Göttingen, FRG, that worldwide there is today more re- than deforestation; would you agree to that?

Kirchner:
I doubt the accuracy of the statement that there is today more re- than deforestation. This is surely right for many regions of Central and Northern Europe, perhaps for the highlands of South America too. But the area of deforestation in the tropical regions (Brazil etc.) is much greater than that of local reforestation.

Hampicke:
The total amount of reforestation in the world and its effects on the atmospheric CO_2 budget are not known with any degree of accuracy. In the industrialized countries of the temperate zone, there have been only small shifts in land use during recent decades so that the area covered by forests has not increased significantly. It is fairly well established, however, that many of these forests are increasing in phytomass so that they constitute a sink for atmospheric CO_2. In the developing countries, numerous reforestation activities are being carried out on a local scale. However, these cannot compensate for the accelerating losses in forest area caused by the spread of shifting cultivation, and for other reasons. Very significant reforestation activities have been reported from China, although it is not known how many of them have been successful.

Gates:
Have you estimated the sensitivity levels of the zonally-averaged model to, say, small randomly-distributed changes of surface albedo, with which an estimate could then be made of the statistical significance of the model's response to a deliberate albedo change?

Potter:
We have introduced randomly perturbed atmospheric temperatures in an earlier version of the model and the model control run was essentially unchanged.

W. Bach, J. Pankrath, and J. Williams (eds.), Interactions of Energy and Climate, 357–361.

Flohn:

Using a similar estimate as Baumgartner for the albedo increase since the Neolithic revolution, and using Manabe's radiative-convective model for the relation between albedo and temperature at the surface, I got similar changes for the last 6000 years (Dahlem Conference 1976). However, one has to be cautious: at that time a not negligible part of the Laurentide Ice-sheet (some 2×10^6 km^2) still existed, and furthermore Professor Oeschger has given here some indications for a CO_2 content which was higher at that time than even now. Thus, any interpretation of paleoclimates should take into account a multitude of natural effects, which have been (and still are) more powerful than anthropogenic effects alone.

Potter:

I agree that to be complete, any model experiment trying to explain temperature changes over the past 10,000 years should incorporate past CO_2 amounts. This study was primarily designed to test certain aspects of the Sagan et al. paper and was not meant to be a comprehensive treatment of all factors that might have contributed to past climate variations.

Gates:

Have you considered making a series of albedo change experiments, say with more and less change than in your present experiment, so that the systematic nature of the model's response could be seen with more confidence?

Potter:

Yes, we have done a similar experiment with an exaggerated albedo perturbation and the model response was similar but with more intense effects.

Bartholomäi:

You showed that there was little difference in the heat island effect of cities during a period of about 40 years and mentioned the city of Karlsruhe. Indeed Karlsruhe did not change very much. However, if you look at a city like Frankfurt, where many low buildings in the center have been replaced by tall ones, which causes a change of ground roughness and waste heat input, there must be a difference in the heat island effect, too.

Oke:

Yes indeed. The relationship between the maximum heat island intensity and sky view factor that I showed (but is not included in the written paper) included results for the growing town of Columbia, Maryland. As the settlement has grown, the heat island has increased because the central area has been continually modified. The example of Karlsruhe was used to illustrate the situation of a city where the central area has remained essentially the same for many years despite increased population and built up-area.

Flohn:
Studies of the role of cities in enhancing the frequency of heavy convective rainfall during summer have also been made in Germany, with similar results. The highest frequency is downwind of the source, at a distance of 5-10 km (Hamburg: Reidat 1970) or 10-15 km downwind of isolated cities (or industrial areas) in the Ruhrgebiet. This seems to occur at the cost of the surrounding areas, where a minimum is observed (Ruhrgebiet), apparently due to a kind of entrainment effect. In Frankfurt and Berlin the urban effect is distorted due to small topographical differences (of 100 and 50 m relative height respectively). The role of condensation nuclei seems to me to be doubtful, because the urban effect can only be shown during summer, while the aerosol content during the cool season is much greater.

Oke:
I had not heard of the possible entrainment effect you refer to and would like to know more of this German research. Your comments regarding the role of nuclei would seem to fit the present consensus as I read it. The role of topography in experiments seeking to validate the existence of any urban enhancement effect is always a problem. It has even been suggested that the slight topographic variations around St. Louis could be sufficient to produce precipitation of the same magnitude as that often attributed to the presence of the city.

Wallèn:
1. I was interested to see that the reduction of incoming short-wave radiation over cities compared to rural areas is now considered to be on the order of only 5%. Is that really true in very polluted cities?
2. Could you comment on influences by cities on duration of sunshine in the same context?

Oke:
1. The number of cities studied recently is too small to draw any specific conclusions. I suspect, however, that the type of pollution is crucial. A city with relatively high gaseous pollution loading may have less impact on solar input than a city with predominantly particulate pollution. Similarly, there will be a very different spectral filtering by the two atmospheres.
2. I am aware that sunshine hours have increased in London, England. There is, however, some discussion as to the reason

and whether this can be wholly attributed to the enforcement of the British Clean Air Act.

Rotty:

First a compliment to Professor Oke on an excellent review of the meteorological impacts of urbanization. It put many factors together in one place and made much sense. There was only one point, that of the water balance, that seemed different from my intuition. Then I realised that your examples are from one type of city and cities like New Orleans or Houston are different. They are indeed drier than the surrounding swampy areas. Have data similar to yours for Montreal etc. been collected for any cities like New Orleans or Houston?

Oke:

I am not aware of direct measurements of the water balance or evapotranspiration in cities such as you mention. Certainly they would appear to fit more closely the old concept of a 'dry' city. Even so I suspect that water loss to the air from such cities will not be significant, especially if urban irrigation is practised.

Potter:

1. Is it water vapor or aerosols that contribute to the reduction in longwave loss from the city?

2. Are you developing or planning to develop a numerical model of the urban climate that might be used to optimize building design or location for optimal energy use?

Oke:

1. It is not clear which is the more important, or indeed if either is especially significant. Our own work has shown that the urban temperature structure is sufficient to support the observed increases of incoming longwave radiation over the city even before considering the role of water vapor or air pollution in affecting the atmospheric emissivity.

2. Not a fully physical-numerical model. The processes acting within the canopy are not sufficiently understood to make such a model realistic at this time, in my opinion. We are, however, developing simple empirical and rule-of-thumb 'models' to aid in generalization.

McKay:

1. Is it not the use of the "sky view factor" as a geometric classification device tantamount to saying that the microclimate of the measurement site is the control? Within a city there would be a large spectrum of such microclimates, and they could

be reproduced in other areas with quite different dimensions and general environment.

2. Experience in snowy environments confirms your remarks about wind deceleration on the windward edge of a city. This process leads to saltation of snowfall such that the winter precipitation on the windward edge exceeds that over the inner city, and presumably there is a deficiency in the zone of acceleration downwind of the city. This is opposite to the summer precipitation effects that you have described.

Oke:

1. Yes, the importance of site geometry points to the role of micro-scale processes in controlling the urban canopy layer heat island rather than local or mesoscale advection. Certainly quite different mixes of site parameters could produce the same microclimate, but under the 'ideal' conditions (calm and clear sky) associated with large heat island intensities, I suspect that radiation geometry (site dimensions) is often the primary control.

2. This is a very interesting observation that I have not seen referred to previously. The situation you describe would certainly fit the strong wind scenario described in my paper. It is of course related to the preferential deposition and accumulation of normal precipitation rather than any enhancement of total precipitation as suggested by the summer results.

PART V

CLIMATIC IMPACTS OF HEAT EMISSION

IMPACT OF WASTE HEAT EMISSIONS IN THE UPPER-RHINE REGION

J. Pankrath

Umweltbundesamt, Bismarckplatz 1, D-1000 Berlin 33

ABSTRACT. The energy supply of industrialized regions is ultimately transformed into waste heat and reaches the atmosphere. In the Upper-Rhine Region (inventory area 27 500 km^2) the mean annual waste heat emission is about 1.6 W/m^2, and thus exceeds the mean global anthropogenic heat release by a factor of 100. The heat emission consists of energy conversion (16 %), major industry (33 %), small industry and domestic consumption (38 %) as well as transportation (13 %); it is released in the form of radiative (17 %), sensible (69 %) and latent heat (14 %). The largest amount of waste heat is released in urban and industrial areas showing flux densities of about 30 W/m^2 on 2 km x 2 km grid areas. The influence on regional climate is studied using a non-hydrostatic mesoscale simulation model, which calculates the topographically influenced wind field.

1. WASTE HEAT RELEASE INTO THE ATMOSPHERE

The economic activity of industrial countries generates environmental pollution, especially gaseous and particulate emissions, and the release of waste heat. Whereas it might be possible to control the release of airborne materials even with increasing economic activities, the release of waste heat can only be reduced by diminishing the consumption of primary energy. Waste heat emissions represent an energy supply to the atmosphere, and we have to ask to what extent the present and the future supply rate may affect weather and climate.

W. Bach, J. Pankrath, and J. Williams (eds.), Interactions of Energy and Climate, 363–381.

The global annual energy consumption is about 8 TW, giving a natural global energy flux density of 0.016 W/m^2, which is less than 0.1 °/oo of the global mean solar radiation at the earth's surface. Tab. 1 and Fig. 1 show, however, that the anthropogenic energy production is much higher than the solar radiation flux over limited industrial areas.

Tab. 1: Some energy flux densities within meteorological scales (2, 11, 14, 18)

Scale	Anthropogenic heat source	Area (km^2)	Annual anthropogenic heat release (W/m^2)
small	Refinery (6 Mio t crude oil/yr)	0.1	4 000
	Nuclear power plant	0.2	16 000
	Steel mill	0.2	530
regional	Fairbanks	37	19
	NY-Manhatten	59	120
	Montreal (1961)	78	99
	Vancouver (1970)	112	19
	Budapest (1970)	113	43
	West Berlin (building area)	234	21
	Munich	310	12
	Hamburg	747	13
	Moscow	878	127
	Chicago	1 800	29
	Los Angeles (city)	3 500	21
	Ruhr Area	6 500	17
	Los Angeles (county)	10 000	7.5
	Upper Rhine Region	27 500	1.6
	Donez Basin	53.00	1.4
	Washington-Boston-Area	87 000	4.4
large	Federal Rep. Germany	248 000	1.36
	Central Western Europe	1 665 000	0.74
	Western Europe	3 946 000	0.38
	USA	9 363 000	0.27
	North America	19 300 000	0.13
	USSR	22 275 000	0.06
global	Continents	135 782 000	0.046
	Globe	510 000 000	0.016

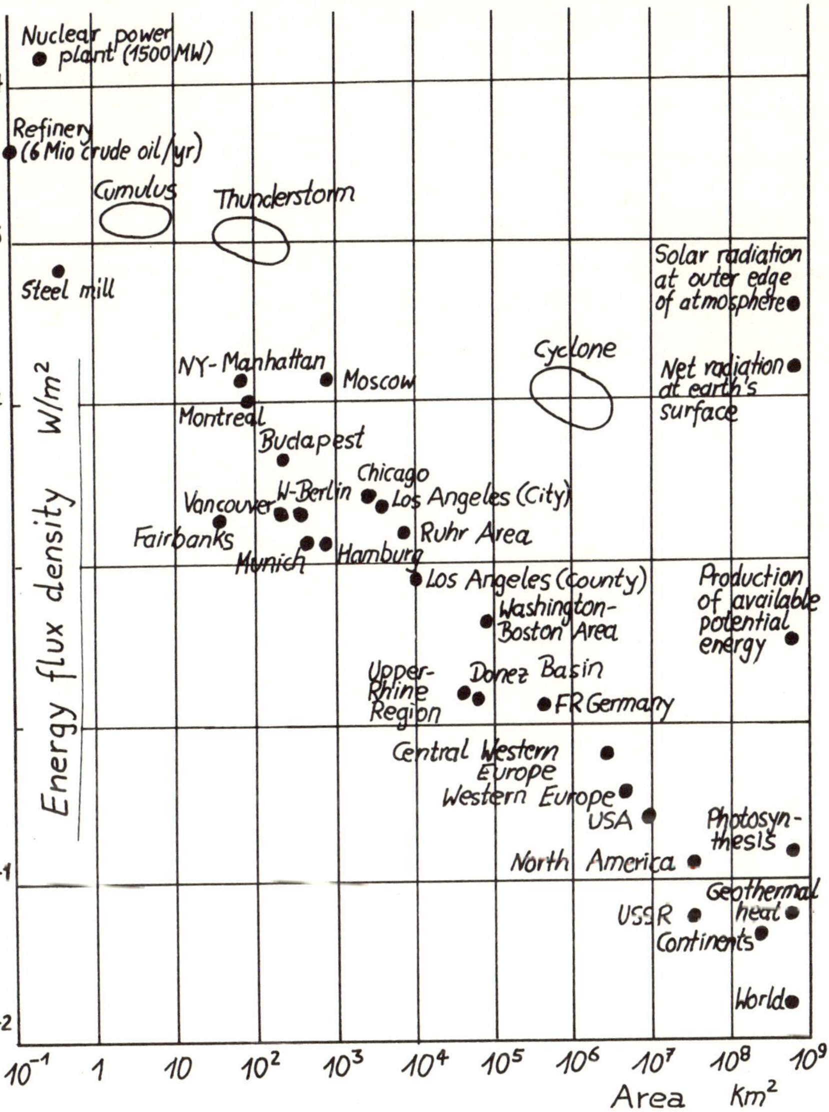

Fig. 1: Comparison of large-, regional- and small scale anthropogenic energy flux densities with some natural global processes (2, 11, 14, 17, 18)

On an urban scale the anthropogenic heat emission is comparable in magnitude to the incoming solar radiation at the ground. The release of waste heat and air pollutants and the change of albedo and roughness cause a local urban climate which differs significantly from its rural surroundings (5, 13, 18, 23).

It is suspected that man-made heat release may already have impacts on the regional (meso) scale weather and climate, if anthropogenic heat addition exceeds about 1 % of the mean global solar radiation at the ground. The Upper-Rhine Region, extending from Basle to Frankfurt, has been selected as an orographically bounded area for the investigation of the effects of waste heat release on regional climate. This area belongs to the most densely populated (327 inhabitants per km^2) and highly industrialized (primary energy supply rate 1.6 W/m^2) areas in Europe.

The "Waste Heat Project in the Upper-Rhine Region", sponsored by the German Federal Environmental Agency by order of the Ministry of the Interior, aims at an evaluation of possible regional impacts of waste heat emissions on weather and climate. The project has four aims:

(1) A waste heat emission inventory.

(2) A mesoscale simulation model which is in the developing stage.

(3) A climatic description of the region.

(4) An evaluation of long time series of climatic parameters such as temperature, area precipitation and sunshine duration.

The results of the project shall form the basis of political decisions regarding for instance, the extent to which measures must be taken in order to avoid undesirable impacts of waste heat release.

2. WASTE HEAT INVENTORY OF THE UPPER-RHINE REGION

2.1 Components of waste heat emission

Waste heat is discharged into the atmosphere primarily in the form of sensible and latent heat and long-wave radiation (Fig. 2).

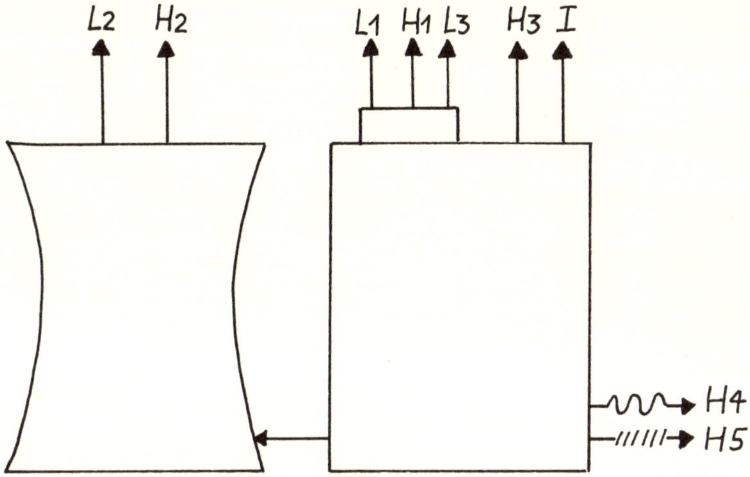

Sesible flux:
H 1 - from flue gas exhaust
H 2 - from cooling devices and product cooling
H 3 - from buildings
H 4 - into water
H 5 - by conduction into the ground

Latent heat flux:
L 1 - from flue gas exhaust
L 2 - from cooling devices and product cooling
L 3 - from drying processes

Radiative heat flux:
I - from buildings

Fig. 2: Major flux components of waste heat emission (3)

The most important waste heat component is the sensible heat flux; substantial contributions are made by gas exhausts, cooling devices and buildings. Latent waste heat is ejected by fossil fuel combustion, wet cooling and evaporation of warmed streams. The contribution of radiation to waste heat release depends on temperature and it can be assumed that the heat of a surface with a temperature of about 100^{0} C is transferred to the air at the same rate as sensible heat and radiation. If temperature rises, the portion of radiative transfer increases, and at about 400^{0} C almost 2/3 of the heat is lost by radiation. Heat conduction is of importance only for heat transfer from buildings and drainage to the ground.

2.2 Categories of waste heat emission

Tab. 2 shows the energy and waste heat balance of the Upper-Rhine Region for the year 1973 by source categories. The share of the energy supply is:

Conversion	16 %
Major industry	33 %
Small industry and domestic consumption	38 %
Transportation	13 %.

The waste heat emission is discharged as:

sensible heat	69 %
latent heat	14 %
radiation	17 %.

Tab. 2: Energy supply and waste heat release rate in the Upper-Rhine Region for the year 1973 (3).

Source category	Energy supply rate			Waste heat release rate into the atmosphere		
	$\frac{W}{m^2}$	GW	%	$\frac{W}{m^2}$	GW	%
Conversion						
- refineries	0.11	2.98	7	0.11	2.98	7
- power plants	0.14	3.81	9	0.05	1.30	3
Major industry						
- point sources	0.26	7.07	16	0.19	5.21	12
- others	0.27	7.35	17	0.28	7.72	18
Small industry and domestic consumption	0.62	16.93	38	0.65	17.86	41
Transportation	0.23	6.32	13	0.24	6.51	14
Total energy supply	1.63	44.45	100			
Net supply of waste heat by rivers				0.08	2.14	5
total waste Heat supply				1.60	43.72	100

2.3 The waste heat emission inventory

The mean annual release of waste heat in the Upper-Rhine Region is about 1.6 W/m^2, which is distributed on grid square elements of 2 km x 2 km as shown in Fig. 3.

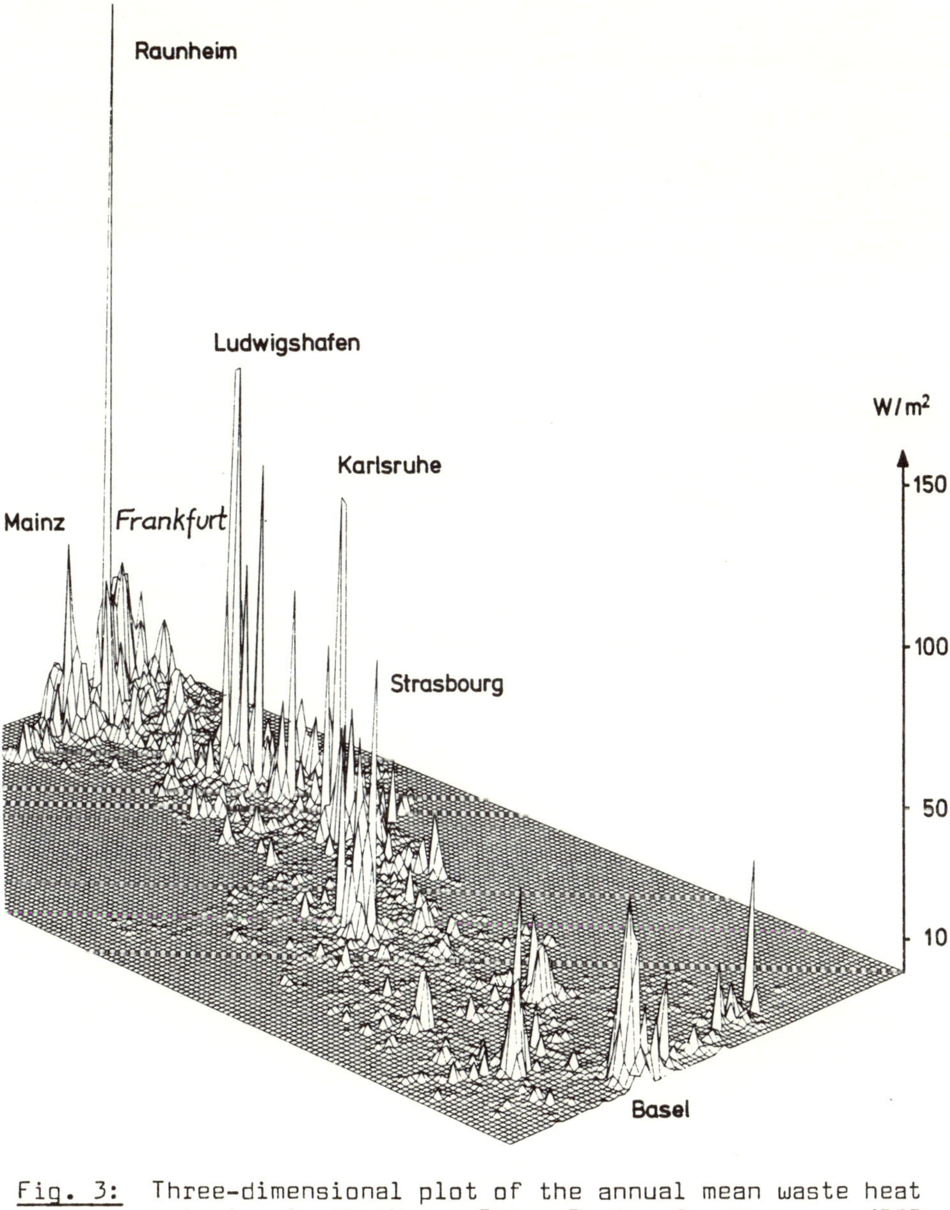

Fig. 3: Three-dimensional plot of the annual mean waste heat emission in the Upper-Rhine Region for the year 1973.

Only 22 % of the inventory area (27 500 km^2) have no significant waste heat emissions. Urban and industrial areas produce most of the heat. The assessment shows that only 3 % of the inventory area emits almost half of the total waste heat resulting in a flux density of more than 10 W/m^2. About 7 % of the grid area releases 70 % of the waste heat of the inventory region with a flux density of about 5 W/m^2. The cities release heat with a flux density of 20 - 30 W/m^2.

2.4 Temporal variation of waste heat emissions

There are representative annual and daily variations of heat emission for each source category which can be applied to the Upper-Rhine Region. The variation of energy supply for different source categories is given in Fig. 4 and Fig. 5. The annual emission for major industry is nearly constant because the basic industry with the greatest share of the industrial energy consumption shows almost no annual variation. For domestic consumption, however, almost 80 % of the energy supply is dependent on temperature which therefore significantly determines the annual variation of this source category. With regard to transportation, the variation of urban traffic can be neglected whereas the traffic outside of urban areas exhibits a significant annual variation.

Yearly and daily variations of important point sources have been determined in particular: Power plants and central heating plants exhibit a typical annual variation, while refineries show no systematic fluctuation. It can be seen that the highest instantaneous value of waste heat release occurs in January between 7.00 and 11.00 a. m., whereas the lowest instantaneous value falls between 11.00 p. m. and 5.00 a. m. in the months June, July and August. The highest and lowest values differ by a factor of 5. The total waste heat flux density of the Upper-Rhine Region increases in winter by about 80 % of the annual mean value to 2.8 W/m^2; in summer the density is reduced by 60 % leading to a value of 0.6 W/m^2. The fluctuations of the mean waste heat release are determined above all by small industry and domestic consumption (Tab. 3), whose share of the total waste heat emission increases from 10 % in summer (night) to 56 % in winter (morning).

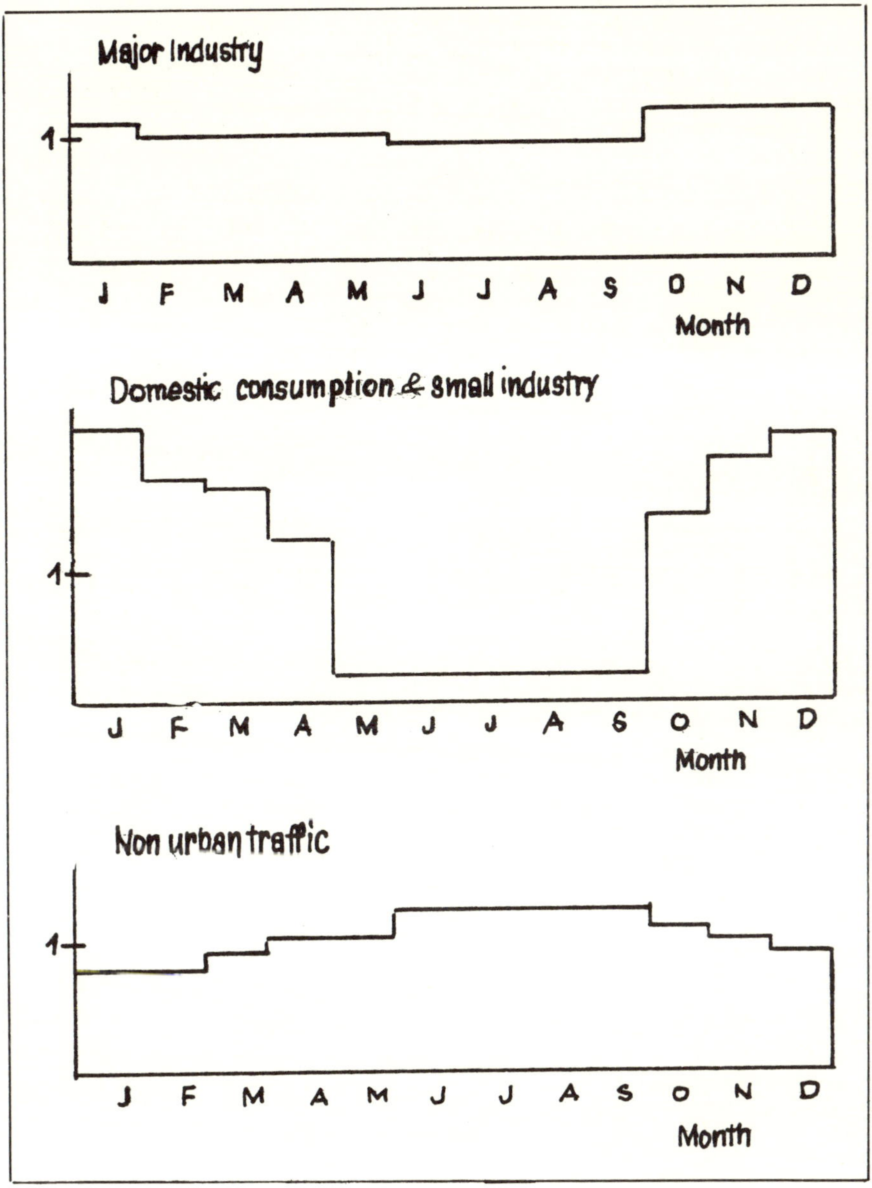

Fig. 4: Variation from annual average of energy consumption for different source categories (3).

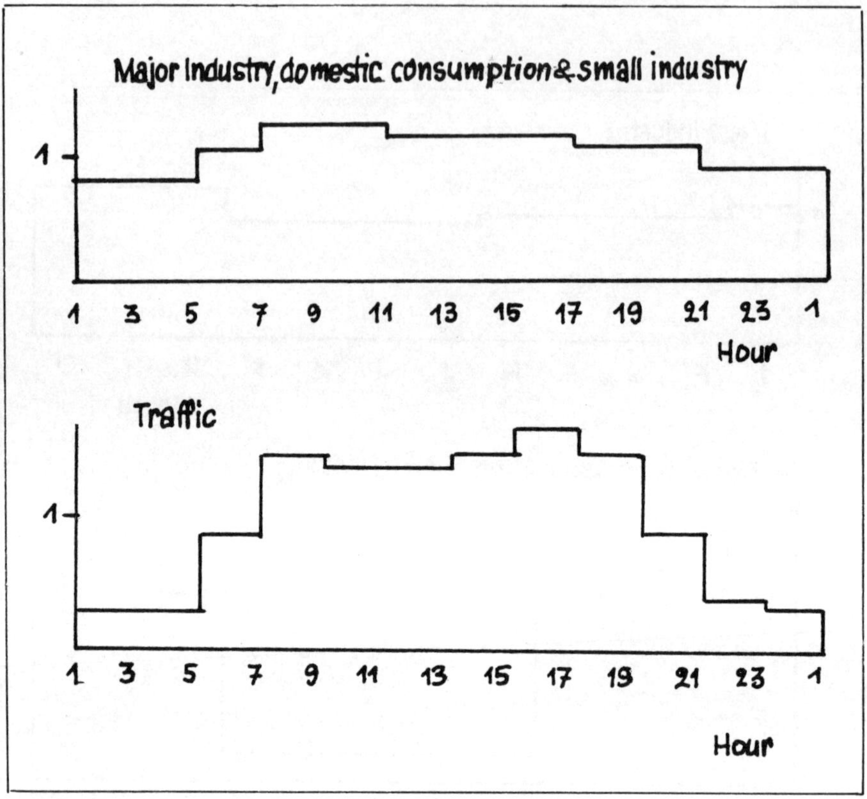

Fig. 5: Variation from daily average of energy consumption for different source categories (3).

Tab. 3: Comparison of winter maximum and summer minimum of the waste heat flux density of the Upper-Rhine Region for the year 1973.

Source category	Waste heat release into the atmosphere (%)		
	annual mean	winter maximum	summer minimum
Conversion			
- refineries	7	4	17
- power plants	3	4	5
Major industry			
- point sources	12	8	20
- other facilities	18	14	25
Small industry and domestic consumption	41	56	10
Transportation	15	11	15
Release from rivers	5	3	9
	100	100	100
Total waste heat supply			
GW	44	78	17
$\frac{W}{m^2}$	1.6	2.8	0.6
Division into waste heat components in %			
radiation	17	19	13
convection	69	70	65
latent heat	14	11	21

2.5 Future waste heat emission

Today the energy conversion rate of the inventory area is 44 GW. For the year 2000 between 66 GW and 91 GW are expected in accordance with the growth of the primary energy consumption of 50 to 100 %. This means an increase of the mean annual waste heat supply from 1.6 W/m^2 to 2.4 and 3.1 W/m^2 respectively.

3. COMPARISON OF ANTHROPOGENIC HEAT EMISSIONS WITH NATURAL HEAT FLUXES

In view of the magnitude of natural heat fluxes, anthropogenic flux densities below about 2.5 W/m^2 are unimportant. More than this amount of waste heat is emitted by only 12 % of the inventory grid area. The grid area of cities have mean waste heat flux densities of up to about 30 W/m^2, which is the same order of magnitude as the natural radiation fluxes, such as solar radiation and net radiation at the ground.

Figure 6 indicates that in winter the anthropogenic heat release can exceed the natural heat supply by solar radiation. This comparison, however, does not permit a statement on the magnitude of impacts due to anthropogenic heat release but it can point to areas where impacts may occur.

4. MESOSCALE SIMULATION MODELING

4.1 Modeling approach

A mesoscale simulation model to investigate the impacts of waste heat release in the Upper-Rhine Region should include the following aspects:

(1) Development of a three-dimensional time-dependent model of air flow in a topographical region;

(2) Incorporation of the temperature and humidity fields including natural heat and water vapour sources;

(3) Incorporation of man-made heat sources.

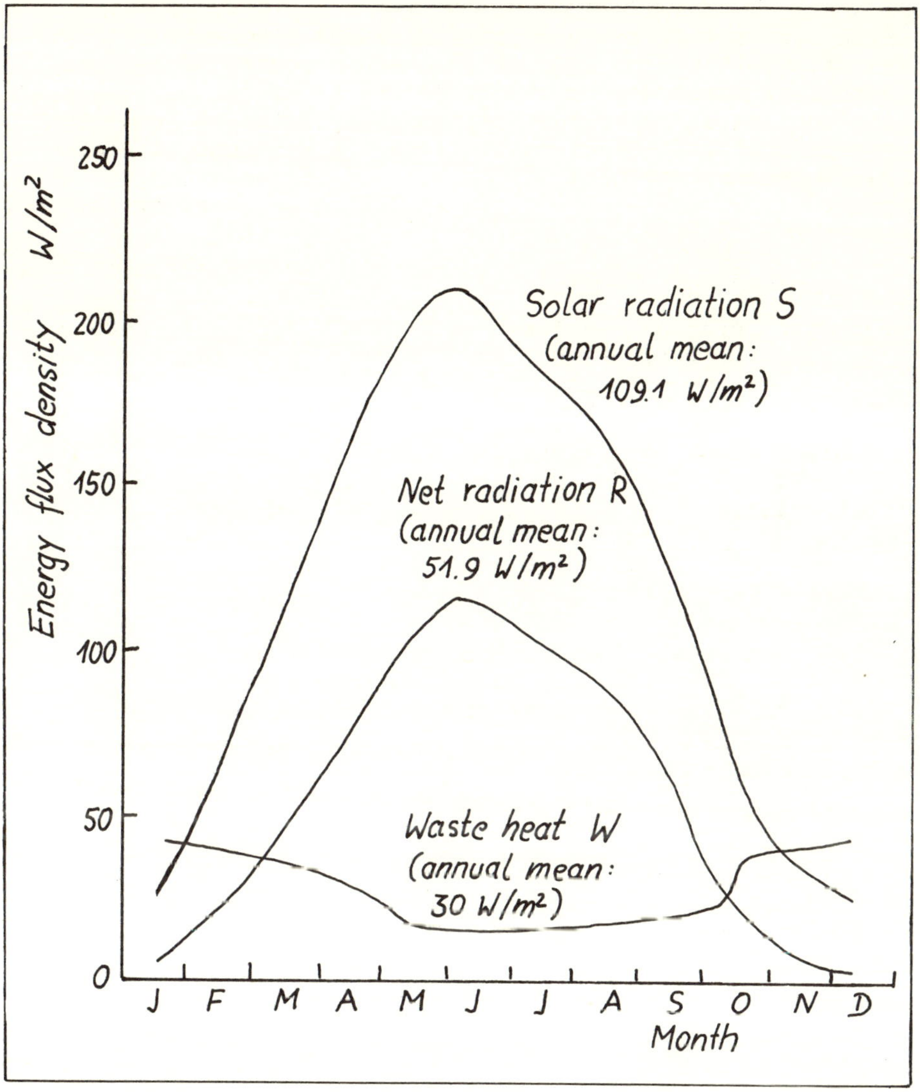

Fig. 6: Comparison of flux density of solar radiation S and net radiation R at Karlsruhe Nuclear Research Center for the year 1973 (7) with anthropogenic heat release W typical for urban areas.

4.2 Non-hydrostatic mesoscale model

Step one has already been performed and step two is in preparation. It can be shown that the hydrostatic approximation is inadequate for investigating the effect of waste heat point sources. Thus the complete system of equations of motion has to be used. In fact. the vertical acceleration becomes important if the horizontal scale approaches the vertical scale of motion (10) and if there are strong horizontal variations in thermal forcing (25).

Tab. 4: Selected publications in the field of mesoscale simulation modeling (D means dimension, h = 0 means without topography)

Hydrostatic	Non-hydrostatic
Shallow water approximation $\nabla \cdot \mathbb{V} = 0$	
PIELKE (20) (3D, h = 0)	
PIELKE & MAHRER (1) (2D, h = 0)	
BORNSTEIN (4) (2D, h = 0)	
MAHRER & PIELKE (15) (2D)	
DIETERLE (6) (2D, h = 0)	
VENKATRAM & VISKANTA (24) (2D, h = 0)	
Anelastic approximation $\nabla \cdot \varrho \mathbb{V} = 0$	
ESTOQUE & BHUMRALKAR (8) (2D, h = 0)	GAL-CHEN & SOMMERVILLE (9) (2D)
PHYSICK (19) (2D, h = 0)	
MAHRER & PIELKE (16) (2D)	
Complete equation system $\nabla \cdot \varrho \mathbb{V} = - \frac{\partial \varrho}{\partial t}$	
KEYSER & ANTHES (12) (3D)	TAPP & WHITE (22) (3D, h = 0)
ANTHES & WARNER (1) (3D)	

The currently used mesoscale models are listed in Tab. 4. As there is no non-hydrostatic simulation model including orography and heat sources, such a model has to be developed.

4.3 Shallow water approximation

The model development for the Upper-Rhine project started with a three-dimensional non-hydrostatic version assuming constant density, i. e. with the shallow water approximation (26). The equations are formulated using an (x, y, η)-coordinate system with

$$\eta = \frac{H - z}{H - h(x, y)}$$

as vertical coordinate, taking into account the height h(x, y) of the terrain; H is the upper boundary of the integration area.

Equations of motion:

$$\frac{du}{dt} - fv + (g\eta + (H-h)\omega^2)\frac{\partial h}{\partial x} = -\frac{1}{\rho}\frac{\partial p}{\partial x} + \\ + \frac{1}{H-h}\left[2\frac{\partial}{\partial x}\left(K_h(H-h)\frac{\partial u}{\partial x}\right) + \frac{\partial}{\partial y}\left(K_h(H-h)\left(\frac{\partial v}{\partial x} + \frac{\partial u}{\partial y}\right)\right) + \frac{1}{H-h}\frac{\partial}{\partial \eta}\left(K_v\frac{\partial u}{\partial \eta}\right)\right]$$

$$\frac{dv}{dt} + fu + (g\eta + (H-h)\omega^2)\frac{\partial h}{\partial y} = -\frac{1}{\rho}\frac{\partial p}{\partial y} + \\ + \frac{1}{H-h}\left[\frac{\partial}{\partial x}\left(K_h(H-h)\left(\frac{\partial v}{\partial x} + \frac{\partial u}{\partial y}\right)\right) + 2\frac{\partial}{\partial y}\left(K_h(H-h)\frac{\partial v}{\partial y}\right) + \frac{1}{H-h}\frac{\partial}{\partial \eta}\left(K_v\frac{\partial v}{\partial \eta}\right)\right]$$

$$\frac{d\omega}{dt} - \frac{g}{H-h} - \frac{2\omega}{H-h}\left(u\frac{\partial h}{\partial x} + v\frac{\partial h}{\partial y}\right) = -\frac{1}{\rho(H-h)^2}\frac{\partial p}{\partial \eta} + \frac{1}{(H-h)^2}\left[\frac{\partial}{\partial x}\left(K_v\frac{\partial u}{\partial \eta}\right) + \frac{\partial}{\partial y}\left(K_v\frac{\partial v}{\partial \eta}\right)\right]$$

Equation of continuity:

$$\frac{\partial u}{\partial x} + \frac{\partial v}{\partial y} + \frac{\partial \omega}{\partial \eta} - \frac{1}{H-h}\left[u\frac{\partial h}{\partial x} + v\frac{\partial h}{\partial y}\right] = 0$$

The values have their common meaning; $\omega = \frac{d}{dt}$ is the generalized vertical velocity. For the horizontal turbulent momentum diffusion coefficient an empirical approach is made:

$$K_h = 2.3\ K_v,$$

where for the vertical turbulent momentum diffusion coefficient K_v the PRANDTL approach

$$K_v = l^2 \left| \frac{\partial v}{\partial z} \right|$$

is used with BLACKADAR's mixing length formula.

Initially the model is in geostrophic balance. At the lower boundary ($\eta = 1$) the velocities u, v, ω are zero. There is no pressure perturbation, if these is no orography; if

$$\nabla h \neq 0, \text{ then } \frac{\partial p}{\partial \eta} = 0.$$

At the upper boundary ($\eta = 0$), the wind is geostrophic, there is no vertical velocity and no pressure perturbation, and all horizontal gradients are zero.

At the lateral boundaries the topography levels out, and the 1-dimensional equations in space have to be solved:

$$\frac{\partial u}{\partial t} = f(v - v_g) + \frac{1}{(H-h)^2} \frac{\partial}{\partial \eta} \left[K_v \frac{\partial u}{\partial \eta} \right]$$

$$\frac{\partial v}{\partial t} = -f(u - u_g) + \frac{1}{(H-h)^2} \frac{\partial}{\partial \eta} \left[K_v \frac{\partial v}{\partial \eta} \right]$$

Additionally the vertical velocity and pressure perturbation are set equal to zero.

To reach a better resolution in layers near the ground, the transformation

$$\xi = \frac{2}{\pi} \arcsin(\eta)$$

has been introduced into the model calculations, which are hence carried out in an (x, y, ξ)-coordinate system.

4.4 Preliminary results of the model calculations

The area of the northern part of the Upper-Rhine Region, the Rhine-Main area of 100 km x 100 km has been selected for model calculations. The horizontal grid width has been chosen to be $\Delta x = \Delta y = 2.5$ km; in the vertical 12 computational levels have been used.

The integration in time is performed by the EULER backward method until a stationary solution is reached.

For the case of a geostrophic wind coming from north-west with a velocity of 20 m/s the following flow pattern has been calculated (Fig. 8).

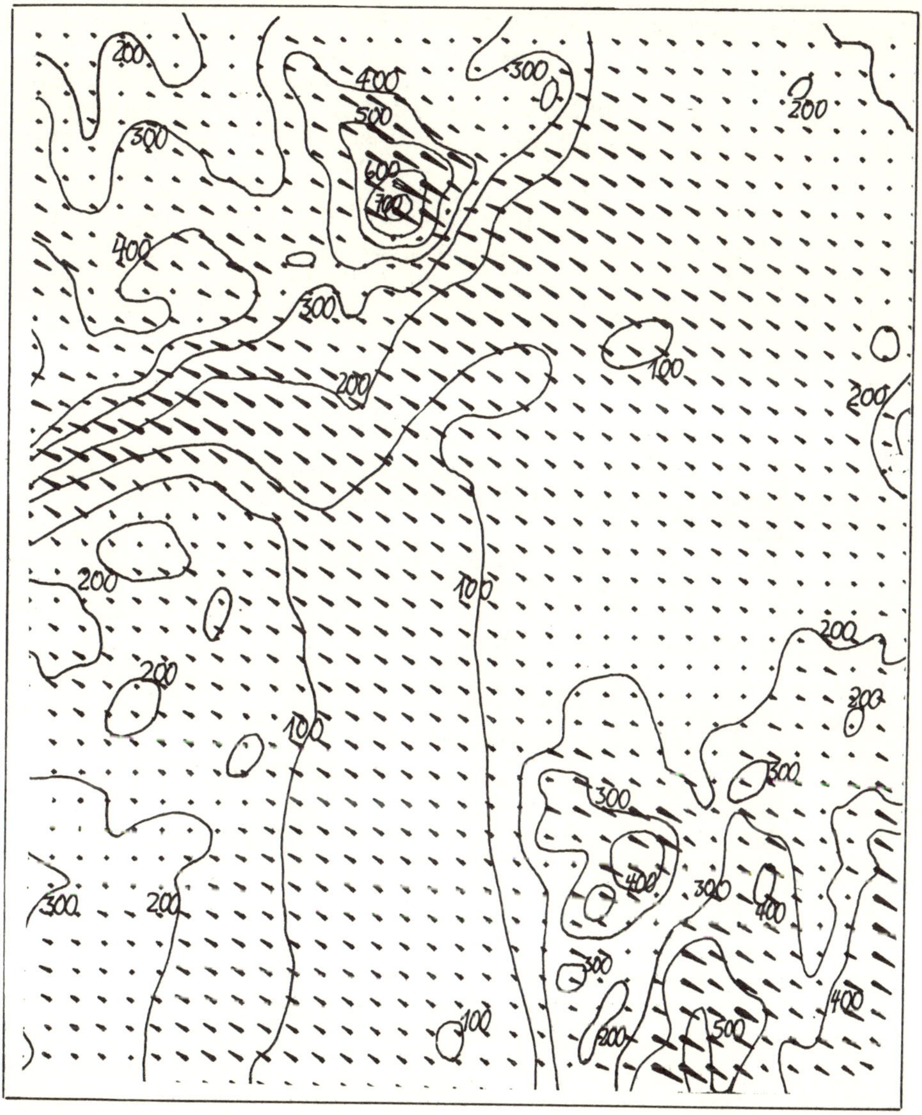

Fig. 8: Topographical contours of the Rhine-Main Region (100 km x 100 km) and vector plot of the deviation of the surface wind (25 m above the ground) from an orographical undisturbed geostrophic flow (north-west, 20 m/s). The grid distance denotes a deviation of 2 m/s.

The results can be summarized as follows:
The flow pattern shows a strikingly low deviation of the wind direction in relation to the orographically undisturbed wind. This effect is due to the application of the shallow water equations which presume a homogeneous atmosphere and exclude vertical accelerations owing to buoyancy. Nevertheless, the calculated, although too small, directional deviations reproduce the influence of the underlying topography on the wind field clearly. Because of a missing density stratification the result of the modeling shows that there is not a flow around but rather over the mountains. The model will be refined in the next project phase by incorporating the thermodynamics which should give more realistic results.

5. OUTLOOK

The Upper-Rhine project will be continued by incorporating temperature and humidity into the mesoscale model as well as by measuring the three-dimensional time-varying structure of the atmosphere. Not until then will it be possible to consider scenarios of waste heat accumulation and their regional climatic impacts. This work shall be a support for political decisions in regional planning.

ACKNOWLEDGEMENT

The "Waste Heat Project Upper-Rhine Region" has been performed under the R & D programme of the Federal Environmental Agency in Berlin (West) of the Federal Republic of Germany by the following institutions:

Nuclear Research Center, Karlsruhe (Project coordinator: Dr. W. Schikarski); German Weather Service, Offenbach; Meteorological Institute of the University Bonn, and Engineering Company Bonnenberg + Drescher, Jülich.

REFERENCES

1. ANTHES, R. A., and WARNER, T. T. 1978:
   Mon. Wea. Rev. 106, 1045 - 1078.
2. BACH, W. 1976:
   Rev. Geophys. Space Phys. 14, 429 - 474.
3. BARTHOLOMÄI, G., and KINZELBACH, W., 1979:
   KfK 2869 UF, 59 p.
4. BORNSTEIN, R. D., 1975:
   J. Appl. Met. 14, 1459 - 1477.

5. CHANDLER, T. J., 1976:
WMO Techn. Note No. 149, Geneva, 61 p.
6. DIETERLE, D., 1976/
IBM Palo Alto Sci. Center, Techn. Rep. No. G 320-3344, 61 p.
7. DILGER, H., NESTER, K., and VOGT, S., 1975:
KfK 2164, 46 p.
8. ESTOQUE, M. A., and BHUMRALKAR, C. M. 1969:
Mon. Wea. Rev. 97, 850 - 859.
9. GAL-CHEN, T., and SOMMERVILLE, R. C. J., 1975:
J. Comput. Phys. 17, 276 - 310.
10. HALTINER, G. J., 1971:
Numerical weather prediction, Wiley, New York.
11. HOSLER, C. L., and LANDSBERG, H. E., 1977:
in Energy and Climate (Ed. Abelson, P. H. and Malone, T. F.) NAS, Washington, D. C. 96 - 105.
12. KEYSER, D., and ANTHES, R. A., 1977:
Mon. Wea. Rev. 105, 1351 - 1371.
13. LANDSBERG, H., 1974:
in: Weather and Climate Modification (Ed. Hess. W. N.), Wiley, New York.
14. LLEWELLYN, R. A., and WASHINGTON, W. M., 1977:
in: Energy and Climate (Ed. Abelson, P. H. and Malone, T. F.), NAS, Washington D. C., 106 - 118.
15. MAHER, Y., and PIELKE, R. A., 1975:
J. Atm. Sci. 32, 2144 - 2155.
16. MAHRER, Y., and PIELKE, R. A., 1977:
Mon. Wea. Rev. 105, 1151 - 1161.
17. MUNN, R. E., and MACHTA, L., 1979:
in: World Climate Conference, WMO-No. 537, WMO Geneva, 170 - 209.
18. OKE, T. R. 1974:
WMO Techn. Note No. 134, WMO Geneva
19. PHYSICK, W., 1976:
J. Atm. Sci. 33, 2107 - 2135.
20. PIELKE, R. A. 1974:
Mon. Wea. Rev. 102, 115 - 139.
21. PIELKE, R. A., and MAHRER, Y., 1975:
J. Atm. Sci. 32, 2288 - 2308.
22. TAPP, M. C., and WHITE, P. W., 1976:
Quart. J. Roy Met. Soc. 102, 277 - 296.
23. TORRANCE, K. E., and SHUM, J. S. W., 1976:
Atm. Env. 10, 329 - 337.
24. VENKATRAM, A., and VISKANTA, R., 1976:
EPA 600/4-76-0.39, 243 p.
25. WIPPERMANN, F., 1977:
Meteorologisches Simulationsmodell Oberrheingebiet, B+D, Nr. 2
26. WIPPERMANN, F., 1977:
Meteorologisches Simulationsmodell Oberrheingebiet, B+D, Nr. 3

LOCAL AND REGIONAL CLIMATIC IMPACTS OF HEAT EMISSION

H. G. Fortak, Free University of Berlin,
1000 Berlin 33, Germany.

ABSTRACT. The impact of anthropogenic heat emission on properties of the planetary boundary layer has been investigated experimentally by air-borne meteorological measurements. These investigations covered a wide range of scales of heat emission: Individual stack plumes, cooling tower plumes, urban thermal plumes, and plumes from area sources in highly industrialized cities. Due to the very rapid dissipation of spatial temperature differences in the developing plumes, point sources of heat emission hardly influence the atmosphere. Moisture emission produces its well known impact on sunshine duration in the immediate vicinity of cooling towers. Thermal plumes originating from heat sources within towns and cities were investigated frequently. However, even in such cases, the spatial temperature differences were extremely small if they could be detected at all. A large number of profile flights in the Upper Rhine area indicated that even in the most industrialized parts of this region the natural spatial variations of meteorological parameters are much larger than the man-made ones. This is not true with respect to the fields of vertical motions, however; the thermal plumes of large industrialized cities are built up by a large number of intense thermal plumes of small size. In many cases the normal thermal activity is enhanced considerably. This fact led to the construction of a simple one-dimensional model for mean vertical velocity fields produced by local heat emission. By comparison with vertical velocities found in natural meteorological phenomena, estimates of man-made heat impacts on the atmosphere and on climate are possible.

W. Bach, J. Pankrath, and J. Williams (eds.), Interactions of Energy and Climate, 383–398.

1. INTRODUCTION

A few years ago there was considerable concern among scientists and decision makers about local and regional impacts of anthropogenic heat emission on weather and climate. It is well known that urban and rural climates differ from each other considerably if a town or a city reaches some critical size. Under certain meteorological conditions, urban heat islands extend to heights of up to 300 meters. Thermal plumes originating from such a heat island may extend horizontally downwind up to 30 kilometers for large cities. These plumes are detectable mainly by observations of visibility rather than by observations of temperature differences between the plume and the surrounding air. In fact, temperature differences of this kind are rather small. This shows that the direct thermal impact of man-made heat emission on the atmosphere is very small and is confined to the lowest layers of the planetary boundary layer. Less information was available about the impact of strong point sources of heat emission. Among these, the heat emission of cooling towers of power plants and its impact on local climate has been discussed for a number of years. Behind such discussions, the problem of power parks played the most important role.

A large number of numerical models for cooling tower plumes as well as for heat islands were developed during recent years. Only a few of them, however, seem to be verified observations. Ground based observations are not likely to reflect what is going on in the atmosphere aloft. In addition, the density of observations in space and in time never can be sufficient for such a validation program. On the other hand, air-borne measurements should be sufficient for most model validation purposes. This was the reason for the author to develop a measurement system consisting of a powered glider and modern system of sensitive meteorological sensors including digital data recording systems. Due to the excellent aerodynamical behaviour of the aircraft, the measurement of vertical motions of the air can be obtained in a fairly simple manner from the vertical motions of the aircraft. Measurements of temperature, humidity, vertical velocity of the air, and of the aircraft's position in space as a function of time enables the investigator to measure this set of variables both in Eulerian and in Lagrangian frames of reference.

Applications of the system to problems of thermal convection, to the problem of plume rise of industrial stacks, to the problem of local climate impacts of cooling towers, and to problems of regional climatology were made between 1973 and 1979. Most of the applicatior were associated with environmental problems, but, with respect to the temporal and spatial structure of thermal convection some basic results were also obtained. The following is a brief outline of results connected with the problem of climatic impacts of anthropogenic heat emission.

2. THE RESEARCH AIRCRAFT AND STRATEGIES OF MEASUREMENTS

The aircraft chosen is a powered glider ASK 16 manufactured by Alexander Schleicher Segelflugzeugbau, Poppenhausen, Germany. Some characteristics are the following: Wing span 16 m, wing area $19 m^2$, maximum wing loading $39.5 kg/m^2$, empty weight 460 kg, maximum gross weight 750 kg, useful load 290 kg, fuel capacity 90 lt, fuel consumption (55% power) 10 lt, power unit VW-Limbach 72 HP, maximum speed 180 km/h, stall speed 75 km/h, rate of climb at sea level 2.5 m/s, gliding ratio 1:25.

Figure 1. The powered glider ASK 16 in flight.

Some reasons for selecting this type of aircraft were the following: High maneuverability, excellent aerodynamic properties, very low stall speed, large range of operating speeds (from 90 km/h to 180 km/h), good gliding performances, sufficiently high rate of climb. In addition, the operating costs are extremely low, the availability of the system is extremely high. Theory and experiments show that the response time to a buoyant plume of mean size is less than 1 s. The resolution of vertical velocities is approximately 25 m, i.e. the aircraft follows the vertical motion of plumes with diameters greater than 25 m and does this with a delay time of less than 1 s. The vertical motion of the air is obtained by feeding all relevant flight data in a suitable computer program.

Meteorological instruments in operation are the following: Pressure: Rosemount REC 831 A 10, REC 1241A-6B; true air temperature: Hartmann und Braun PT 110 and Rosemount REC 102EBJ; Humidity: Lyman Alpha and Väisälä Humicap. In addition, acceleration, pitch and roll are stored together with the other variables on magnetic tape as well as on a four-channel recorder.

Due to specific characteristics of the whole system (sensors and recording on magnetic tape) the highest horizontal resolution for most of the variables is 2.5 m, for some 5 m. Here an operating speed of 90 km/h is taken as lowest safe operating speed.

The strategy of measurement depends on the meteorological system to be investigated. The individual life history of a thermal was investigated in a Lagrangian frame of reference. A soaring glider marked the thermal, encircling the center of the thermal with constant rate of climb. The powered glider flew horizontal traverses through the center of the thermal with various headings. Normally, the mean rate of climb of the powered glider was approximately the same as that of the soaring glider. Statistical properties of horizontally distributed thermals were obtained by flying traverses very frequently above a part of the ground where thermal activity could be expected. With regard to thermal plumes, traverses were flown at many heights within a cross section perpendicular to the direction of the mean wind vector. Normally, up to three cross sections could be done during the maximal recording time of the tape recorder. Typical source distances of such cross sections were: 0.5 km, 1.5 km, and 3.0 km. Sometimes the aircraft was tracked by radar, in most other cases ground navigation was sufficient to fly accurately within a chosen vertical cross section. With regard to problems of regional climatology, horizontal traverses were flown over greater distances, therefore, only a limited vertical resolution could be achieved. During such campaigns an additional instrumented aircraft did the vertical soundings.

3. LOCAL IMPACTS RESULTING FROM POINT SOURCES

The first example of cross section measurements in a thermal plume may be the case of a power station stack near Wilhelmshaven. The stack height is 275 m, the power 700 MW, the exit temperature of the effluent gases 130°C. The thermal plume was investigated in cross sections with source distances up to 6 km under various meteorological conditions. The nearest cross section had a source distance of 600 m. At the same time, sodar measurements of the plume covered the range between 100 m and 600 m source distance. A typical traverse at a distance of 600 m is shown in Figure 2. Temperature as well as humidity recordings show bifurcation of the plume at a height of 100 m above the stack. The original temperature difference between effluent gases and surrounding air (119 C) has decreased to less than 2 C. This is referred to as "rapid temperature dissipation". At source distances greater than 3 km, only tenths of a degree could be observed. The resolution of all sensors allowed, however, the plume to be tracked downwind up to source distances of 6 km. This way, a well known plume rise formula of Briggs was verified.

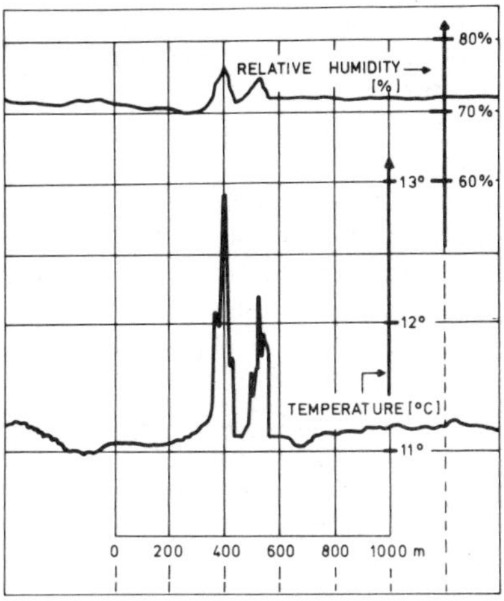

Figure 2. Typical temperature and humidity records on a traverse in a cross section 600 m downwind of the stack (physical height 275m) of a 700 MW power station near Wilhelmshaven. Height above the ground was 375 m. Wind velocity in this height approximately 10m/s, stratification was near neutral. August 28, 1979, 1830 GMT.

It is obvious that there will be no impact on the atmosphere by this type of anthropogenic heat emission. Even in cases of power stations in the order of 3,000 MW, no impact could be observed.

Starting in 1974, many field studies were made to investigate the physical structure and the local impact on climate of cooling tower plumes. Within one vertical cross section perpendicular to the mean wind direction up to 20 traverses were flown. During two hours of recording time three such cross sections with a total number of nearly 60 traverses were investigated. This led to a fairly complete three-dimensional physical picture of such cooling tower plumes.

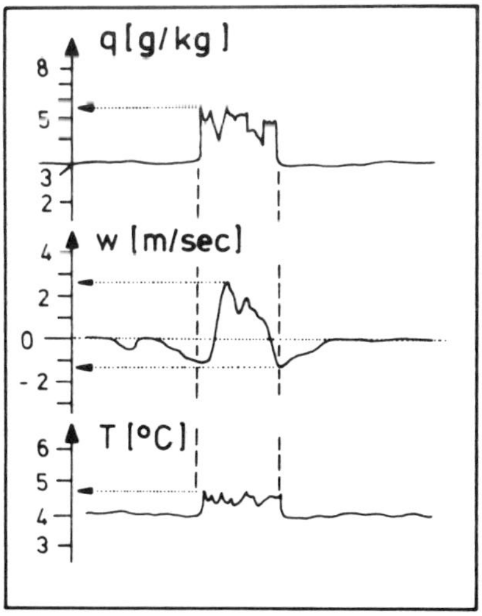

Figure 3. Typical temperature, humidity, and vertical motion records in a cross section 850 m downwind of the cooling tower (physical height 135 m) of a 600 MW power station near Meppen. Height above the ground was 500 m. February 18, 1975, 1535 GMT.

Characteristics of almost all such records are the following: Temperature and humidity assume the internal plume values within a distance of approximately 10 m when entering the plume. The same steep gradient is found when the aircraft is leaving the plume. Near the center of the plume horizontal vorticity lines produce circulations within the vertical cross section that result in maximum downward velocity near the regions of steepest gradients of temperature and humidity. In all cases, the difference between the air within the plume and the surrounding air is in the order of tenths of a degree.

Figure 4 summarizes the results of such flights: The upper part shows the typical records of vertical velocity, temperature and humidity along a selected traverse according to Figure 3. The lower part of Figure 4 shows the result of many measurements within a cross section. The visible plume is found in the upper part of the plume, the main entrainment is found at the bottom of the plume. This indicates that below the plume air from the surroundings was always found. Ground based measurements only occasionally showed short term influences of the plume. Apart from the shading of the visible plume, there certainly is no impact on local climate at the ground. The same is true for the atmosphere above the ground where the thermal effects are very small (1).

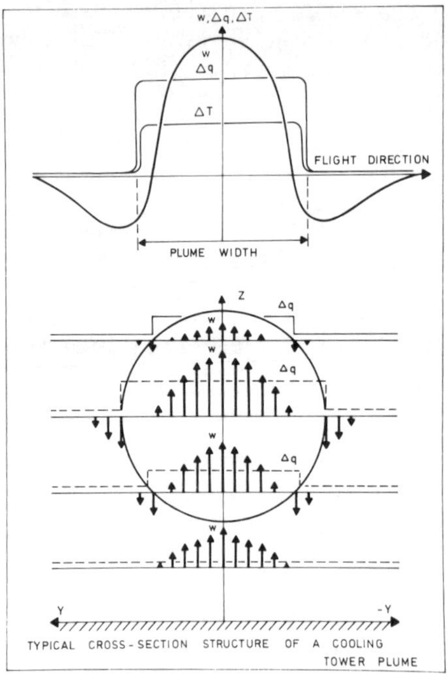

Figure 4. Typical structure of a cooling tower plume in a cross section perpendicular to the mean wind direction. The upper part shows records of temperature humidity, and vertical velocity along a traverse. The lower part shows the distribution of these variables within a cross section.

4. REGIONAL IMPACTS

Increasing urbanization and industrialization in certain regions led to considerable concern about regional impacts on climate resulting from anthropogenic heat release. The region of the Upper Rhine Valley attracted the interest of scientists and politicians most in this respect. On the one hand, this region does not have a favourable climate even if there were no anthropogenic activity. On the other hand, urbanization and industrialization develop very rapidly especially in this region. Therefore a research project was sponsored by the German "Umweltbundesamt" and started in late 1976. Based on a detailed heat source inventory (2) a numerical model for the simulation of the undisturbed climate as well as for the disturbed climate is expected to give the answer to the question of the impact on climate of the anthropogenic heat emission. In addition, a program of field measurements, ground based and air borne, was launched to investigate the present day climate of the region, and further to estimate the present day local impacts of area sources.

Up to now, the author has conducted six field studies in this region. Besides the powered glider ASK 16, another light aircraft (KLEMM 107 C) was used for vertical soundings. Figure 5 shows the arrangement of vertical cross sections on a map of the region. The region of investigation around Freiburg (FR) is typical for a broad valley (40 km width) with fairly high mountain ridges at

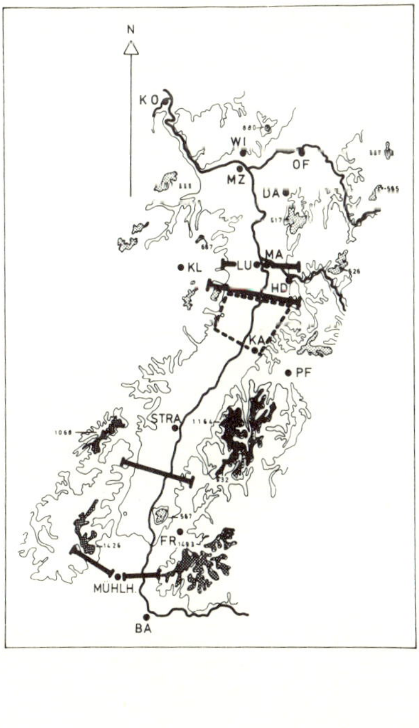

Figure 5. Map of the Upper Rhine valley. Shown are the cities (starting from the bottom): BA=Basel, MÜHLH.=Mühlhausen, FR=Freiburg, STRA=Straßburg, HD=Heidelberg, LU=Ludwigshafen, MA=Mannheim, DA=Darmstadt, MZ=Mainz, WI=Wisbaden. The Pfälzer Wald is located around KL=Kaiserslautern and has a typical height of 500 m, the Schwarzwald is located between PF=Pforzheim and Basel. The Vogesen are located opposite to the Schwarzwald.

both sides (Schwarzwald and Vogesen). Typical of this region is a much smaller amount of urbanization and industrialization compared with the second main region of investigation around and south of Mannheim-Ludwigshafen. Here the mountain ridges are less pronounced but here we find a region of most intense urbanization and industrialization.

Main traverses were made south of Mannheim-Ludwigshafen with Speyer in the center of the traverse and south of Freiburg with Mülhausen in the center of this traverse. These traverses were considered to be suitable to study more normal, i.e. rural conditions in that region. Figure 6 shows typical records of temperature and humidity along traverses at different heights. Within the planetary boundary layer we find a thermal asymmetry between the east and the west side of the valley. Humidity shows the expected opposite behaviour. The temperature difference may be 2°C and more. This happens at all times at which thermal activity is possible in the region. An explanation for this thermal asymmetry, which is found to be much more pronounced in

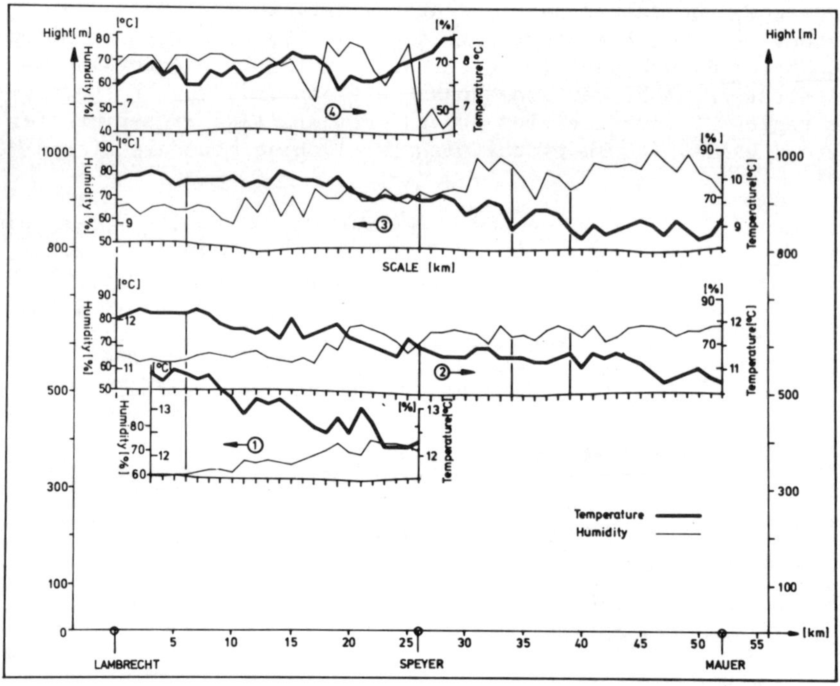

Figure 6. Typical records of temperature and of humidity in a vertical cross section of the Upper Rhine Valley south of Mannheim-Ludwigshafen. Lambrecht is located at the west side of the valley, Mauer at the east side.

the region around Freiburg, could be given as follows: During the morning hours thermal convection starts much earlier at the eastward sloping parts of the mountain ridges, i.e. at the Pfälzer Wald and at the Vogesen. Here, the morning inversion is destroyed very early, whereas the inversion above the Rhine Valley prevents convection for most of the morning hours.

Such situations are found during the year much more often in the region than believed. As a matter of fact, the statistics of all results of flights within the same cross section show the thermal asymmetry during the morning hours as Figure 7 depicts it. Here the deviations of temperature from the mean value of a horizontal traverse are drawn. Even in the average, the atmosphere at the west side of the valley is about 1 C warmer than at the east side. Each model for the simulation of the climate of the region should be able to simulate thermal convection and its spatial and temporal distribution in order to simulate the mentioned thermal asymmetry of the region. We will see later, that man-made heating of the atmosphere is much smaller than this natural spatial asymmetry.

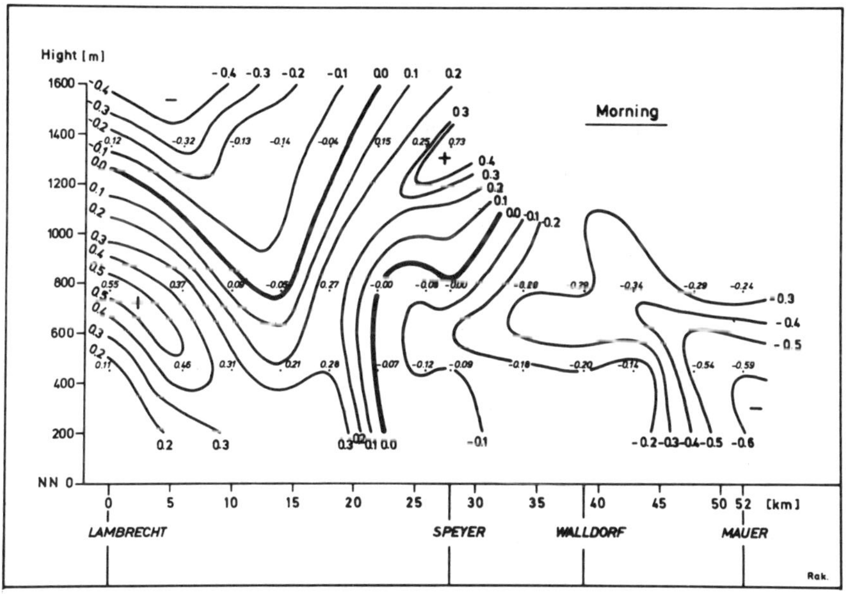

Figure 7. Deviations of temperature from the mean value along a traverse for the morning hours taken from all flights during the years 1977/79. The cross section is located south of Mannheim-Ludwigshafen.

The mentioned thermal asymmetry during the morning hours is seen to be not uniform at the sides of the valley as Figure 8 depicts. Near Neustadt/W convection is most intense and leads to high temperatures approaching the slopes of the Pfälzer Wald. The cities of the region, e.g. Karlsruhe, produce no thermal effect in the atmosphere at a height of 650 m above the ground. Although it was a very cold day in the middle of March, we did not expect anthropogenic thermal effects at such heights above the ground. It was interesting, however, that in this and other cases there was large spatial variability of temperature and humidity under natural conditions. The onset of thermal convection at noon all over the region smooths horizontal differences out. However, due to different properties of the ground, even thermal convection shows a spatial distribution of its intensity which may be so pronounced that even during the time around noon spatial differences of temperature occur.

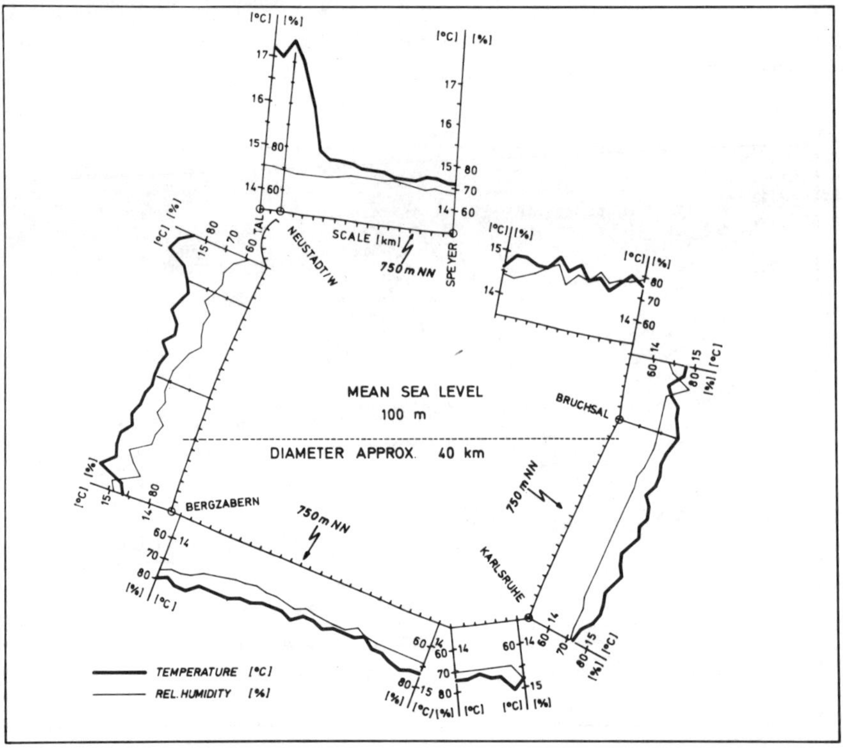

Figure 8. Records of temperature and humidity along horizontal traverses in constant height above mean sea level. The moutain ridge of the Pfälzer Wald is located between Bergzabern and Neustadt/W. The east side of the valley consists of a hilly region of moderate heights.

During all field studies conducted so far, it has been hard to find any anthropogenic effects of heat emission within the atmosphere in the region. Some flights at approximately 300 m above the ground during March 1978, however, indicated such effects produced by comparatively small towns very clearly. Figure 9 shows that this is the case over the town of Bruchsal and over Weingarten. The warming is on the order of half a degree. At the same time, however, spatial differences of more than 1.5°C were found along the southern traverse due to the physical properties of the underlying ground (forest between Karlsruhe and Bergzabern first, then dry fields without vegetation). It was surprising that the city of Karlsruhe did not produce any effect of anthropogenic heating.

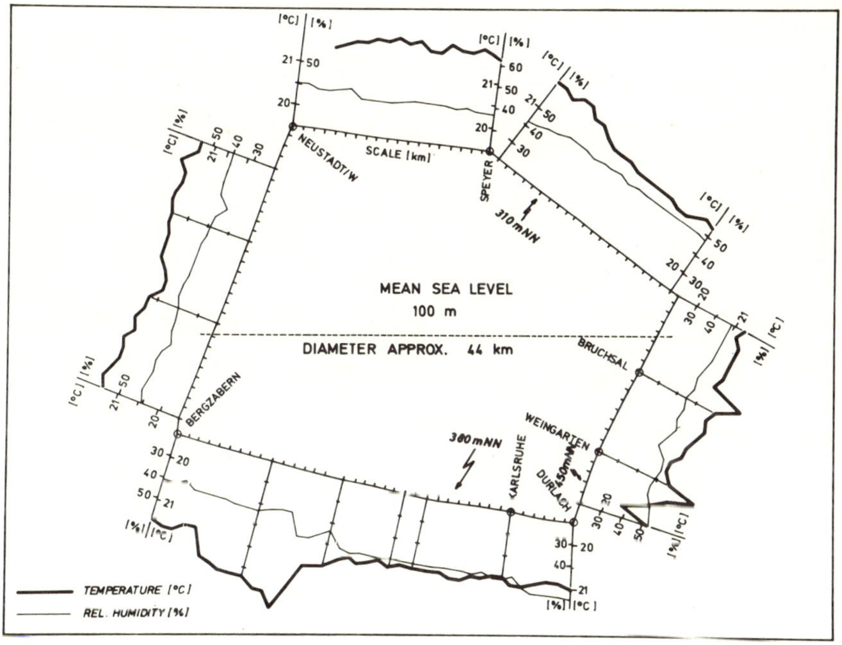

Figure 9. Same as Figure 8, flight time in the late afternoon.

Similar flights crossed the highly industrialized cities of Mannheim and Ludwigshafen. During the same period of cold but sunny days in March natural thermal convection was enhanced drastically around noon. Although the field of vertical motion was influenced this way, the fields of temperature and of humidity were not. Therefore, heat emission resulted in a build up of an intense field of vertical motion that may lead to the build up of secondary cloud formation during other periods of the year.

5. ESTIMATES OF IMPACT BY ANALOGY WITH NATURAL PHENOMENA

Mainly with regard to the problem of energy centers (EC) and their impact on local and regional climate, a simple one-dimensional model was developed that should be able to predict - among others - vertical motions within the atmosphere induced by heat release near the ground. The impact of heat release on the atmosphere then could be estimated simply by comparing man-made fields of vertical motion with those found in connection with typical atmospheric phenomena. With respect to ECs the question arises: At what power densities and spacing of cooling towers do ECs become undesirable from a meteorological/climatological point of view?

In what follows, we compare the state of the unperturbed atmosphere with the state at which heat and water vapor has been added. The thermal power P (MW) of waste heat induces a vertical flux of heated air in a vertical column of radius r. The height of the bottom is z_B, the vertical extent of the cylinder is l_o (see Figure 10). The emitted heat is continuously transported vertically through this cylinder. The entrainment/detrainment mechanism is not taken into account, the most unfavourable case of a windless weather situation is assumed.

With density ρ, vertical velocity w, potential of gravity $\phi=gz$, and enthalpy h, the equations of continuity and of total energy for the air passing the volume V of the cylinder are given by

$$\frac{d}{dz}(\rho w) = 0, \qquad \frac{d}{dz}(\rho w(w^2/2 + gz + h)) = 0 \tag{1}$$

For the unperturbed air, e.g. for the surrounding of the cylinder the index "o" will indicate this state. There, the equation for total energy is given by

$$\frac{d}{dz}(gz + h_o) = 0 \tag{2}$$

Combining equations (1) and (2) and introducing the difference $\Delta h = h - h_o$ of enthalpy between the two states, the equation for total energy (1) reads as

$$\frac{d}{dz}(\rho w(w^2/2 + \Delta h)) = 0 \tag{3}$$

This equation is integrated over the volume of the cylinder successively as follows:

$$\iint_{F_o} df(\rho w(w^2/2 + \Delta h))\Big|_{z_B + l_o} = \iint_{F_o} df(\rho w(w^2/2 + \Delta h))\Big|_{z_B} = P$$

where P (MW) is the total power of the source, further,

an integration over the vertical extension of the cylinder gives

$$\int_{z_B}^{z_B+l_o} dz \iint_{F_o} df(\, \varrho w(\, w^2/2 + \Delta h\,)) \approx \overline{w} \iiint_G dV \varrho \Delta h = \overline{w} \Delta H = l_o P \qquad (4)$$

Here, $w^2/2 \ll \Delta h$ was assumed. The characteristic vertical velocity $\overline{w}$ is the mean of w within the cylinder of Volume $V=F_o l_o$. The residence time of heated air in the volume is given by $t=l_o/\overline{w}$.

The total enthalpy H of the volume can be written as

$$H = \iiint_G dV \varrho h = M_d h_d(\overline{T}) + M_v h_v(\overline{T}) + M_l h_l(\overline{T})$$

The masses of dry air, water vapor, and liquid water are M_d, M_v, and M_l respectively; the corresponding partial specific enthalpies are written in terms of the mean temperature of the volume. Introducing the latent heat of vaporization L and the mean heat capacity

$$M_o c_{p,o} = M_d c_{p,d} + M_{v,o} c_{p,v} + M_{l,o} c_l$$

an excellent approximation for ΔH is given

$$\Delta H = M_o c_{p,o} \Delta \overline{T} + L(\overline{T}) \Delta M_v \qquad (5)$$

Mass and enthalpy of the unperturbed atmosphere are given by $M_o = \overline{\varrho}_o F_o l_o$ and $H_o = M_o c_{p,o} \overline{T}_o$ respectively.

Now, a characteristic vertical acceleration $\overline{A}$ and an associated buoyancy flux $\overline{E}$ are introduced according to

$$\overline{A} = g \Delta H / H_o, \quad \overline{E} = F_o \overline{w}\, \overline{A} \qquad (6)$$

and ΔH is eliminated in equation (4). The resulting simple relationship between power P and buoyancy flux $\overline{E}$ then is given by

$$P = \frac{H_o}{F_o l_o} \cdot \frac{\overline{E}}{g} \qquad (7)$$

Now, the acceleration $A = dw/dt = w \frac{\partial w}{\partial z} = \frac{\partial}{\partial z}(w^2/2)$ is integrated over the lower part of the cylinder

$$\frac{1}{F_o} \int_{z_B}^{z_B+l_o/2} dz \iint_{F_o} df \frac{\partial}{\partial z}(w^2/2) = \overline{w}^2/2 - w_B^2/2 = \frac{l_o}{2} \overline{A}$$

Utilizing the equation of continuity $\overline{\varrho}\, \overline{w} = \varrho_B w_B$ and introducing the parameter $a = \varrho_B^2/(\varrho_B^2 - \overline{\varrho}^2)$ the relationship between $\overline{A}$ and $\overline{w}$

is given by $\overline{w}^2 = al_o\overline{A}$. The buoyancy flux $\overline{E}$ is given by

$$\overline{E} = \frac{F_o}{al_o}\,\overline{w}^3 \tag{8}$$

Together with equation (7) the final relationship between power and characteristic vertical motion is obtained as

$$P = \frac{H_o}{agl_o^2}\,\overline{w}^3 \tag{9}$$

Approximating ΔH by $M_o c_{p,o} \Delta\overline{T}$, it is seen that $\Delta\overline{T} \sim P^{2/3}$. Since $\overline{T}_o$ as well as M_o can be obtained for a given volume and since the same is true for the densities ϱ and ϱ_B, all interesting parameters can be calculated in terms of the power P of the heat source.

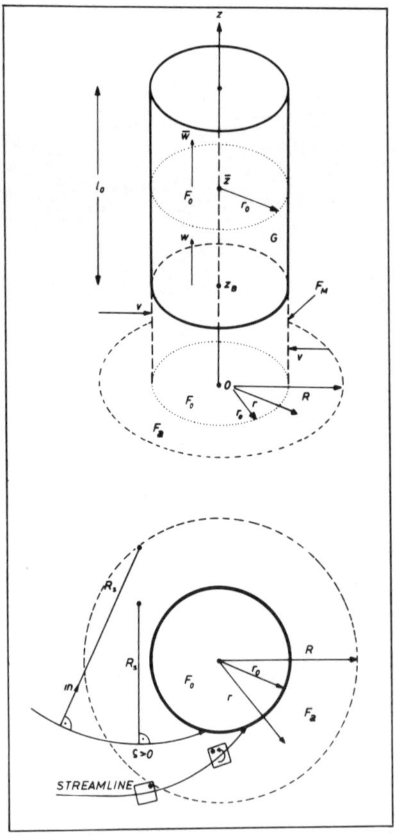

Figure 10. Explanation of the symbols used in the text. The entrainment takes place below the bottom of the cylinder at $z = z_B$. The lower part of the Figure shows a possible case of mass inflow from a certain region around the cylinder near the ground.

The above theory is used to construct a diagram shown in Figure 11, where the radius r_o and the area F_o of the vertical air column are plotted on the abscissa, and the vertical heat emission P (MW_{th}) is plotted on the ordinate. The slanting broken lines give the vertical velocities $\overline{w}$ of the rising air. The slanting solid lines give the specific vertical heat flux densities $S = P/F_o$ (W/m^2). The spacing of cooling towers is depicted in the upper left corner of the Figure.

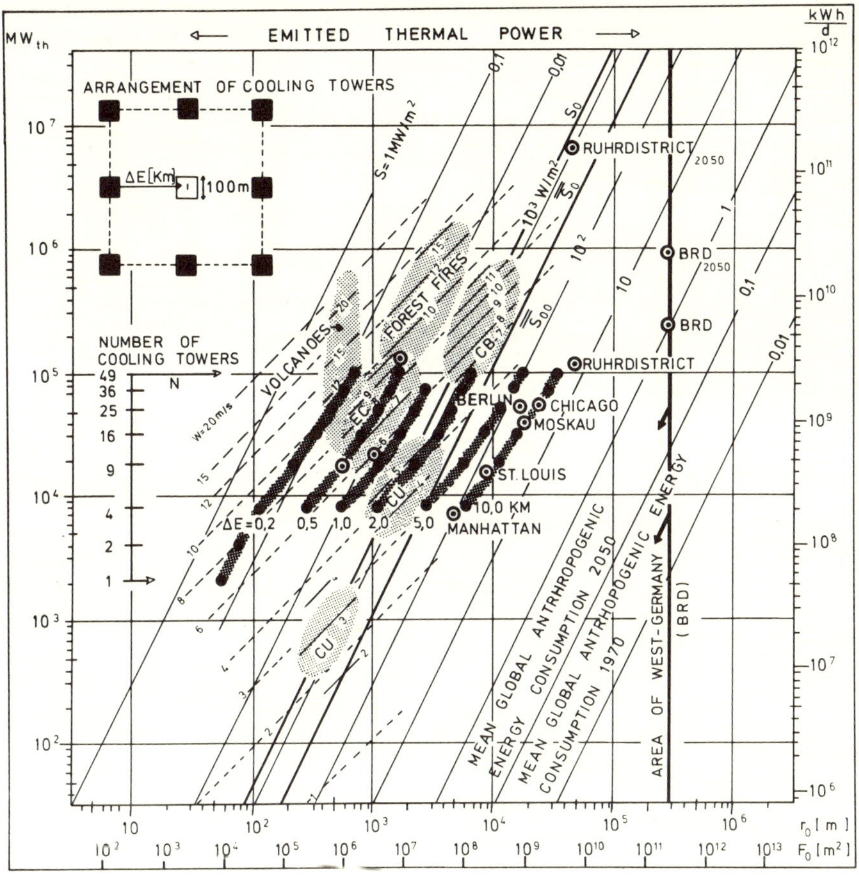

Figure 11. Characteristic vertical velocities in the atmosphere as function of the emitted thermal power at the bottom of a vertical column of air and as a function of the geometrical structure of the column itself. The number and the spacing of cooling towers is varied and the corresponding vertical velocities are comparable with those estimated for natural phenomena. Data are taken from (3), (4), (5) and (6).

It can be seen that ECs of 10^4 to 10^5 MW_{th} and vertical velocities of 6 to 10 m/s have characteristics that fall close to regions in the diagram where the effects of very intense natural phenomena are located in the diagram. In order to avoid the effects of these phenomena it is clear that the larger the ECs the greater the spacing between the cooling towers must be. From a meteorological point of view, large ECs are undesirable.

Concluding these investigations, something should be mentioned about the inflow velocities at the bottom of the heated volume of air. It would be desirable to express this velocity also in terms of the thermal power P. A simple attempt to do so is given as follows. Referring to Figure 10 the equation of continuity requires

$$\pi r_o^2 \rho_B w_B = \pi r_o^2 \bar{\rho} \bar{w} = 2\pi r_o z_B \rho_o v(r_o) \text{ i.e. } v(r_o) = \frac{r_o \bar{\rho}}{2 z_B \rho_o} \bar{w} \quad (10)$$

Here $\bar{w}$ is known as function of the thermal power according to equation (9). The velocity $v(r_o)$ is thought to be a vertical mean value. The same is assumed with regard to the density ρ_o. Equation (10) is already the required relationship between power P and mean inflow velocity. Of some interest is, however, also the distribution of the horizontal wind velocity outside the column near the ground. A simple application of the equation of continuity written in cylindrical coordinates leads to

$$v(r) = \frac{r_o}{r} v(r_o) - \frac{1}{2z_B} \frac{\bar{\rho}}{\rho_B} \frac{F_o}{F_a} \frac{r^2 - r_o^2}{r} \bar{w}$$

where F_a is given by $\pi R^2 - F_o$.

REFERENCES

(1) Fortak, H. G.: 1975, Dreidimensionale Vermessung von Kühlturmfahnen. KFA Jülich, JÜL-1250 (Parts 3 and 4)

(2) Bartholomäi, G. und W. Kinzelbach: 1980, Das Abwärmekataster Oberrheingebiet. KfK Karlsruhe, KfK 2869 UF

(3) Hanna, S. R. and F. A. Gifford: 1975, Meteorological Effects of Energy Dissipation at Large Power Parks. Bull. Amer. Met. Soc. 56, p. 1069 - 1076

(4) Hanna, S. R. and S. D. Swisher: 1971, Meteorological Effects of the Heat and Moisture Produced by Man. Nuclear Safety, 12, p. 114 - 122

(5) Bourne, A. G.: 1964, Birth of an Island. Discovery, 25, p. 16 - 19

(6) Häfele, W.: 1973, Energy Systems. Proc. IIASA Planning Conference on Energy Systems. Laxenburg, Austria

COMPUTER SIMULATION STUDIES OF THE REGIONAL AND GLOBAL CLIMATIC IMPACTS OF WASTE HEAT EMISSION

Robert M. Chervin

National Center for Atmospheric Research*
Boulder, Colorado 80307, U.S.A.

ABSTRACT. A series of computer experiments was performed with a global version of the NCAR atmospheric general circulation model to assess the potential climatic impact (in terms of altered climatic means) of a hypothetical 90 W m^{-2} waste heat emission (or thermal pollution) released from a projected eastern United States megalopolis circa 2000 A.D. To provide an indication of the possible seasonal range of thermal pollution climatic impacts, separate experiments were performed with the model in its January and July configurations containing the identical prescribed change. It was found that, when the inherent variability of the model's climatic means in the control populations was taken into account, the waste heat emissions produced large and statistically significant changes in the megalopolis region with respect to mean temperature, vertical velocity, precipitation and soil moisture. However, considerable differences in the model response for the different seasons were obvious for these climatically important fields.

1. INTRODUCTION

During the past decade, computer climate models of widely varying complexities have been used to assess the potential climatic impact from altered boundary conditions, atmospheric composition or external conditions. The basic premise in any of these computer experiments is that the sensitivity of the model to such prescribed changes bears some resemblance to the

*The National Center for Atmospheric Research is sponsored by the National Science Foundation.

W. Bach, J. Pankrath, and J. Williams (eds.), Interactions of Energy and Climate, 399–415.

the sensitivity of the actual climate system. Naturally the validity of this premise is enhanced by the complexity of the model in terms of the incorporation of appropriate physical processes and the degree of spatial resolution. Also, the more complex models permit assessment of climatic impact in terms of a broader spectrum of climate measures as determined by different physical variables, different statistical moments and different degrees of temporal or spatial compression. Included in this latter class of climate models are the three-dimensional atmospheric circulation models, commonly called general circulation models (GCMs), which have been developed by various university groups or national laboratories throughout the world.

Within the context of the methodology described above, a number of modeling groups have used GCMs to study the possible climatic effects of man-made thermal energy (or thermal pollution). A series of GCM experiments (1,2) was performed at the National Center for Atmospheric Research (NCAR) using different geographical distributions of thermal energy added to all non-ocean regions. These distributions were based on different projections of future population growth with further assumptions on the concomitant energy requirements (3). A summary of these early efforts is presented in a National Academy of Sciences Report on energy and climate (4).

Another series of thermal pollution GCM experiments (5-8) was carried out by the International Institute for Applied Systems Analysis (IIASA) with a model (9) provided by the British Meteorological Office. These experiments investigated the atmospheric response to the waste heat emitted from hypothetical ocean energy parks and hence differed from the early NCAR experiments by the comparatively localized nature of the surface heat source which was restricted to small ocean areas as opposed to large continental regions.

Recently, a second series of thermal pollution experiments (10) was performed with a version of the NCAR GCM. The intent of this study was to assess the potential climatic impact of a hypothetical 90 W m^{-2} of waste heat released from a projected eastern United States megalopolis circa 2000 A.D. (11). It was also assumed that this megalopolis would have the same energy requirements and produce the same thermal energy output in terms of energy density as present-day Manhattan Island. Both the magnitude and areal extent of the heat emission were exaggerated in the experiments to enhance the possibility of obtaining a statistically significant response in the model while taking into account the inherent variability of the model's climatic means in the control population. Furthermore, to provide an indication of the possible seasonal range of thermal pollution climatic impacts, separate experiments were performed with the model in its January and July configurations containing the same additional surface heating distribution. The results and conclusions of this recent study are summarized in this paper with emphasis on the analysis of regional climate change in the vicinity of the prescribed change and globally averaged climate change. More details may be found in the original paper (10).

2. MODEL DESCRIPTION

The atmospheric model used in this study is global in domain, with a 5° latitude-longitude horizontal grid and six 3 km thick vertical layers. The computational procedures for solving the finite difference approximations to the governing equations and the included parameterizations of subgrid-scale physical processes are described in several publications and reports (12-14). Dynamically interactive physical processes in the model are solar and infrared radiation, cloudiness and the horizontal and vertical diffusion of sensible heat, moisture and momentum. Also included are large-scale precipitation dependent on local relative humidity and small-scale precipitation consistent with the convective latent heat release predicted by a cumulus parameterization (15-16). Ocean surface temperatures and solar declination are held constant in time at the appropriate climatological values for the particular month in question (i.e., January or July). Continental surface temperatures are calculated from a surface energy balance which includes time-dependent snow cover and soil moisture as prescribed by a parameterized surface hydrology. Orography is included, consistent with the horizontal resolution.

3. EXPERIMENTAL DESIGN

To provide a standard about which the model's thermal pollution response can be measured (and its statistical significance assessed), sampled climate ensembles of the basic January and July climatic states were generated from sets of extended simulations with identical external and boundary conditions but with initial conditions differing by small random perturbations. These individual climatic states were defined as 60-day means with the averaging interval commencing 30 days after the initial time to produce more stable climatological statistics (17) and independent states (18). Following this procedure, five independent time-averaged samples were used for each model configuration (January and July) to estimate the ensemble average (i.e., the typical climatic mean state) and standard deviation (i.e., the unforced inherent climate variability) of the control climates simulated by the model (10).

The relatively coarse vertical and horizontal resolution of the model required the simple approximation that the impact of the waste heat emission of 90 W m^{-2} is transmitted to the atmosphere solely as a prescribed change in the surface energy balance in the form of an additional heat source. (Ordinarily, waste heat is not included in any form as an input parameter in the standard model control experiments.) Individual January and July thermal pollution experiments were performed in a parallel fashion to the control population starting at the same initial time but with the inclusion of a uniform 90 W m^{-2} heat source as an additional contribution to the surface energy balance for the eighteen grid points (each representing a 5° latitude-longitude area) contained in the megalopolis region shown in Figure 1. No other changes were prescribed in the model for thermal pollution experiments in an attempt to isolate the climatic impact produced by a single large source. The climatic means for the thermal pollution populations were determined by single 60-day averages for each of the January and July model configurations computed over the same time-span as for the control populations. In this way, a quasi-equilibrium, independent atmospheric state is generated for each month compatible with the prescribed change in surface forcing.

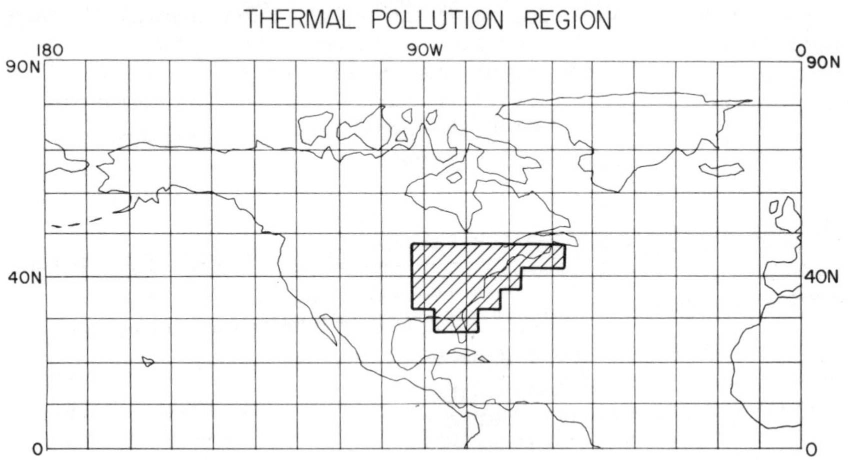

Figure 1. Prescribed change region (cross-hatched) in which thermal pollution (added thermal energy) is 90 W m^{-2}.

In the analysis that follows in the next two sections, the model's response to the waste heat emission for each month, or prescribed change response (19) Δ_{60}, is evaluated in terms of the difference between the 60-day mean from each prescribed change experiment and the estimated ensemble average of the 60-day means from the corresponding control population. Δ_{60} was computed for several key climatic variables for both grid points and globally averaged climate measures. To obtain an objective measure of the statistical significance of any apparent difference in climatic means, a normalized response $r = \Delta_{60}/\sigma_{60}$ is defined for any particular grid point or globally averaged climate measure, where σ_{60} is the standard deviation of 60-day means as estimated from the sample of five control cases for the appropriate month (20). r can be related to the test variate used in the Student's t-test and the resulting significance test is of the null hypothesis that no change in a particular climatic mean was caused by the addition of thermal energy in the model. For the purpose of this study, a two-sided t-test was used with an *a priori* significance criterion $|r| \geq 3$, implying an approximately 5% significance level (or probability of sampling error) in rejecting the hypothesis that a nonzero prescribed change response is merely the result of random fluctuations and not the waste heat emission. Consequently, areas with significantly altered climatic means are shown stippled in the regional response charts displayed in the next section. The additional question of thermal pollution climatic impact in terms of other climate measures such as second moment statistics (e.g., the variability of climatic means or time-lagged autocorrelations) was not addressed in this study but could be considered with the generation of sampled ensembles of prescribed change experiments.

4. REGIONAL RESPONSE

The regional responses for January and July are presented here for the north-west quarter-sphere (i.e., the western half of the northern hemisphere) for the following climatically important model fields: ground surface temperature ($\overline{T_G}$), 1.5 km atmospheric temperature ($\overline{T}_{1.5\ km}$), 3 km vertical velocity ($\overline{W}_{3\ km}$), precipitation and soil moisture. This quarter-sphere is the region of concentration since little evidence was found of significant responses elsewhere (10). The corresponding first and second moment statistics for these five fields in the control populations were previously displayed and discussed (10).

4.1 Ground Surface Temperature

The January prescribed change response for $\overline{T}_G$ is shown in Figure 2. As to be expected, considerable warming takes place at the surface in the prescribed change region with a maximum of 12°C. Significant changes occur primarily in the megalopolis region itself. The large differences of 4°C or more in the vicinity of Greenland are comparable to the inherent variability in the control population and hence cannot be attributed to the thermal pollution from the eastern United States.

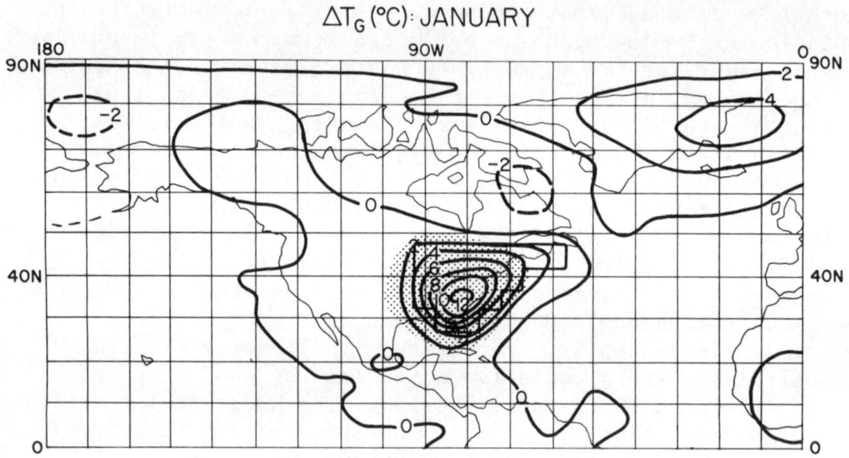

Figure 2. January prescribed change response for ground surface temperature. The stippled areas indicate statistically significant differences at the 5% level. The prescribed change region is outlined with a heavy line.

Figure 3 shows the July Δ_{60} for $\overline{T}_G$. Although the same waste heat emission level of 90 W m^{-2} was added to the January and July thermal pollution experiments, the maximum response in July is substantially smaller. This difference in response is, in part, a consequence of the thermal pollution component playing a greater role in the surface energy balance in January because of the relatively smaller contribution of the solar flux in January versus July for the megalopolis region. In contrast to the January response, the area of statistically significant changes extends to western Canada. Smaller values of Δ_{60} are significant in July because of the reduced inherent variability in the simulated July climatic means (10).

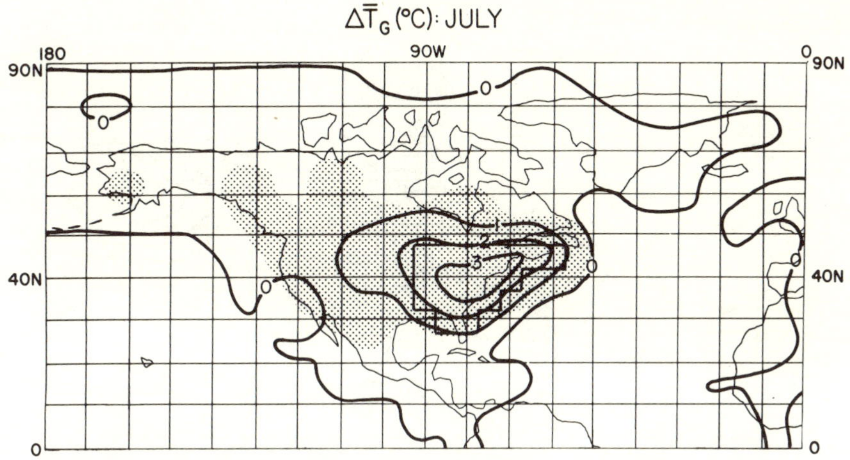

Figure 3. July prescribed change response for ground surface temperature. The stippled areas indicate statistically significant differences at the 5% level. The prescribed change region is outlined with a heavy line.

4.2 1.5 km Temperature

The January prescribed change response for $\overline{T}_{1.5\ km}$ is shown in Figure 4. This response pattern in the first free atmosphere layer is quite similar to the pattern for $\overline{T}_G$ except that the maximum change in the prescribed change region is reduced by a factor of 4. As at the ground, statistically significant responses are confined to the megalopolis region and the 4°C difference over Greenland is comparable to the inherent variability.

Figure 5 shows the July Δ_{60} for $\overline{T}_{1.5\ km}$. This pattern also resembles the corresponding surface temperature response pattern (Figure 3) but with the magnitude reduced only by a factor of approximately 2. At 1.5 km the area of statistically significant response also extends far west of the prescribed change region. Another factor in the seasonal differences in the model response is that the atmosphere above the megalopolis is more stable during January and that the thermal pollution causes relatively large changes in the simulated surface inversion. During July, this surface inversion does not exist in the mean, and the convective parameterization in the model distributes more sensible heat in the vertical. These model results are consistent with the earlier conjectures of other investigators (21).

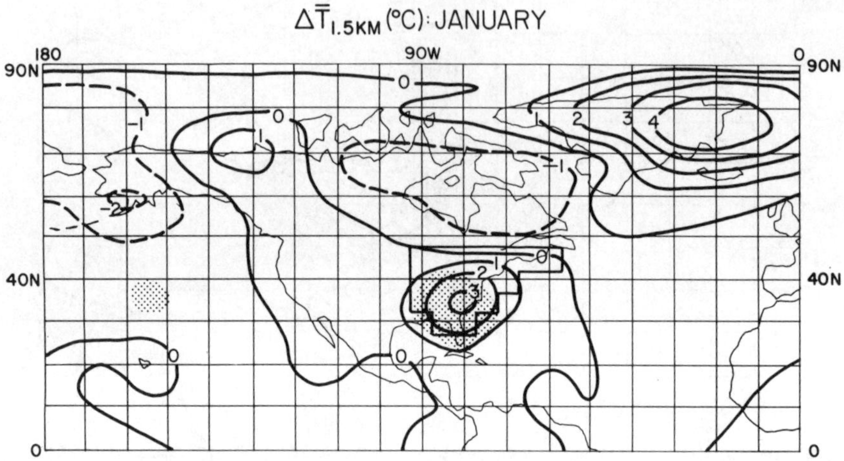

Figure 4. January prescribed change response for 1.5 km temperature. The stippled areas indicate statistically significant differences at the 5% level. The prescribed change region is outlined with a heavy line.

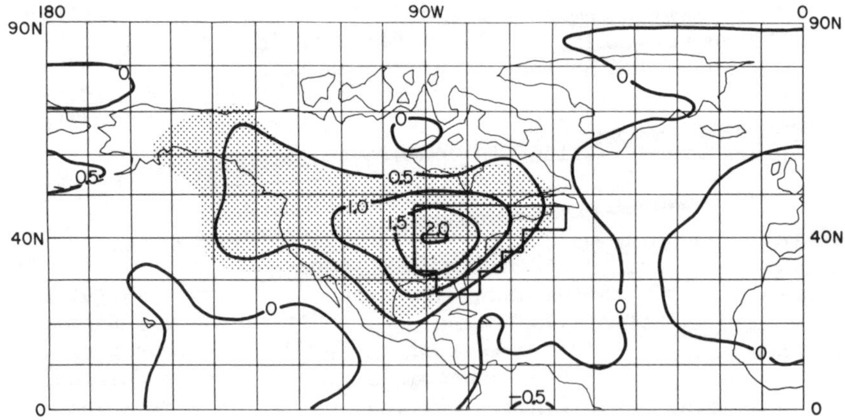

Figure 5. July prescribed change response for 1.5 km temperature. The stippled areas indicate statistically significant differences at the 5% level. The prescribed change region is outlined with a heavy line.

4.3 3 km Vertical Velocity

The January prescribed change response for $\overline{W}_{3\ km}$ indicating the dynamic coupling between the surface heating and the atmosphere is shown in Figure 6. Δ_{60} generally describes a direct thermally driven circulation with increased upward motion over the megalopolis region and increased downward motion surrounding it. This increased upward motion is statistically significant but the increased downward motion is significant only immediately upstream from the prescribed change region. The general response pattern over the continental United States is the same as the spatial pattern of the $\overline{W}_{3\ km}$ field in the control population (10), but is of opposite sign. Hence, in this case, the thermal pollution acts to reduce or counteract the spatial asymmetries in vertical dynamics. The large changes in the higher latitudes are comparable to the model's inherent variability.

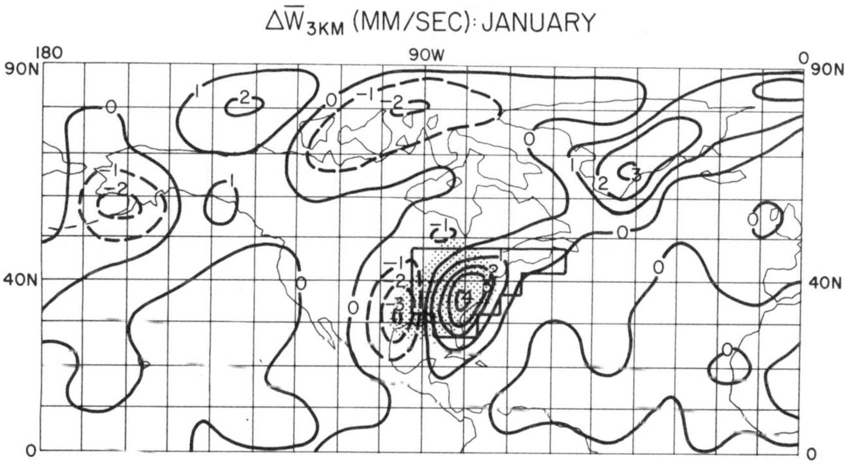

Figure 6. January prescribed change response for 3 km vertical velocity. The stippled areas indicate statistically significant differences at the 5% level. The prescribed change region is outlined with a heavy line.

Figure 7 shows the July Δ_{60} for $\overline{W}_{3\ km}$. A direct thermally driven circulation is apparent over the prescribed change region with increased downward motion generally surrounding it, although the magnitude of this circulation is several times smaller than in January. Nevertheless, there is an area of significantly increased upward motion within the megalopolis region as well as an area of significant reduction of upward motion of the west. The high latitude changes cannot be judged statistically significant by the criterion set forth in Section 3.

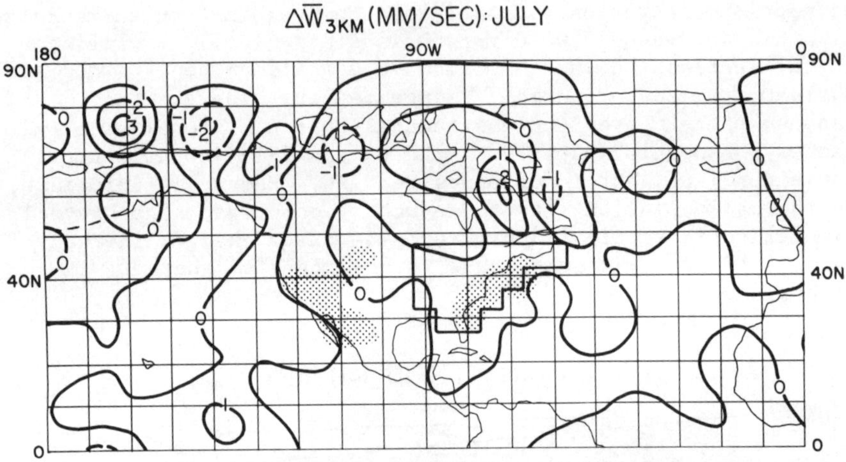

Figure 7. July prescribed change response for 3 km vertical velocity. The stippled areas indicate statistically significant differences at the 5% level. The prescribed change region is outlined with a heavy line.

4.4 Precipitation

The January prescribed change response for the precipitation field, which is an integral measure in height of the various condensation processes in the model, is shown in Figure 8. The spatial structure of the precipitation response is quite similar to the vertical velocity response (see Figure 6) with increased precipitation in the eastern United States (over the megalopolis region), generally reduced precipitation in the central and increased precipitation in the west. However, only the change over the prescribed change region is significantly different from zero. The stippled areas over Greenland and North Africa appear to be isolated individual grid point events without any direct relationship to the prescribed thermal pollution.

Figure 9 shows the July Δ_{60} for the precipitation field. As in January, the July precipitation response shows a three-cell structure across the continental United States, but with less

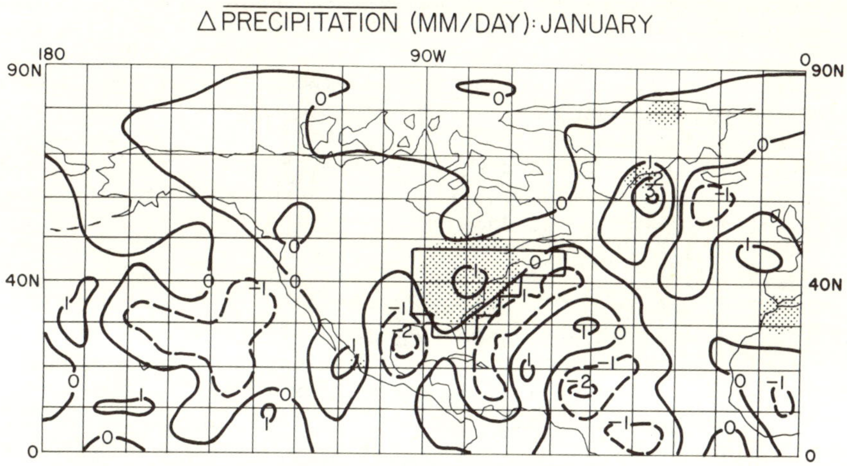

Figure 8. January prescribed change response for precipitation. The stippled areas indicate statistically significant differences at the 5% level. The prescribed change region is outlined with a heavy line.

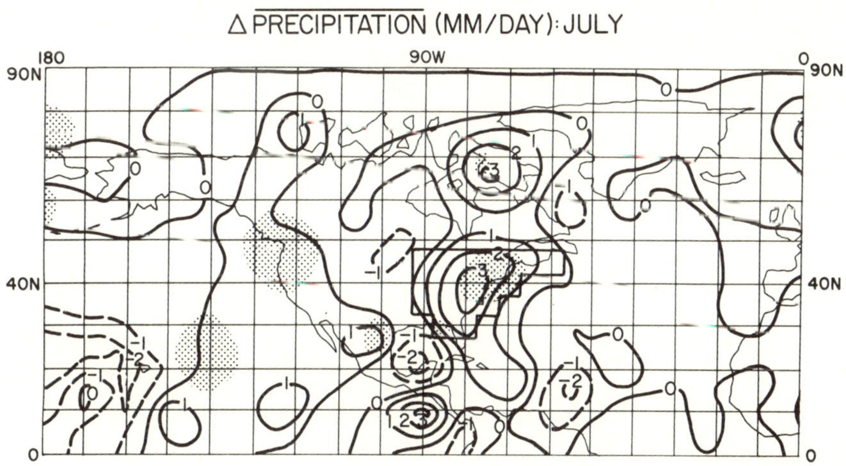

Figure 9. July prescribed change response for precipitation. The stippled areas indicate statistically significant differences at the 5% level. The prescribed change region is outlined with a heavy line.

correspondence to the vertical velocity response. The significant increase over the prescribed change region is three times greater than the January precipitation increase in spite of the smaller surface temperature response. This enhanced precipitation response is due to the higher temperature in the July control population climatic means in conjunction with a greater moisture content and reduced stability in the July atmosphere. Also, there are areas of statistically significant increased precipitation over the western United States which were not apparent in the January thermal pollution experiment.

4.5 Soil Moisture

The January prescribed change response for the soil moisture field, which in the model parameterization is an integral measure in the time of the net surface water flux (i.e., the difference between precipitation and evaporation rates), is shown in Figure 10. The major feature of Δ_{60} is a significant drying out of the soil in the megalopolis region in spite of the increased precipitation (see Figure 8). This reduced soil moisture is primarily due to increased evaporation from the higher ground surface temperature.

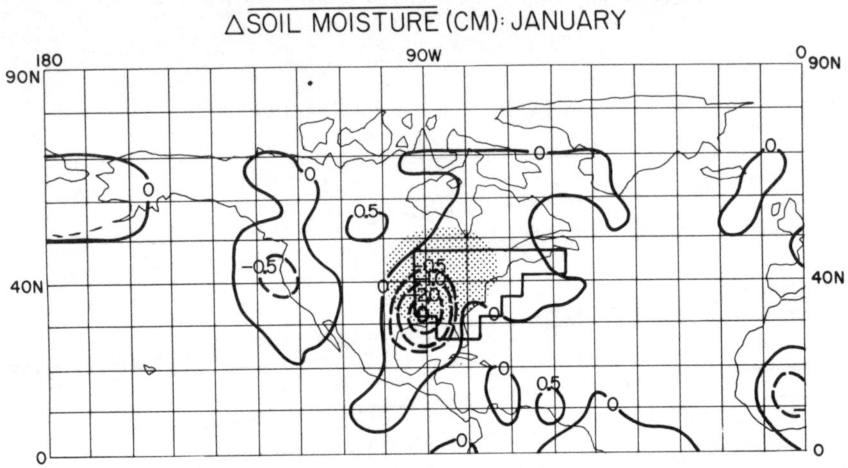

Figure 10. January prescribed change response for soil moisture. The stippled areas indicate statistically significant differences at the 5% level. The prescribed change region is outlined with a heavy line.

Figure 11 shows the July Δ_{60} for the soil moisture field. Statistically significant differences are found over a large fraction of the North American continent. The soil moisture decreases by as much as 2 cm on the western edge of the prescribed change region and increases by 0.5 to 1 cm in the northern and eastern part of North America. This increase within the megalopolis region can be accounted for by the increase in precipitation while the reduction in the Mississippi Valley is caused by both a decrease in precipitation and increased surface temperature due to the thermal pollution.

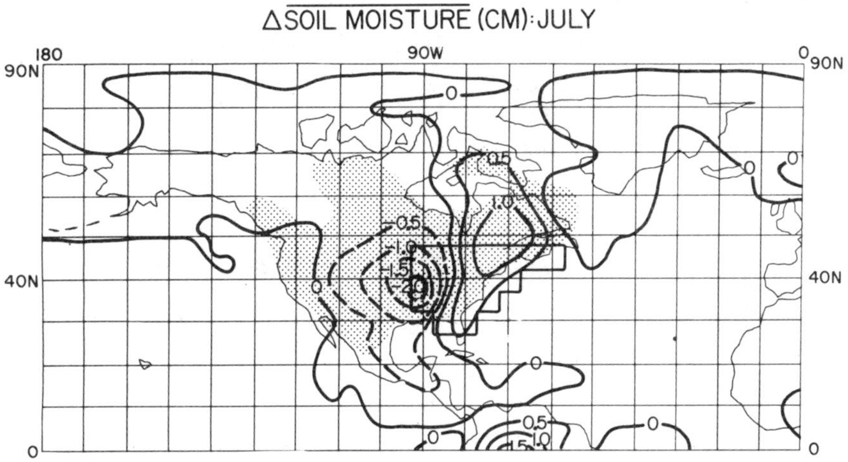

Figure 11. July prescribed change response for soil moisture. The stippled areas indicate statistically significant differences at the 5% level. The prescribed change region is outlined with a heavy line.

5. GLOBALLY AVERAGED RESPONSE

The possibility of globally averaged climate changes being produced by waste heat emissions confined to a relatively small localized region is considered in this section. Table 1 shows the January and July 60-day mean globally averaged precipitation changes and the temperature changes at the ground surface, at the anemometer height layer ($\overline{T}_A$) and within each of the six model layers. To evaluate the statistical significance of these changes, the estimated standard deviations (σ_{60}) of these globally averaged statistics as well as the normalized responses (r) are included in Table 1.

The change in globally averaged precipitation is positive in both experiments but is not significantly different from zero according to the significance criterion ($|r| \geq 3$) from Section 3.

Table 1. Globally averaged responses for January and July thermal pollution experiments. (Units for temperature: °C; units for precipitation $\overline{P}$: mm day^{-1}.)

	January			July		
	Δ_{60}	σ_{60}	$r=\Delta_{60}/\sigma_{60}$	Δ_{60}	σ_{60}	$r=\Delta_{60}/\sigma_{60}$
$\overline{P}$	0.0412	0.0275	1.500	0.0271	0.0669	0.41
$\overline{T}_G$	0.0239	0.1420	0.168	0.1130	0.0433	2.61
$\overline{T}_A$	-0.0090	0.1590	-0.006	0.0736	0.0612	1.20
$\overline{T}_{1.5km}$	-0.0582	0.1820	-0.320	0.0470	0.0814	0.58
$\overline{T}_{4.5km}$	-0.0001	0.0671	-0.001	0.1060	0.1050	1.01
$\overline{T}_{7.5km}$	0.0440	0.0690	0.638	0.1120	0.0635	1.76
$\overline{T}_{10.5km}$	0.0066	0.0575	0.115	0.2670	0.1190	2.25
$\overline{T}_{13.5km}$	-0.0354	0.0545	-0.650	0.1790	0.0822	2.18
$\overline{T}_{16.5km}$	-0.0377	0.0522	-0.722	0.1890	0.0783	2.42

Although $|\Delta_{60}|$ is typically larger in July than in January, all the globally averaged temperature responses fail the significance criterion and hence the apparent differences cannot be attributed to the waste heat emissions. Thus, while large and significant regional responses were found in the thermal pollution experiments, the globally averaged climate is not affected in a statistically significant fashion by the addition of thermal pollution in the prescribed change region.

6. SUMMARY AND CONCLUDING REMARKS

In computer simulation studies with January and July versions of the NCAR atmospheric general circulation model it was found that the addition of thermal pollution resulting from a hypothetical eastern United States megalopolis produced large and statistically significant changes in the prescribed change region in terms of mean temperature, vertical velocity, precipitation and soil moisture. As to be expected, the model's prescribed change response differed considerably in the January and July thermal pollution experiments. The maximum ground surface temperature change in the prescribed change region was 12°C in January and only 3°C in July. In general, statistically significant increases in temperature and precipitation resulted over the prescribed change region although the magnitudes were markedly different in the two months. The sign of the soil moisture response was primarily determined by the magnitude of the surface evaporation rate which in turn was a function of the surface temperature change. Over most of the prescribed change region, the soil dried out in January and became wetter in July. However, neither experiment produced any evidence of either a statistically significant downstream "teleconnection" over the Atlantic Ocean or Europe.

Of course, these experimental results should in no way be construed as a definitive assessment of the climatic impact of the waste heat released for a projected eastern United States megalopolis. The experiments do constitute valid climate sensitivity tests and thereby establish a baseline for the sensitivity of a typical atmospheric general circulation model to such forcing. Furthermore, these experiments clearly demonstrate the seasonal dependence of the model's sensitivity, a conclusion that will likely be appropriate for other climatic impact studies. Finally, the experimental results and methods presented here should be useful for later comparison purposes and also should provide guidance to those planning similar experiments in the future with different and physically more complete climate models.

REFERENCES

(1) Washington, W.M.: 1971, in *Inadvertent Climate Modification--Study of Man's Impact on Climate*, W.H. Matthews et al., Eds., MIT Press, pp. 265-276; also in *Air and Water Pollution*, W.E. Britten et al., Eds., Colorado Associated University Press, pp. 599-613.

(2) Washington, W.M.: 1972, J. Appl. Meteor. 11, pp. 768-772.

(3) Weinberg, A.M., and Hammond, R.P.: 1970, Amer. Sci. 58, pp. 412-418.

(4) Llewellyn, R.A., and Washington, W.M.: 1977, in *Energy and Climate, Studies in Geophysics*, National Academy of Sciences, Washington, D.C., pp. 106-118.

(5) Murphy, A.H., Gilchrist, A., Häfele, W., Krömer, G., and Williams, J.: 1976, *The impact of waste heat release on simulated global climate. RM-76-79,* International Institute for Applied Systems Analysis, Laxenburg, Austria, 27 pp.

(6) Williams, J., Krömer, G., and Gilchrist, A.: 1977, *Further studies of the impact of waste heat release on simulated global climate: Part I. RM-77-15,* International Institute for Applied Systems Analysis, Laxenburg, Austria, 22 pp.

(7) Williams, J., Krömer, G., and Gilchrist, A.: 1977, *Further studies of the impact of waste heat release on simulated global climate: Part II. RM-77-34,* International Institute for Applied Systems Analysis, Laxenburg, Austria, 33 pp.

(8) Williams, J., Krömer, G., and Gilchrist, A.: 1979, J. Appl. Meteor. 18, pp. 1501-1511.

(9) Corby, G.A., Gilchrist, A., and Newson, R.L.: 1972, Quart. J. Roy. Meteor. Soc. 98, pp. 809-832.

(10) Washington, W.M., and Chervin, R.M.: 1979, J. Appl. Meteor. 18, pp. 3-18.

(11) Sundquist, J.L.: 1975, *Dispersing Population: What America Can Learn from Europe,* Brookings Institution, Washington, D.C., pp. 24-32.

(12) Kasahara, A., and Washington, W.M.: 1971, J. Atmos. Sci. 28, pp. 657-701.

(13) Washington, W.M., Otto-Bliesner, B., and Williamson, G.: 1977, *January and July Simulation Experiments with the 2.5° Latitude-Longitude Version of the NCAR General Circulation Model.* NCAR/TN-123+STR, National Center for Atmospheric Research, Boulder, CO, Vol. 1, Text, 39 pp.; Vol. 2, Figures, 61 pp.

(14) Washington, W.M., and Williamson, D.L.: 1977, in *Methods in Computational Physics,* Vol. 17, *General Circulation Models of the Atmosphere,* J. Chang, Ed., Academic Press, pp. 111-172.

(15) Kuo, H.L.: 1965, J. Atmos. Sci. 22, pp. 40-63.

(16) Krishnamurti, T.N., and Moxim, W.J.: 1971, J. Appl. Meteor. 10, pp. 3-13.

(17) Chervin, R.M., Gates, W.L., and Schneider, S.H.: 1974, J. Atmos. Sci. 31, pp. 2216-2219.

(18) Chervin, R.M., and Schneider, S.H.: 1976, J. Atmos. Sci. 33, pp. 391-404.

(19) Chervin, R.M., Washington, W.M., and Schneider, S.H.: 1976, J. Atmos. Sci. 33, pp. 413-423.

(20) Chervin, R.M., and Schneider, S.H.: 1976, J. Atmos. Sci. 33, pp. 405-412.

(21) Schneider, S.H., and Dennett, R.D.: 1975, Ambio 4, pp. 65-74.

DISCUSSION

Laurmann:
A few years ago the possibility of induction of major convective instabilities, or even tornados, by large scale power parks was of concern. Could you tell us the present status of this problem?

Fortak:
Under light wind conditions almost all cooling tower plumes penetrate through inversion layers within the planetary boundary layer or above it. Convective instability has been found over large power stations even when the surroundings showed slightly stable stratification. Even above industrialized cities, convection is enhanced leading to a thermal stratification that differs from that of the surroundings in that it is more unstable. Systematic investigations of this kind will be continued in the upper Rhine VAlley during the next three years.

Rotty:
Professor Oke, in describing the effects of urbanization, used the parameter of energy flux density. Have you been able to make any quantitative correlations of vertical velocity with the energy flux density above cooling towers?

Fortak:
The fairly simple model presented in the paper leads to the relationship

$$S = \frac{1}{\alpha l_o} \frac{\bar{p}_o c_p}{Rg} \bar{w}^3$$

Here S is energy flux density (power/area) and $\bar{w}$ is the characteristic vertical velocity in the spatial center of the vertical plume. Further, p_o is pressure at the characteristic height, c_p is specific heat, R is the gas constant, g is the gravity constant, and l_o is the vertical extent of the plume. Measurements indicate that this relationship is true for more or less all of the heat emissions from "point sources". In the case of area sources, heat transport is carried out by many individual convective cells. In that case it is difficult to define characteristic properties of the integrated thermal plume.

Flohn:
Have you ever observed, in the Ruhr district, a merger of the convective system on a scale of, say 5-10000 km^2? I know of such

W. Bach, J. Pankrath, and J. Williams (eds.), Interactions of Energy and Climate, 416–419.

(natural!) systems only in the Tropics, at Lake Victoria (70,000 km^2) and above Borneo (750,000 km^2) where cumulonimbus systems regularly develop due to diurnal land-sea circulations.

Fortak:
A merger of convective systems has not been observed in the Ruhr district so far. Instead, normal convection with a spacing of about 1 to 3 km is enhanced in that area. Only the normal scale development in the vertical is observed in all cases of measurements, i.e. some smaller cells starting in the lower boundary layer merge in the upper part, building up larger cells.

Oke:
1. Does your anthropogenic heat flux include a latent as well as a sensible heat component? The addition of water vapour has been shown to add a considerable buoyancy flux (e.g. Bornstein and Tam).
2. I am concerned about the ability of such models (which do not have very full boundary layer characteristics) to realistically simulate surface temperatures. Even models constructed specifically for the boundary layer can produce either warming or cooling with an increase in energy input because of the intricate partitioning of latent and sensible heat (e.g. excessive evaporative cooling can suppress the sensible heat flux). Therefore I have doubts about a model in which the boundary layer processes are handled relatively simply.

Chervin:
1. As a very crude first approximation, the waste heat emission was included in the model simply as a constant heat source at the ground surface. No energy-related water vapor flux was added at the surface. However, the evaporation rate in the model did change in response to the inserted thermal pollution heat source consistent with the surface temperature change. This "secondary response" turned out to be a key factor in the sign of the soil moisture response.
2. While the boundary layer representation in the model was relatively crude, the surface energy balance relationship was was more complete. I agree that any possible ground surface temperature change is a consequence of the relative magnitudes of the various changes in the individual components of the surface energy balance, and these components are highly dependent on the boundary layer formulation in the model. Your concerns can be allayed only by a repetition of the experiments with a model featuring a more detailed boundary layer parameterization. I feel obligated to repeat my statement in the conclusions that these experiments should in no way be considered as a definitive assessment of the climatic impact of the given regional

distribution of thermal pollution. They do, however, establish a base line against which other GCM experiments may be compared.

McKay:
Did the model allow for the existence of snow cover and melt energy? How did it manage soil moisture in frozen ground and under snow? Did it account for the control of snow and dormancy on evapotranspiration?

Chervin:
Time dependent snow cover was computed in the model from a parameterized surface hydrology. The melting and sublimation of snow or ice were included as components in the model's surface energy balance. Snow-covered soil moisture was not permitted to interact directly with the atmosphere in the model. Of course, once the snow completely melted, the resulting soil moisture would contribute to the surface evaporation and latent heat flux. The physical process of evapotranspiration through a plant canopy cover was not included explicitly in the model.

Flohn:
In spite of the need to obtain a signal large enough to overcome the model's variability, is it really realistic to consider a 450 TW-model for regional thermal pollution, about 20 times larger than the global scenarios which have been discussed here by Rotty and Sassin?

Chervin:
Of course, 90 Wm^2 of thermal pollution distributed uniformly over an area of approximately 5 million km^2 is not realistic and was not intended as a prediction of a future situation. Performing climate sensitivity experiments with such extreme prescribed changes is a standard experimental design procedure to obtain a statistically significant response with limited sampled climate ensembles and to identify more easily the key physical processes responsible for the model response. Then, armed with the information and experience gained from a "super anomaly" experiment, more realistic experiments can be performed with a greater chance of recovering a significant response.

Page:
There is an interaction between climate and human activity. If the temperature in winter rises 12°C, then heating demand will fall considerably. If the summer temperature rises by 3°C, the air conditioning load will rise drastically. In one case the feedback is negative, and the other positive. What is the actual monthly pattern of demand in Manhattan at the moment? The responses seem

proportional to the ratio 90 W/m^2 divided by mean monthly surface irradiation.

Chervin:
I agree that a feedback would exist from a climate change, like the temperature changes produced in these model experiments, back to human activity. However, "open loop" experiments serve to isolate the impact of human activity on climate and to identify key physical processes. Results from "closed loop" experiments, while possibly more realistic, would be more difficult to analyse and are beyond the scope of this investigation. I do not have any information on the present seasonal variation in the energy demand of Manhattan Island. The experiments reported here examined the seasonal dependence of the GCM's sensitivity to the same prescribed (constant in time) change in surface forcing. No attempt was made at an actual prediction of climate change in response to time-varying waste heat emission. The temperature response may be proportional to the suggested ratio of surface energy balance components but such a relationship would definitely not hold for the precipitation or soil moisture response.

PART VI

RISK – BENEFIT AND COST-BENEFIT ASSESSMENT

COMPARISON OF RISKS AND BENEFITS AMONG DIFFERENT ENERGY SYSTEMS

S.C. Black and F. Niehaus

Joint IAEA/IIASA Risk Assessment Project, International Atomic Energy Agency, P.O. Box 100, A-1400 Vienna.

ABSTRACT. The public and occupational health effects of five centralized electricity generating systems are compared assuming each type supplies one GWa(e) of energy. Results show large uncertainties in estimates especially with regard to coal. The health effects from nuclear power are comparatively well understood. Quantification of effects in monetary units, as required by a cost/benefit analysis, indicates that these external costs are in the range of 1% or less for all energy systems considered. It is suggested that cost-effectiveness analysis (marginal cost/benefit analysis) provides a better tool for decisions on risk reduction measures. A more complex measure of risk is needed which includes additional parameters, particularly the uncertainty of the risk estimates.

1. INTRODUCTION

Decisions on the future mix of the available energy sources should consider information on their risks and benefits. With increasing size of technologies and their implementation on a truly global scale, particular attention is being paid to their side-effects. However, this new order of magnitude has led to the fact that the structure of consequences becomes more complex and less accessible to intuitive holistic judgments. Thus, there arose a need for methodologies which, while using all available information for an assessment, maintained transparency in each step. For this purpose, it is important to keep two types of data separate:

a. the technical data on the physical variables;
b. evaluations with regard to the planning objectives.

W. Bach, J. Pankrath, and J. Williams (eds.), Interactions of Energy and Climate, 421–436.

The technical data which describe risks from the use of a technology, both for planned operation and unplanned occurrences, are:

- events (e.g., emissions, accident sequences, wind direction, etc.) and their probability of occurrence;
- consequences of these events (e.g., health effects, property damage, etc.);
- the distribution of consequences within the population affected;
- and uncertainties in these estimates.

These four categories of technical data establish the "objective" part of any assessment of risks, although they will often have a certain degree of subjectivity introduced by judgments of the technical expert. However, the combination rule for aggregation of these data is not specified by any natural law. Therefore, there is no such thing as an objective unified measure of risk. Any aggregation of these data is part of an evaluation process and thereby subjective.

Two categories of methods may be utilized to obtain these technical data:

The first category applies to *frequent events* of well established technologies which may be identified and quantified by analyzing historical data with the help of statistical methods. The extreme case of course is the continuous release of pollutants from normal operation. The crucial issue for the significance of such results is the availability of a reliable data source. Problems arise through the definition of test population versus control population, the sampling procedure, background noise, correlation versus causal relation and competing risks. In spite of these problems, statistical methods are often thought to lead to the most reliable results because they are based on actual experience. In reality, however, these problems are often insurmountable. This is demonstrated by the fact, as will be seen later, that the highest uncertainties in analyzing risk of energy systems are in the coal system, which has the longest history of the commercial energy carriers.

In the past it was possible to base the development of a technology on trial and error, which is equivalent to hypothesis and experiment in science. With the implementation of technical systems on a truly global scale, man entered the domain of "hypotheticality" (1), where it is no longer acceptable (as applied to the complete system) to correct a hypothesis by the outcome of an experiment. This applies to accidents in large modern technical installations such as LNG terminals or nuclear power plants, as well as to routine emissions from well established technologies such as CO_2 emissions from fossil fuel power plants. Therefore, a second category of methods has to be used for analysis of *low frequency risks* (2). These are predictive in principle and usually rely on an extrapolation of statistical information. Such methods include fault and event trees and computer simulation models which use failure rates of the components of a facility, such as pipes, valves, vessels, etc. Information on the reliability of system elements is usually available because they are in common use. Performance of special equipment has to be extrapolated from this data source. The behavior of the total system is then synthesized. Since a large number of event paths are possible, those which contribute significantly to the risk have to be identified and are estimated with regard to probability and consequence distributions. Generally such results follow smoothed curves of decreasing probabilities of occurrence with increasing magnitude of consequences. It has been suggested that this total density function is important in describing risks (3, 4). However, these methods have certain shortcomings: One limitation of such analyses is that only those failures and event sequences can be included which experts can imagine. In this context, especially, common mode failures pose severe problems. Another limitation is the fact that human failures, especially under stress, are difficult to quantify (3). Most such studies are therefore essentially based on extrapolation techniques requiring expert judgment. Since the probability of many event sequences is as low as 10^{-6} per year, there will never be a base of historical data on which to decide whether results of such an analysis are correct or not.

2. TECHNICAL APPROACHES TO RISK EVALUATION

Three basic methods are presently used to decide on acceptable levels of risk. Each of these methods has its advantages and shortcomings, and will be discussed in the following.

2.1. Putting Risk into Context

Constant danger of morbidity and mortality is an attribute of life. The risk may be caused by normal biological processes (ageing), illness, hazard of the natural environment (earthquake), or by human activities such as driving a car. When new technologies are to be introduced it seems reasonable as a first rough approach to compare these new risks to those which are already generally accepted.

Thus, this approach tries to establish routine criteria based on historical data disregarding any information about the benefit to society. Results of such an analysis can be found in the literature (e.g., 3, 5, 6).

For modern large-scale technologies much concern has been expressed about the potential for large accidents. Therefore, Figure 1 compares some natural and man-made risks by means of a diagram, plotting frequency versus magnitude distributions. Data for natural hazards like earthquakes and hurricanes, and for man-made hazards like explosions, air crashes, or chlorine releases, are based on statistical analysis of past experience. The curves for nuclear power plants in the USA and for the petro-chemical facilities installed on Canvey Island, UK, were estimated in (3, 7) using event- and fault-tree analysis.

Figure 1 shows that the risk from Canvey Island is about two orders of magnitude lower than many other risks for accidents causing less than ten deaths, whereas it reaches the range of these risks for large accidents. The risk of 100 NPP located in the USA at the currently used sites is estimated to be two to four orders of magnitude lower than the risk posed by Canvey Island. However, this immediately demonstrates one of the problems in comparing this type of risk data. Clearly, the risk depends on the number of facilities so that a common basis for comparison has to be found.

2.2. Comparison of Risk and Benefit

Once it has been established that a risk is not out of proportion with other risks, a common basis for a comparison of technologies can be achieved by normalizing their risks to a common measure of benefits which they provide to society. For the following comparison of electricity production systems, the benefits of producing a unit of electricity has been chosen. Additional benefits are then given by differing costs of electricity production and some attributes which are difficult to quantify such as security of supply, diversification, etc. The latter are not considered here. Those costs accounted for by the utility are known as internal costs and are reflected in the price for a kWh of electricity (8).

Some recent studies outlined the problems of comparing very different technologies (i.e., centralized versus decentralized systems, transportation systems versus electricity production, etc.). This paper concentrates on a comparison of electricity production by centralized systems. The following technologies have been considered:

Coal: The analysis is based on data from the US where, in 1975, 55% of the coal was produced in surface mines and 45% by deep mining. On the average, the coal is transported for 500 km, 65% by rail, 21% by truck and 13% by barge. The 2 x 700 MWe plant is assumed to operate on a load factor of 71% with 38% efficiency. Flue gas desulfurisation is considered to be equivalent to a 0.5%

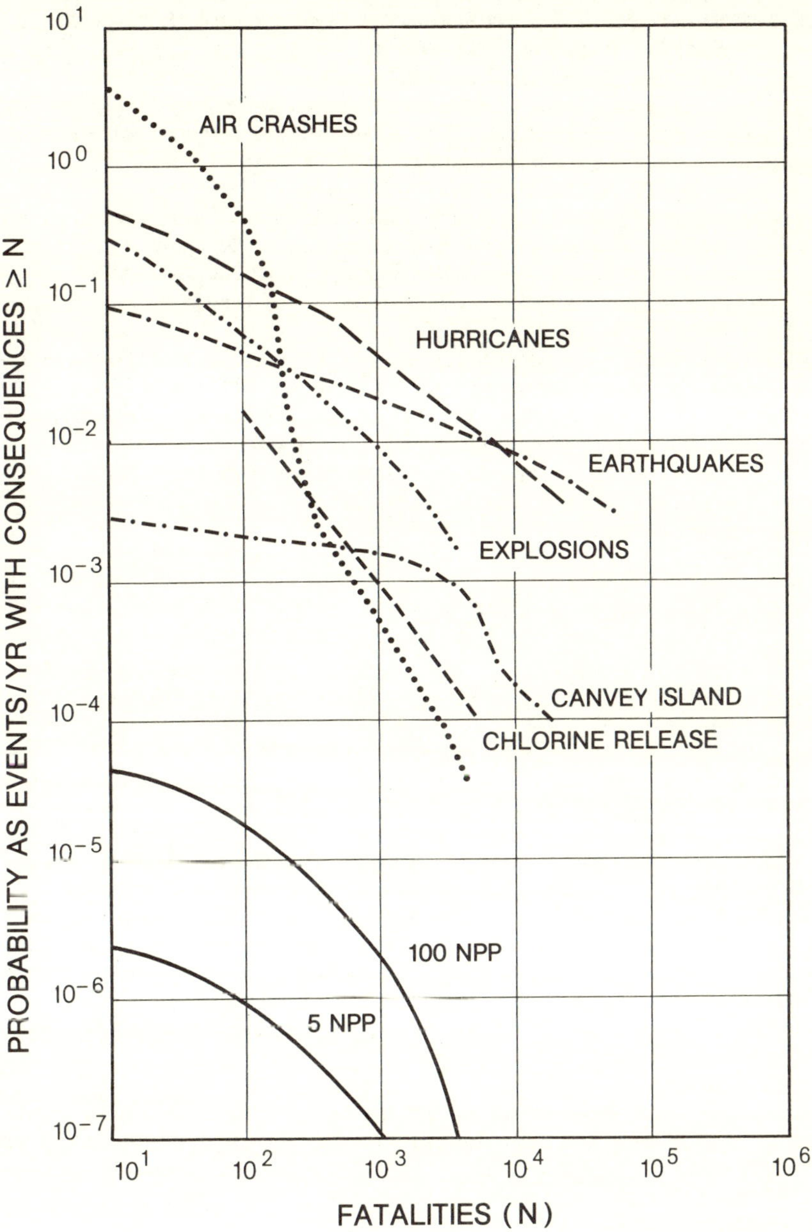

FIGURE 1: Frequency as events per year causing fatalities ≥ N for some man-caused and natural events together with estimates for 5 petro-chemical plants on Canvey Island, for 100 Nuclear Power Plants, and for 5 NPP.

sulfur content in coal. Public health effects are based on a stack height of 305 m and consider an average population of $2.2 \cdot 10^6$ people within a distance of 80 km.

Oil: A plant of similar size and operating parameters, fuelled by residual oil, is used. The sulfur content is assumed to be 0.2%. The fuel is transported by tanker.

Gas: The SO_2 emissions are assumed to be 24 t/GWa(e). This may be compared to the 31,000 t/GWa(e) estimated for coal with 0.5% sulfur. The gas is transported by pipeline and data on pipeline accidents have been allocated accordingly.

LWR: Results are based on a 1350 MW(e) light water reactor operating at a load factor of 0.74 assuming 33% efficiency. The fuel cycle considers reprocessing. Health effects of storage of nuclear waste were assumed to be close to zero and were not considered. No credit was given for the recovered plutonium.

STEC: For solar thermal electric conversion systems, 100 MW(e) (peak) plants are considered with 6 h rock and oil storage operating at a 54% capacity factor. Thus, 18.5 plants are needed to supply 1 GWa(e)/year. No back-up systems have been considered. Based on most favourable conditions in the South-Western desert of the US with an average annual insolation of 8 kWh/m^2 x day, each 100 MW(e) plant would consist of 28,600 heliostats with 30.4 m^2 surface. In total, each plant would cover 3.5 km^2 of land. Extra costs have been assumed for dry cooling towers.

All plants are assumed to have a lifetime of 30 years.

Based on (9 - 28) Table 1 summarizes the results of estimated health effects for the above described energy systems. Data have been normalized for the production of 1 GWa(e). The Table also gives the range of estimates which can be found in the literature and, for occupational effects, the results from (9) for the FRG. These different sets of data have been given to emphasize the uncertainties involved in these estimates and to point to national differences.

Results have been kept separate for occupational and public health effects and for accidental death, accidental injuries and fatal diseases. Diseases which have led to temporary or permanent disabilities were not considered since the data base is insufficient.

Results are listed for the various steps of the fuel cycle. Transportation includes both the transport of fuel and the transport of materials for the construction of the plant and the supporting equipment of the fuel cycle. Construction includes

TABLE 1: Effects from accidents and diseases from supply of 1 GWa(e) (8.76×10^9 kWh) (9 - 28)

OCCUPATIONAL	Accident Injuries (in MDL)			Accident Deaths			Fatal Diseases		
	This Study	Literature	FRG	This Study	Literature	FRG	This Study	Literature	FRG
COAL									
Fuel Supply	3020-3090	1490-8900	5410	0.66 -0.73	0.45 -2.71	3.66	0.097	0.008-6.4	2.77
Transport	710-880	19-4460	1550	0.351-0.381	0.073-3.07	2.89			
Normal Operation	213	150-2400	81	0.016	0.013-0.46	0.04			
Construction	240	-	92	0.055	-	0.1			
TOTALS	4200-4400	1660-14000	7130	1.08 -1.18	0.54 -6.2	6.7	0.10	0.008-6.4	2.8
OIL									
Fuel Supply	2640	430-6500	41	0.387	0.027-1.7	0.04			
Transport	470	53-750	936	0.068	0.04 -1.8	1.8			0.026
Normal Operation	130	50-171	59	0.013	0.013-0.052	0.04			
Construction	230	-	76	0.054	-	0.07			
TOTALS	3500	530-7520	1110	0.52	0.08 -3.6	2.0			0.03
GAS									
Fuel Supply	1770	200-3100	41	0.22	0.036-0.4	0.04			
Transport	100	44-193	936	0.02	0.026-0.034	1.80			
Normal Operation	140	68-210	59	0.012	0.012-0.053	0.04			
Construction	210	-	76	0.049	-	0.07			
TOTALS	2200	310-3500	1110	0.30	0.07 -0.49	2.0			
LWR									
Fuel & Reprocess	190	180-1130	71	0.094	0.071-0.57	0.091	0.28-0.3	0.003-0.67	0.187
Transport	7	3-47	1	0.002	0.0026-0.015	0.003	0.0006	0.0008-0.115	0.0016
Normal Operation	140	44-140	8	0.014	0.013-0.02	0.012	0.13-0.18	0.024-0.14	
Construction	235	-	190	0.055	-	0.12			
TOTALS	570	230-1370	190	0.17	0.087-0.60	0.23	0.42-0.49	0.028-0.92	0.19
SOLAR THERMAL									
Material Supply	290	900-1200		0.055	0.18				
Transport Mat.	82	630-2380		0.021	0.77 -2.52				
Construct Plant	4620	5990-7780		1.07	1.91				
Construct Storage	30	130-8300		0.012	0.038-0.91				
Normal Operation	3450	2660		0.96	0.92				
TOTALS	8500	10330-22300		2.12	3.8 -6.4				

TABLE 1: Continued

PUBLIC	Accident Injuries (in MDL) This Study	Accident Injuries (in MDL) Literature	Accident Injuries (in MDL) FRG	Accident Deaths This Study	Accident Deaths Literature	Accident Deaths FRG	Fatal Diseases This Study	Fatal Diseases Literature	Fatal Diseases FRG
COAL									
Fuel Supply								1.3 -14	
Transport	610-740	60-765		0.57 -1.38	0.15 -1.9				
Normal Operation							3.2 -22	0.1 -140	
Construction							0.008 -0.059		
TOTALS	610-740	60-765		0.57 -1.38	0.15 -1.9		3.2 -22	1.4 -150	
OIL									
Fuel Supply									
Transport	2			0.0008-0.0019					
Normal Operation							1.1 -7.5	0.8 -140	
Construction							0.005 -0.034		
TOTALS	2			0.0008-0.0019			1.1 -7.5	0.8 -140	
GAS									
Fuel Supply	1770			0.15					
Transport	1000			0.20					
Normal Operation							0.0025-0.017		
Construction							0.0028-0.019		
TOTALS	2800			0.35			0.005 -0.036		
LWR									
Fuel & Reprocess							0.073	0.001 -0.13	
Transport	6	8		0.003 -0.006	0.006-0.012		0.0006	0.0002 -0.0012	
Normal Operation							0.19	0.013 -0.23	
Construction							0.008 -0.057		
TOTALS	6	8		0.003 -0.006	0.006-0.012		0.27 -0.32	0.014 -0.36	
SOLAR THERMAL									
Material Supply							0.07 -0.47	0.02 -0.03	
Transport Mat.	77	112		0.03 -0.08	0.6 -1.4				
Construct Plant									
Construct Storage									
Normal Operation									
TOTALS	77	112		0.03 -0.08	0.6 -1.4		0.07 -0.47	0.02 -0.03	

provision of materials and equipment for the plant and for the adequate share of the fuel cycle.

The data in Table 1 represent the current status of ongoing investigations. The following remarks are therefore meant to stress the uncertainties of the estimates and to point to the limitations of this paper.

Coal. With regard to earlier studies (15,19) data have been significantly reduced for occupational fatal diseases as a result of new data available after the US Health and Safety Act of 1969. The given estimate represents optimistic extrapolation of the available data to modern large coal mines. Estimates of public fatal diseases are based on two epidemiological studies. The lower boundary uses the median value from Lave and the upper boundary is based on the lower (linear) results from Winkelstein (15). It should be noted that Hamilton (10) gives a range from 36 to 240 public fatal diseases based on 100 currently used sites in the US and integrating over all distances (not only up to 80 km).

Data on emissions of trace elements have not been incorporated since the respective dose-effect relationships are not yet understood. The effects from emissions of benz(a)pyrene (about 6% increase of lung cancer per 1 ng/m^3 (29)) have not been included.

Radioactive emissions from coal-fired plants heavily depend on the coal used and have been estimated to be in the same order of magnitude as those from pressurized water reactors. Thus, they are negligible compared to the variations in natural background radiation exposure and were thus not included in Table 1.

The risks of climatic change caused by emissions of CO_2 from combustion of fossil fuels are not yet quantifiable and have not been incorporated in Table 1.

Oil. A recent study of Norwegian off-shore oil drilling activities in the North Sea indicates a total of 82 occupational deaths for 28,700 to 49,200 man-years of labor during the time period from 1966 to 1978. Preliminary allocation of the various activities confirms the upper bounds given in Table 1.

Gas. The occupational risks of gas supply from the North Sea would be considerably higher (about a factor of five) than listed in the Table. The low frequency/high consequence risk of destroying large fixed off-shore platforms (up to 200 people) has not been included. Also the risk of destroying population centers by gas clouds from LNG tanks (7) has not been considered.

LWR. Effects are based on a risk factor of 2 x 10^{-4}/man-rem which includes cancers plus genetic effects for all generations (30). Data for fuel supply and reprocessing include 0.004 to 0.018 lung cancers in miners based on experience (31). Low-frequency hypothetical reactor accidents (3) are not included.

Solar. Data are assumed to be optimistic. New calculations using Input/Output methodology indicate generally higher risks for construction of all energy systems (32). This analysis is not yet completed.

The data from Table 1 show a definite ranking for various effects. However, the structure of risk posed by these technologies is very dissimilar. Therefore, a first step of evaluation aggregates injuries and fatalities by assuming that one death is equivalent to 6,000 lost man-days. The results of this aggregation, which contains a value judgment, are displayed in Figure 2. As mentioned above, only fatal occupational diseases have been included. It should be noted that permanent disabilities among coal miners are about a factor of ten higher than the number of deaths listed in Table 1.

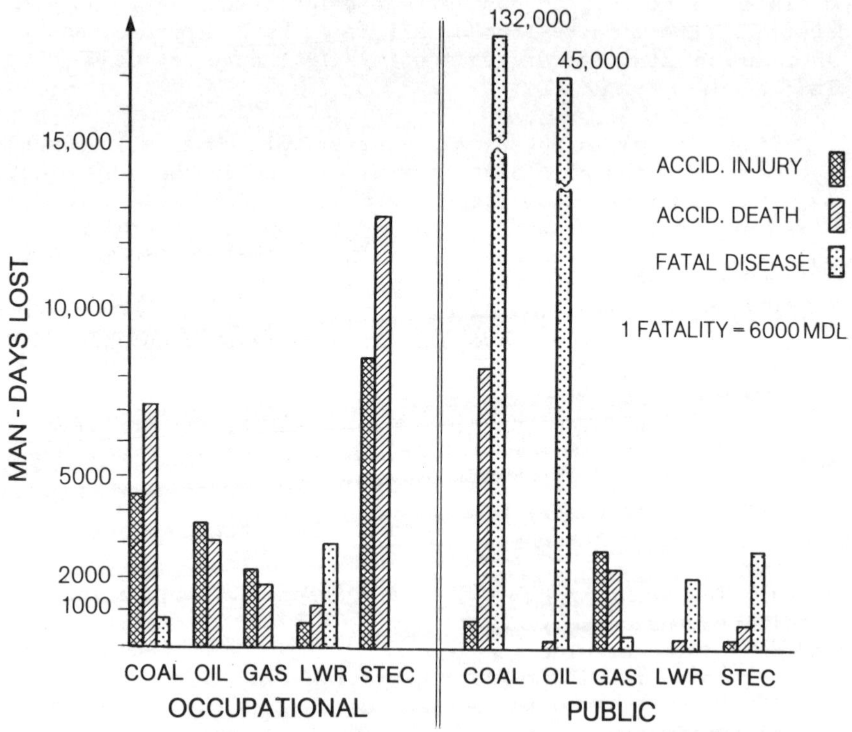

FIGURE 2: Occupational and public man-days lost per GWa(e) for each energy system

The final step in a cost/benefit analysis requires that health effects and benefits be quantified in common units, usually monetary terms. This poses the difficult question of assigning a monetary value to a human life and to work-days lost. For the calculations given in Table 2, a value of $ 300,000 has been chosen for a life and $ 100 per work-day lost. It should be noted that such values somewhat alter the relations as were given in Figure 2. The equivalence of 6,000 man-days per fatal effect would lead to a value of $ 50 per man-day. A factor of two has been introduced because of expenses for medical treatment (33). The limitations of such an evaluation based on the human capital approach have been outlined in (34).

TABLE 2: Costs of electricity production (million US $ in 1977/GWa)

	Coal	Oil	Gas	LWR	STEC
Capital	158	132	132	228	580-710
O & M	17.5	8.8	8.8	17.5	98
Fuel	114	228	199	61.3	50 (1)
Abatement	79	(2)	(2)	1.9 (3)	140-175 (4)
Internal Costs	368	369	340	309	865-1030
Occupational Effects	0.77-0.82	0.51	0.31	0.23-0.26	1.49
Public Effects	1.2 - 7.1	0.33-2.25	0.39-0.4	0.08-0.1	0.04-0.17
Total health costs	2.0 - 7.9	0.8 -2.8	0.7	0.31-0.36	1.5 - 1.7
% of internal cost	0.5 - 2.1	0.2 - 0.8	0.2	0.1 - 0.12*	0.17

(1) Cost of electricity transmission, 2000 km to load center.
(2) Oil and gas desulfurization at refinery, natural gas price still controlled in 1977.
(3) Near-zero Rad-waste system, LWR includes decommissioning waste disposal.
(4) Extra cost for dry cooling towers required in desert areas.

* Major accidents would add $ 40,000 or 0.01% to this.

Table 2 shows the high degree of internalization of environmental costs (abatement costs) (8). Health effects as quantified by this procedure are generally in the range of about one percent or less of internal costs. From this table it is obvious that a cost/benefit comparison is dominated by internal cost. The quantified health effects are less than the uncertainties of estimates of capital or fuel costs.

2.3. Cost-Effectiveness Analysis (Marginal Cost/Benefit Analysis)

Comparing risks and benefits as described above allows for a comparison of technical options in terms of their aggregated effects on society. However, the technology may be made safer at higher economic costs, introducing the concept of "how safe is safe enough". Recognizing the limited resources of society which may be allocated for safety, it seems to be reasonable to distribute them for risk reduction to the most cost-effective alternatives. Such a procedure does not lead to an absolute limit for an acceptable risk but to standard values for the marginal cost of risk reduction. In the nuclear field, an example is given by assigning a monetary value to the reduction of one man-rem (e.g., \$ 1,000/man-rem reduction) (35). Marginal risk reduction costs range from less than \$ 30,000 per life saved to more than \$ 10^9. It appears that low probability/high consequence accidents are associated with higher marginal costs per life. The quoted value of \$ 1,000 per man-rem reduction implies assigning a value of about \$ 5 million to a human life.

Generally, it has been assumed that any standard value to be chosen for cost-effectiveness studies would represent an arbitrary trade-off between the two objectives of achieving highest safety and minimizing economic costs, given the risk could be reduced below any given value with sufficient expenditures. However, it has to be recognized that the production of safety equipment itself is associated with occupational and public risk. Details of such an analysis were described in (32). The conclusion was drawn that the production of \$ 30 million of safety equipment would result in one equivalent fatality.

In addition, about 1,400 man-years of labor requirements would be needed to produce the equipment. It is suggested that such information on marginal cost of risk reduction be used for decisions on standard values.

3. CONCLUSIONS

Comparing the risks of electricity-producing systems, assuming their benefit is to supply 1 GWa(e), indicates that combined public and occupational effects, quantified in monetary units, represent only a percent or so of the total costs. It would be difficult to decide, empirically, how much should be spent to continue reducing risks. It is suggested that a cost-effectiveness analysis may be useful because the health-effects costs of safety equipment can be calculated and used to balance the health-effects prevented.

Either of the above methods can provide only a portion of the data necessary for decision-makers. Other factors, such as security of supply, strength of the economy, the industrial base, etc., need to be considered but have not been addressed herein.

Remarkable are the large uncertainties involved in some of the estimates. Also several effects have not yet been included since an adequate data base is not available. These considerations certainly tend to increase the range of uncertainties as listed in this paper. Thus, it seems to be necessary to develop a more complex measure of risk which would include these uncertainties. A review of suggested methodologies (most of them based on multi-attribute utility theory) has been given in (36).

REFERENCES

(1) Häfele, W.: 1975, Hypotheticality and the New Challenges: The Pathfinder Role of Nuclear Energy. *Anticipation*, 20, pp. 15-23.

(2) Gumbel, E.J.: 1938, *Statistics of Extremes*. Columbia University Press, New York.

(3) U.S. Nuclear Regulatory Commission: 1975, *Reactor Safety Study: An Assessment of Accident Risks in U.S. Commerical Nuclear Power Plants*. WASH-1400 (NUREG-75/014), Washington, D.C.

(4) Farmer, F.R.: 1967, Siting Criteria - A New Approach. *Proc. of a Symposium on Containment and Siting of Nuclear Power Plants*. IAEA-SM-89/34, International Atomic Energy Agency, Vienna, Austria.

(5) Pochin, E.: 1977, *Colloquium on the Pyscho-Sociological Implications of the Developments of the Nuclear Industry*. Edited by Societe Française de Radioprotection, Paris.

(6) Wilson, R.: 1980, *Hazards of Daily Life*. Presented at the Colloquium about Different Energies, organized by the Societe Française d'Energie Nucleaire. 24-26 January, 1980, Paris, France. To be published in proceedings.

(7) Canvey Island Study: 1978, *Canvey: Summary of an Investigation of Potential Hazards from Operations in the Canvey Island/Thurrock Area*. ISBN 0 11 883203. Health and Safety Executive, London, UK.

(8) Rossin, A.D., and Rieck, T.A.: 1978, Economics of Nuclear Power. *Science*, 201, pp. 582-589.

(9) Battelle Institute: 1976, *Comparison of Health Effects of Various Electricity Supply Technologies and First Attempt at Classifying Risks of Nuclear Energy*. Progress Report 200/1. Frankfurt/Main: Battelle Institute.

(10) Hamilton, L.D.: 1977, Alternative Sources and Health. In: *CRC Forum on Energy*. Report BNL-24852, Brookhaven National Laboratory, N.Y.

(11) Morris, S.C., Moskowitz, P.D., Sevian, W.A., Silberstein, S., and Hamilton, L.D.: 1979, *Coal Conversion Technologies: Analysis of Some Health and Environmental Effects*. Brookhaven National Laboratory, N.Y.

(12) Hamilton, L.D.: 1979, *Areas of Uncertainty in Estimates of Health Risks*. Paper presented at the Symposium on Energy and Human Health, Pittsburgh, Penn., February, 1979.

(13) Lave, L.B.: 1976, *Health Effects of Electricity Generation from Coal, Oil, and Nuclear Fuel*. In: H. Ashley, R.L. Rudman, and C. Whipple (eds.), *Energy and the Environment: A Risk Benefit Approach*. Pergamon Press, New York.

(14) Barrager, S.M., Judd, B.R., and North, D.W.: 1976, *Economic and Social Costs of Coal and Nuclear Electric Generation*. Decision Analysis Department, SRI.

(15) Hamilton, L.D., and Manne, A.S.: 1977, *Health and Economic Costs of Alternative Energy Sources*. Presented at the International Conference on Nuclear Power and Its Fuel Cycle; International Atomic Energy Agency, Salzburg, Austria, May, 1977.

(16) Lave, L.B., and Freeburg, L.C.: 1973, Health Effects of Electricity Generation from Coal, Oil, and Nuclear Fuel. *Nuclear Safety*, 14, pp. 409-428.

(17) U.S. Nuclear Regulatory Commission: 1974, *Comparative Risk-Cost-Benefit Study of Alternative Sources of Electrical Energy*. WASH-1224, Washington, D.C.

(18) Comar, C.L., and Sagan, L.A.: 1976, Health Effects of Energy Production and Conversion. *Annual Rev. Energy*, 1, pp. 580-600.

(19) U.S. Environmental Protection Agency: 1977, *Accidents and Unscheduled Events Associated with Non-Nuclear Energy Resources and Technology*. Report EPA-600/7-77-016, Washington, D.C.

(20) Redfern, J.: 1978, Hazards of Non-Nuclear Energy. *Nature*, 272, pp. 487.

(21) Inhaber, H.: 1978, *Risk of Energy*. Report AECB-1119/Rev.2, Atomic Energy Control Board, Canada.

(22) Morris, S.C., Novak, K.M., and Hamilton, L.D.: 1979, *Health Effects of Coal in the National Energy Plan*. Brookhaven National Laboratory, N.Y.

(23) Hamilton, L.D.: 1979, *Assessing the Health and Environmental Costs of Energy Production and Use*. International Atomic Energy Agency Seminar on Comparative Health Impacts of Nuclear and Alternative Sources of Energy. Trieste, Italy, 7-11 May, 1979.

(24) United Nations Environmental Program: 1979, *Impacts of Production and Use of Energy, I: Fossil Fuels*. UNEP Report ERS-1-79.

(25) Caputo, R.: 1977, Comparative Assessment of Orbital and Terrestrial Central Power Plants. *Proc. of the 12th IECEC.*

(26) Myers, D.K., and Newcombe, H.B.: 1979, *Nuclear Power and Low-Level Radiation Hazards*. AECL-6432, Toronto, Canada.

(27) Holdren, J.P., Anderson, K., Gleick, P.H., Mintzer, I., Morris, G., and Smith, K.R.: 1979, *Risk of Renewable Energy Sources: A Critique of the Inhaber Report*. Report # ERG-79-3, University of California, Berkley.

(28) Hamilton, L.D., and Morris, S.C.: 1974, *Health Effects of Fossil Fuel Power Plants*. 8th Midyear Topical Symposium, Health Physics Society, CONF-74/0/8, NTIS, USA.

(29) Vohra, K.G.: 1977, *A Perspective on the Radiation Protection Problem and Risk Analysis for the Nuclear Era*. Presented at the International Conference on Nuclear Power and Its Fuel Cycle; International Atomic Energy Agency, Salzburg, Austria, May, 1977.

(30) ICRP 26: 1977, *Recommendation of the International Commission on Radiological Protection*.

(31) UNSCEAR: 1977, *Sources and Effects of Ionizing Radiation*. Report to the General Assembly.

(32) Black, S.C., and Niehaus, F.: 1980, How Safe is "Too" Safe? To be published in the *International Atomic Energy Agency Bulletin* (in press).

(33) Sagan, L.: 1976, *Public Health Aspects of Energy Systems*. In: H. Ashley, R.L. Rudman, and C. Whipple (eds.), *Energy and the Environment*. Pergamon Press, New York.

(34) Linnerooth, J.: 1979, The Value of Human Life: A Review of the Models. *Economic Inquiry*, 17, pp. 52-74.

(35) U.S. Nuclear Regulatory Commission: 1975, *Radioactive Material in Light-Waster-Cooled Nuclear Power Reactor Effluents. Proposed Regulations to Title 10, Part 50*. Vol. 40, No. 87, Washington, D.C.

(36) Schaefer, R.E.: 1978, *What Are We Talking About When We Talk About "Risk"? A Critical Survey of Risk and Risk Preference Theories*. IIASA RM-78-69, International Institute for Applied Systems Analysis, Laxenburg, Austria.

ASSESSING THE IMPORTANCE OF CO_2-INDUCED CLIMATIC CHANGES USING RISK-BENEFIT ANALYSIS

J.A. LAURMANN

Stanford University, California

ABSTRACT. A decision analytic approach is described for estimating the economic impact of CO_2 induced climatic change, allowing for probabilistic features that arise from uncertainty in the magnitude of global climatic change. Since we have no means for assessing the error of the economic cost calculations, we have to regard the results as illustrative only, though the general impression that CO_2 doubling indeed produces major stress is confirmed. Thus, the analysis predicts a 3% per annum economic loss relative to gross world product (GWP), roughly the rate of growth of the GWP, by the year 2020 for a 3% per annum exponential growth of CO_2 in the atmosphere, with catastrophic losses occuring shortly thereafter. Some general results that we believe are independent of the fidelity of the analysis are readily apparent. Thus, we find that the question of the need for immediate action to curb fossil fuel use in order to avoid CO_2 induced disaster is a function of two primary factors: 1) the validity of the long market penetration time concept as applied to new energy sources, and 2) the nature and magnitude of social discounting of future events.

1. BACKGROUND

Anthropogenic release of carbon dioxide into the atmosphere from fossil fuel based energy use is now recognised as resulting in a potentially serious environmental threat. It is manifest in a global increase of temperature that could have important impacts within a fifty year period, though the uncertainty in this climate change projection is very large. Since the appreciation some three years ago of the importance of the problem, a substantial and growing international effort has been mounted to research the

W. Bach, J. Pankrath, and J. Williams (eds.), Interactions of Energy and Climate, 437–460.

numerous scientific questions that need to be answered. Even so, it is recognised that a decade or more will elapse before the uncertainties that now plague the climate change projections can be reduced to scientifically acceptable norms.

Were it not for two additional complications it would no doubt be satisfactory to await the outcome of this long range research program before considering the implications of the potential CO_2-induced climate change on energy use policy. Whether we can or cannot afford this luxury depends on

A) the effect of large uncertainties of the projections, resulting in a finite probability of a much earlier onset of significant effects than the average estimate gives;
B) the long time period required to transfer global energy production to a non-carbon based fuel, if the CO_2 threat is found to be sufficiently serious and remedial actions are called for.

In an earlier study (1) we considered the ramifications of the first of these problems in some detail and concluded that quantification of costs applying decision analytic methodology was a reasonable approach for assessing the seriousness of the environmental threat and its implications for present day policy. This paper represents a first attempt to analyse the CO_2 problem along such lines by evaluating climate change costs and by comparing them with some rudimentary estimates of remedial measure costs. The method can allow for the effects of uncertain knowledge, both in estimates of climatic change and in its economic impact. We have incorporated the former into our calculations, but since we have very little notion of the accuracy of the cost estimation, the latter has been omitted. This could prove to be a serious deficiency of the analysis, and hence we cannot be assured of the quantitative validity of our results; at this stage they should therefore be regarded as illustrative only.

In the past we have also discussed the second of the above major problems (2); it affects the conclusions of this study in a major way, as will be covered in section 5.

1.1. Basis for the cost calculations

Following reference (1), we have assumed a global mean temperature rise of 2.5°C for doubling of atmospheric CO_2 levels above the pre-industrial value; this is ½°C less than the value selected in the latest National Academy of Sciences review of the CO_2 problem (3). The factor of two probable error associated with this figure by the NAS review panel is the same as we had used, and results in the probabilistic estimates of global temperature rise shown in figure 1. We have assumed a 3% per annum rate of rise in atmospheric CO_2 level, which, for constant fractional

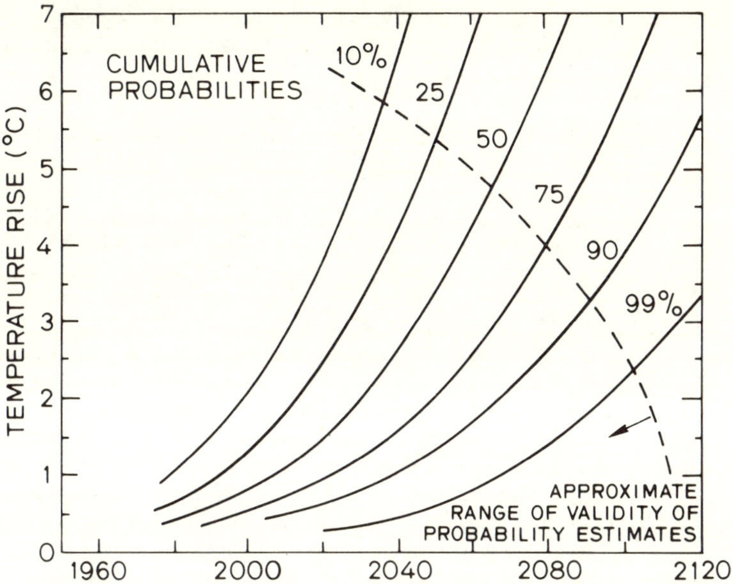

FIGURE 1. CUMULATIVE PROBABILITIES OF GLOBAL TEMPERATURE CHANGE

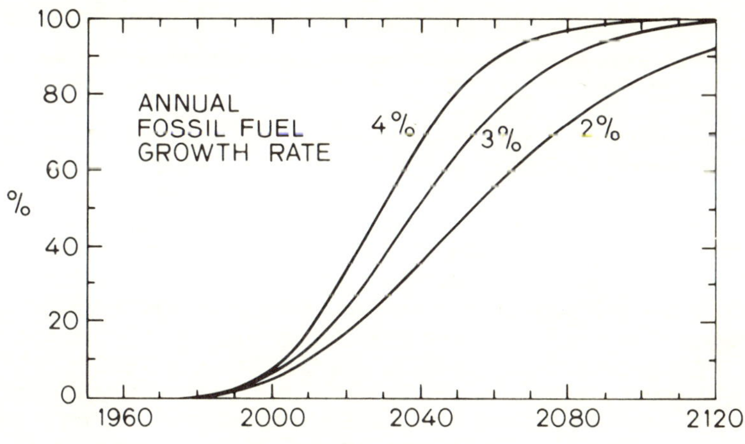

FIGURE 2. CUMULATIVE PROBABILITY OF A 2.5 °C TEMPERATURE RISE

atmospheric retention of anthropogenically released CO_2, means a 3% p.a. rise in atmospheric CO_2. A 3% annual increase in fossil fuel use over the next fifty years represents a mean for several projections made through 1978 (see (1)), though current estimates seem to be decreasing. Figure 2 illustrates the effect of changing the CO_2 release rate; reduction to 2% per annum only postpones the CO_2 doubling date by some 16 years.

The effect of carbon cycle modeling error for projections to mid 21st century we regard as minor, due essentially to their linear behavior over this time frame and their fitting to empirical data (1). However, the possibly large biospheric contribution to CO_2 release from deforestation could considerably change the magnitude of the climatic warming if deforestation rates or fuel use decrease substantially in the future. Such a possibility is not discussed in this paper; we hope that current efforts to more accurately quantify deforestation rates will shortly bear fruit.

The cost estimates we shall describe all assume a 3% p.a. rise rate in atmospheric CO_2 concentration, which we shall ascribe to a 3% p.a. rate of increase in fossil fuel burning. Note that under these conditions we predict the 50% probability for a rise in global temperature of 2.5°C or above (corresponding to doubling of atmospheric CO_2) to occur by the year 2038 (see figure 1).

2. CLIMATIC CHANGE IMPACTS

Although enhancement of atmospheric CO_2 concentration alters photosynthetic processes and plant growth, its major harmful effects are thought to arise from increased global temperatures and a more vigorous hydrologic cycle. The impact cost estimates for this type of change that we shall give are essentially based on extrapolations from the effects of present day climate variability, so that radical and hence more costly climatic shifts are not envisaged. However, warnings of very dramatic effects of climatic warming have been given, including removal of all Arctic sea ice, with major, but unknown effects on northern hemisphere circulation, and also the possible surging of the West Antarctic ice sheet (4), precipitated by minor warming of the upper layers of the ocean. Impact costs of such an occurrence have been made recentl (5), but since we have no idea as to its probability of happening, we have no means for introducing this catastrophic event into the decision analytic approach that we use.

In computing the economic impacts of even relatively minor climatic changes, we are faced with two serious deficiencies. Firstly, climatic change costs have been very little investigated, whatever their cause, so we are dealing with a near complete absenc of information directly applicable to CO_2 impact calculations.

Secondly, climate is highly regional in character, yet the only quantitative CO_2 induced climate change estimates that we have are at best zonal averages, and a translation of this type of information into local or regional data is a difficult matter. The numerical values of climate change costs that we will present thus have to be regarded quite tentatively. However, because of the exponential growth of CO_2 and its impacts, and because of the sweeping nature of the global threat, we believe that even rough estimates can guide us in assessing its policy implications.

The dollar cost figures for CO_2- induced climate change impacts that we shall give are all interpolations or extrapolations from cost calculations made for the CIAP study (Climatic Impact Assessment Program) (6). Unfortunately, the research performed for that program was all for assumed decreases in world temperature, and our analysis for an increased temperature condition suffers thereby.

2.1. The world economy

The CO_2/climate problem is global both in its causes and in its impacts; its remediation similarly involves international actions. If only for these reasons, it is necessary to analyse the problem globally. Indeed, we find that, with the present lack of information on regional climatic changes, we are quite unable to make cost assessments with any reliability for individual nations, and only world totals can be quoted with a modicum of confidence. National figures could give a rather different impression, since in a global warming no doubt some would benefit from the change, whilst others would lose, further complicating the pursuit of CO_2 control policies, if ever they need to be contemplated.

In quantification for the decision analysis we thus require a global measure of the seriousness of the impact; both gross world product and total world assets provide such a reference norm. In extrapolating to future conditions it is customary to allow for both inflationary effects on a dollar numeraire and for economic growth via a discount factor, so some idea of these features is needed. Table 1 gives the figures we have used for U.S. and world assets and gross product for 1971, the arbitrary date we have chosen for our base year. Differences in definitions of gross or net product and in what constitutes total wealth or assets, together with data accumulation difficulties, results in variability in the reports of economic statistics we have used for producing table 1, as we discuss in reference (7).

We should point out that the asset values listed in table 1 specifically exclude inherent (unimproved) land values, in contrast to some other uses of the term. We have chosen this definition since we regard climatic effects on asset values to

Table 1. U.S. and world accounts

Assumed value for 1971 in billions of 1971 U.S. dollars		
U.S. gross national product	1,052	ratio = 3.24
Gross world product	3,410	
U.S. gross national wealth	2,945	
World total assets	9,542	

Growth rates, percent per annum		
U.S. GNP	in current $	in constant $
1948-66	6.4	4.0
1966-69	8.3	3.8
1969-73	8.9	3.6
U.S. GNW		
1948-66	6.0	3.3
1966-69	8.6	4.0
1969-73	9.4	3.8

arise from the need to re-capitalise the investment in land development and infrastructure that can result from a change in the local environment. It is interesting to note that with this defintio it only requires some three years of complete national commitment of the national budget to reconstruct a devastated country. Or, alternatively, we may say that a third of the nation's economic effort is constantly spent in refurbishing the system and in promoting growth (though the latter is a very small portion of the total).

2.2 Climate change cost estimation

We have already commented on the absence of directly applicable information on climate change costs and the resultant estimation difficulties. The CIAP generated results we use are also incomplete in their estimates for a global cooling, so our extrapolations suffer further from this cause. Reference (7) goes further into these problems. All in all we therefore need to emphas that the cost figures we shall present should be used primarily as guides to policy assessment, and not taken as quantitatively reliable.

We shall first treat the effect of climate change on capital assets. Niedercorn (8), as part of the CIAP study, estimate this for one sector of the U.S. economy, namely, agriculture of the corn belt. He computed the total re-investment required to shift agricultural production to physically new locations following a 5°F fall in global temperature. These costs do not reflect any change in productivity that might result from the move; the latter we place in another catagory - operational losses - that we shall cover later. Neidercorn estimates an increased investment of \$27.1 x 10^9 (1971 \$) with a median growth rate assumption, but spread over 25 years, between 2000 and 2025.

In order to use this result for the CO_2 case, we need first to extrapolate to total U.S. capital investment losses, then to a world figure. We do this by picturing changing climatic conditions to result in ever shifting geographical locations for optimal human activities, both in the agricultural, the industrial and in other economically significant sectors. Niedercorn's result gives just a portion of the costs of such adjustments, and we expand this to the entire U.S. system by factoring it by the ratio of total U.S. capital assets to assets in the corn belt. Extrapolation to world scale is then accomplished by multiplying by a factor of 3.42 (table 1).

We next have to translate these figures for a step change in climatic conditions into the exponentially growing CO_2 case. To do this we need to assume a functional form for the relationship between costs and temperature change, and a power law is the simplest that could be applicable. Niedercorn makes an estimate for costs of a 1°C change in his model, and we find that this fits the law

$$C = 596\,\Delta T^{1.35} \quad \text{per annum} \tag{1}$$

for costs in billions of 1971 dollars, with the temperature rise, ΔT, in degrees Celsius. Our own estimate is for a rather higher rate of cost increase with temperature, and we believe a quadratic law is to be preferred:

$$C = 308\,\Delta T^{2} \quad \text{per annum} \tag{2}$$

so that costs are less than (1) for small temperature changes, but the same for a 2.78°C (5°F) change.

The importance of non-linear growth of costs with temperature rise needs to be emphasised as a significant characteristic of climatic change impacts. Modest changes - say less than 1°C in a global average - are easily and often accomodated (see the discussion in section 6), but changes of 3°- 4°C correspond to glacial-interglacial events, with quite radical effects on habitability of different sectors of the earth; a 10°C change could well eradicate the human species.

We have used eq.(1) & (2), together with projections of CO_2 induced temperature change, to estimate capital investment losses due to changing climate, and the results are shown in figure 3. The curves refer to annual losses and apply to a 3% per annum growth in atmospheric CO_2 level. Note that the absolute dollar values refer to world asset status as of 1971 and not to the date given on the abscissa. To compute actual dollar costs for a

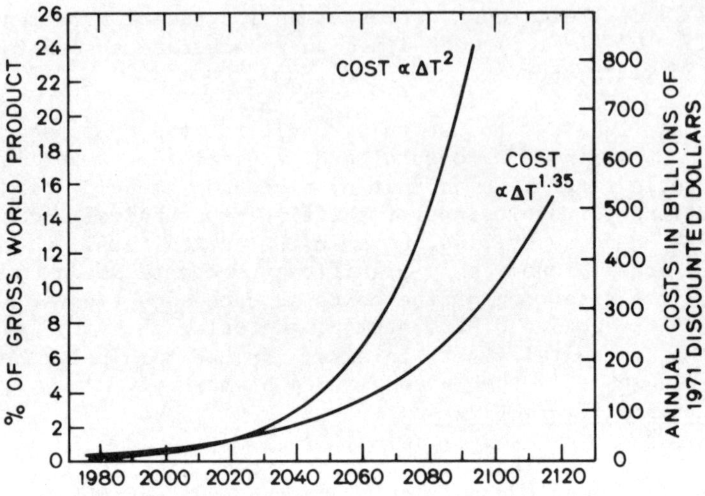

FIGURE 3. ANNUAL CAPITAL ASSET LOSS FOR CO_2 GROWTH AT 3% PER ANNUM

a particular year assumptions need to be made about the growth rate of the world economy, as well as dollar value changes. Although projecting fifty years into the future is not a charge easily accepted by economists, a reasonable guess is for a 3% per annum growth in gross world product (for example as given in (6)), which, if the world economy behaves as has the U.S., implies an approximate equal rate of growth in total world assets.

Rather than trying to deal with the difficulties of estimating future economic growth rates and inflation effects on the dollar, we can eliminate their consideration by appropriate normalisation of the absolute costs. By measuring costs relative to gross capital assets or to gross capital asset growth rate we automatically allow for inflationary effects and in large measure for changes in economic growth rate that might occur. This follows once we make the reasonable assumption that capital stock asset losses due to climate change vary in proportion to the size of the assets and that capital stock grows linearly with gross world product growth. Under these conditions the ratio of capital asset loss to total assets and the ratio of annual capital loss to GWP become functions of only the atmospheric CO_2 level and its rate of change. The left hand ordinate in figure 3 expresses cost in such a dimensionless form.

In using the CIAP enumeration of capital loss impacts we have assumed symmetry in response to positive and negative temperatu changes. In view of the model we have assumed of shifting belts of climatic optima for economic activity, such an assumption has the ring of plausibility. By contrast, operational cost changes attendant upon climatic change are known to be asymmetric with regard to temperature sign. This is best documented - as are most climatic costs - for the agricultural sector (for a recent extended analysis see (9)). Use of the CIAP generated data is in this case therefore much more tentative, and in fact fitting a parabola between temperature changes of $-1^\circ C$ and $+5^\circ C$ gives the formula

$$C = 31.8\,\Delta T(\Delta T - 1) \quad \text{per annum} \qquad (3)$$

for operational costs in billions of dollars (1971 \$). For $\Delta T = 2.5^\circ C$ (our estimate for temperature change due to doubling of CO_2 levels) equation (3) yields a figure of $\$119 \times 10^9$ p.a.

Comparing figure 3 with formula (3) we see that, at the doubling condition (2038), operational costs are of the same order of magnitude as the rate of capital investment loss. However, we place less faith in the reliability of equation (3) than for the corresponding asset equation (2), and our decision analytic approach will be based solely on the latter. To this extent the climatic impact effects will be assessed optimistically.

2.3 The effect of uncertainty on costs

Considering just the capital investment loss due to climate change, we can use the probabilistic results for temperature increase with time, as depicted in figure 1, to calculate the effect of this source of uncertainty on costs. Figure 4 plots the cumulative probability of capital investment loss as a function of time for an assumed 3% per annum rise rate in atmospheric CO_2 level, or in fossil fuel burning rate if deforestation effects can be ignored. The 50% cumulative probability curve corresponds to the result plotted in figure 3 for a quadratic cost-temperature relationship. The large uncertainty in date for a given impact shown in figure 4 is important for correct handling of the decision analysis described in section 4.

3. REMEDIAL COSTS

If the climate cost calculations presented above are believed, at some stage in the CO_2 build up process we can expect remediation measures to become less costly than climate change costs. There are three general approaches for reduction of CO_2 levels. Most obvious is the reduction of fossil fuel use (or wood burning), with the substitution of non-carbon based fuels. The alternative energy source might be nuclear, the cost of which is comparable to fossil fuel usage, or more exotic, such as solar or biomass. The latter are not only more costly than fossil fuels for large scale production, but require technological development if to

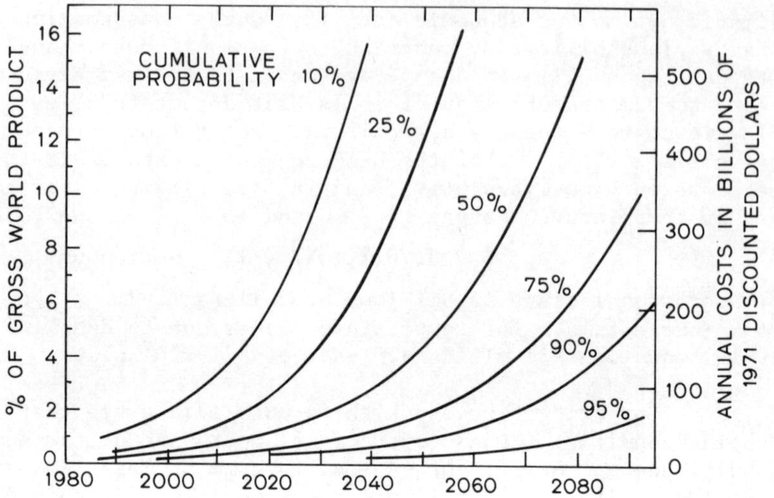

FIGURE 4. CLIMATE CHANGE COSTS

be employed on a scale sufficient to replace fossil fuels within a few decades. The second alternative is to remove CO_2 directly from the atmosphere. Conversion to biomass and its storage has been suggested (10), as has the use of wind energy (11). Neither feasibility nor costs of such radical measures have been sufficiently analysed for inclusion in this study.

A third expedient is to remove carbon dioxide from the combustion effluents before it is emitted to the atmosphere. Following an original suggestion by Marchetti (12), two estimates have been made for stack removal of CO_2 using scrubbers (13,14). When transcribed to our figures for world fossil fuel use - 228 Quads in 1975 (see reference (1) for an accounting of past fossil fuel use and its projection) - we find approximate agreement between these estimates for operational costs of scrubbing of about $\$110 \times 10^9$ p.a. for 1971 costs in 1971 dollars, assuming all fossil fuel generated can be so treated. Reference (7) gives details. The estimates of capital investment costs do not agree; however, since these appear to be smaller than operational costs, we have ignored them in presenting our final CO_2 removal cost results. These are shown in figures 5 and 6, assuming 50% removal of all additional fossil fuel generated CO_2 starting at two dates, 2000 and 2020, which correspond approximately to the installation of 100% effective scrubbers on all new industrial fossil fuel using plants as of these dates. We have assumed continuous growth of fossil fuel use at 3% per annum. The costs depicted are the sum of scrubbing costs and the median value of climate change costs (the 50% curve in figure

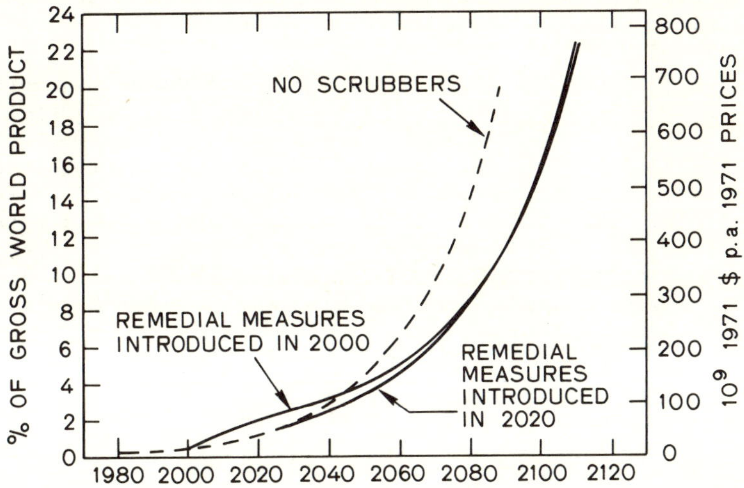

FIGURE 5. NET ANNUAL COSTS WITH 50% CO_2 REMOVAL

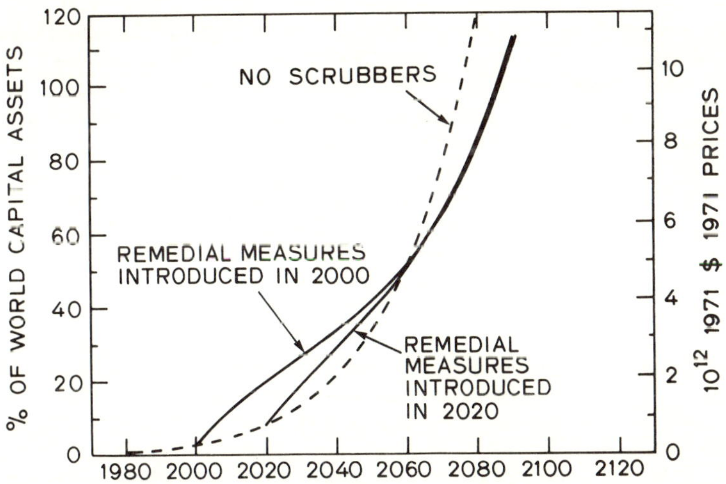

FIGURE 6. NET ASSET LOSSES WITH 50% CO_2 REMOVAL

4). We conclude from these results that the cost of scrubbing is very high, even compared with climate change costs, until quite late in the CO_2 build up process, and its introduction prior to the CO_2 doubling time (2038) is not cost effective.

Nordhaus (15,16) has estimated the additional energy production costs in a strategy of substitution of non-carbon based energy sources for fossil fuels, aimed at keeping atmospheric CO_2 concentrations below a chosen level. He uses a market allocation model combined with a carbon cycle model in order to investigate control strategies for minimizing impact costs in keeping the increase in CO_2 level below 50%, 100% and 200% of the pre-industrial value. This prescription does not minimize total costs, since the climatic change costs are omitted. Table 2 is taken from two of

Table 2. Discounted costs for CO_2 reduction strategy

Maximum allowed increase of CO_2 relative to 1860	200%	100%	50%	Ref.
Total discounted cost in billions of 1975 $s	30 4	87 87	540 540	(15) (16)
Total discounted cost relative to total discounted GWP	.06% .004%	.12% .08%	.81% .48%	(15) (16)

Nordhaus' papers and gives discounted costs for the three emission control strategies he selected. The source of discrepency in the values quoted is not known. Note that the costs are not annualized, but are totals. In order to make estimates of the former we need to know both his discount rate (10%) and his assumptions regarding GWP growth, and the latter cannot be readily allowed for from the information in his published material. Direct comparison of the dollar costs in table 2 with our climate change costs (section 2) cannot therefore be made, though his values for the ratio of costs to discounted GWP should accord approximately with the normalized form of climate costs that we have employed. In any event, it is evident that the energy transition costs calculated by Nordhaus are relatively small compared with climate change costs at the doubling (100% increase) and the tripling (200% increase) levels, and quite modest compared with the cost of scrubbers.

One of the main conclusions Nordhaus draws from his results is that the optimum control strategy that he finds differs little in its energy use path from the uncontrolled path (no CO_2 reduction policy) until quite close to the date for reaching the critical CO_2 level with no controls. For example, to avoid CO_2 doubling - without remedial measures expected to occur around 2040 - requires implementation of severe controls only some twenty years earlier, about forty years from now. This behavior is in sharp contrast to the long time lags predicted according to market penetration time ideas, as we shall discuss further in section 5. The source of the market penetration time lag, whatever it may be, does not seem to be included in the macro-modeling employed by Nordhaus.

In addition to this discrepancy as a possible source of error in Nordhaus' conclusions, his analysis, in failing to include climate change costs, thereby also errs by leaving out the important probabilistic element of cost uncertainty. By including climate change costs and their probable error distribution in a decision analytic approach, the importance of this feature for the timing of remedial strategies can be brought out, as we shall see in the following.

4. THE DECISION ANALYTIC APPROACH

Decision anlysis can deal with two of the critical features of decision making under uncertainty that are present in the CO_2 problem. The first concerns the probabilistic element that results from the large uncertainty in climate change prediction models and the second relates to the definition of the degree of seriousness of the threat of climate change. By formulating "value" or "disbenefit" in quantitative terms, a rational prescription for the latter can be given, and at the same time, under a set of axioms developed by Von Neuman and Morgenstern (see, for example (17)), we can logically allow for uncertainty in drawing conclusions.

By assigning a specific probability distribution to the possible climatic states that might result from increasing CO_2 levels, we convert our problem of decision making under uncertainty to one of decision making under risk, provided we can also assign a measure of utility to the climatic states. The cost estimates given in section 2 go part way toward this goal. However, the resultant dollar numeraire for value is clearly inadequate. In the first instance we require a reference level for significance of the impact costs, and, as we have suggested in section 2.2, we find this to be conveniently provided in the CO_2 problem by gross world product or total world assets. Secondly, we must allow for the fact that the disutility of climatic change costs increases disproportionally with increasing cost, so that high costs at low risk weigh more heavily in the decision making process than medium costs at medium risk, i.e. the utility function chosen to represent value must be risk averse. For our specific situation it is impossible to define the function with any degree of certainty (the decision analysis practice of discovering the appropriate measure by interrogation of the decision maker involved is obviously impossible in our situation). The best we can do is to choose a common analytic form that has been used and check for its sensitivity. We have in fact selected the exponential utility function

$$U = - \exp - W/\kappa W_o \qquad (4)$$

where W stands either for net world product (i.e. GWP minus annual climatic change costs) or net world assets (assets minus capital losses from climate). The critical parameter κ relates to the point at which aversion becomes strong, and is thus closely linked to world assets or GWP, W_o. In the calculations that follow we have selected κ to correspond to 5%, 10% or 50% of W_o.

The utility function (4) as it stands cannot yet be a correct representation of value in our problem. Thus, it assigns rapidly increasing weight to costs as they approach the value of GWP or world assets, yet it is clear that it is impossible for costs to exceed either of these two functions. In fact, once the expectation of climate impact costs rising to levels that completely disrupt the system is reached, one becomes indifferent to the possibility that costs may become even larger - the degree of bankruptcy is irrelevant. We thus need to level out the exponential form (4) above some cost value, and we have done this by setting

$$U = - \exp \frac{\kappa - 1}{\kappa} \quad , \quad W < (1-\kappa) W_o \qquad (5)$$

this corresponding to a situation in which climatic change costs C exceed κW_o. The stipulation also eliminates infinite values of expected utility that would otherwise result from the forms (1) and (2) that we assumed for cost dependence on temperature change.

In the decision analytic approach quantification of the disbenefit of a particular climatic change, as depicted for example in figure 1, is obtained by calculating the expected value of the utility function (4) for the distribution of uncertainty we have assigned, initially to the temperature response, figure 1, then to dollar costs, figure 4.

For a final quantification of benefit or disbenefit we return from the measure U, as defined by equations (4) and (5), to a dollar numeraire by defining the "certain equivalent", W_{eq}, in the relationship

$$\langle U \rangle = - \exp - W_{eq}/\kappa W_o \qquad (6)$$

More commonly, we deal with the certain equivalent loss variable, C_{eq}, defined by

$$\langle U \rangle = - \exp - [W_o - C_{eq}]/\kappa W_o \qquad (7)$$

The above risk aversion formulation leads to estimates of certain equivalent costs of climate change as a function of time, i.e. we have eliminated the probabilistic aspect of the problem though we should repeat that these expected value estimates still contain uncertainties that result from the gross assumptions we had to make in estimating costs. If we had been able to assign error values to the latter they could also have been incorporated in our framework of analysis and the value of the results thereby enhanced.

The certain equivalent loss values that are obtained in this way can be combined with estimates of costs of remedial strategies, in order to develop minimum cost paths. At this stage in our perception of the nature of the CO_2 problem, rather than elaborate in this way, we are more concerned with developing a better understanding of its broader features, those that serve best to define energy policy options and reveal weaknesses in logic of the assessment process. In this paper we have therefore not attempted to develop optimum strategies for CO_2 impact reduction, but have restricted ourselves to an assessment of the significance of the risk averse cost calculations as defined above, together with their modification by discounting the future. Since the events under discussion are far distant, the latter can have a significant effect on present-day policy regarding the threat of CO_2 induced climate change.

4.1. Risk averse cost calculations

We again treat the case of 3% per annum rate of increase of atmospheric CO_2 levels and assume a normal distribution of error in the logarithm of the temperature increase, with a standard deviation of ln 2. The numerical integrations that yield the certain equivalent annual capital loss rates have been carried through for three choices of the parameter κ : 0.05, 0.1 and 0.5, corresponding to critical climate cost levels of 5%, 10% and 50% of the (unperturbed) gross world product. A rough interpretation of the significance of these figures is that the impact costs have by then become large enough to result in world economic collapse.

Results are shown in figure 7; the curve labelled "expected value" corresponds to the no risk aversion case, or, equivalently, the case when $\kappa = \infty$. The "median value" cost curve represents the 50% cumulative probability result (figure 1) when no provision is made for the effects of uncertainty. The latter gives a date of 2038 for CO_2 doubling with a 2.5°C temperature rise, the rate of capital asset loss from this climatic change amounts approximately to 3% of gross world product. Figure 7 shows that the effect of uncertainty is to advance attainment of this cost level by some twenty years (with a slight dependence on choice of utility function), thus making the problem more severe. Past this time the sensitivity of costs to assumptions regarding risk aversion increases rapidly, though this feature may not be important for general policy conclusions, as we shall discuss later.

Figure 8 presents results analagous to those in figure 7 but for accumulated asset losses as a function of time. In this

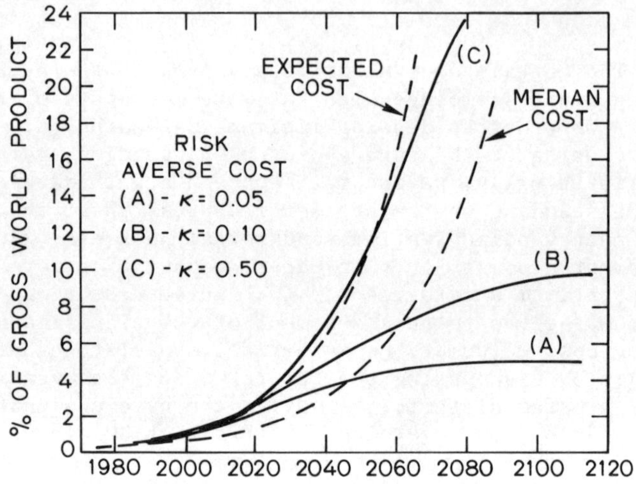

FIGURE 7.
ANNUAL CAPITAL LOSS RATE WITH RISK AVERSION.

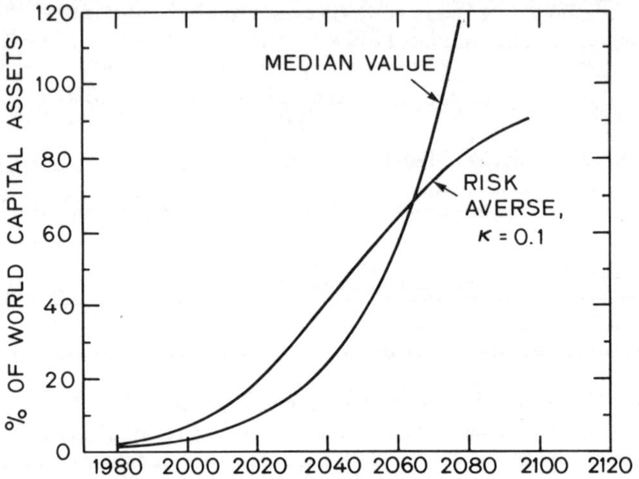

FIGURE 8. ACCUMULATED CAPITAL ASSET LOSSES WITH RISK AVERSION.

instance we have computed risk averse predictions only for the median value for κ of 0.1. From it we see that total asset loss values are even less sensitive to assumptions than are the annual loss rates. The no risk averse cost for CO_2 doubling, slated to occur in 2038, is about 20% of world capital assets; the effect of risk is to advance the date for such a cost by 15 years. Note that from figure 7, no matter what assumptions are made regarding the utility function, the loss estimates vary by no more than a factor of two at any time for which the results make any sense (very high values of losses, approaching the 100% level, are obviously impossible, and the results of the analysis should not be taken seriously under these circumstances).

4.2. Discounting

It is standard practice in economic analysis and in evaluation of future marketing conditions to apply discount rates to future costs or incomes. Managerial or policy decisions that are concerned with future outcomes become more and more sensitive to choice of discount rate as the outcomes become more distant. Any usual choice of discount rate (ranging upward from 5% p.a.) applied to events fifty years away (such as the CO_2 doubling date) results in such a large discount factor that economic analyses tend to ignore such events. Climate change costs, however, are so large that we cannot assume that the same logic applies, and we need still to estimate the present worth of future costs.

Economic discount factors have to allow for inflationary effects and for investment interest rates, and it is essentially impossible to estimate these over a fifty year period. However, we have argued that these factors are also reflected directly in growth of gross national product or gross world product, and hence in total asset growth. Thus, by normalizing with respect to GWP or total world assets we can largly eliminate the need to consider economic discounting in assessing the present day significance of future climatic impact costs. In addition to economic discounting we must still consider <u>social</u> discounting, and here we have to treat a little understood psychological attribute that is very difficult to quantify. This is often discussed as a question of the ethics of intergenerational valuation, but discounting the future can also be regarded as a way to allow for unknown future eventualities that cannot be explicitly taken into account. Such externalities certainly exist for the bounded domain of analysis we have used for the CO_2 problem, and whether a suitable social discounting rate can be ascribed to our problem without considering a complete range of alternative future and the role of the CO_2/climate change issue in each of them is doubtful.

As an example of the effects of discounting we have plotted in figure 9 the rate of capital investment loss due to warming for

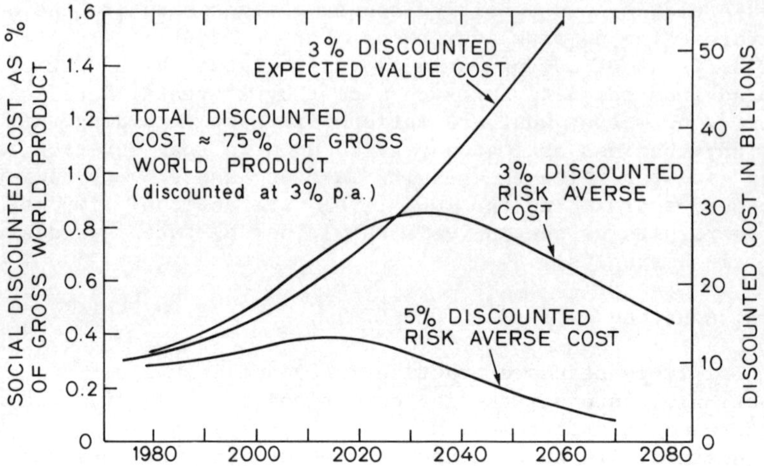

FIGURE 9. EFFECT OF DISCOUNTING THE FUTURE ON CLIMATIC CHANGE COSTS

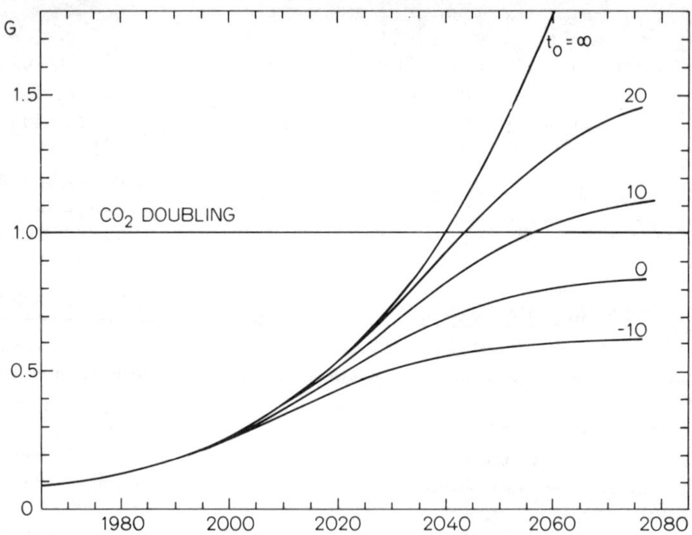

FIGURE 10. EFFECT OF MARKET PENETRATION TIME LAG ON CO_2 GROWTH

two discount rates, 3% and 5% per annum. The 3% figure corresponds to the value of "utility time preference" discount rate suggested by Arrow (18). The results show the importance of discounting the future in assessing the present day seriousness of the CO_2 threat. Thus, although we find that major economic consequences occur within fifty years, when discounted - even at the rather low rate of 3% p.a.- the costs are considerably reduced, with a maximum present day cost of only .8% of GWP, compared with a non-discounted cost of 5% GWP at the CO_2 doubling date.

An immediate reaction to the relatively small discounted costs shown in figure 9 is that they demonstrate the unimportance of CO_2 growth as a current issue. However, and quite apart from the controversy that exists concerning the use of social discounting, one can argue that discount factors should be applied to at least some aspects of gross world product. For example, a general depression that reduced energy use would also reduce GWP, thereby increasing the socially discounted measure of percentage costs plotted in figure 9. Other externalities, such as a new environmental threat that could reduce the weight attached to CO_2 impact, might have no effect on GWP. In general we would recommend that, with our present lack of appreciation of the nature of the forces affecting social discounting of future events, decisions should not be based on blind application of a particular social discount rate.

5. MARKET PENETRATION TIMES

As we have seen, it is likely that we have a grace period of some fifty years before the climatic effects of CO_2 growth become major. However, if it is deemed necessary to resort to a fundamental change in energy use policy to avert the potential disaster, the question of substitution time to convert to a non-carbon fuel base arises. According to Marchetti (19) at least 50 years is required for such a process, implying that policy steps should be taken today to avoid the climatic impacts we are predicting could occur in the next century. We have used the logistic function form for market penetration (see (2)) to compute atmospheric CO_2 levels as a function of time when a new non-carbon based fuel is introduced in the year 1975 + t_o (with a 1% share of the energy market), replacing a completely fossil fuel system. We have assumed the best case situation of a fifty year market penetration time (the time it takes the new energy source fraction to rise from 1% to 50% of the market) and an overall 3% per annum growth in energy use. In figure 10 we have plotted the resultant growth G in fractional increase in atmospheric CO_2 concentration relative to the 1860 norm. The figure shows the near impossibility, under the above assumptions, of reducing fossil fuel with sufficient rapidity to avoid CO_2 doubling, except via the use of nuclear fuel (which already has a significant share of energy production, corresponding roughly to the curve marked $t_o = -10$).

It is thus clear that market penetration times that are as long as Marchetti claims result in lags that force us to take the CO_2/climate impact question very seriously. Even so, questions have been raised as to the legitimacy of projections using the market penetration time concept for energy sources, as we have discussed elsewhere (2), and there are disagreements as to whether we should be using it for analysing the CO_2 problem. This issue is central to policy questions and has to be settled before policy decisions can be reached.

6. COST COMPARISONS

Figure 11 assembles some of the cost calculations made in sections 4 and 5. The climatic change costs pertain to an assumed 3% per annum growth in fossil fuel use (more accurately, in CO_2 release rate to the atmosphere) at the CO_2 level doubling date of 2038. The expected global temperature rise is then 2.5°C, of the order of the value during the hypsithermal period, perhaps 7000 years ago, as we have indicated with the temperature scale on the right hand ordinate in figure 11. Costs of both capital asset losses and of increased operational expenditures are shown, though, as we discussed earlier, we believe the former to be more reliable than the latter.

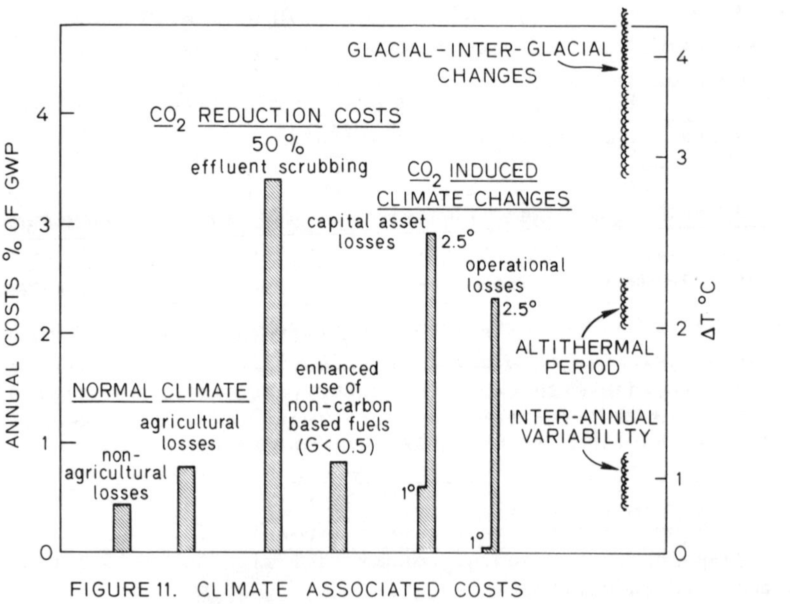

FIGURE 11. CLIMATE ASSOCIATED COSTS

The two types of costs for remedial action covered in section 3 are included in the figure; scrubbing costs are for the 2038 date (figure 6) and for 50% removal of CO_2 released. Nordhaus' estimate of costs for replacing fossil fuels by non-carbon based fuels is for maintenance of CO_2 levels under 150% of the pre-industrial value. We should point out that is based on <u>total discounted</u> costs and <u>total discounted</u> GWP (table 2), and is thus only an approximation to the measure of costs we have used. Transcription of Nordhaus' values requires more information than we have available. However, it is clear that his estimate of costs for reducing CO_2 growth is much less than the cost we anticipate for climatic changes from unrestrained CO_2 release. By contrast, stack scrubbing costs are very high and of dubious merit as a CO_2 control measure.

Figure 11 also shows our estimate of climate change costs for a 1°C rise in global temperature (projected to occur in the year 2008). These are of the order of ½% of the GWP, or some $\$20 \times 10^9$ per year, and may seem large in absolute measure. But they are less than the estimates made by Thompson (20) for the costs of natural variability of climate, indicated on the left hand side of figure 11. We note that the latter are to be associated with the interannual variability of climate, which is also of the order of 1°C. Thus the cost of an upward shift of global temperatures by 1°C will not have a major impact, and its effect will likely be lost in the variability of interannual climate related costs that arise from natural causes.

The final feature included in figure 11 gives an indication of the range of temperature change occuring between glacial and interglacial periods, shown in the upper right. An atmospheric concentration of CO_2 three or four times the pre-industrial value is required to reach this temperature range, a state likely to be reached only some 30 years after the doubling date, i.e. around 2070. Our knowledge from paleoclimatology of the ecological shifts associated with glacial-interglacial periods indicates that these levels of CO_2 increase would indeed produce traumatic results and be difficult for our civilization to bear.

The intercomparisons provided by figure 11 bring out the following main points:

(1) The exponentially growing characteristic of the CO_2 problem means that its climatic costs remain unimportant up to the natural climate variability level - around 1% of GWP, occuring in the first decade of the 21st century - but thereafter rise rapidly to dominance by the CO_2 doubling date of 2038, with costs of the order of 3% of GWP, and become overwhelming by the quadrupling date, about 2070.

(2) Remedial measure costs vary widely; stack scrubbing for removal of CO_2 is much more expensive than replacement of

fossil by a non-carbon based fuel, the latter costs being modest compared with those resulting from any major climate change.

(3) Since gross world product growth projections lie between 3 and 4% per annum, we can expect climatic impact costs to have serious economic consequences worldwide by the doubling date, if not before.

These conclusions do not allow for uncertainities present in the climatic change estimates; the risk averse calculations presented in section 4 take care of this. There we find, referring to figure 7, that expected climatic change costs reach 3% of GWP by 2025, just 45 years from now, instead of in 2038. The three widely differing risk averse formulations change the date only slightly (2021, 2024 and 2033 for κ = 0.5, 0.1 and 0.05, respectively). Thus, if we accept a 3% level of cost as serious, decisions are relatively insensitive to the particular choice of utility function. Note that a 3% of GWP cost effectively reduces world economic growth to zero, since projected GWP growth is also about 3% p.a. in the next century.

7. CONCLUSIONS AND POLICY IMPLICATIONS

Results of the cost analyses discussed above lead to the following main conclusions:

A) Estimates of climatic impact costs of increasing atmospheric CO_2 concentrations confirm the often reported seriousness of CO_2 build up within a fifty year period. At a 3% per annum release rate of CO_2, capital loss rates from CO_2 induced warming are of the order of 3% of the gross world product in year 2025, reducing world economic growth to zero.

B) This conclusion is not sensitive to a particular choice of utility function or risk aversion formulation, but the probabilistic aspect of the decision analysis, arising from uncertainties in climatic response to increased CO_2 levels, advances the effective CO_2 doubling date by some 15 years, from 2038 to 2025.

C) Social discounting of the future can considerably reduce the present day significance of the anticipated costs. Uncertainty regarding an appropriate magnitude for the discount factor and disagreement on means for allowing for social discounting makes it difficult to incorporate this feature into the decision analysis.

A cost/benefit analysis weighing the attractiveness of alternative CO_2 control strategies has not been carried out, but would clearly be sensitive to item C). We anticipate, however, that the marginal benefit of any optimum control strategy in (socially) discounted dollars would not be very large (compared with natural climate variability costs), so that it is not clear what effect publication of the results of such a study would have. A clear case in favor of acting today to alleviate the potential CO_2 problem would thus appear to be difficult to make, unless market penetration times for new energy sources are indeed as long as has been suggested. We thus regard resolution of the market penetration time question as central to the issue of immediacy of the need for a CO_2 control policy. If market penetration time is not a barrier, replacement of fossil fuel use by a non-carbon based energy source, as postulated by Nordhaus, would appear to be the appropriate strategy - one in which the cost of delay (especially when discounted) would be small.

If it is concluded that market penetration times can be fifty years or more, a cost/benefit analysis that allows for such long lags can still be carried out for various CO_2 control paths. Even in this case we envisage that the importance attached to the results thus obtained will depend on individual and subjective attitudes with regard to social discounting of future events. In order to obtain a concensus on the present day gravity of the problem, appropriate means for dissemination of the results of the analysis would be needed. The story would have to be presented clearly and fairly so as to bring out the probabilistic and discounting aspects of the problem; perhaps then reactions could be culled and a basis for determining a course for policy actions established.

REFERENCES

(1) J.A. Laurmann: Climate Change from Fossil Fuel Generated CO_2 and Energy Policy. Environment International, in press.

(2) J.A. Laurmann: Market Penetration Characteristics for Energy Production and Atmospheric Carbon Dioxide Growth. Science 205, pp. 896-898 1979.

(3) National Academy of Sciences: Carbon Dioxide and Climate: A Scientific Assessment. NAS, Washington D.C. 1979.

(4) J.H. Mercer: West Antarctic Ice Sheet and CO_2 Greenhouse Effect: A Threat of Disaster. Nature 271, p. 321 1978.

(5) S.H. Schneider: So What if Climate Changes? AAAS meeting preprint, Houston Jan. 1979.

(6) CIAP: Monograph #6; Economic and Social Measures of Biologic and Climatic Change. Ed: A.J. Grobecker. DOT-TST-75-56 1975.

(7) J.A. Laurmann: Economic Impact of CO_2-induced Climatic Change. Draft report, Electric Power Research Institute, July 1979.

(8) J.H. Niedercorn: The Capital Costs of Climatically Induced Shifts: the Example of the American Corn Belt. In "The Urban Costs of Climate Modification" Ed: T.A. Ferrar. John Wiley, New York 1976.

(9) W.A. Gasser. Climatic Change to the Year 2000 and Possible Impacts on World Agriculture. AAAS-DOE workshop on Environmental and Societal Consequences of Possible CO_2 Induced Climate Change. Annapolis, April 1979.

(10) F.J. Dyson and G. Marland: Technical Fixes for the Climatic Effects of CO_2. Workshop on the Global Effects of Carbon Dioxide from Fossil Fuels. Miami Beach, March 1977.

(11) R. Rotty: personal communication.

(12) C. Marchetti: On Geoengineering and the CO_2 Problem. Climatic Change, 1 p. 59 1977.

(13) C. Mustacchi, P. Armenante and V. Cena: Carbon Dioxide Disposal in the Ocean. In "Carbon Dioxide, Climate and Society". Ed: J. Williams. Pergamon Press, 1978.

(14) Radian Corp: An Assessment of the Implications of Increased CO_2 Emissions and their Control. March 1978

(15) W.D. Nordhaus: Economic Growth and Climate: the Carbon Dioxide Problem. Preprint, annual meeting Am. Econ. Ass. Sept. 1976.

(16) W.D. Nordhaus: Strategies for the Control of Carbon Dioxide. Cowles Foundation Discussion Paper # 443, Jan. 1977.

(17) R.L. Keeney and H. Raiffa: "Decisions with Multiple Objectives" John Wiley, New York, 1976.

(18) K.J. Arrow: The Rate of Discount of Long-Term Public Investment. In "Energy and the Environment" Pergamon Press, 1976.

(19) C. Marchetti: Multicompetition and the Diffusion of New Technology in the Energy System. Chem. Econ. & Eng. Rev. 11 pp. 7-13 1979.

(20) J.C. Thompson: In "Living with Climate Change. Phase II" Mitre Corp. 1976.

ACKNOWLEDGMENT. The material described in this paper is in the main abstracted from a draft report prepared for the Elecric Power Researc Institute (7). However, the viewpoints expressed are those of the author and are not necessarily endorsed by the Electric Power Researc Institute.

ADDENDUM. Since completion of the paper, it has come to the author's attention that the calculation by Niedercorn that were used for maki the present estimates, were based on incorrect specification of the influence of climate on U.S. corn belt location. The magnitude of er resulting from our use of the Niedercorn results is not known.

THE COSTS OF CLIMATE IMPACTS

Walter Orr Roberts

Program on Food, Climate and the World's Future,
Aspen Institute for Humanistic Studies,
1229 University Avenue, Boulder, Colorado, USA

ABSTRACT. The surest prospect for future world climate patterns is that they will differ from present ones. What is uncertain is how much, and exactly in what way in different geographical regions. The anthropogenic CO_2 increase will probably exceed the unknown forcing functions of "natural" climate change within 30 to 60 years. It is not unlikely that by AD 2040 the world's climate, driven by the CO_2 increase, will enter a domain warmer than any within the past few million years.

The costs of averting this climate change or of absorbing its impact are likely to be huge, even though today imponderable. Not least among these are intangible and unquantifiable costs associated with changes in human values and the quality of everyday life for future generations.

As Donald Michael and others have said, we live in an emerging society and the evolution of such a society, on a 25-50 year time scale, cannot be predicted from the characteristics of the many components from which it will be built. Thus, economists are powerless to predict, beyond the short term, fiscal policies to stabilize the cost of gold, to predict future population growth rates, or to assess the costs of accommodating to a warmer climate 60 years hence, when the very nature and dynamic interactions of the emergent society are uncertain.

It is probable, however, that food and agriculture will be the sectors most strongly impacted. Better arid land management, increased use of irrigation, water and soil conservation, and the use of yet-to-be-developed genetic strains of food plants and animals will feature among the costly new strategies most

W. Bach, J. Pankrath, and J. Williams (eds.), Interactions of Energy and Climate, 461–468.

promising of benefits. Moreover, these strategies will bring benefits that are entirely independent of their worth in minimizing the impact of climate changes.

1. FUTURE CLIMATE WILL BE DIFFERENT

Perhaps the most constant feature of climate is its tendency to change. On every time and spatial scale the weather and climate vary. In my home in Boulder, Colorado this autumn and winter has been characterized by cold weather and frequent deep snowfall. Yet in the winter of 1976-77 Colorado and all of the west of the U.S. suffered drought and unusually warm weather. The summer of 1972 was a period of severe heat and drought in the Moscow region of the USSR. These fluctuations are basically unexplained, but history tells us that they have characterized the climate from the dawn of history. Not only are these changes large from year to year and from decade to decade, but there are warm centuries and cold, frigid millenia and mild ones. There are extreme ice ages and there are long interglacial spans with no year-round terrestrial ice.

The impact of these changes has been traced by many authors. In ancient times as today, human societies have been vulnerable to the natural vagaries of weather and climate. Garcia[1] has argued, moreover, that climate often triggers unnecessary human suffering in societies rendered precarious by exploitative patterns of land use and by agricultural development policies that favor cash benefits for the rich over food security for the poor. Nonetheless, all social systems, whether socialist or capitalist, advanced or developing, are subjected to great costs when the "normal" climate extremes occur.

The magnitude of a climate fluctuation can be measured meaningfully only in terms of its human impact. There will be, for example, a large impact if a region suffers a 30% decrease in growing season rainfall in a crop area that is already marginally too dry and too hot, like the High Plains of USA. If the same rainfall deficiency occurs in a region with predominantly irrigated agriculture, the adverse impact may be slight, or the effect even beneficial. A few decades of climate that is 1 C warmer may produce a nil effect, a substantial benefit, or a severe loss depending on many things. It will be troublous if the temperature change produces a large decrease in soil moisture through its effect on the large scale wind circulation causing decreased rainfall. A robust and healthy social system, with food and energy reserves, insurance schemes, and other safeguards can withstand large climate anomalies safely; a precarious one can suffer disaster under the same stress.

It is customary among climatologists to separate out long-term trends of climate from shorter fluctuations. The latter are usually regarded as part of the "noise" of the climatic system, with the trends perhaps being driven by some external forcing factor such as, for example, a postulated change in the sun's emission. However it is, to my mind, a very unclear distinction. Within a few years, however, it seems likely that we will see great progress in understanding climate changes on all scales of time and space. We will better know what is signal and what is noise. Only in recent years have large scientific resources been brought to bear on climatology, and only in the near past have we had the powerful computers that now permit us to use appropriately complex interactive simulation models for climate research. The capacity of regions and nations to benefit from this new knowledge will also be dependent on their social and political characteristics.

Whatever else we speculate, however, it can be safely expected that the climate of the future will be markedly different from that of the present in most regions of the world. Thus, the resilience of the social, political and moral fibre of a nation will markedly affect the ability of its people to "weather" the changes, and this resilience will be a major factor in determining the cost of climate impacts.

2. HUMAN INFLUENCES WILL SOON DOMINATE

Most emerging research in climatology points convincingly toward the conclusion that human influences on the radiative balance of our planet will soon produce climate changes that exceed the recent range of "normal" fluctuations. Foremost among the factors expected to influence the climate is the steadily mounting atmospheric burden of anthropogenic CO_2 produced mainly by the steadily mounting use of fossil fuel. It is expected to produce a mean global warming of 2 to 3 C by about AD 2030 or 2050. Such a warming, if it occurs, will exceed any since Homo Sapiens emerged to begin the long trek to mastery of the globe's natural resources. Moreover, other human influences are thought to be additive rather than subtractive. They will augment, rather than ameliorate, the warming. The temperature change by itself is not highly important. The most significant effects are likely to be the changes in precipitation distribution coming about from the changed wind circulation that results from the warming.

A wide range of authoritative authors express the general concern of the scientific community on this matter. The Soviet radiation expert M.I. Budyko (2) says: "The development of civilization, an exceptionally rapid event on the time scale of geological history, has radically changed the outlook for the further existence of the biosphere." Robert M. White [3] of the U.S. National Academy of Sciences, and President-Elect of the University Corporation for Atmospheric Research, indicates that these potential climate changes "could have a pervasive impact upon the nations of the world and may require unprecedented forms of international actions to deal with them effectively." George M. Woodwell [4] of the Marine Biological Laboratory at Woods Hole Oceanographic Institution, expresses this opinion: "Carbon dioxide, until now an apparently innocuous trace gas in the atmosphere, may be moving rapidly toward a central role as a major threat to the present world order."

3. IT WILL NOT BE POSSIBLE TO AVERT THE WARMING

The driving force for the CO_2 buildup is the human appetite for energy. The world's petroleum and natural gas reserves are rapidly being consumed, and the major other source of fossil fuel is coal. Coal can be burned directly for energy, or it can be converted to "synfuel" that is more readily transported and used. In either form, it produces CO_2 as it is used, and it does so at a faster rate per energy unit than natural gas or petroleum. The best present projections lead me to expect, by the year AD 2040 that the following circumstance will prevail for the world:

population:	2x present
food demand:	3x present
energy usage:	4x present

To meet such food and fuel consumption patterns will place severe demands on world societies, even without a climate change. It seems unlikely that the food needs and wants can be satisfied without increased use of energy and fertilizer, though all possible encouragement should be given to research or improved techniques for energy-conserving food production. The total energy demand is probably understated, if the development of the Third World progresses as all humane persons hope. The widespread fear of the health risks--both present and future--of nuclear power are likely to restrict its use as a replacement for fossil fuels until the latter half of the next century. Biomass production of fuel and the direct use of solar energy are also probably not going to meet the needs at reasonable cost before AD 2050.

Thus, coal is likely to be the fuel of the near future. And with it will come increased atmospheric CO_2. Though disposal of CO_2 from coal usage has been advocated by some[5], most experts are not optimistic about the feasibility or cost-effectiveness. It seems, therefore, most likely that the warming will come, and with it major changes of global climates, differing from region to region. There will be some who benefit and some who suffer, in probability. However, the losers will probably outnumber the winners. As the Panel on Climate Variation of the U.S. National Academy of Sciences stated recently:[6] "A major climatic change would force economic and social adjustments on a world scale, because the global patterns of food production and population that have evolved are implicitly dependent on the climate of the present century.......This dependence of the nation's welfare, as well as that of the international community as a whole, should serve as a warning signal that we simply cannot afford to be unprepared for either a natural or a manmade climate catastrophe."

The essential step for local, national and international communities, thus, appears to be to develop new strategies to accommodate to a change of climate that will come to us slowly over the next half-century, that involves scientific uncertainties, and that requires considerations of lead times beyond those we normally consider in planning.

4. STRATEGIES FOR A VIABLE FUTURE WILL BE COSTLY

Assessing the costs of various policy options to meet far future military, environmental, or economic threats is a difficult business. Most of the risk-benefit cost assessments that are offered to us seem, to me at least, to be fuzzy and uncertain. Moreover, as admirable as are their goals, the actual steps taken in political bodies of the world do not seem to take much account of the quasi-quantitative products of the applied systems analysts. Rather, the decisions appear to be based on more intuitive considerations, and especially on short-term projections of risks and benefits. The welfare of the next generations seems distant, abstract, and highly discounted.

The advanced countries of the world do very well at solving rather straightforward scientific-technical problems, when we place high priority on the goals. The accomplishments in space illustrate this. We do rather badly, however, when real social innovation is a precondition for achieving gains. Meeting the costs of climate impact is a domain clearly demanding far-reaching social innovation.

I find it convenient to divide cost considerations of climate impacts into two classes depending on the immediacy with which we should pay them. The first class, which we must deal with at once, is the easier. It includes costs that we must incur to increase the certainty of the scientific and technical knowledge about the risk. In this class are heightened priorities for the pursuit of complex mathematical modelling of the causes of climate changes and fluctuations. Also included is the establishment of improved monitoring of the progress of the climate, with attention to parameters not usually considered, but that are suggested by our present speculations about human and natural causes of climate change. This means such things as the aerosol composition of air pollution, the depletion rate of atmospheric ozone, the behavior of atmospheric electricity as a function of time, and, of course, the suspected prime culprit, carbon dioxide. Because of the magnitude of the unexpected impact of climate change on human societies, we must be willing to build some redundancy into our research and monitoring, and we must gamble on pursuing leads that are possibly false ones, to assure that we don't overlook important leads that now appear somewhat improbable. The attention given the World Climate Program testifies to our seriousness about moving ahead on this class of costs. But the real test will be when we commit the necessary monies to the programs, and enter into the essential international cooperations to make the programs meaningful and comprehensive.

The second class of costs is tougher to assess and still more difficult to decide to pay. It involves social innovation, long lead times, and serious uncertainties about the scientific nature of the risks. Moreover, it involves planning today for social innovations for a future society whose characteristics are uncertain, and cannot be predicted from the present knowledge of the components from which that future society will be built. Donald N. Michael (7) and others have called attention to the difficulties of technology assessment in an emerging world society. The evolution of such a society cannot be predicted with any security on the 25-50 year time scale over which we expect climate changes to become important stresses. Economists cannot predict the growth of capital or the cost of money at such lead times. Military experts cannot foretell the character of war nor the probability of disarmament 50 years hence. Yet we face the need to assess future potential costs that justify present expenditures to start us on the way to the social innovations that will reduce our losses in this distant future.

Among the costs we must assess are such things as what it will take to bring about the social and political reforms in all countries that will build resilience in the face of climate stresses. How far can we go toward a new international order--not just economic--that will advance equity and freedom and the quality of life? What should we pay now to take the first steps? How much are we prepared to sacrifice today in order to exercise a heightened responsibility toward future generations in this uncertain emerging world? Are we willing to reduce our energy usage? What are we ready to forego in order to elevate moral and philosophical and humanistic values to a more revered place in political decision-making?

One cost that will be levied against the scientific community has been underlined by Thomas H. Moss[8]. He points out that "science will be thrust into the center stage of basic social and philosophical debates" and that this will probably require "loss of some of the cherished anarchy, independence and detachment of the scientific enterprise." It seems clear that, in a changed world order, scientific and technological expertise will have higher value than ever before. Almost certainly the scientist will have to become more directly the servant of social need. As we move closer to the limit of the carrying capacity of the planet, in terms of meeting the wants and needs of increasing world populations, it becomes ever more important to manage our environment and resources more scientifically and more skillfully.

A cheerful note in all this, however, is that many of the things that will be mandated for human societies by climate change will be good for us to do, quite aside from considerations of climate. Robert Schware [9] indicates: "climate-related policies can increasingly be viewed in terms of reinforced motivations to get policy-makers to do what they should (and may want to) do anyway." He cites several examples. Reforestation may alleviate the CO_2 problem, and it will also improve the environment and economies of the forested regions. Interdisciplinary research on the CO_2 problem will help break down the barriers separating disciplines.

To me, however, the most important reinforced incentives will come in the fields likely to be most impacted by climate fluctuations or changes, namely agriculture and water management. It will be costly to develop better arid land management for agriculture and forestry. But the gains will be substantial, quite aside from considerations of improved resilience against climate vagaries. The same is true of water and soil conservation. Expanded effort to develop genetic strains of food and fibre crops that will resist bad climate will also yield improved protection against disease, and more nutritious and desirable food crops. Efforts to improve food processing and storage to prevent climate damage will also result in better food availability even in normal years, and also greater resilience in the face of disasters other than those resulting from climate.

We are, I am sure, only at the very beginning of the long pathway that will lead us to a better understanding of the cost of climate impacts. The problems we face seem at times intractable. The political world often appears chaotic and irrational, particularly when we view the incredible trillions of dollars we spend to prepare for a world war we earnestly hope to deter. To be sure, the emergent society is only dimly visible to us today. But I am confident that we can start down the path, and that its mainstream and its byways will, for the costs we pay, bring benefits that we are as powerless to predict today as we are the problems.

References

(1) R. Garcia, Nature Pleads Not Guilty, Pergamon (New York), In Press

(2) M. I. Budyko, Climatic Changes, published by American Geophysical Union, Waverly Press, Inc., Baltimore 1977

(3) R. M. White "Climate at the Millenium," World Climate Conference: Extended Summaries of Papers Presented at the Conference, World Meteorological Organization, Geneva 1979

(4) G. M. Woodwell, "The Carbon Dioxide Question," Scientific American, Vol. 238, no. 1, January 1978

(5) C. Marchetti, "On Geoengineering and The CO_2 Problem," Climatic Change, 1, 59-68, 1977

(6) U.S. GARP Committee, Understanding Climatic Change: A Program for Action, National Academy of Sciences (Washington) 1975

(7) D. N. Michael "Technology Assessment in an Emerging World," Technological Forecasting and Social Change, II, 189-195 (Elsevier) 1978

(8) T. H. Moss, private communication of a draft paper for future publication in a policy journal.

(9) R. Schware, IIASA Task Force paper, "The Nature of Climate and Society Research," Laxenburg, February 4-6, 1980

DISCUSSION

Schiefer:

1. When considering risks and costs of decisions, you have also to take into account the risks and costs that these decisions will cause in the future. Used only in LWR, uranium will be in short supply soon. Thus the nuclear energy strategy must lead to the fast breeder reactor. But for the FBR and the reprocessing of nuclear waste there are now no risk and cost data available. So an overall risk and cost calculation of the nuclear energy strategy is not possible, and a calculation for only one component can lead to decision errors.
2. Nuclear waste nowadays is mostly stored in temporal deposits. The problems of final storage are not yet solved, or the solutions have not been realised on a large enough scale. So the risk and cost calculations for the light water reactor system alone is not complete, and so the results can not be compared with those of other energy systems.

Niehaus:

1. The development of LWRs is a useful technology even if it does not lead to fast breeder reactors. However, I agree that because of the limited uranium resources, LWRs can not solve the problems of long term energy supply.
2. In my opinion there exist several safe and cheap concepts for the disposal of highly radioactive waste. The problem is that of choosing the best out of a set of good solutions. The risk, especially per unit of electricity production, will be comparatively very low. The technology of reprocessing is well understood and data on health impacts are reliable upper estimates.

Zimmermeyer:

I found your presentation very impressive but I disagree with your data on environmental health diseases and deaths. These data are mainly based on the epidemiological studies of Lave and Seskin. However, they have been criticized very much by American and German epidemiologists. It seems to them and to me that the risks in this field have been overestimated by a factor of about 10. Because this gives the biggest contribution to the overall risks of coal, so that they seem to be too large by an order of magnitude, the studies of Lave and Seskin have never been used as a basis for setting national standards.
Secondly, I should like to mention that the data on SO_2-emissions are too large. In Germany the standard is now 2.75kg/MWh referring to 850mg/m^3. That leads to a percentage of 0,2-0,3

W. Bach, J. Pankrath, and J. Williams (eds.), Interactions of Energy and Climate, 469–471.

instead of 0,5. Therefore the calculations by Hamilton are at least wrong for the FRG.

Niehaus:
In my opinion the study done by Lave represents the best data available at present. I also pointed out that we assume a 5% probability that there will be no effects at all, and also a 5% probability that the effects will be larger than indicated. The data are based on the US situation, they will be different in each country depending on the coal and the standards applied, e.g. accident and disease data will be higher for the FRG (as pointed out in the paper).

Waltz:
Since the climate may also be cooling for natural reasons, temperature increases are generally growth-enhancing for agriculture and atmospheric CO_2 increases are also growth-enhancing for agriculture, your capital transformation costs seem highly uncertain and very likely overstated. The final two points, indeed, suggest that CO_2 would yield a net contribution of "natural climatic resource capital" to society. Have you considered these possibilities and what difference do you think they might make to your results?

Laurmann:
I accept in principle the conceptual possibilities outlined by Dr. Waltz. In particular the operational costs induced by potential climatic change, both as estimated in CIAP and in later studies of the US Department of Agriculture, show a net benefit from small amounts of global warming. However, it is clear that a sufficiently large increase in global temperatures must result in economic loss. The operational cost estimates, not highlighted in the paper, reflect this situation for CO_2 doubling conditions, but are even less reliable than the figures calculated for capital loss rates. As for the latter, I can only suggest that, if we think of the socio-economic system as being optimally adapted to its immediate environment, any change in climate, be it reflected in an increase or a decrease in temperature, will result in reduced economic efficiency. Finally, regarding your request to have me estimate the change in costs that might result from a proper allowance for the effects you identify, if I knew the answer, the ingredient, in the form of an estimate of the probable error distribution, would be available for making a complete and final decision analytic estimate, and except for the question of social discounting, our problem would be solved. This is, indeed, a target at which we should be aiming.

Quirk:

I have a comment on all three papers in this session. One source of energy is conservation. Dr. Page has shown that for each house in Hamburg that you bring up to the Scandinavian standard you would free up to 3 kw of energy generation. There would be industrial health costs, and also public health costs as a result of decreased ventilation. I would be very curious to see where this would fit on the graphs of Dr. Niehaus.

I would be very interested in seeing a calculation of the costs of conservation for avoiding the CO_2 doubling. Could this be added to the graphs of Dr. Laurmann?

I found the talk of Dr. Roberts very inspiring. If the developed world were to conserve energy and raw materials there would be more resources available for the developing countries. The United States in particular has a large potential for conservation. By bringing our per person energy use down to the Western European average, we could stop importing oil, and we would have an energy surplus. Would not this be one of the most important steps that could be taken to help the developing world, whether or not there is a large economic impact from increased carbon dioxide?

PART VII

POSSIBLE PREVENTIVE MEASURES

REDUCTION OF FOSSIL FUEL USE AND ADOPTION OF ALTERNATIVE ENERGY SOURCES

G. Egberts, A. Voss

Programmgruppe Systemforschung und Technologische Entwicklung, Kernforschungsanlage Jülich GmbH, Postfach 1913, D-5170 Jülich 1, West Germany

ABSTRACT. Reduction of carbon dioxide emissions means a much more radical change in the energy supply structure than any possibly successful strategy to replace scarce petroleum could ever be. Moreover, the vulnerability of oil supplies leads to problem solutions, which even aggravate the carbon dioxide problem. Therefore, any improvements must heavily rely on forced energy saving and enhanced use of renewable and nuclear energy. Although the theoretical potential to substitute carbon - and hydrocarbon - fuels has the rather high value of about 2/3 of final energy demand, the technical and economical feasibility of alternative fuels must be judged much more pessimistically. Some recent results of an analysis sponsored by the IEA show, that nuclear and renewable energy supplies may only be able to sustain about 50% of the growth or primary energy demand of the IEA-countries. Thus, if we want to be sure of having enough energy to sustain our lifestyle, we must not expect reductions of carbon dioxide emissions. This can only be achieved by huge efforts in energy saving, connected with a full utilization of the nuclear and renewable resources.

1. INTRODUCTION

The energy supply is by far the greatest man-made source of carbon dioxide. It is obvious, therefore, that the growing energy needs are responsible for the slow but steady increase of the equilibrium concentration of carbon dioxide in the atmosphere. Thus, if we do not want to further perturb the carbon dioxide cycle, the logical procedure would be a considerable reduction of carbon use for energy supplies. This implies

W. Bach, J. Pankrath, and J. Williams (eds.), Interactions of Energy and Climate, 473–481.

a radical change of our energy system to non-fossil sources, whether they are renewable or not. Such a change would be forced anyway, sooner or later, by the finite extent of fossil energy resources.

2. ALTERNATIVE ENERGY SYSTEMS FOR REDUCTION OF FOSSIL FUEL USE

One might doubt to what extent such a substitution process could be feasible in a foreseeable time period. The experiences after the oil crisis justify a certain scepticism. If we did not manage to substitute scarce and expensive oil to a satisfactory extent in the past, how can we expect to make progress to substitute not only for oil but for all fossil fuels in the future? Even worse, if we can reach a considerable reduction of oil imports, it seems most likely that the carbon dioxide situation becomes even more aggravated. The reason is that a reduction of oil product consumption will only partially be achieved by energy saving. A great deal will be met by substitutes, such as electricity or products from coal conversion. But these are produced much less efficiently and usually from sources with a higher carbon content. This is shown in Figure 1 for motor fuel, useful energy for heating and electricity (1) :

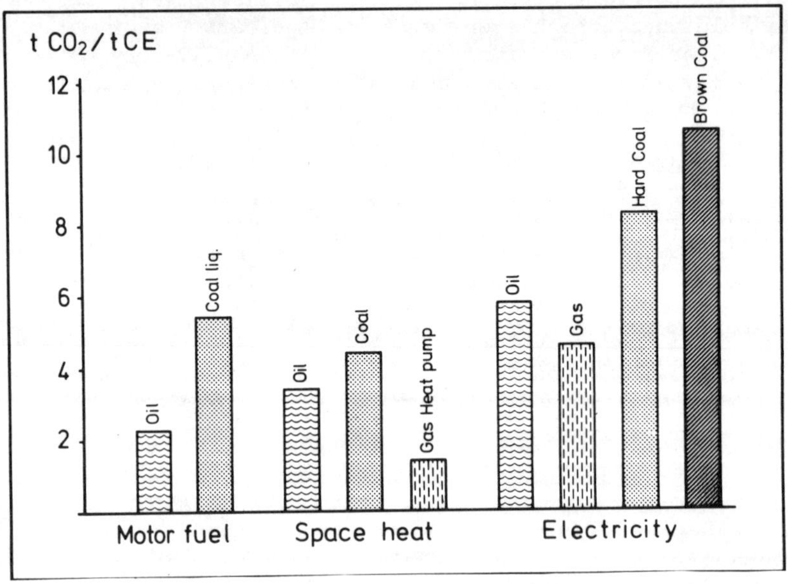

Fig. 1: Carbon dioxide emissions to produce selected final energy forms

A substitution of oil by coal products leads to a considerable increase of carbon dioxide emissions, in some cases more than two times as much. Natural gas supplies are also rather restricted and gas prices are more or less tied to oil prices. So, if we want to save oil and reduce carbon dioxide, we cannot stick to conventional measures. Instead new technologies and new sources must be utilized to a great extent. These are energy saving, renewable and nuclear energy.

One of the most important advantages of fossil fuels, especially oil products, is the very high specific energy content of the fuel. This has economic and technical advantages for transportation and storage of energy. Thus a shift away from fossil fuel will be accompanied by a far reaching change of the infrastructure for transportation and distribution of energy. Technically, a large part of the final energy demand based on fossil fuels at present could be met by electricity produced from nuclear or renewable sources. But the needs for energy storage will require a certain fraction of chemically bound energy anyway. This must mainly be supplied by hydrogen in a real post-carbon future. If we add local hot water distribution systems for heating, fed by coupled production plants and synthetic hydrocarbons based on biomass, we get the following picture of an imaginary post-carbon energy supply structure:

Energy Sources:

-Solar
-Biomass
-Geothermal
-Nuclear
-Energy Saving

Energy Carriers:

-Electricity
-Hot Water
-Hydrogen
-Hydrocarbons from Biomass

In principle all energy services could be met by those secondary energy carriers. Electricity is able to meet all energy demands in the same way as hydrogen does. Hot water is restricted to space heating and warm water generation, whereas fuels from biomass, such as ethanol or methanol, could mainly be used for individual transportation. Going one step further one can imagine a more distant future mainly based on nuclear fission or fusion, where practically all energy demands are met by electricity or hydrogen.

3. POSSIBILITIES OF FOSSIL FUEL SUBSTITUTION IN THE OECD COUNTRIES

If we restrict ourselves to the foreseeable future, we can surmise that we will not be able to substitute carbon-based fuels in the sectors of non-energy use and individual transportation to a satisfactory degree. Since these sectors are heavily dependent on fluid hydrocarbons for both chemical and physical reasons, we can expect that a substitution, if possible, will concentrate on the other sectors of energy demand. A glance at the present structure of energy demand and supply of the OECD countries shows that demands for non-energy use and transportation form only about 7% and 25% respectively of the total final energy demand. Thus, from the viewpoint of the technical nature of the energy utilization systems, about two thirds of the final energy demand could be from non-fossil sources. The question is, whether we are able to develop such sources to the necessary degree.

Considering the energy resources we are relying on at present, our task becomes obvious; we have to substitute oil, but not by a regression to older and less efficient technologies, such as coal conversion. Since such a strategy would lead to a considerable aggravation of the carbon dioxide problem, we are forced to look forward for new unconventional problem solutions. One of the most attractive ways out of this dilemma could be the direct conversion of solar light into hydrogen and hydrocarbons by means of biological catalysts. Genetic engineering, the scientific and technological challenge of the next decades, is likely to provide an important contribution to a stable, efficient and clean solution to the energy problem. However, this area of research has just been opened, everything is still in the laboratory stage, but the problems we are faced with are urgent at the moment (2).

To get an idea of the possible development of the energy system to a reasonable degree in a reasonable time frame, reference can be made to a study performed by the International Energy Agency. This work has been carried out in co-operation with the member countries of the agency in order to elaborate common recommendations for a co-ordinated research and development policy within the OECD to replace the scarce oil (3).

The resulting quantities and structures of future energy supplies of this study shall be taken here as an example of a conceivable or possible future of energy supplies in the western hemisphere. The general objective of this study was to estimate the possibilities for oil substitution within a general economic framework but without any care about environmental factors.

Using a linear optimization model of energy supply and utilization, the development over the next forty years of all national energy economies involved in that project has been investigated individually (4). For a certain, well defined set of scenarios only those system developments are determined, which lead to minimal system costs of energy supply over the whole time period of consideration. The optimal mix of technologies and energy carriers resulting from such calculations can give valuable information about the relative virtues and shortcomings of energy technologies. This may be used as a guideline for a goal-orientated technology policy. The starting point of the calculations is the year 1980. The structure of the 15 OECD-countries, which participated in the project so far is shown in Table 1:

Table 1: Projected energy supply structure of 15 IEA-countries
1EJ = 34.12 Mio t Coal Equivalent

	Coal	Oil	Gas	Hydro/ Nuclear	Electricity
Primary Consumption	31,0	75,0	28,0	16.0	
Input to Conversion/Losses	22,0	13,5	2,5	16.0	
Final Consumption	9,0	61,6	25,5		15,0
Traffic		30,5			0,5
Non energy use		6,5	1,5		
Industry	8,0	11,0	14,5		7,0
Residential & Commercial	1,0	13,5	9,5		7,5

These 15 countries, namely USA, Japan, Canada, New Zealand and western Europe except France, Portugal and the Netherlands, form, at present, about two thirds of the oil demand of non-communist countries. It can be seen that mineral oil covers about 50% of the primary energy demand. Coal and gas have a fraction of about 20% each.

The assumptions for an unperturbed development yield the following optimized supply structure for the year 2000 (5). The scenario shown in Figure 2 is based on the rather optimistic assumption of an average growth of final energy demand of about 2% per annum. This growth is predominantly met by coal and nuclear energy, whereas the use of oil and gas remains approximately constant over time. So within this framework the dependence on crude oil is reduced from about 50% now to about one third in the next 20 years. This is achieved by the relative substitution of oil products in the end use, especially the industrial and residential sectors. No liquid substitute can gain a market share.

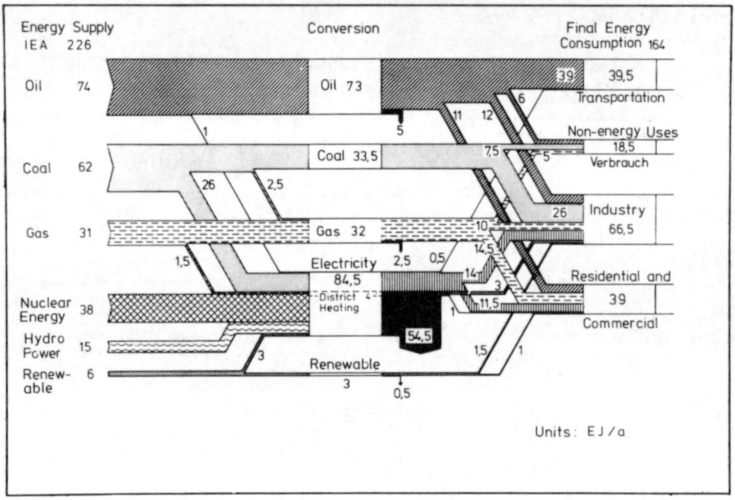

Fig. 2: Optimized energy supply structure of 15 IEA-countries for the year 2000, - Base case -

Since a constant level of oil consumption over time does not seem very satisfactory, a set of strategy runs assuming a forced reduction of oil imports has been performed. The system configuration in the year 2000 over the whole planning period is shown in Figure 3.

Here, oil inputs are reduced to about one quarter of the total primary energy demand. The substitution is mainly made by coal, nuclear and renewables. A small amount of coal liquefaction enters the market first. But again the main part of oil saving and substitution is achieved in the consumption sectors. Here, heat pumps and industrial co-generation have the greatest importance as technologies for oil substitution.

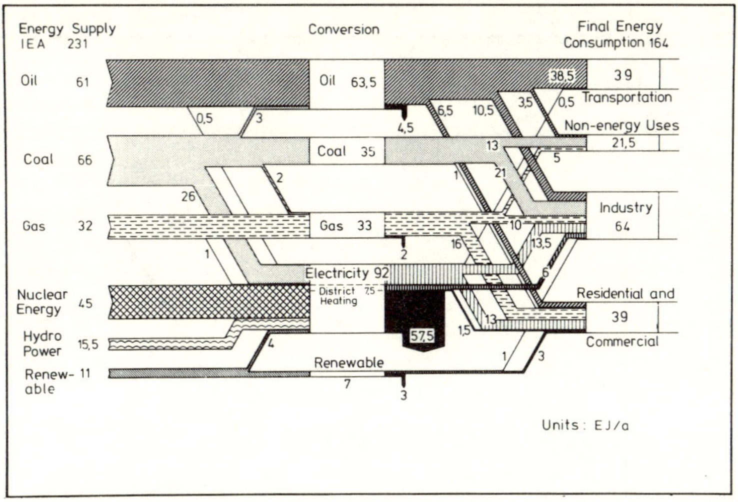

Fig. 3: Optimized energy supply structure of 15 IEA-countries for the year 2000, - Forced oil saving -

Figure 4 gives an idea of the impacts of those scenarios on the carbon dioxide cycle. These are the total carbon dioxide emissions of the energy supply systems we are dealing with in billion of tons per year. The base case shows an increase of emissions of about 70% over the planning period. The case of forced oil reduction shows a slight reduction in carbon dioxide emissions too. This is due to the fact that in our model calculations energy saving turned out to be more important for oil saving than substitution by synthetic fuels. As a comparison the hypothetical cases are drawn, where all nuclear and all renewables are replaced by coal. These two upper lines can give an idea of the relative importance of both nuclear and renewable energy for the carbon dioxide emissions.

4. CONCLUSIONS

Thus, we can conclude that an improvement of the carbon dioxide situation is rather unlikely. As this diagram shows, even with optimistic assumptions about energy demand growth and the availability of alternative sources, emissions turned out to show a considerable growth. Thus, even if nuclear and renewable energy can help us a great deal, none of these sources are sufficient. Success can only be obtained by the use of all measures and it seems that the most important one will be an extensive energy saving by more efficient equipment and a new lifestyle, conscious of scarcities.

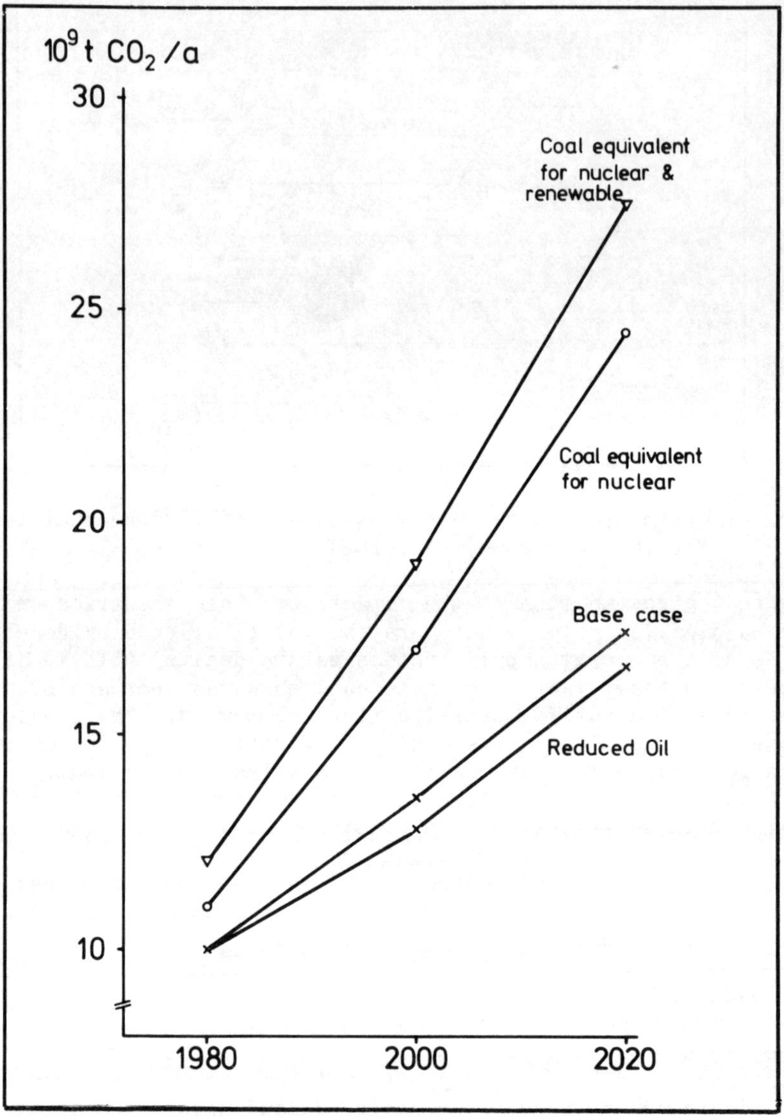

Fig. 4: Carbon dioxide emissions resulting from model calculations for 15 IEA-countries

REFERENCES

1. F. Niehaus
Langzeitaspekte der Umweltbelastung durch Energieerzeugung: CO_2 und H^3 Jül - 1165; Jülich 1975

2. P. Böger
Photobiologische Umwandlung der Sonnenenergie
Naturwissenschaften 65, pp. 407 - 412
Springer-Verlag 1978

3. Annual Report on Energy Research, Development and Demonstration - Activities of the IEA 1977-1978
IEA, Paris 1979

4. G. Egberts
MARKAL - A multi-period linear optimization model of the energy supply system
Jülich 1980 (to be published)

5. Beckurts, K.-H.
Einsatzmöglichkeiten und Substitutionspotential der Kernenergie
Deutsches Atomforum; Mainz 1980

CONSERVATION PRACTICES AND INCREASED EFFICIENCY OF ENERGY CONVERSION AND USAGE

H. Schaefer

Research Institute for Energy Utilization and Technology, Technical University of Munich, 8000 Munich 50, F.R.G.

ABSTRACT. Society cannot do without modern energy technology if it is to maintain the present structures of human relations and the standard of living reached so far. It is therefore imperative that energy supplies be secured at reasonable cost and with a minimum of environmental impact.

In view of the present situation in West Germany and the energy supply pattern it is obvious that a more economic, efficient and sensible use of energy should be demanded.

Possibilities to make more of resources are partially of a non-technical, more of a technical, and mostly of an energy nature. In general, no new technologies need to be introduced, but existing knowlegde and well-tried techniques will have to be coordinated and integrated if we are to meet the following objectives:

- to avoid unnecessary consumption of net energy;
- to reduce specific consumption of net energy for some applications;
- to reduce primary energy needs related to net energy needs;
- to increase utilisation of renewable energy sources;
- to recover energy wherever technically and economically feasible.

Substantial latent potential for energy conservation can be liberated if we succeed in explaining to the layman the physical and technical concepts which enable him to access the true value of each type of net energy. This would also make him realize that there is a way of operating systems that will reduce the specific energy cost without sacrificing productivity or comfort.

W. Bach, J. Pankrath, and J. Williams (eds.), Interactions of Energy and Climate, 483–494.

1. INTRODUCTION

The need to use energy more efficiently, i.e., more economically and wisely, on technical and ecological reasons, is no new demand. In antiquity, the wasteful use of wood had disastrous implications which are still evidenced nowadays in the ancient cities of Miletus, Priene and Ephesus in Asia Minor. England's forests were only saved from complete clear-cutting by the invention of the steam engine which accelerated the expansion of the coal-mining industry. Ever since man made energy technology his servant, to make himself independant of his environment, many efforts were undertaken to make better use of energy; in particular the past two centuries and the technical advances they bore have been of great importance in this respect. Restraints were abandoned to use fossil fuels in their original form, as gas works, refineries and power plants helped to supply the consumer with refined energy which was more efficient and ecologically reliable to use, increasing comfort at the same time.

The three main reasons underlying the demand for energy conservation are the following:

(i) At present more than 90 per cent of the world's total energy demand rests on fossil fuels which are non-renewable and limited in quantity. Taking the global resources of fossil fuels as 100 % the share of oil is about 16 percent and that of natural gas about 10 percent. However, the share of oil and natural gas accounts for 65 per cent of the world's total energy consumption. A break away from worldwide dependence on liquid energy sources is therefore a prime necessity.

(ii) Any use of energy has an ecological impact, which does not necessarily have to be adverse. Thus the thermal load of conurbations resulting from the total energy consumption nowadays contributes towards higher air exchange rates; and in our degree of latitude it also means that urban areas are blessed with more sunshine duration during the winter months than the countryside (1) . Adverse effects, however, do impose constraints which, in many cases, are still difficult to quantify.

(iii) Although fossil and nuclear fuels, as well as solar energy are abundant, they are limited in their availability which will make the cost of meeting energy demands rise considerably.

Directions where to use energy more rationally can be found by an analysis of the energy balance of the Federal Republic of Germany. The main components of this energy balance are pointed out in Figure 1 with all data based on the year 1977 (2).

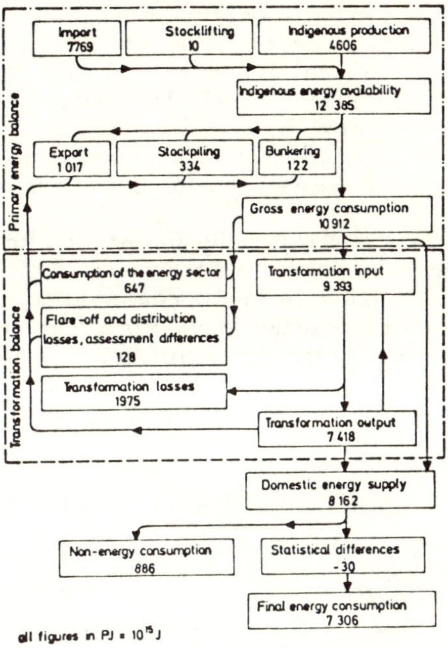

Figure 1 : Energy balance Scheme for the Federal Republic of Germany in 1977

Energy needs and losses in the energy sector account for some 28 per cent of the primary energy consumed, with the needs and losses in generation and distribution of electricity making up three-quarters. As 7 % of the primary energy consumption was used for energetic purposes, 65 % of the primary energy was the final consumption on electricity, solid, fluid and gaseous fuels. Energy conversion losses in final consumption amount to approximately 38 per cent of the consumed primary energy, in other words, they accounted for 1 1/2 times the losses in the energy sector. Viewed in this light, it is evident that rationalisation must concentrate on final consumption, all the more since improved efficiencies in the energy sector can only be realized on a small scale, and despite high cost. There are, however, much easier and simpler ways to increase efficiencies in final consumption and energy application. What is more is that considerably lower efficiencies are to be expected in the energy sector if we are to turn to new technologies, such as the gasification and liquefaction of coal, to reduce oil dependence. These reduced efficiencies can hardly be offset against the advantages inherent in using primary energy for power generation, such as in improved cogeneration of heat and power or the construction of binary systems as in combined gas-steam turbine systems.

Looking at the breakdown of final energy consumption into applications (Figure 2), it becomes clear that heat plays a dominating role. Space heating at 39 per cent, and process heat at some 36 per cent, together account for three-fourths of the total demand of final energy demand in the Federal Republic of Germany in 1977. Transportation takes 20 per cent of the remaining and approx. 4 per cent is used for meeting the demand for drives in industries, households, commercial and agriculture sectors (3). The share of final energy used for lighting is about 1 per cent.

This breakdown clearly underlines the top priority to be given to space heating, followed by process heat and the energy required in the transportation sector when it comes to taking conservation action. Next in line is the mechanical energy needed in industry, households and the other sectors. Any intervention in the light consumption field could, at best, have a signalling effect.

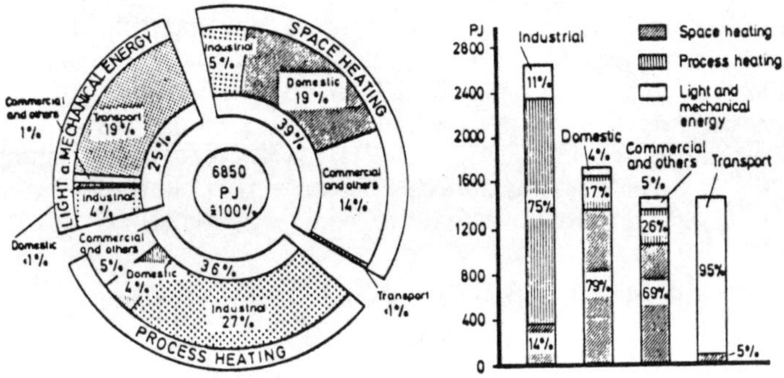

Figure 2 : Final energy demand 1975 Fed. Republic of Germany

Ways towards energy conservation come in five main groups, which are discussed in the following.

2. AVOIDING UNDUE CONSUMPTION

Energy is consumed unnecessarily whenever machinery and plants are operated with idle running, whenever rooms are overheated and undue quantities of hot water are tapped, in other words, in all those cases where the generation of net energy does neither improve production nor create additional services or enhance comfort. Little can be done in this field technically, but rather a lot can be achieved by informing the public at large about fundamental concepts. Light and its importance as the source of all organic life is normally overrated, at least subconsciously. This also goes for power and mechanical work, which, because of our associating it with muscle power, is also rated high. This contrasts strongly with heat which is extremly underrated, because we are only able to perceive temperatures but not quantities of heat. No one seems to realise that one kilowatt-hour can be used to keep a 60 W bulb on for some 17 hours, or to raise 400 litres of water to six times the height of the Cologne Cathedral, but that one kilowatt-hour would be just enough for a two-minute shower!

In view of the necessary information and education in this field, school education plays a most important role. What is desirable is less disparagement of technology and an improvement in the observation and teaching of the physical and technical world. At the same time, the mass media ought to aim at a more objective and matter-of-fact presentation of the problem.

3. REDUCING SPECIFIC NET ENERGY DEMAND

It is mainly technical action that could help to cut the net energy demand in certain applications. Good heat insulation in buildings, coupled with easy-to-control heating systems, scale down the need for heat. The same goes for an increased passive utilisation of solar energy by a sun orientated positioning of buildings and their main window areas. Thus each architecture and space enclosing materials with great heat capacity will reduce the need for cooling and/or eliminate this need altogether. Aerodynamicly styled vehicles and reduced vehicle weight will deflate the specific need in some parts of the transportation sector. Optimum energy technologies should be singled out. For instance, mechanical instead of thermal drying will cause the energy demand per kilogramme of extracted water to drop to one-hundredth. Other examples include vacuum evaporation, pressure cooking, glueing instead of welding, material recycling, non-cutting instead of cutting shaping.

The net energy need could also be reduced if more passengers were to use public transport instead of their own private car. To transport 50 persons over a distance of one kilometer it takes 3.8 MJ of primary energy if they all go in their own car. If this need were to equal 100 per cent, the use of a public transport bus would bring the percentage down to about 7 per cent, and the tramway to 40 per cent, with utilisation being 50 and 23 per cent, respectively (4).

Recycling material can equally contribute to energy savings. The cumulated energy required to make a euro-glass bottle of 0.5 litre will drop by 2.5 per cent in power consumption if the share of breakage is increased from 17 per cent to 36 per cent. Simultaneously, fuel consumption would drop by 11 per cent, and the total final energy consumption would fall by 10 per cent.

This raises the question of packaging, which is very often undervalued in terms of cumulated energy that goes into making a product. Studies at the Institute for Energy Technology and Power Plant Engineering (IfE,

Technical University of Munich) by Flaschar (5) have shown that the energy that goes into packaging of yoghurt, packed in a 175 g PVC pot with aluminium seal, accounts for 41 per cent of the total cumulated primary energy requirement of 14,700 kJ/kg. Further down the production chain is raw milk production, at 32 per cent; and the energy cost consumption of a dairy comes third in line, although, intuitively, one would have estimated the dairy's share to be much higher.

A change of material can also lead to substantial reductions in energy consumption. For instance, the 3,500 kJ of energy needed to produce a 0.33 litre tin with aluminium lid can be lowered by 37 per cent to 2,200 kJ for the same tin with a tin lid (6).

Despite all such savings there may be substitutes which introduce unfavourable effects. Replacing copper as conductive material by aluminium in electrical engineering applications will raise the cumulated energy consumption by a factor of 1.5, with equal conductivity.

4. IMPROVING EFFICIENCIES

Design improvements of plant and machinery and in particular the integration of modern control techniques will in many cases lead to a considerable reduction of specific energy consumption. This is true of household appliances; for instance, washing machines where new technologies are expected to bring about savings of 25 per cent. It also applies to single-purpose appliances, such as coffee makers which use only half as much energy to prepare the same potful of coffee on a cooker with a kettle, both being multi-purpose appliances.

In the field of road transport, considerable savings are anticipated long-term: electronic ignition control systems, fuel injection and load-dependent cylinder cut-off. Including design modifications and weight reduction, the petrol demand of cars could almost be halved, and that of trucks be cut to 70 per cent (7).

Another case in point of conservation is the co-generation of power and heat, which has been applied in the Federal Republic of Germany for many years. This provides a sound basis for interesting further developments in this field. This type of co-generation almost doubles primary energy efficiencies. Power stations using gas-driven combustion engines constitute a significant additional alternative to conventional thermal power plants using back pressure steam or gas turbines (Figure 3).

The principal source of conservation of energy lies in the advanced operating mode and in the optimum utilisation of existing energy technologies. In all energy sectors, a major part of the energy consumption is not a function of load, but one of size and type of the plant or machinery. This also means that the specific consumption drops as the load rises. Operating plant and machinery to full capacity and avoiding idlerunning will improve the efficiency on a large scale; this goes for washing machines in households, forging furnaces, passenger cars, boilers, and aircraft alike. Important is the matching of installed system loadings to actual needs. Overdimensioned drives in particular are not only unduly high in investment cost, but also cause an increased energy consumption.

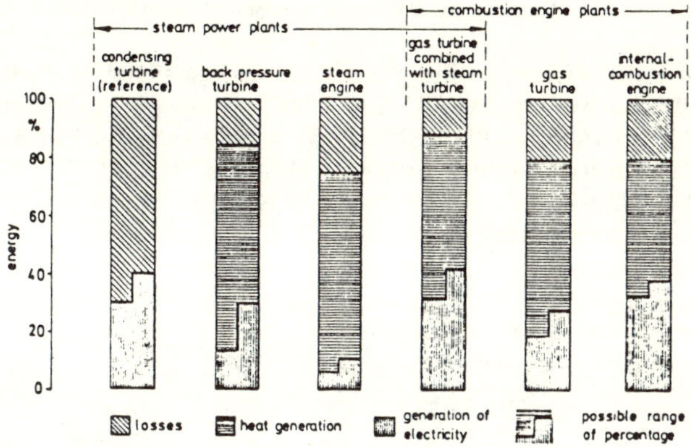

Figure 3 : Combined generation of electricity and heat energy balances of different types of plants

5. ENERGY RECOVERY

Only in electric traction and cars containing a flywheel mechanical energy during braking can be recovered. All other kinds of energy recovery rely on the recovery of heat.

Such methods are by no means new. Leonardo da Vinci already designed a 'waste gas turbine' for his roasting spit, and the Bavarian King Ludwig II had a similar system fitted at the kitchen of the Neuschwanstein Castle. In industry, heat recovery has been practised in many different ways and for a long time; even so, more development work will be needed. What is more important, however, is to introduce the well-known and tried systems in all those fields where heat recovery has not yet become a household term. While it is common practice, for instance, to use heat exchangers and heat pumps to recover the heat in air-conditioned buildings, the translation of such techniques into the household and small consumer applications calls for modifications and even new developments because here the professional control and maintenance of the systems will be lacking.

Problems in heat recovery arise whenever the temperature level of the waste heat is too low. In this case a heat pump is introduced to exploit the heat flux. It needs to be noted, however, that at present, the maximum temperature level is approximately 60 °C, which makes it a valid proposition in heating systems that can be run at low temperatures, such as floor, wall and ceiling heating systems.

Another problem rests in the disparity between the time and place of waste heat supply on the one hand, and the demand of low temperature heat on the other hand. The installation of additional storages could ease the situation to some extent by balancing out supply and demand.

The recovery and use of waste heat may also have an ecological impact. As Figure 4 shows in principle the heat in domestic waste water is utilised through heat pumps (with these waste waters containing about 15 per cent of the total energy consumed in a household) (8), the temperature of the waste water will be reduced and therefore the quality of purification in sewage works will be lowered.

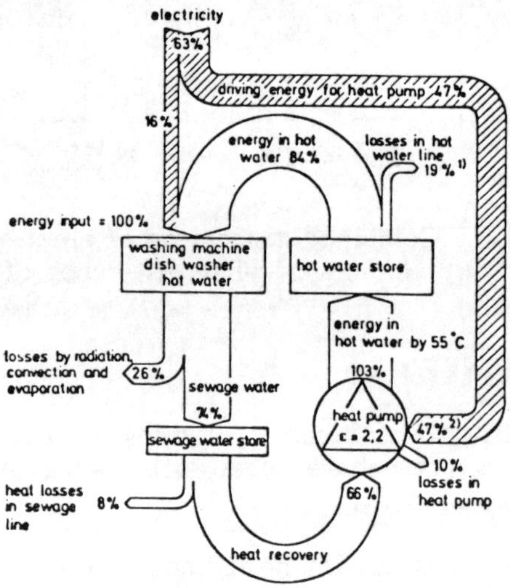

1) without regard of circulation losses
2) without auxiliary energy consumption

figured in percentage of the energy input

Figure 4 : Flow diagram for heat recovered from domestic sewage water

Here, as in other cases, any intervention in one part of the overall energy grid will have to be checked against adverse impacts in other parts of the system. Then one will have to investigate how much effects can be mitigated. In the above case, structural modifications, heat recovery systems and optimum utilisation of the by-product biogas, will contribute towards an improvement of the sewage plant energy balance, so that lower intake temperatures will not affect the purification process.

6. USING RENEWABLE ENERGY SOURCES

Renewable sources of energy are for the most part direct or indirect solar energy, such as, the energy in wind and flowing streams, ground heat and chemically bound biomass energy. In addition to that, there is geothermal energy flowing from the hot interior of the earth to the surface, energies from the gravitational attraction between Earth and Moon, and the waste heat from anthropogenic energy consumption. Several ways are given to use this energy, and although there are technical and economic constraints today many of them are most likely to be removed in the future.

Renewables, besides nuclear fuels and possibly useful nuclear fusion energy, are of prime importance in replacing fossil fuels in the long term.

It should be noted, that, in contrast to fossil fuels, these sources depend upon technologies based on a growing need of auxiliary energy; and it is electric energy that comes into consideration most. In oil central heating systems, the electric energy that powers the burner, pumps, and controls accounts for 1 to 1.5 per cent of the total heat output. This compares with about 10 per cent in heating systems that use solar collectors. Moreover, the space and materials required per unit of power will have to be higher than in conventional systems since the power densities of natural renewable sources are low. The environmental impact, due to occur with large-scale utilisation of renewable energy sources, will also have to be taken into account. A few cases in point are the shift in the radiation balance through solar energy use, the formation of cavities in the interior of the earth as a result of contraction following the extraction of geothermal energy, the change of the Earth's rotation caused by tidal power plants, a reduction in air exchange and water vapour transport following the exploitation of wind energy by wind power stations, and the creation of cold air sinks in the case of air heat pumps.

Solar collectors could best be used for heating swimming pools; for making hot water, however, they are hardly economical up to now. If solar energy is to be used in a heating system, this would also require a heat pump, in general, often an 'energy roof' which combines the principle of a solar collector with that of a large surface air-heat exchanger, might turn out to be more economical than flat collectors (9).

For heat pumps, ground water or soil will be the most favoured source of heat, because both these sources will be subject to only slight seasonal changes in temperature. Applications of these sources are limited for ecological and water conservation reasons on the one hand, and the large space requirement on the other hand. This means that in future, heat pumps preferably will use air as a heat source since this is available in abundance. Its inherent disadvantage of requiring more heat when temperatures fall will be overcome by designing suitable bi-valent systems.

The only tested drive available for small-capacity heat pumps is the electric motor; for the larger capacities, a combustion engine may be a very interesting proposition, especially when gas-fired.

Biomass is already being used to some extent in energy technology, such as in the incineration of refuse in thermal plants and biogas being used in sewage works. A lot remains to be done in this field, and methods of straw combustion and pyrolytic processes are presently under investigation. Biogas production could also be increased. As regards the energy situation in West Germany, prime attention should be devoted to the use of organic wastes; since conditions do not favour energy farming in this country.

7. PROBLEMS AND CONSTRAINTS OF ENERGY CONSERVATION

The fact that energy conservation has become the subject of vigorous debates points to the great challenge in pushing the technical advance in this direction. However, in spite of all distrust in technology and experts, expectations in 'overnight' improvements are often much too great. More often than not, it is believed that one can do anything with technology and that it is an easy thing to develop an idea, and within ever shorter periods of time see it through to a well-tested, reliable, safe, maintenance-free and low-cost product; ignoring, of course, that in fact technical advances take considerable time to mature from first conception to final demonstration. It is the mere necessity to carry out field tests, in addition to analytical studies and laboratory tests, that calls for a period of several, if not many, years.

As for anything else, there is a price to pay for energy conservation. Even the plethora of possibilities available to meet the energy demand more efficiently by making better use of existing energy systems does require much to stimulate the information process, actuating a re-thinking process on the part of the individual consumer.

In many cases the more efficient energy system will require more material cost than the conventional system; in other words, energy is substituted by capital. Moreover, technology is getting more and more complex in general, and its fundamental concepts less and less clear to the layman, just as to the expert. Accordingly, repair and maintenance require increasingly qualified and specially trained staff, and hence are turning into a significant cost factor.

As with most other technologies used for energy conservation, the electric energy used for auxiliary and control functions is a substantial factor in the total balance. No matter whether such systems are used for recovering waste heat or converting direct or indirect solar energy, they will require a larger share of auxiliary energy in relation to the net energy supplied.

In terms of national economy, the assessment of energy conservation measures presupposes relevant data, and therefore causes problems of practicality. In heat recovery technologies, it has been next to impossible to supply statistical evidence on their effect upon the final and primary energy balance of, say, the Federal Republic of Germany. This problem is amplified with the use of environmental energy. In many cases it is impossible to precisely quantify the energy savings because they are not only determinded by the efficiency of systems and components, but also by their correct design and operation. A large-capacity electric motor will be more efficient on rated duty than a smaller one. However, its efficiency may be substantially lower under partial load than that of a smaller motor made to match the load. It is very often ignored that improved efficiencies of system components do not add to, but multiply, savings. In other words, a 15 per cent reduction of consumption obtained with a better boiler, and another 15 per cent reduction obtained with an improved control of the connected heating systems, will not cut the overall savings by 30 per cent, but by 22.5 per cent. Questionable are all those statements on savings and percentages if the basis to which they relate and the conditions under which they apply, fail to appear.

8. CONCLUSIONS

The efforts not only recently made in the technological and scientific field have led to the commercialisation of a number of technologies and economic alternatives which reduce the specific energy demand. Also, all realistic paths should be taken simultaneously. Certainly, there will be no way to 'eternal energy bliss', and urgent warnings should be raised against monolithic solutions.

In all sectors, future action should rely on the energy-conscious and sensible behaviour of the individual who has a full understanding of the world of physics. An awareness needs to be created among consumers of the importance of such purpose-oriented behaviour. The necessary information on fundamental energy concepts is to protect the citizen against misconceptions, exaggerated expectations, false decision-making, and it will enable him to sufficiently judge strategies and policies of energy questions. Not cheap showmanship on economic and/or political grounds, but realistic and sensible problem solutions make us advance.

REFERENCES

(1) Geiger B.: Methodik zur Strukturentwicklung des urbanen Energiebedarfs für die Analyse ökologischer Auswirkungen. Dissertation, Technische Universität München, 1978

(2) Energiebilanzen der Bundesrepublik Deutschland. Verlags- und Wirtschaftsgesellschaft der Elektrizitätswerke mbH, Frankfurt. Jahrgang 1974 bis 1978

(3) Technologien zur Einsparung von Energie, Band Haushalt und Kleinverbrauch. Forschungsvorhaben ET 5012 des Bundesministers für Forschung und Technologie, bearbeitet durch: Forschungsstelle für Energiewirtschaft, München, FfE-Schriften 77-02, München, Nov. 1977

(4) Technologien zur Einsparung von Energie, Band Verkehr. Forschungsvorhaben ET 5012 des Bundesministers für Forschung und Technologie, bearbeitet durch: Forschungsstelle für Energiewirtschaft, München, FfE-Schriften 77-03, München, 1977

(5) Flaschar, W.: Mikro- und makroanalytische Methoden zur Ermittlung des spezifischen kumulierten Energieverbrauchs zum Herstellen von Verbrauchsgütern. Dissertation, Technische Universität München, 1979

(6) Schaefer, H.: Energetische Bewertung von Verpackungsmaterialien. Vortrag anläßlich der Internationalen Ausstellung UPAKOVKA, Moskau, 1979.

(7) Helling, J.: Energiesparen im Straßenverkehr. Vortrag anläßlich der Internationalen Automobil-Ausstellung, Frankfurt-Main, 1979

(8) Zentrale Wärmerückgewinnung aus dem Warmwasserverbrauch in Mehrfamilienhäusern. Forschungsvorhaben ET 5012 des Bundesministers für Forschung und Technologie, bearbeitet durch: Forschungsstelle für Energiewirtschaft, München; RWE Anwendungstechnik, Essen

(9) Stoy, B.: Wunschenergie Sonne. Energie-Verlag, Heidelberg, 1977

THE COLLECTION, DISPOSAL, AND STORAGE OF CARBON DIOXIDE[1]

C. F. Baes, Jr., S. E. Beall, and D. W. Lee

Oak Ridge National Laboratory,
Oak Ridge, Tennessee 37830, U.S.A.

G. Marland

Institute for Energy Analysis, Oak Ridge Associated Universities, Oak Ridge, Tennessee 37830, U.S.A.

ABSTRACT. To scrub CO_2 from the stack gas of a coal fired plant by the most efficient means presently available will require about 43% of the combustion energy of the coal. While appreciable improvement might be realized by burning the coal in pure oxygen, any method of recovery from stack gas will use a substantial fraction of the energy content of the fuel. The best option for disposal is in the ocean, in the form of a negatively-buoyant plume of concentrated solution that would carry the CO_2 to deep water and give retention times of many hundreds of years. Alternatively, liquid CO_2 could be injected at great depth, or blocks of the solid hydrate or dry ice might be released at the ocean surface. The most efficient option for the future might be floating power stations that use cold, deep seawater both for condenser cooling and CO_2 disposal. Depleted oil and gas fields could be used as a near-term storage option. Storage as buried biomass or as a solid near the South Pole seem the least attractive of the options considered.

1. INTRODUCTION

The carbon dioxide problem (1-3) may be briefly stated as follows: The amount of fossil carbon ultimately available for use as fuel by mankind may be more than ten times the amount of carbon presently in the atmosphere as carbon dioxide. A large fraction of this fuel may be consumed in the next hundred years. Because the

W. Bach, J. Pankrath, and J. Williams (eds.), Interactions of Energy and Climate, 495–519.

resulting rate of release of carbon dioxide to the atmosphere may far exceed the rate at which the remainder of the natural carbon cycle can take it up, the atmospheric carbon dioxide concentration may well increase several fold. The consequent warming effect on the world's climate from the "greenhouse effect" could considerably exceed the natural rate and extent of change, posing a threat to mankind that may truly be unprecedented.

The difficulty of dealing with this threat may also be without precedent: The potential magnitude of the future impact on climate from increasing atmospheric carbon dioxide is almost matched by its uncertainty. The average warming from the "greenhouse effect" may be as little as 1.5° or as much as 4.5°C per doubling of the carbon dioxide concentration (4). Prediction of the fraction of fossil carbon that will remain in the atmosphere in the future requires a more detailed knowledge of the carbon cycle than we are likely to achieve soon. Prediction of the consequent climate changes and the economic, political, and ecological impacts is even more uncertain. The climate changes produced by excess carbon dioxide could appear, grow and become unacceptable before effective countermeasures--if indeed such are possible--could be taken. Thus, in spite of the uncertainties and the difficulties involved, early use of preventive measures may be the only available means of avoiding unacceptable consequences. Finally, world-wide cooperation must somehow be achieved if control of the release of carbon dioxide is to be effective.

In order to deal with this intractable situation it would seem prudent to: (i) work vigorously to reduce the important uncertainties involved in assessing the magnitude of this threat, particularly those associated with predicting climatic changes; (ii) continue development of non-fossil energy sources; (iii) establish mechanisms for the international control of fossil fuel use if such should become necessary; and as a last resort (iv) be prepared to dispose of carbon dioxide if this should ever be necessary.

It is our purpose here to consider this last resort. Starting from the premise that man could someday determine a maximum tolerable rate of release of anthropogenic carbon dioxide to the atmosphere, we ask whether there is some technical strategy that would allow a higher rate of use of fossil fuels, accompanied by some means of disposal of the excess carbon dioxide that would keep the atmospheric release rate within the established limit.

Our goal has been to examine ways in which carbon dioxide from fossil fuels might be collected, converted to an appropriate form for transport and disposal, and then placed in some environment where it could be retained for a period of time comparable to the mixing time of the deep oceans, 500 years or longer. The deep oceans are the only reservoir with sufficient capacity to accommodate the large amount of fossil carbon ultimately to be injected into the terrestrial carbon cycle, and the rate of take-up will probably be determined by the rate of circulation between the deep ocean and the well-mixed surface layer.

We will discuss the technical feasibility, energy cost, and monetary cost of the options that appear the most promising. Previous such efforts have not given reason for optimism: Marchetti (5) first gave serious consideration to methods of collection and disposal, suggesting that liquid carbon dioxide could be piped from power plant scrubbers to injection points where deep-water, such as the Mediterranean Outflow Water, originates. Later (6) he estimated that carbon dioxide collection could increase the cost of electricity by 30%. Dyson (7) suggested a massive tree-planting program to buy time to shift to non-fossil energy sources. Dyson and Marland (8) considered this and previously suggested options further and concluded that while technical fixes were feasible in theory, they would be exceedingly expensive and would require unprecedented global cooperation. Steinberg and Albanese (9) have systematically analyzed the methodology of carbon dioxide collection and discussed the more promising options. While Wilkinson (10) and Gribbin (11) seem to retain hope that suitable "technical fixes" can be designed if needed, Cooper (12) concludes that the institutional obstacles to the requisite international cooperation are "almost insurmountable."

The generally pessimistic conclusions drawn from previous studies have narrowed the scope of the present effort. For example, we did not consider further the option of direct absorption from the atmosphere by some chemical processing facility, since Albanese and Steinberg (13) have shown clearly that the energy costs alone for concentration from such a dilute source would be prohibitive. Rather we have considered primarily collection from concentrated sources such as large power plants. While at present these account for only a third of the fossil carbon released worldwide and only a sixth of that in the United States, such sources are likely to become the major contributors in the future, especially if a threat from increasing carbon dioxide in the atmosphere is perceived. We do consider briefly the scheme of Dyson and Marland (8) for removing carbon dioxide from the atmosphere by natural photosynthesis, despite the immense scale of the planting required. While we consider various disposal options, we emphasize ocean disposal as the most promising choice.

2. COLLECTING AND PROCESSING CARBON DIOXIDE

In this section we consider the collection and processing of carbon dioxide from concentrated sources by mechanical-chemical methods such as scrubbing and compressing. If processing from concentrated sources should prove not to be technically attractive, certainly collection from dilute sources will be even less attractive.

2.1. Concentrated Sources

In the United States in 1978 there were fossil-fired power stations with a total capacity of 400,000 MW which generated 1.65×10^{12} kWh of electricity (14). Approximately 60% of this was coal fired. For the world the total fossil-fired capacity in 1978 was approximately 1,300,000 MW and about 6.5×10^{12} kWh of electricity were generated.

We can make an estimate of the amount of carbon dioxide released from these utility plants by assuming that all were fired with U.N. standard coal (7000 kcal/kg). Combustion of one kg of this coal produces 2.55 kg of carbon dioxide and, at a plant efficiency of 38%, 3.09 kWh of electricity. For the U.S. the output of carbon dioxide for 1978 would then be 1.36 Gt, and for the world 5.36 Gt. For various reasons, in particular the small size and remote locations of many utility plants, probably no more than 50% of these emissions could be treated, so the total world "controllable" carbon dioxide for 1978 was about 2.7 Gt, containing 0.73 Gt carbon. This is a small fraction of the estimated total emissions from fossil fuels and cement production of 5.2 Gt of carbon for that year (15).

2.2. Carbon Dioxide Scrubbing

A wide variety of methods has been developed for scrubbing carbon dioxide from flue gas or more concentrated streams. In a review Strelzoff (16) listed thirteen licensed processes. The energy costs and dollar costs of collecting carbon dioxide from stack gas by present technology have been estimated by Pappano et al (17) and Rump et al (18) for large scale operations (in the latter case, 270 metric tons of CO_2/hr) for enhanced oil recovery. In Table 1 the energy costs are compared with the energy derived by burning the coal (again U.N. standard coal). To scrub and strip 90% of the CO_2 produced is seen to require 43% of the combustion energy of the coal. The dollar cost is estimated as $20 per metric ton of carbon dioxide collected if the price of energy is $7.14/Gcal ($1.80/million BTU).

The estimated energy cost for compressing the CO_2 to 150 bars and transporting it 40 km are small by comparison. The estimated energy costs of conversion to dry ice or carbon dioxide hydrate are appreciable, 13 and 17% respectively of the energy value of the coal. The dollar costs of these steps are proportionately higher because much of the energy consumed is electrical.

Table 1. Cost of Collecting and Processing Carbon Dioxide from Stack Gas

	Energy Cost Gcal/Mg CO_2	Per Cent of Combustion Energy of Coal[a]	Monetary Cost $/Mg CO_2
Capture CO_2[b]	1.31	43	20
Compress to 150 bar and transport 40 km[b]	0.10	3	4
Liquefy CO_2[c]	0.12	4	4
Solidify CO_2[d]			
To Dry Ice	0.36	13	14
To $CO_2 \cdot 6H_2O$	0.47	17	17

[a]Assuming one Mg (metric ton) of coal yields 2.55 Mg of CO_2 and 7.0 Gcal of heat, the boiler efficiency for steam is 90%, and the plant efficiency for electricity is 38%.

[b]Based on results of Rump et al (ref.18) for a 290 Mg/hr monoethanolamine (MEA) process that captures 90% of the carbon dioxide.

[c]Econergy Associates (private communication, 1979).

[d]Calculated from data of Stickney (ref. 29).

2.3. Combined Carbon Dioxide-Sulfur Dioxide Scrubbing

We explored the possibility that carbon dioxide might be removed from stack gas in conjunction with sulfur dioxide removal, thus taking advantage of the sulfur dioxide scrubbing systems that will be required for coal-burning power plants in the United States. If both sulfur dioxide and carbon dioxide could be removed with the same equipment, energy and dollar costs might be lower than for separate systems. With the help of consultants A.A. Pappano[2] and Econergy Associates[3] we examined several scrubbing systems, both regenerative (carbonate, amine and bisulfite) and non-regenerative (lime and caustic soda). A fundamental problem is the great difference in the concentration of the two gases. The absorption characteristics of the various solvents do not match the great excess of carbon dioxide, which results in one gas or the other being removed inefficiently. In addition, unwanted chemical reactions between the gases and the solvents require replacement of the solvent and the disposal of even larger amounts of waste solids than with sulfur dioxide scrubbing alone. The high cost of chemicals is another problem. The consultants and we concluded independently that simultaneous removal of sulfur dioxide and carbon dioxide is not attractive with existing technology.

2.4. Burning Coal with Pure Oxygen

Marchetti (6) has suggested that coal be burned in pure oxygen to produce a nearly pure mixture of carbon dioxide and water, thus avoiding the costly scrubbing - stripping method of concentration. Table 2 summarizes the comparable costs of this alternative.

Table 2. Combustion in Oxygen vs Combustion in Air

	Energy Cost Gcal/Mg CO_2		Per Cent of Combustion Energy of Coal[a]	
	Plant Efficiency for Elect. Prod.			
Combustion in Oxygen	(38%	42%)	(38%	42%)
Preparation of pure oxygen (22 bar)	0.88	0.80	32.2	29.5
Compress CO_2 to 150 bar and transport 40 km	0.13	0.12	4.7	4.3
	1.01	0.92	36.9	33.8
Useful (net) electricity produced			24.0	28.0
Combustion in Air	(38%)		(38%)	
Capture from normal stack gas and transport 40 km (from Table 1)	1.41		46.1	
Useful (net) electricity produced			20.5	

[a]From data supplied by P. Koster, Linde Div., Union Carbide Corp: 441 kWh of mostly electrical energy required per metric ton (Mg) of oxygen; 2.26 Mg oxygen per Mg coal; the same coal (U.N. Standard) assumed in Table 1.

The separation and compression of oxygen from air (presently done on a large scale for the steel industry) is of course a major cost, requiring 441 kWh of mostly electrical energy per metric ton of oxygen. Since 2.26 metric tons of oxygen are required to burn one metric ton of U.N. standard coal, this amounts to 2.25 Gcal per metric ton of coal (assuming 38% efficiency for the production of electricity). Thus the energy cost of oxygen production amounts to 0.88 Gcal/metric ton of carbon dioxide, considerably less than the 1.31 Gcal/metric ton estimated for recovering carbon dioxide from normal flue gas.

Moreover, it is reasonable to expect there will be an increase in plant efficiency from (i) the reduction of heat loss to the stack gas, (ii) the elimination of combustion air fans, and (iii) perhaps an increase in combustion temperature. If the overall efficiency for the production of electricity could be improved from 38% to 42%, this would reduce the energy costs of oxygen production and CO_2 compression, and leave a larger fraction of the energy of the fuel for useful electricity (Table 2).

The flue gas from an oxygen-fed plant would contain about 45% CO_2, 40% H_2O, unused oxygen, and small amounts of sulfur dioxide and nitrogen. It would be at a temperature in the range 260 to 500°C, making recovery of the heat economically attractive. The heat of vaporization of the water could be recovered upon cooling before the remaining gas (75% CO_2) is compressed. The cost of compression should be about a third more than for normal flue gas, but still only about 0.13 Gcal/metric ton CO_2, approximately 4.3% of the combustion energy of the coal.

Thus we estimate (Table 2) the total energy cost of carbon dioxide collection from an oxygen-fed plant as 0.92 Gcal/metric ton CO_2, 34% of the energy available in the coal. Since this is considerably less than the energy required by the normal scrub and strip process discussed earlier, it appears to us that Marchetti's idea results in a significant improvement. An additional advantage of combustion in pure oxygen is that there may be no need for the separate recovery of sulfur dioxide. Provided the additional environmental impact is not significant, it could be disposed of together with the carbon dioxide.

3. OCEAN DISPOSAL

As previously noted, the most promising place to dispose of carbon dioxide is in the deep ocean since it is there that most of the atmospheric excess is destined ultimately to go. If carbon dioxide can be delivered there, a retention time of many hundreds of years, perhaps more than a thousand years, could be achieved because of the low rate of mixing between the deep waters and surface waters of the ocean. Of course if the carbon dioxide were to react at an appreciable rate with basic minerals such as calcium carbonate in the ocean sediments, even longer retention times would be realized.

As will be seen, the most attractive means of such disposal appears to be dissolving the compressed gas in seawater to produce a liquid stream with a density greater than seawater and then injecting this at a depth in the ocean sufficient to prevent bubble formation. We will consider the mixing properties of the negatively buoyant plume of carbon dioxide-enriched seawater that would be formed as a result. We will consider briefly the suitability of the liquid and solid forms of carbon dioxide for disposal in the ocean. Finally we will examine the availability of offshore locations that might be suitable for such disposal.

3.1. Concentrated Solutions of Carbon Dioxide

The solubility of carbon dioxide in seawater up to the liquefaction pressure of CO_2 at various temperatures (19-21) is shown in Figure 1. Its presence increases the density of seawater by an amount given by its apparent molar volume in aqueous solutions. Measurements of the latter quantity in water and in seawater fall in the range 30 to 35 cm^3 (22). Adopting the upper limit we calculate conservatively that at the maximum solubility, carbon dioxide should increase the density of seawater by 0.008 g/cm^3 or more. Such a solution would produce plumes of substantial negative buoyancy, and hence might sink to great depths without excessive mixing with the surrounding water.

To consider this point further, let us imagine that seawater is taken for use as a disposal medium from some point in the ocean at which the temperature is t_1 and the density is ρ_1. It is pumped through a gas-liquid contactor at elevated pressure which dissolves carbon dioxide to a concentration C, the density rises to ρ_2 and because of the heat of solution the temperature rises to t_2. Finally the stream is discharged at a point in the ocean where the temperature is t_3 and its density is ρ_3. The temperature t_2 can be estimated from the approximation

$$t_2 = t_1 + 5 \times C ,$$

(C is in moles/kg and t in °C), based on a heat of solution of 5 kcal/mole for carbon dioxide. The density difference between the waste stream at the discharge point and the surrounding water can be calculated as the sum of two terms,

$$\rho_2 - \rho_3 = [\ \rho(t_2,0) - \rho(t_3,0)\ ] + [\ \rho(t_2,C) - \rho(t_2,0)\ ] .$$

The first term on the right is the density change caused by the

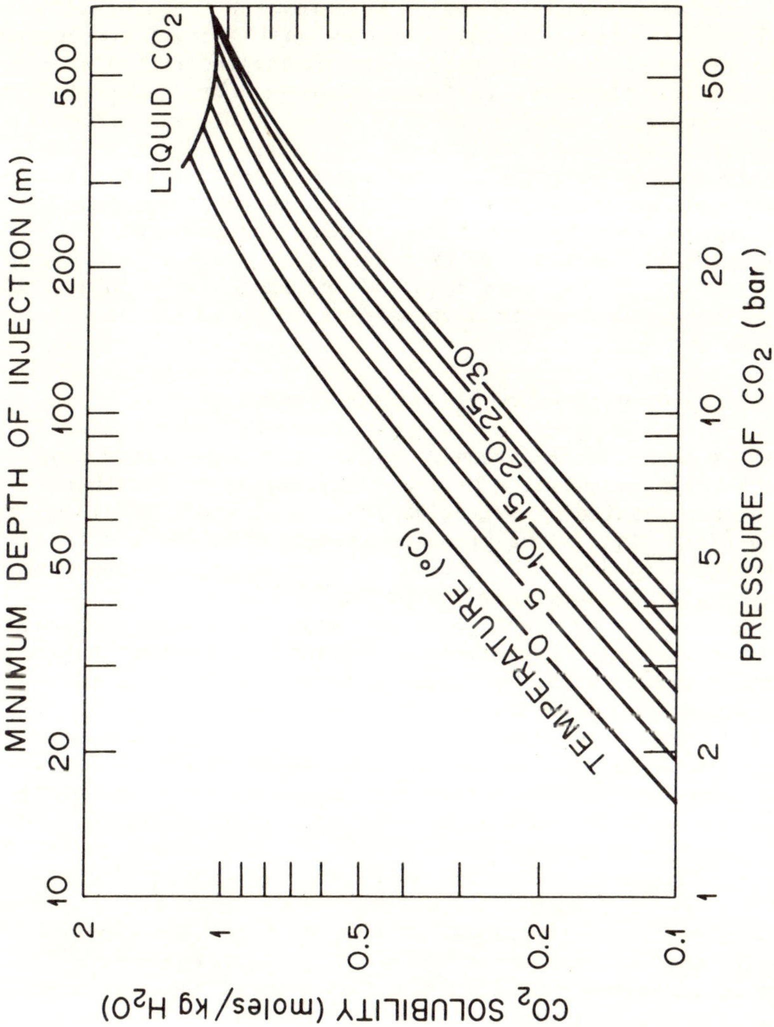

Figure 1. The solubility of carbon dioxide in seawater and the minimum depth for its injection without bubble formation (19, 20).

temperature difference $t_2 - t_3$ and can be determined accurately from the known temperature coefficients of seawater density. The second term is the density increase produced by introducing the carbon dioxide and can be estimated from the approximation

$$\rho(t_2,C) - \rho(t_2,0) = 0.0072 \times C ,$$

based on a molar volume of 35 cm^3. It turns out (Figure 2) that the density difference, $\rho_2 - \rho_3$, which we wish to maximize depends primarily on two quantities, the inlet temperature (which should be minimized) and the concentration of carbon dioxide C (which should be maximized).

Of course, the design of a seawater disposal system will involve many considerations which may affect the depth of the inlet and discharge points, and the concentration and temperature of the discharge stream. If the resulting density increment of the discharge stream is smaller than required for adequate disposal, it may be possible to evaporate some water from the disposal stream to increase its density.

3.2. The Behavior of Negatively Buoyant Plumes

The behavior of a jet of carbon dioxide-enriched seawater discharged to stagnant seawater of lower density below the thermocline was investigated by use of an analytical model developed by one of us (23). This model produces results that compare favorably with existing data and numerical codes for thermal buoyant jets and should be useful for the present application. A 1000 MW(e) coal-fired power plant produces about 230 kg/sec of carbon dioxide. If this were dissolved in seawater to a concentration of 1 mole/kg, 5200 kg/sec of seawater with a substantially increased density would be discharged. An ocean disposal facility dedicated to the disposal of carbon dioxide from ten such plants was considered with a total discharge of 50,000 kg/sec. The difference in density between the discharge and the surrounding water was taken to be 0.006 g/cm^3.

The discharge is considered to have three distinct zones (Figure 3): (i) a zone of flow establishment where the velocity profile evolves from its initial flat ("top hat") distribution to one that is nearly Gaussian in shape, (ii) a jet zone where the behavior is dominated by the initial momentum of the discharge, and (iii) a plume zone where behavior is controlled by the buoyancy of the discharge. The nature of the flow in each zone is highly dependent on the physical nature of the receiving water (density stratification, salinity variation, currents, and bathymetry). For our calculations we assumed that the temperature and salinity of the receiving water were uniform and that there were no ambient currents present. This last condition is the most important.

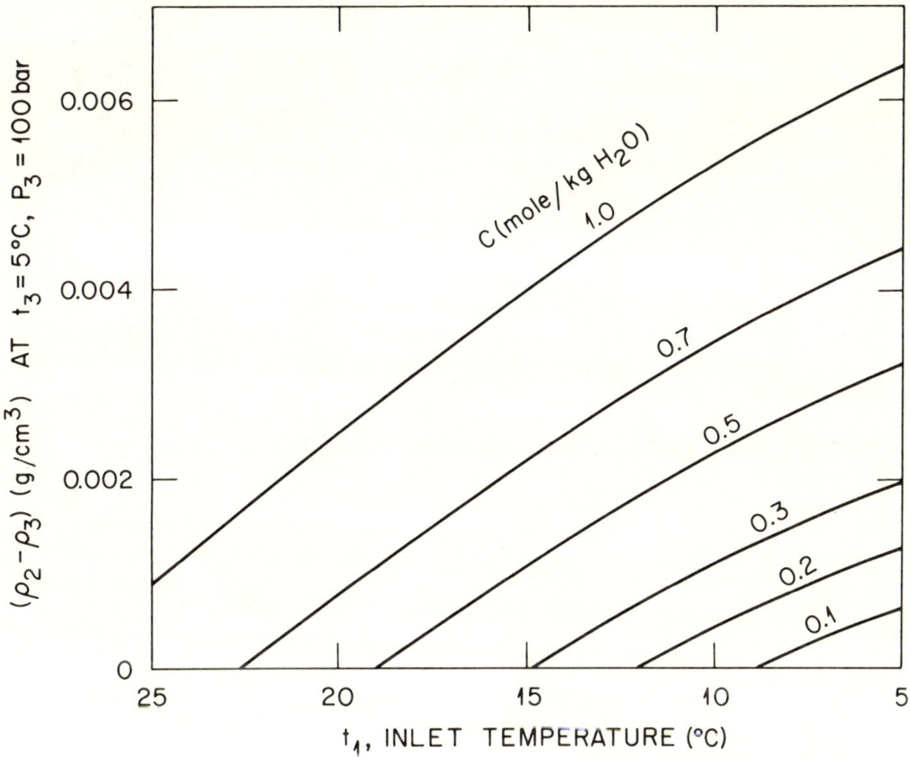

Figure 2. The excess density of a stream of seawater which has been taken at a temperature t_1, has C moles/kg of carbon dioxide dissolved in it, and is discharged into a water mass of temperature 5° C at a depth of 1000 m.

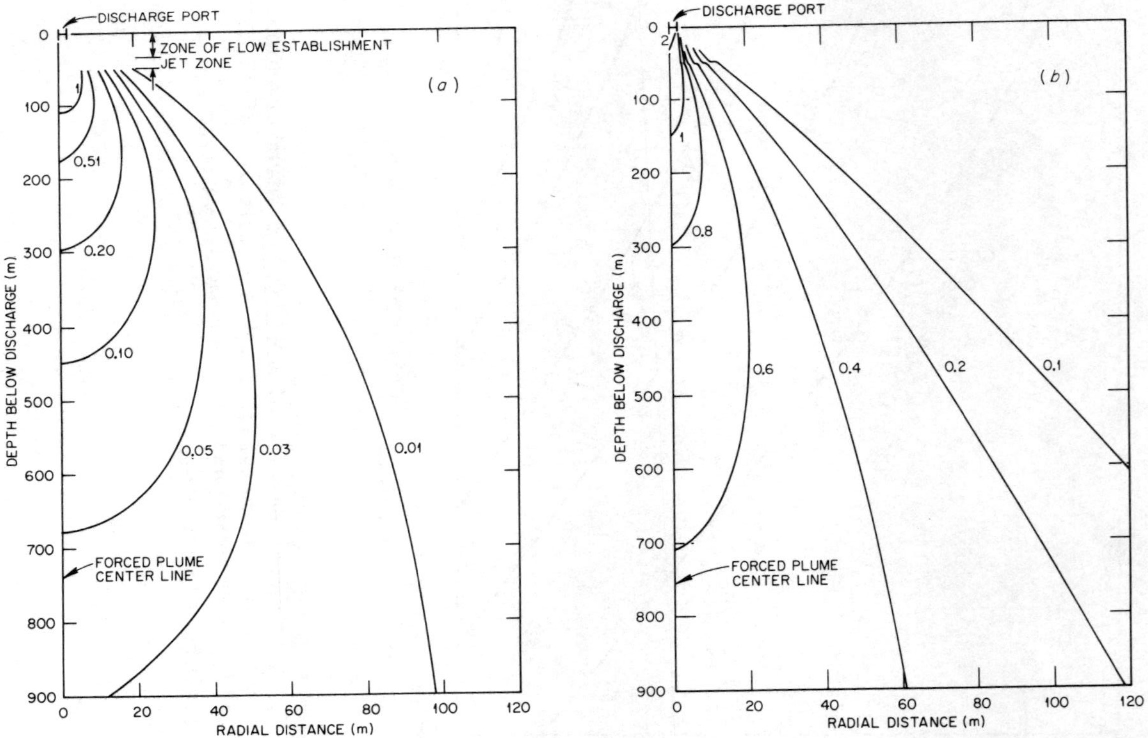

Figure 3. Profiles of (a) excess density (g/l) and (b) vertical velocity (m/s) in a negatively buoyant plume. The portion of the plume shown contains 90% of the carbon dioxide discharged. The initial density excess is 6 g/l, produced by a CO_2 concentration of 1 mole/kg; the discharge rate is 50,000 kg/s; the jet diameter is 5.55 m and the jet velocity is 2 m/s at the point of injection.

The analytical modeling of the discharge (Table 3) utilized separate formulations for each flow zone. The zone boundaries were matched by requiring continuity of centerline velocity and density excess. The matching procedure determined the virtual source of each zone, the length of each zone, the decay of velocity and concentration along the centerline, the velocity profiles in the jet zone and the plume zone, and the concentration profiles in the plume zone.

Table 3. Characteristics of Carbon Dioxide Enriched Discharge Plume in Seawater[a]

Notation:

- z = distance below discharge port
- D = diameter of discharge port
- U = discharge velocity at discharge port
- ρ_2 = density of effluent at discharge port
- ρ_3 = density of ambient seawater at discharge port
- g = gravitational constant
- z_p = location of virtual source of plume

Plume half-width in jet zone:	$0.106(z + 0.4D)$
Plume half-width in plume zone:	$0.130(z - z_p)$
Length of jet zone:	$2.54 \left(\frac{U^2 D}{g}\right)^{\frac{1}{2}} \left(\frac{\rho_2 - \rho_3}{\rho_3}\right)^{-\frac{1}{2}}$
Centerline dilution in jet zone:	$0.166 \left(\frac{z}{D} + 0.4\right)$
Centerline dilution in plume zone:	$0.916[g(\rho_2 - \rho_3)]^{1/3} (D^2 U)^{-2/3} (z - z_p)^{5/3}$
Virtual source for plume (z_p):	$-0.4D + 0.029\left(\frac{D^4 U^5}{g}\right)^{\frac{1}{2}} \left(\frac{\rho 2 - \rho 3}{\rho 3}\right)^{-\frac{1}{2}}$

[a]From ref. 23.

Density and velocity profiles for the plume, calculated for an initial density excess of 0.006 g/cm^3 and a discharge rate of 50,000 kg/sec, are presented in Figure 3. The density profiles shown include 90% of the discharged carbon dioxide. Upon reaching the ocean floor the plume would continue to disperse along the bottom. Reaction of the dissolved carbon dioxide with the carbonate ion in seawater to form bicarbonate ion would be determined by the rate at which the carbonate ion is supplied by the seawater diluting the plume. Since the concentration of carbonate in deep seawater is about 0.0001 moles/kg, this reaction would become appreciable only after the original waste stream had been diluted more than 1000 fold. At the bottom, reaction of the dissolved carbon dioxide with calcium carbonate may also occur at an appreciable rate.

The width of the plume zone proved insensitive to the density excess of the jet in the range 0.002 to 0.009 g/cm^3 and to its discharge rate in the range 5000 to 500,000 kg/sec. The variables that influenced the length of the jet zone and the degree of dilution along the centerline were, in order of their importance, the discharge flow rate, the discharge port diameter, the magnitude of the density excess, and the depth of the receiving water body. In particular, the dilution along the centerline varied directly as the 1/3 power of the density excess and inversely as the 2/3 power of discharge velocity (Table 3).

The model results indicated that dilution could be reduced by use of a line of discharge ports producing plumes that interact at depth. Plume interaction is poorly understood, however, and the appropriate spacing between adjacent discharge ports can be specified only qualitatively.

3.3. Liquid Carbon Dioxide

In Figure 4 we see that although the density of liquid carbon dioxide (24) would be less than that of seawater at moderate depths, it is much more compressible and has a much higher thermal coefficient of expansion. With increasing depth the density of the liquid increases quite rapidly and should exceed that of seawater of similar temperature at about 3000 m. Thus, if the liquid were piped to a somewhat greater depth it could be expected to form a negatively buoyant plume and descend the remaining distance to the bottom.

3.4. Carbon Dioxide Hydrate

The hydrate of carbon dioxide has a clathrate structure (25) with a density of 1.12 g/cm^3, greater than that of seawater. If large blocks of this solid were dropped into the sea they would sink rapidly. Once such a block reached a depth of about 500 m where the hydrostatic pressure is about 50 bars and the temperature is less than 12° C, gas bubble formation from the decomposition of the hydrate would cease (26). The decomposition product thereafter would be a carbon dioxide-rich seawater solution more dense than the surrounding water. Depending upon the reaction rates,

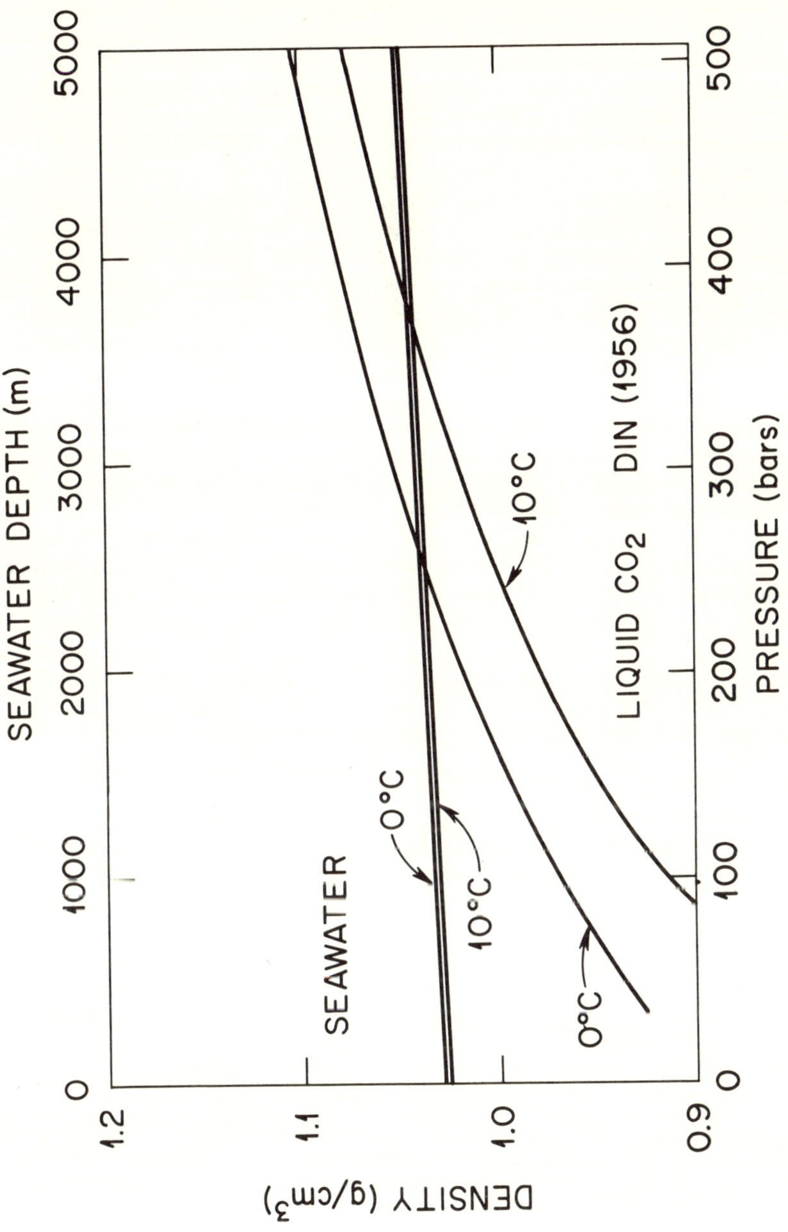

Figure 4. Comparison of the densities of liquid carbon dioxide and seawater as a function of pressure (depth) and temperature.

it might be practical to prepare blocks of the hydrate large enough to reach the bottom before an appreciable fraction had been consumed.

3.5. Dry Ice

This form of carbon dioxide has the greatest density, about 1.5 g/cm^3, and could be economically transported by water. Large blocks would sink rapidly and, again, once below a depth of about 500 m, bubbling would cease and the reaction products--the hydrate, or a carbon dioxide-rich solution--should be denser than the surrounding water. This form of carbon dioxide might survive the trip to the ocean floor best of all.

3.6. Potential Disposal Sites

Obviously, desirable ocean disposal sites should be near shore and areas of energy demand; yet they should have depths greater than 1000 m, the usual lower limit of the thermocline, and be relatively free of strong currents and stratification. Such sites are limited in number, and the transmission of electricity from an off-shore plant or the transport of carbon dioxide to an off-shore disposal site will involve substantial distances in most cases.

As an example we have considered a "smoothed" East Coast of the United States, one that runs from Miami along the seaward side of the near-shore islands and past bay mouths some 3100 km to the Canadian border (Table 4). Over this distance there are only 40 km of shoreline, near Cape Hatteras, for which 1000 m depths are less than 50 km off-shore. There are another 230 km of shoreline for which 1000 m depths are between 50 and 100 km off-shore (27). In the limited areas where the continental shelf is narrow, strong currents are often present in the adjacent deep water. South of Cape Canaveral the Straits of Florida reach depths of 700 to 800 m but contain strong currents that reach to the bottom. North of Cape Hatteras the continental shelf is quite wide. At Atlantic City, NJ, for example, one must go 135 km off-shore to reach depths of 200 m, although the slope there is quite steep and an additional 10 km reaches water 1000 m deep. A similar situation is found off the midsection of Long Island.

The U.S. West Coast is more attractive, with a narrower continental shelf and consequently much less distance from shore to deep water, but it is adjacent to a limited fraction of the U.S. power market.

Table 4. Distribution of Water Depths Off U. S. East Coast

Distance Off-shore (km)	Water Depth (m) 200	500	1000	1500
		Shoreline (km)[a]		
< 10	110	0	0	0
10 - 25	45	0	0	0
25 - 50	130	215	40	35
50 - 100	785	610	230	210
100 - 150	1325	960	685	570
> 150	705	1315	2145	2285
Total Shoreline	3100	3100	3100	3100

[a] Smoothed outermost shoreline from Miami, Florida to the Canadian border. Shore configuration and off-shore bathymetry were taken from ref. 27.

4. FLOATING POWER STATIONS

In view of the advantages of disposal of carbon dioxide in the deep ocean, we have explored the feasibility of a floating coal-fired power station. At first we considered such a plant combined with an ocean thermal energy conversion (OTEC) plant, using the exit (warm) cooling water stream of the coal plant to feed the OTEC evaporators, and the cooling water discharge from the OTEC plant for the disposal of carbon dioxide from the coal plant. The discharge for a typical OTEC plant (e.g., 380,000 kg/s for a 100 MW(c) TRW design) would be so large compared to the volume of seawater required to produce a concentrated waste stream from a coal-fired plant (e.g., as little as 5200 kg/s for a 1000 MW(e) plant), it was soon obvious that there was no significant advantage of coupling the two kinds of plants for the purpose of carbon dioxide disposal. While we abandoned coupling, we retained the idea of using cold deep water for carbon dioxide disposal and for optimum condenser cooling. The carbon dioxide-enriched stream would be discharged at a sufficiently greater depth to avoid mixing with the inlet stream. The seawater requirements for the carbon dioxide discharge are so modest that only 10% of the cooling water flow would be needed to carry away the carbon dioxide produced by the plant.

We pursued the engineering of this option with the help of Leonard Seglin of Econergy Associates[3]. The concept is shown schematically in Figure 5. The coal-fired plant could be mounted

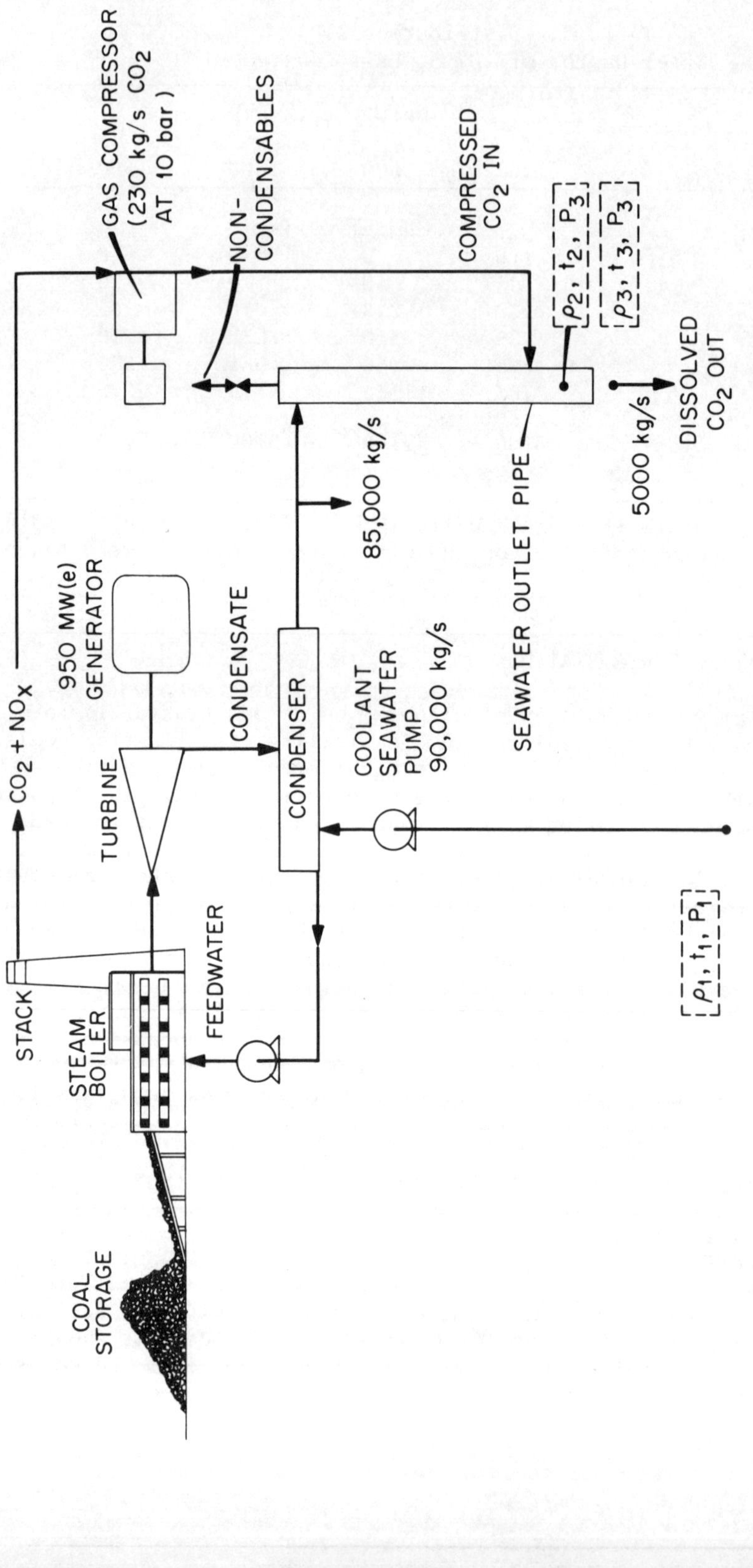

Figure 5. Flowsheet for a floating coal-fired power plant with CO_2 disposal to the ocean.

on a barge anchored at a deep water site, similar to a proposed OTEC plant, with inlet and discharge pipes extending to depths of optimum cooling-water temperature and of optimum carbon-dioxide stream discharge. If pure oxygen were used for combustion, the dry, filtered stream of 75% pure carbon dioxide (along with sulfur dioxide and other minor constituents) could be compressed directly.

The water flow and the pressure in the scrubber, shown in Figure 5 as 5000 kg/s and 10 bar, would be optimized with other design variables, such as the cooling-water inlet temperature and the height of the scrubber required to dissolve the carbon dioxide and produce a discharge stream with a sufficient density excess for effective disposal. The scrubber was assumed to be a pipe without packing, 16 m in diameter and 300 m long, and with a porous membrane at the gas inlet (lower end) to produce a stream of small bubbles. Calculation suggested that with a water velocity down the column of 0.07 m/s and a rising gas velocity of 0.03 m/s the carbon dioxide would dissolve in a very short distance (within 2 m) above the inlet. The function of most of the length of the carbon-dioxide discharge system is to provide the hydrostatic head necessary to dissolve the carbon dioxide from the compressed gas. Sulfur dioxide would be removed as well and if pure oxygen were used for combustion there would be little non-condensable gas (<10%) exiting from the top of the scrubber.

The capital and operating costs estimated for several cases are shown in Table 5. These results suggest an important idea: Since the energy and dollar costs of oxygen-firing overwhelm the other costs, and since the carbon dioxide dissolves so easily, why not simply compress the filtered flue gas from an air-fired plant? While in this case five times more gas must be compressed and the energy cost of compression oven for shallow injection is 360 MW, or about 36% of the electrical output of the plant, this appears the most attractive option. Its feasibility will depend upon the actual injection depth required to produce a plume of sufficient negative buoyancy to insure deep-water disposal.

5. OTHER STORAGE POSSIBILITIES

5.1. Depleted Gas Fields

Gas fields provide a large subterranean volume that could be used for the storage of carbon dioxide as the natural gas is depleted. If we assume in the U.S. that gas fields become depleted at a rate near the rate of gas consumption, then the annual capacity for carbon dioxide storage at the original gas pressure would be about 0.55×10^{12} m^3/yr, or about 1.1 Gt/yr. This is not much less than the present U.S. production of carbon dioxide from

Table 5. Dollar and Power Requirements for Processing and Disposal of 229 kg/s (30,300 lbs/min) of CO_2 from a Floating Power Plant[a]

System	Dollar (1979) Requirements, Millions		Power Requirements, MW(e) or MW(t)		Useful (Net) Electricity	
	Capital	Oper./yr	O_2 Plant	Proc. & Disp.	% Combustion Energy of Coal	Power MW(e)
Gaseous CO_2 (99% pure) scrubbed from flue gas, compressed to 150 bar discharged 1000 m below ocean surface	148	225	0	1210(t)	21	539
Gaseous CO_2 (75% pure) compressed to 33 bar, discharged 330 m below ocean surface	287[b]	159	322(e)	68(e)	23 27[c]	610 715[c]
Dry flue gas (19% CO_2) compressed to 33 bar, discharged at 330 m below ocean surface	235	224	0	361(e)	24	639

[a] Assuming the conditions given in Tables 1 and 2. The efficiency for electricity production is 38% in all cases but one (footnote *c*).

[b] Includes cost of oxygen plant at $240 million.

[c] At 42% plant efficiency for electricity production.

coal-fired power plants (about 1.3 Gt for 1978). For the world, the natural gas already consumed as fuel or flared is approximately 24 x 10^{12} m^3 (28) and the present rate of withdrawal is 1.3 x 10^{12} m^3/yr. These quantities suggest that gas wells already depleted could contain about 48 Gt of carbon dioxide and that new capacity becomes available at a rate of 2.6 Gt/yr. Thus we conclude that abandoned gas fields could store a significant fraction of the world's production of carbon dioxide from fossil fuels for some time into the future.

5.2. Depleted Oil Fields

Estimates of recoverable world resources of oil are approximately 2 x 10^{12} bbls (28), which is equivalent to a volume of about 0.32 x 10^{12} m^3. The volume of carbon dioxide that could be contained in the same space is of course much larger because of the compressibility of the gas and its solubility in the oil that remains in a depleted field. Carbon dioxide may prove valuable for stimulating the production of low yield oil wells. The U.S. Department of Energy and several oil companies are considering carbon dioxide miscible flooding as a method of enhancing oil production. Econergy Associates estimates that 0.42 metric tons of carbon dioxide could be "consumed" in the unreleased oil and in the gas space of the field for each barrel of enhanced oil recovered. If so, and if carbon dioxide miscible flooding were used for one half of present U. S. oil production, the carbon dioxide from all U. S. coal-fired utility plants could be consumed.

5.3. The South Pole

The mean annual temperature at the South Pole is -50°C and some locations near the South Pole remain at temperatures below -20°C year round. The decomposition pressure of carbon dioxide hydrate does not fall below one atmosphere until it is cooled to about -54°C; hence, like dry ice, its long-term storage near the South Pole would require venting and the storage time would depend upon the rate at which heat reaches it. While it might be possible to store the hydrate or dry ice for hundreds of years in a large, well-insulated repository located near the South Pole, the problems of preparing such a facility and transporting the materials to it would be formidable.

5.4. Buried Plant Material

Dyson and Marland (8) examined the problem of storing carbon as standing crops of biomass and found that 18 Gt/yr of carbon dioxide could be withdrawn from the atmosphere with 7 x 10^6 km^2 of land (about the size of Australia) planted in sycamore trees. If such cultivated biomass could be prevented from decaying or otherwise being oxidized back to carbon dioxide, a large storage rate might be achieved. Burial in excavations or abandoned mines within the cultivation areas seems the only plausible way to do this, considering the cost of long distance transport. As Dyson and Marland point out, a significant part of the energy cost of this option would be the fertilizer required. Using the cultivated biomass for fuel to replace fossil fuels appears to be a preferable way of reducing the net release of carbon dioxide to the atmosphere.

6. SUMMARY AND CONCLUSIONS

If society should ever decide that unacceptable changes in climate are imminent because of the accumulation of carbon dioxide in the atmosphere from the use of fossil fuels, it may be necessary to collect and dispose of carbon dioxide unless or until alternative energy sources can be substituted. At present it appears that such disposal would consume a large fraction of the energy derived from the fossil fuel. Attention here has been focused on large carbon dioxide sources where collection seems most likely to be practical.

To scrub carbon dioxide from the flue gas of a coal-fired power station by the most efficient method presently available would require about 43% of the combustion energy of the coal (Table 1). Simultaneous CO_2-SO_2 scrubbing does not appear practical because of the great difference in the concentrations of the two gases. If, as Marchetti (6) has suggested, the coal were burned in pure oxygen to avoid the scrubbing step, at least 30% of the combustion energy would be required to supply the oxygen (Table 2).

After recovery the best option for disposal of carbon dioxide is in the deep ocean where the low rate of circulation should give a retention time of hundreds of years. Injection could be in the form of a concentrated seawater solution and at a depth sufficient to prevent bubble formation (Figure 1). Such a stream could be denser than seawater (Figure 2) and form a negatively buoyant plume that might convey much of the CO_2 to the bottom (Figure 3). Alternatively, liquid carbon dioxide could be injected into the ocean, but at greater depths to achieve the needed density (Figure 4). Or blocks of solid carbon dioxide hydrate or dry ice released at the ocean surface might reach deep water quickly enough for effective disposal. Sites for ocean disposal near shore and energy markets, and with sufficient depth are not abundant (Table 4).

The most efficient option for the future might be floating, coal-fired power plants that use cold deep seawater for both condenser cooling and carbon dioxide disposal (Figure 5). In particular, it might be practical simply to compress the flue gas and scrub it with seawater at a sufficient depth to insure effective disposal (Table 5). As a near-term option, depleted oil and gas fields could be used for disposal of carbon dioxide. Storage near the South Pole or as buried biomass are perhaps the least attractive of the alternatives considered.

ACKNOWLEDGEMENTS

We are indebted to W. L. Fulkerson of the ORNL Energy Division and to R. M. Rotty of the ORAU Institute for Energy Analysis for valuable discussions during this study.

FOOTNOTES

[1]Research sponsored by the Division of Environmental Control Technology, Office of Environmental Compliance and Overview, U.S. Department of Energy, under contract W-7405-eng-26 with Union Carbide Corporation.

[2]Chemical Engineering Department, University of West Virginia, Morgantown, West Virginia.

[3]799 Broadway, New York, New York.

REFERENCES

1. Baes, C. F., Jr., Goeller, H. E., Olson, J. S., and Rotty, R. M.: 1977, *Carbon Dioxide and Climate: The Uncontrolled Experiment*, American Scientist 65, pp. 310-320.

2. Elliott, W. P., and Machta, L.: 1979, *Workshop on the Global Effects of Carbon Dioxide from Fossil Fuels*, U.S. Department of Energy Report CONF-770385.

3. Kellogg, W. W.: 1979, *Influences of Mankind on Climate*, in Annual Reviews of Earth and Planetary Sciences, Annual Reviews Inc., Palo Alto, CA, v. 7, pp. 63-92.

4. Charney, J. G.: 1979, *Carbon Dioxide and Climate: A Scientific Assessment*, Report of Ad Hoc Study Group on Carbon Dioxide and Climate, J. G. Charney Chairman, National Academy of Sciences, Washington, D.C.

5. Marchetti, C.: 1976, *On Geoengineering and the CO_2 Problem*, International Institute for Applied Systems Analysis, Research Memorandum RM-76-17.

6. Marchetti, C.: 1978, *Constructive Solutions to the CO_2 Problem,* International Institute for Applied Systems Analysis, unpublished manuscript.

7. Dyson, F. J.: 1977, *Can We Control Carbon Dioxide in the Atmosphere?* Energy 2, pp. 287-291.

8. Dyson, F. J., and Marland, G.: 1979, *Technical Fixes for the Climatic Effects of CO_2*, in *Workshop on the Global Effects of Carbon Dioxide from Fossil Fuels*. W. P. Elliott and L. Machta (eds.), U. S. Department of Energy Report, CONF-770385, pp. 111-118.

9. Steinberg, M., and Albanese, A. S.: 1978, *Environmental Control Technology for Atmospheric Carbon Dioxide. Progress Report No. 2,* Brookhaven National Laboratory Report BNL 50877, April.

10. Wilkinson, J. F.: 1977, *An Exercise in Unnecessary Fright,* Coal Age, September, p. 444.

11. Gribbin, J.: 1978, *Carbon Dioxide and Climate: Too Much Heat Clouds Debate,* Energy Policy 6, pp. 314-319.

12. Cooper, C. F.: 1978, *What Might Man-induced Climate Change Mean?* Foreign Affairs 56, pp. 500-520.

13. Albanese, A. S., and Steinberg, M.: 1979, *Environmental Control Technology for Atmospheric Carbon Dioxide, Final Report,* Brookhaven National Laboratory Report BNL 51116, September.

14. Department of Energy: 1979, Annual Report to Congress, 1978, Vol. 2, DOE/EIA 0173/2, April.

15. Rotty, R. M.: 1980, *Past and Future Emissions of Carbon Dioxide,* Experientia, in press.

16. Strelzoff, S.: 1975, *Choosing the Optimum-CO_2 Removal System,* Chemical Engineering, September 15, pp. 115-120.

17. Pappano, A. W., Sears, J. T., and Parr, W. R.: 1976, *Availability and Economics of CO_2 for Enhanced Oil Recovery in Appalachia,* U. S. Department of Energy Report TID-22787.

18. Rump, W. M., Hare, M, and Porter, R. E.: 1977, *The Supply of Carbon Dioxide for Enhanced Oil Recovery,* Pullman-Kellogg Report FE-2515-10, December, for the U. S. Department of Energy.

19. Markham, A. E., and Kobe, K. A.: 1941, *The Solubility of Carbon Dioxide and Nitrous Oxide in Aqueous Salt Solutions*, J. American Chemical Society 63, pp. 449-454.

20. Linke, W. F.: 1958, *Solubilities* 1, D. Van Nostrand Company, Inc., New York.

21. Vukalovich, M. P., and Altunin, V. V.: 1968, *Thermodynamic Properties of Carbon Dioxide*, Collet's Ltd., London.

22. Bradshaw, A.: 1973, *The Effect of Carbon Dioxide on the Specific Volume of Seawater*, Limnology and Oceanography 18, pp. 95-105.

23. Lee, D. W.: 1980, *An Analytical Model for a Vertical Buoyant Jet*, Oak Ridge National Laboratory Report, ORNL TM 7140.

24. Din, F.: 1956, *Thermodynamic Functions of Gases, Vol. 1*, Butterworth Scientific Publications, London, p. 133.

25. Miller, S. L., and Smythe, W. D.: 1970, *Carbon Dioxide Clathrate in the Martian Ice Cap*, Science 170, pp. 531-533.

26. Quinn, E. L., and Jones, C. L.: 1936, *Carbon Dioxide*, Reinhold Publishing Corp., New York, p. 113.

27. From maps in Uchupi, E.: 1968, *Atlantic Continental Shelf and Slope of the United States: Physiography*, U. S. Geological Survey Professional Paper 529-C.

28. World Energy Conference: 1978, *World Energy Resources 1985-2020*, IPC Science and Technology Press, Guildford, U.K.

29. Stickney, W. J.: 1932, Refrigeration Engineering 23, p. 334, cited in *Chemical Engineer's Handbook*, Perry, J. H., Ed., 3rd ed., McGraw-Hill Book Company, New York, p. 1700.

ENVIRONMENTAL CONTROL TECHNOLOGY FOR ATMOSPHERIC CARBON DIOXIDE

Meyer Steinberg and Anthony S. Albanese

Process Sciences Division,
Department of Energy and Environment
Brookhaven National Laboratory, Upton, New York 11973

ABSTRACT. The impact of fossil fuel use in the United States on worldwide CO_2 emissions and the impact of increased coal utilization on CO_2 emission rates are assessed. The aspects of CO_2 control are discussed as well as the available CO_2 control points (CO_2 removal sites).

The primary factor affecting the practicability of a CO_2 control system is the energy required by the control system. Removal of CO_2 from the atmosphere or from the ocean surface water is energetically not feasible with a fossil energy source. More CO_2 is generated than is recovered. Removal from the stacks of fossil fuel power plants requires the least amount of energy and allows use of a fossil energy source. Estimates of the energy required to capture and recover CO_2 from coal-fired power plant stacks by various processes are presented. The effect on the reduction in overall plant efficiency is determined.

Although capture and recovery of CO_2 are important considerations in the overall scheme of CO_2 control, disposal or reuse of recovered CO_2 is the greatest problem in the CO_2 control chain. Of the several options considered, deep ocean storage appears to be a possibility.

Two control scenarios are evaluated, one based on the absorption of CO_2 contained in power plant flue gas by seawater; the other, based on absorption of CO_2 by MEA. Captured CO_2 is injected into the deep ocean in both cases. The analyses indicate that capture and disposal by seawater is energetically not feasible, whereas capture and disposal using

W. Bach, J. Pankrath, and J. Williams (eds.), Interactions of Energy and Climate, 521–551.

MEA is a possibility. However, the economic penalties of CO_2 control are significant. At a CO_2 removal efficiency of 50% the power generation efficiency of a conventional coal fired plant would be reduced from 34% to 25% and the cost is expected to double. For 90% removal the efficiency is reduced to between 15% and 6% and the power generation cost would increase by a factor of from 4 to 7.

The use of non-fossil energy sources, such as hydroelectric, nuclear or solar energy is considered as an alternative for limiting and controlling CO_2 emissions resulting from fossil energy usage.

I. INTRODUCTION

(1) Global CO_2 Problem

During the last one hundred years or so, the release of fossil carbon as CO_2 has increased at an exponential rate as shown in Figure 1.[1] As a result, the concentration of CO_2

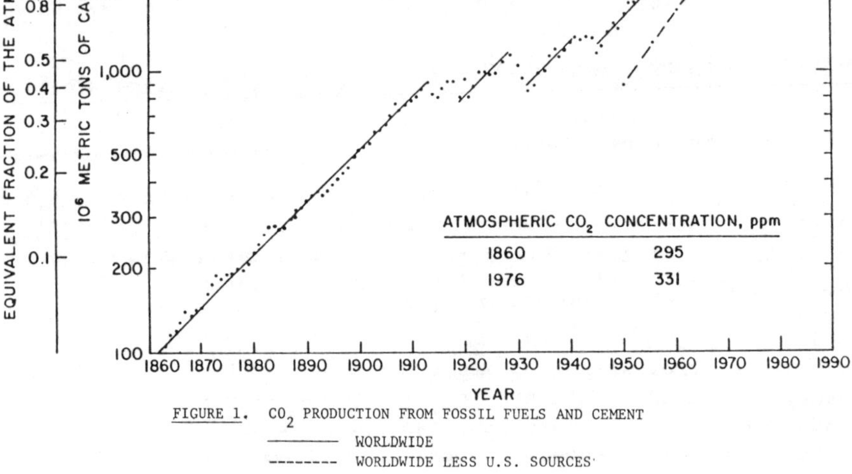

FIGURE 1. CO_2 PRODUCTION FROM FOSSIL FUELS AND CEMENT

———— WORLDWIDE

-------- WORLDWIDE LESS U.S. SOURCES

contained in the atmosphere has grown by about 12%, increasing from about 295 ppm by volume in 1860 (pre-Industrial Revolution) to the current level of 331 ppm. Except for short periods during World Wars I and II and the Depression, the release rate of fossil carbon has increased at the rate of about 4.3% per year.[2] At present, with no CO_2 emission controls, about 5 x 10^9 metric tons of carbon per year in the form of CO_2 are emitted to the atmosphere by the burning of fossil fuels from worldwide sources. About 50% of this carbon can be accounted for by the increased CO_2 concentration in the atmosphere. The remaining carbon is presumably absorbed by the oceans and by the land biomass; however, the actual quantities of CO_2 absorbed by the land biomass and by the oceans are still undetermined. Deforestation considerations tend to complicate the issue even further.

It is predicted that the concentration of atmospheric CO_2 relative to the pre-industrial value could increase by a factor of from about 2 (low-use case) to about 5 (high-use case) during the next one hundred years.[2] The low-use case corresponds to a fossil fuel growth rate of 2% per year until the year 2025, followed by a symmetrical decrease as alternative and renewable energy sources become more available and the use of fossil fuel is discouraged. The high use case corresponds to an initial annual growth rate of 4.3% with subsequent reductions in proportion to the ultimate fossil fuel supply that has been consumed.

Of primary concern is the warming ("greenhouse effect") which may be produced by these increased CO_2 concentrations due to the absorption by CO_2 of a portion of the infrared radiation returning to space from the earth. The amount of warming produced by a given CO_2 concentration increase is estimated at 1° to 5^oK per doubling of the CO_2 concentration. For the low-use case described above, this corresponds to a mean global temperature rise of from about 1° to 5^oK over the next one hundred years and from about 2° to 10^oK for the high-use case. A temperature rise of 1° or 2^oK during this period may be acceptable; however, a rise of 10^oK would probably cause catastrophic climatic and environmental effects.

Increased concentrations of particulate matter in the global air mass might produce a cooling effect by backscatter of incoming solar radiation and thus counteract, the warming produced by increasing CO_2 levels. However, uncertainties in this effect and increased reductions in particulate emissions make control by this method of questionable value.

Since there is a chance that naturally occurring and/or anthropogenic opposing factors will not just counter the effects of increasing CO_2 levels, it may be necessary to limit industrial CO_2 emissions. It is thus of interest to examine the technical feasibility and economic costs of CO_2 control should the need arise for regulating the quantity of CO_2 contained in the atmosphere.

(2) Impact of U.S. Emissions

In order to determine the impact of fossil fuel use in the United States on overall (worldwide) CO_2 emissions, we have superimposed several points on Figure 1, based on U.S. fuel use data for the years 1950 through 1976.[1,3] The points were obtained by subtracting the estimated emissions of carbon (as CO_2) by U.S. sources from the corresponding worldwide emissions for each of the above years, and thus indicate the quantities of carbon that would have been emitted if the U.S. had not burned any fossil fuels during these years. The reduction in carbon (and thus in CO_2) by this hypothetical non-use of fossil fuels in the U.S. ranges from 0.32 ppm in 1950 to 0.64 ppm in 1973 (and in 1976). However, during this period the percentage of the total yearly emissions contributed by the U.S. decreased from 43% in 1950 to 27% in 1976. Furthermore, it is projected that the contribution to global CO_2 production by the U.S. and Canada will decrease to about 8% by the year 2025 due to use of alternative energy sources and lower rate of energy growth demand. Developing nations will account for 36% of the CO_2 production, the USSR and Eastern Europe 27%, Asia 19%, Western Europe 7%, and Japan and Australia 3%.[1] Thus, it appears that a worldwide effort would be required to effectively control the level of atmospheric CO_2.

(3) Impact of Alternative Energy Sources

Table 1 contains the source distribution of the carbon (CO_2) produced from major worldwide sources for 1976.[1] Using heating values of 18,000 BTU/lb of contained carbon for coal, 22,400 BTU/lb carbon for fuel oil and 31,500 BTU/lb carbon for natural gas, and assuming that all fossil energy was produced by the burning of solid coal, the carbon released to the atmosphere would have been roughly 6 x 10^9 metric tons or about 1.2 times the quantity actually emitted. Energy differences required in mining versus drilling, preparation, transportation, etc. and in combustion efficiencies were not taken into account. If we had assumed that the natural gas and petroleum consumed in 1976 had been replaced by coal derived gaseous and liquid fuels, the quantity of carbon released to the atmosphere would have been greater than 6 x 10^9 metric tons due to inefficiencies in the gasification and liquefaction processes. Thus, added reliance on coal for energy production tends to compound the atmospheric release rate of CO_2,

TABLE 1

WORLD-WIDE CO_2 EMISSIONS BY SOURCE
ALL ENTRIES IN 10^6 METRIC TONS CARBON AS CO_2

Source	1976	1976 Distribution %
Coal	1676	33.2
Lignite	260	5.1
Crude petroleum & N.G. liquids	2224	44.1
Natural gas	675	13.4
Natural gas flared	113	2.2
Cement	99	2.0
Total carbon as CO_2	5047	
Incremental atm CO_2 conc. (ppm)	2.38	
U.S. emissions contribution	0.64	

depending upon the extent to which coal displaces oil and natural gas and future energy use patterns.

The expected reduction in the U.S. contribution to worldwide atmospheric CO_2 emissions is partially attributed to the development and implementation of alternative energy sources, which includes, nuclear (fission and fusion) geothermal and solar. The impact of these alternatives in the U.S. depends to a great extent on four factors: (1) the demand growth for energy in the U.S. (2) the availability of the technology (3) the capital and production costs and (4) the supply of world petroleum sources and (5) the demand of the developing countries for petroleum and coal resources.

Although uncertainties in all of these factors make it extremely difficult to estimate the impact of substitute energy sources, nevertheless some projections can be made for the U.S. in the next 50 to 100 years as listed in Table 2. Thus, if the energy demand no more than doubles in the next 50 years, and the fossil fractional use decreases in half, the U.S. would then contribute approximately the same amount as at present to the world atmospheric CO_2 emissions, as indicated by the CO_2 Index in the lower half of Table 2. However, the other countries would exceed their present contributions as they develop technologically at a higher rate. Eventually even

TABLE 2

PROJECTED DISTRIBUTION OF ALTERNATE ENERGY SOURCES FOR THE U.S. IN THE NEXT 50 TO 100 YEARS

Source	Percent of Demand 1975	2025	2075
Fossil (coal, oil and gas)	90%	50%	30%
Nuclear (Fusion and Fission)	5	20	30
Solar	0	20	30
Hydro, Geothermal and others	5	10	10
Total	100	100	100

U.S. AND WORLD ENERGY GROWTH AND DISTRIBUTION OF ALTERNATIVE SOURCES

Region	1975 Quads	Fraction Fossil	CO_2 Index*	2025 Quads	Fraction Fossil	CO_2 Index*
U.S.	75	0.90	67.5	150	0.50	75
Other Nations	175	1.00	175.0	1100	0.84	925
World (total)	250	0.97	242.5	1250	0.80	1000

*CO_2 index is product of Quads consumed and fossil fuel fraction.

these countries will acquire alternative energy sources and the CO_2 emissions may then become stabilized.

Table 3 shows the quantities of CO_2 produced per unit energy released for various naturally occurring and synthetic fuels. In determining the CO_2 generated by synthetic fuels such as substitute natural gas (SNG), the CO_2 generated in the production process is included. It is noted that natural gas produces about one third the amount of CO_2 per unit energy released in comparison to SNG derived from coal, and less than half the CO_2 from the direct burning of coal. Synthetic liquids and gases from coal generate the largest amount of CO_2 per unit of energy released. Of course there is no CO_2 generation if non-fossil energy sources are used to produce hydrogen as a synthetic fuel.

II. CO_2 CONTROL OVERVIEW

Perhaps the most significant difference between the control of CO_2 in comparison to other pollutants is the vast amount of CO_2 generated by the activities of mankind--currently about 20 x 10^9 tons per year from the use of fossil fuels. Effective control would therefore require an extremely large-scale application of technology and the expenditure of large amounts of energy and capital. Methods for removing and

TABLE 3

CO_2 GENERATION FOR VARIOUS FUELS

Fuel type	Chemical Formula	Higher heating value Btu/lb	CO_2 generated lb CO_2/lb fuel	CO_2 generated lb CO_2/1000 Btu	Energy generated kWh(e)/lb CO_2 generated[(1)]
Coke (ashless C)	C	14,000	3.67	0.26	
Bituminous coal	$CH_{0.85}O_{0.1}$	12,700	2.59	0.20	0.55[(2)]
Fuel oil & gasoline (petrol. dist.)	$(CH_2)_n$	19,600	3.14	0.16	0.70
Benzene	C_6H_6	18,000	3.38	0.19	
Acetylene	C_2H_2	21,500	3.38	0.16	
Methanol (wood alc)	CH_3OH	10,000	1.38	0.14	
Methanol (syn. from coal)	CH_3OH	10,000	3.3	0.33[(3)]	0.34
Biomass (wood, cellulose)	$(CH_2O)_n$	8,000	1.47	0.18	
Carbon monoxide	CO	4,350	1.57	0.36	0.31
Natural gas	CH_4	24,000	2.75	0.115	0.97
SNG from coal gasification	CH_4	24,000	7.9	0.33[(4)]	0.34
Synthetic liquid fuels (by coal hydrogenation)	—	—	—	0.34[(5)]	
Synthetic liquid & gaseous fuels (mixed products from SRC-II process)	—	—	—	0.26[(6)]	
SNG coal & non-fossil hydrogen	CH_4	24,100	2.75	0.11	1.07
Hydrogen generated by:	H_2	61,000	0	0	
Coal gasification	H_2	61,000	16.5	0.27	
Natural gas reforming	H_2	61,000	7.0	0.11	
Non-fossil energy					
Hydro, solar, nuclear-electrolytic	H_2	61,000	0	0	
Geothermal-electrolytic	H_2	61,000	2.44	0.04	

(1) Assuming a power plant efficiency of 38%.
(2) For a power plant with FGC, assuming power plant efficiency is reduced from 38% to 35%, the kWh(e)/lb CO_2 generated is reduced to 0.50.
(3) Estimated from ref. 14.
(4) Estimated from ref. 15 and 16.
(5) Estimated from ref. 17.
(6) Estimated from ref. 17 and 18.

recovering CO_2 from gases and liquids have been practiced in industry and are currently available,[4]. However, because there was little need for reuse, experience with storage and disposal techniques is very limited.

Although there are several key factors in assessing the practicability of a prospective control methodology, the most important consideration is the energy required by the control system. As this energy increases, assuming that the energy required to drive the control system is furnished by fossil fuel, overall power plant efficiency decreases and the quantity of CO_2 emitted to the atmosphere per KWH(e) generated increases. A point could be reached where the only effect produced by the CO_2 control system is a decrease in power plant efficiency with no net reduction in CO_2 emissions.

In a coal-fired power plant as shown in Table 3 about 0.55 KWH of electricity is produced per pound of CO_2 generated. Table 4 indicates in a general manner, the effect of CO_2 control system energy requirements and CO_2 collection efficiency on the overall power plant efficiency and on net CO_2 emissions.

TABLE 4

EFFECT OF CO_2 CONTROL SYSTEM ENERGY REQUIREMENTS ON POWER PLANT EFFICIENCY

Energy required by CO_2 control system kWh(e)/lb CO_2	CO_2 removal efficiency, %	Net power plant efficiency %	lbs CO_2 discharged to atmosphere per net kWh(e) generated
0	100	38.0	0
0.1	100	31.0	0
	75	32.8	0.53
	50	34.5	1.01
	25	36.3	1.44
	0	38.0	1.83
0.2	100	24.1	0
	75	27.6	0.63
	50	31.0	1.12
	25	34.5	1.51
	0	38.C	1.83
0.3	100	17.1	0
	75	22.4	0.78
	50	27.6	1.26
	25	32.8	1.59
	0	38.0	1.83
0.4	100	10.2	0
	75	17.1	1.01
	50	24.1	1.44
	25	31.0	1.68
	0	38.0	1.83
0.5	100	3.2	0
	75	11.9	1.46
	50	20.6	1.69
	25	29.3	1.78
	0	38.0	1.83
0.55	100	0	0
	75	9.5	1.83
	50	19.0	1.83
	25	28.5	1.83
	0	38.0	1.83

Three potential control points (CO_2 removal sites) are available for regulating the concentration of CO_2 in the atmosphere: (1) the atmosphere itself, (2) the surface waters of the oceans, and [3] the stacks of fossil fuel power plants. The latter sites have the advantage of providing a source of relatively high CO_2 concentration, while the first two provide more flexibility for site selection and simpler disposal or recovery and reuse logistics. A choice of control point limits the available choices of CO_2 removal processes and will reflect on the size of operation, the number of required sites, energy requirements, the comparative economics, and the overall feasibility of operation.

The CO_2 concentration ranges of the three control points are listed in Table 5. Since the concentration of CO_2 in flue gas is significantly greater than that in the atmosphere or the oceans, the energy required to remove CO_2 from flue gas should

TABLE 5

CO_2 CONCENTRATIONS AT CONTROL POINTS AND MINIMUM CO_2 SEPARATION ENERGIES

Control point	CO_2 concentration		Min. separation energy[(1)]
	% by vol	% by wt	kWh/lb CO_2
(1) Atmosphere	331×10^{-4}	502×10^{-4}	0.0570
(2) Ocean	—	100×10^{-4} (2)	0.0570
(3) Fossil fueled power plant stacks	7-20	11-28	0.0259-0.0179

(1) Minimum energy based on free energy of mixing: $\Delta F = RT\Sigma n_i \ln(P_i/P_i^o)$, where P_i^o and P_i are the initial and final partial pressures of the ith species, respectively. R is the gas constant, T is temperature in degrees absolute, and n_i is the number of moles of the ith species.

(2) ΣCO_2 (total carbon dioxide content) = $C_{H_2CO_3} + C_{CO_2(aq)} + C_{HCO_3(T)} + C_{CO_3(T)}$.

be appreciably lower than that for the other two control points. The minimum theoretical energy (the thermodynamic free energy) to separate CO_2 from each of the three sources are also listed in Table 5. Based on these theoretical values, it would take about three times more energy to separate CO_2 from the atmosphere than from flue gas.

Although removal or capture of CO_2 from a particular source is an important consideration in devising a CO_2 control system, CO_2 disposal (storage) and reuse potentials are equally or perhaps more important, for unless a control scenario produces a net near term decrease in the rate of buildup of atmospheric CO_2 it will be of no value. This means that the captured CO_2 must be stored or recycled until large scale non-fossil energy sources become available.

A block diagram showing the relationship between the various aspects of CO_2 control is presented in Figure 2.

III. CO_2 REMOVAL AND RECOVERY

(1) Atmosphere and Oceans

The present concentration of CO_2 contained in the atmosphere is about 331 ppm by volume. Therefore, in order to

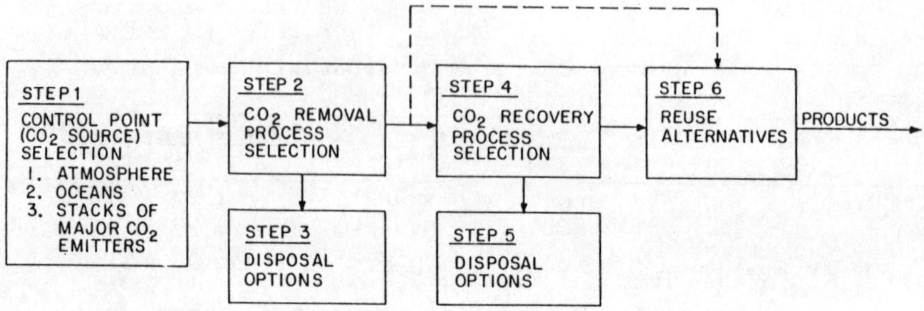

FIGURE 2. LOGIC DIAGRAM FOR CO_2 CONTROL

remove one mole of CO_2 from the atmosphere a minimum of 3021 moles of air must be processed. This is equivalent to about one pound of CO_2 per 2000 pounds of air. The equation for calculating the air horsepower of a fan can be expressed as follows[5]

$$\text{Air h.p.} = 157 \times 10^{-6}\, Q\, P \;,$$

where Q is the volume handled in cfm and P is the pressure drop across the fan in inches of water. For air at standard conditions, this reduces to:

$$\text{Air h.p.} = 34.9 \times 10^{-6}\, W\, P \;,$$

where W equals the mass flow rate of air in pounds per hour. Assuming a fan efficiency of 75%, and converting to kilowatts yields

$$\text{kW fan} = 34.7 \times 10^{-6}\, W\, P \;.$$

Dividing by pounds of CO_2 and assuming 2,000 pounds of air per pound of CO_2, yields the following equation:

$$\frac{\text{kWh(e)}}{\text{lb } CO_2} = 0.0694 \;\; P \;.$$

Assuming an overall power plant efficiency of 38% and a coal heating value of 18,000 Btu per pound of contained carbon, the pounds of CO_2 produced per KWH(e) generated amounts to 1.83. Setting 1/1.83 (0.5464) equal to the latter equation P is found to equal 7.9 inches of water. Thus, if the pressure drop across a CO_2 absorption or adsorption unit is equal to about 8 in. of water (assuming a 100% removal efficiency and electricity generation by the combustion of coal), as much CO_2 will be generated in supplying electricity to the air-handling fan as will be removed from the atmosphere by the CO_2 removal system. For electricity generation by the combustion of natural gas instead of coal, P maximum calculates to 14 in. of water. Since a maximum allowable pressure drop of from 8 to 14 in. of water is a rather stringent constraint to impose on such equipment (actually the pressure drop would have to be much less than the calculated maximum allowable for the process to be of value), and since there are other system energy requirements to satisfy, e.g., sorbent pumping energy in absorption systems, it is unlikely that a practicable system can be designed for removing CO_2 from the atmosphere, using a fossil energy power source. Note that Steinberg and Dang[6] determined that the energy required to remove CO_2 from the atmosphere by refrigeration techniques was greater (9.4 kWh(e)/lb CO_2) than that required for removal by atmospheric absorption or adsorption processes (0.4 kWh(e)/lb CO_2). It was also determined that the energy required to remove CO_2 from the ocean was at a minimum equal to or greater than that required for removal from the atmosphere. It is concluded that for the removal of CO_2 from the atmosphere or the ocean, a non-fossil fuel source would be necessary.

(2) Flue Gas

The concentration of CO_2 in fossil fuel power plant flue gas is about 500 times greater than its concentration in the atmosphere, therefore, CO_2 removal should be greatly facilitated. Furthermore, commercial processes, i.e., absorption/desorption processes, are currently available for removing CO_2 from flue gas and other gaseous mixtures. For example, in the manufacture of liquid CO_2 and dry ice, CO_2 resulting from the burning of a carbonaceous material is absorbed from flue gas with sodium carbonate, aqueous potassium carbonate, or amines. CO_2 removal systems are also employed in the production of synthesis gases for ammonia, methanol, and in hydrogen production.

Estimates of the energy requirements for removing CO_2 from flue gas are given in Table 6. Values range from 0.15 kWh(e)/lb CO_2 for 50% absorption with monoethanolamine (MEA) to 0.8 kWh(e)/lb CO_2 for 90% absorption with seawater. The most

TABLE 6

ENERGY REQUIRED TO REMOVE AND RECOVER CO_2 (GASEOUS) FROM FLUE GAS

Process	CO_2 removal efficiency %	kWh(e)/lb CO_2[(1,2,3)] recovered
1. Amine (MEA) absorption/Stripping	50 90	0.15 0.27
2. Potassium carbonate absorption/stripping	90	0.32
3. Molecular sieves	90	0.40
4. Refrigeration	90	0.40
5. Seawater absorption[(4)]	60 90	0.60 0.80

(1) Energy required to recover CO_2 (as a gas) from flue gas of coal-fired power plant subsequent to SO_2 removal. Inlet conditions to CO_2 removal process are 120°F and 14.7 psia, assuming flue gas is saturated with water vapor and free of SO_2. Reheat of flue gas accounted for by SO_2 system.

(2) All energy requirements have been converted to the equivalent amount of electrical energy.

(3) To liquefy the recovered CO_2 gas, an additional 0.05 kWh(e)/lb CO_2 is required; to solidify the CO_2 gas, about 0.17 kWh(e)/lb CO_2 is required.

(4) Including deep ocean disposal of captured CO_2.

effective solvent appears to be MEA; however, even the use of MEA will significantly reduce overall power plant efficiency.

It should be pointed out that only about 30% of the total CO_2 released by the U.S. today is located at central stationary electric power plants. The rest comes from smaller stationary industrial and residential boiler and heating systems and from transportation sources (automobiles and railroads).

IV. CO_2 DISPOSAL (STORAGE)

Four ways of disposing of captured CO_2 have been considered: (1) injection in the deep ocean, (2) terrestrial burial (3) storage in plants and trees, (4) conversion of CO_2 to useful products and (5) extraterrestrial disposal. The latter method is considered impractical because of the large fuel to payload ratio to lift a unit mass into earth orbit, thus generating more CO_2 than can be disposed of.

(1) Injection in the Deep Ocean

In the two-box model, the world ocean is represented by two reservoirs, separated by a thermocline. One reservoir consists of surface layers to a depth of about 100 meters; the other, of deeper layers from the 100-meter depth to the ocean floor. Mixing is assumed to occur rapidly within each box or reservoir, but exchange between the two boxes proceeds slowly. Thus, the deep ocean is a potential sink for CO_2 since there is no direct exchange between the atmosphere and deep ocean waters. Only the surface layers are in CO_2 equilibrium with the atmosphere. The main discrepancy between the simple two-box model and the real ocean is that in some regions (e.g., Arctic, Antarctic, and Gibraltar) convection currents extend from the ocean surface to the ocean bottom.

One proposal for disposing of collected CO_2 is to inject it into the sea at the Strait of Gibraltar.[7] Presumably, the current would carry it to the deeper ocean waters where it would remain for centuries. Another proposal is to inject liquid CO_2 into the deep sea at a density greater than that of deep sea water via a long pipe.[24] The CO_2 would sink to the bottom of the ocean forming a liquid CO_2 lake within the ocean. CO_2 could also be injected into the deep ocean as a gas, provided the injection depth is deep enough to permit dissolution of the CO_2 before it rises to the upper layers.

The current release rate of fossil carbon to the atmosphere is about 5×10^{15} g/yr, whereas the current carbon content of the deep ocean is estimated to be at 0.373×10^{20} g. Therefore, if all this fossil carbon was captured and deposited in the deep ocean, its carbon content over a one year period, would increase by only about 0.013%. However, the effects of this injected carbon on the deep ocean ecosystem have not been assessed.

(2) Terrestrial Burial

Storage in abandoned mine shafts or wells and fixation in natural clays are two options in this category which can be considered. The latter has been dismissed as impractical because of the large quantities of clays that would have to be handled, even if they were available and effective in fixing CO_2. In the case of below ground storage the rate at which CO_2 would re-enter the atmosphere is an important consideration as well as the available storage capacity of the below-ground cavities. A related disposal scheme is to use captured CO_2 for enhanced oil recovery, in which case, the CO_2 would be injected into depleted oil wells to effect additional oil recovery. Captured CO_2 would thus have an economic value. However, the effectiveness of such a method in controlling atmospheric CO_2 does not only depend on economics but upon (1) the energy requirements of the CO_2 removal and disposal system, (2) how much CO_2 can be injected into depleted wells, and (3) the length of time the CO_2 would remain underground. These factors have not been studied in enough detail to draw any firm conclusions.

(3) Storage in Plants and Trees

It has been suggested that in the event of a CO_2 worldwide emergency, corrective measures to halt or reverse the growth of atmospheric CO_2 could be instituted by planting fast-growing long-lived trees which are left unharvested and/or by growing short-lived plants which are converted to humus or allowed to accumulate in artificial peat bogs. However, this action would be limited by the availability of fertilizer.[8] Furthermore, additional CO_2 would be generated by the reforestation activities and in supplying energy for increased fertilizer production, which would reduce the effectiveness of the method. Further analysis is required to rationalize this method.

(4) CO_2 Conversion to Useful Products

Although a number of commercial CO_2 products can be made for the economy, e.g., dry ice, or soda ash, the quantity of these products consumed yearly is extremely limited in comparison to the quantity of fossil CO_2 emitted to the atmosphere. Therefore, atmospheric CO_2 levels will not be significantly affected by the way in which these products are made. Furthermore, these products eventually return the CO_2 back to the atmosphere.

A more variable possibility for reducing the buildup of CO_2 in the atmosphere, is to recover and convert CO_2 to synthetic liquid and gaseous carbonaceous fuels using a non-fossil energy source such as nuclear, solar or hydro energy.[6,9,10] Another alternative is to combine a non-fossil energy source with coal to produce synthetic fuel. This would tend to reduce the rate of increase of CO_2 emissions by a factor of up to 3 compared to producing hydrocarbon synfuels from coal alone. The latter is a step toward complete substitution of fossil with non-fossil energy.

V. REMOVAL AND DEEP OCEAN DISPOSAL OF FLUE GAS CO_2 USING SEAWATER

(1) System Description

In light of the previous considerations, control point 3 (power plant stacks) has been selected as the initial control point of interest. Accordingly, a scenario based on the absorption of CO_2 generated by coal-fired power plants has been

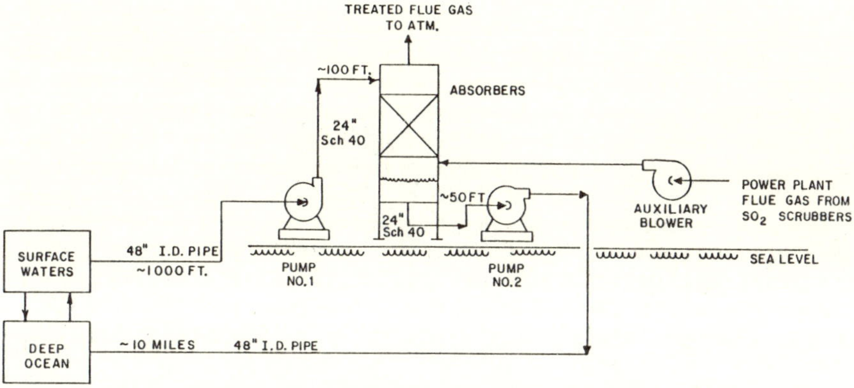

FIGURE 3. FLOWSHEET FOR CO_2 ABSORPTION DISPOSAL FROM POWER PLANT FLUE GAS USING SEAWATER.

formulated. A process diagram of the system is presented in Figure 3. Seawater from the surface of the ocean is pumped to the top of a CO_2 absorber (packed tower). The seawater flows downward through the packing by gravity, absorbing CO_2 from the power plant flue gas which flows countercurrent to the seawater. The water exits at the bottom of the absorber and is pumped to the deep ocean while the treated flue gas is discharged to the atmosphere. The flue gas is fed to the absorber, subsequent to SO_2 removal, by an auxiliary blower. Such a system is perhaps the simplest for one step absorption and disposal of CO_2. No CO_2 desorption step is required and no additional handling of recovered CO_2; the captured CO_2 is discharged directly into the ocean reservoir. Furthermore, seawater is readily available in large quantities at no cost. However, the scenario does impose constraints on power plant location.

In evaluating this subject, it was assumed that the power plant under consideration has a capacity of 200 MWe, an efficiency of about 34% (subsequent to SO_2 removal), is located near the U.S. coastline at sea level, and that deep ocean waters are accessible at a distance of about 10 miles from the power plant site.

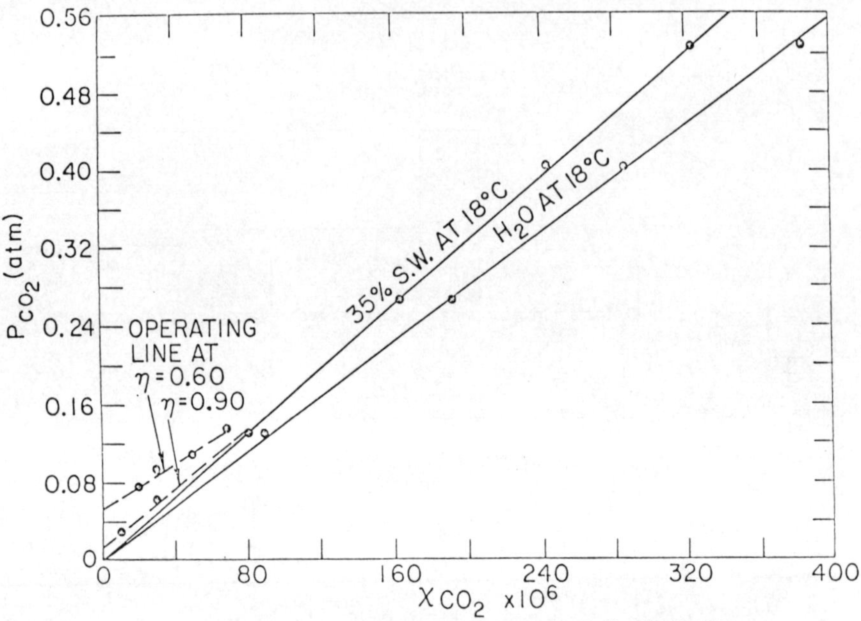

FIGURE 4. EQUILIBRIUM PARTIAL PRESSURE OF CO_2 VERSUS MOLE FRACTION IN WATER AND IN SEAWATER (3.5% SALINITY)

(2) Process and Absorber Design

Flue gas composition was determined assuming 20% excess combustion air. A calculation of the design[11] of the absorption tower was based on CO_2 solubility data from Riley and Skirrow[12] and the transfer unit method described by Sherwood and Pigford.[13] Two cases were determined; one for a CO_2 removal efficiency (R) of 0.6 (60%) and the other for a removal efficiency of 0.9. Figure 4 shows the plots of CO_2 partial pressure versus mole fraction for pure water and for seawater with the two absorber operating lines indicated. The absorber tower size was determined from the liquid to gas mass flow ratio and the pressure drop correlation for two inch Raschig Ring packing. A summary of the design and operating characteristics for the two removal efficiency cases is given in Table 7. It is seen that the pressure drop is 15 inches and 70 inches water respectively. The system would require between 84 to 90 and 100 to 107 towers of 20 ft diameter each for each of the cases. For the 60% removal case the packed height is 23 ft while for 90% efficiency the tower height is 130 ft. The equipment and consequently the capital investment becomes formidable.

TABLE 7

SUMMARY OF CO_2 ABSORBER DESIGN DATA
AND RESULTS FOR CO_2 REMOVAL EFFICIENCIES
OF 60 AND 90 PERCENT
SEAWATER ABSORPTION/DISPOSAL

	CO_2 removal efficiency	
	60%	90%
x_1, mf $^*CO_2$ in outlet seawater	$70x10^{-6}$	$80x10^{-6}$
x_2, mf CO_2 in inlet seawater	$0.2x10^{-6}$	$0.2x10^{-6}$
y_1, mf CO_2 in inlet gas	0.1340	0.1340
y_2, mf CO_2 in outlet gas	0.0536	0.0134
N_{OL}, number of overall liquid-phase transfer units	3.3	16.3
m, slope of equilibrium line	1620	1620
L_m/G_m, molar ratio of liquid to gas	1217-1493	1533-1981
L/G, mass ratio of liquid to gas	754-927	949-1271
G, mass velocity of gas, lb/(sec) (ft^2)	0.020-0.018	0.017-0.014
H_{OL}, height of overall liquid-phase transfer unit, feet	7	8
h, packing height, feet	23	130
ΔP/foot packing, inches water	0.5	0.5
ΔP/tower, inches water	15	70
Tower diameter, feet	20	20
Overall tower height, feet (approx.)	42	150
No. towers	84-90	100-107

*Mt = male fraction

(3) Energy Requirement

A summary of energy requirement is given in Table 8. The major energy requirement, as would be expected, is due to the seawater pumping load. A relatively minor quantity of energy is required to pump the flue gas through the absorbers. It should be mentioned that the high horsepower and thus energy requirements of the seawater pumps are due to the large quantities of seawater (about $3x10^6$ gpm) that must be supplied to and exhausted from the CO_2 absorption towers (because of the limited solubility of CO_2 in seawater). Friction loss (for the selected pipe sizes) and static head requirements are not excessive (about 100 psi overall). Furthermore, static head is dictated by absorber height and not by the depth to which the spent seawater must be pumped because the head against which the spent seawater is pumped is balanced by the static head gain of the spent seawater in flowing from sea level to the deep ocean. However, the deeper the injection point the greater the friction loss (for a given pipe size and flow rate) because a greater length of pipe is required.

TABLE 8

SUMMARY OF ENERGY REQUIREMENTS FOR CO_2
REMOVAL EFFICIENCIES OF 60 AND 90 PERCENT
SEAWATER ABSORPTION/DISPOSAL

	Energy requirements, kWh(e)/Lb CO_2 captured 60%	90%
Seawater pump #1	0.2235	0.4839
Seawater pump #2	0.3688	0.3087
Flue gas exhaust blower	0.0045	0.0136
Total	0.5968	0.8062

For each of the CO_2 removal efficiencies considered, more CO_2 would be produced in supplying electricity to the control system (assuming coal derived power and electrically driven equipment) than would be extracted from the power plant flue gas. For the 60% efficiency case, about 0.6 kWh(e)/lb CO_2 captured is required and for the 90% case, about 0.8 kWh(e)/lb CO_2 captured. At an overall power plant efficiency of 34% and a coal heating value of 18,000 BTU per pound of contained carbon, only 0.5 kWh of electricity is produced per pound of CO_2 generated. Thus for a fossil fueled power source more power is required to remove CO_2 than is generated. Therefore, this sytem is not feasible. The only alternative is to use a non-fossil energy source to produce the power for CO_2 removal. However, it can be argued that the non-fossil source could replace the fossil fuel plant in the first place.

VI. REMOVAL OF CO_2 FROM FLUE GAS USING MEA, FOLLOWED BY DEEP OCEAN DISPOSAL

(1) MEA Absorption/Stripping

Of the several processes evaluated for removing CO_2 from stack gases (Table 6), monoethanolamine (MEA) absorption/stripping requires the least amount of energy per unit of CO_2 recovered. MEA absorbs CO_2 rapidly and efficiently at low partial pressures and is therefore well suited for removing CO_2

from power plant flue gas. Impurities such as COS, CS_2, HCN and SO_2 react non-regenerably with MEA so SO_2 must be removed from the flue gas prior to treatment with MEA. Flue gas does not normally contain COS, CS_2 or HCN.

A flow sheet of the MEA process for CO_2 absorption and recovery from power plant flue gas is shown in Figure 5.[19,20]

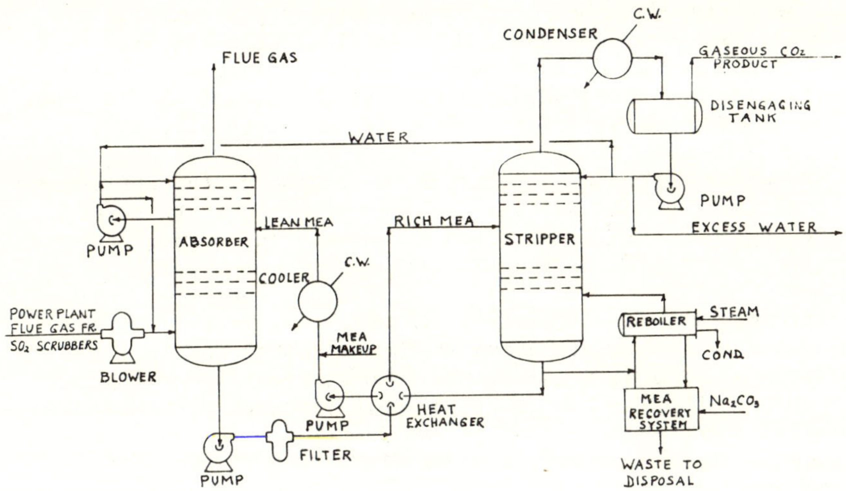

FIGURE 5. FLOWSHEET FOR CO_2 ABSORPTION RECOVERY FROM POWER PLANT FLUE GAS USING MEA.

Flue gas at 120°F and 14.7 psia (water saturated) from SO_2 removal is fed to the suction of a blower which compresses the flue gas to about 16 psia and feeds it to the MEA absorber. As the flue gas passes through the absorber CO_2 is removed by reaction with lean MEA solution, flowing countercurrent to the flue gas, in accordance with the following reaction:

$$HO(CH_2)_2NH_2 + H_2O + CO_2 \rightleftharpoons HO(CH_2)_2NH_3HCO_3$$

The reaction takes place to the right (forward direction) at temperatures between about 100 and 175°F.

Spent MEA solution is pumped from the bottom of the absorber through a filter, for removal of solid impurities, to the top of the regenerator. Before entering the regenerator, the cold spent MEA solution is heated by heat exchange with the hot regenerated (lean) solution from the bottom of the regeneration. Carbon dioxide is stripped from the MEA as the solution flows downward through the regenerator. A steam heated reboiler supplies the neccessary heat for regeneration. The bottom of the regenerator is maintained at about 250°F. Elevated temperature combined with steam stripping favors the reverse of the absorption reaction.

The stripped carbon dioxide exits the regenerator at the top and flows to a condenser for removal of water vapor. Most of the condensed water is returned to the generator as reflux, a portion of the remainder is pumped to the top of the absorber to reduce MEA losses. Excess water is pumped to disposal.

Regenerated MEA solution from the bottom of the stripping tower is first cooled by heat exchange with spent MEA solution flowing to the stripper and then by water. The cooled lean solution is then fed to the top of the absorber tower.

A reclaimer for recovering MEA which has reacted with residual flue gas SO_2 is shown at the bottom of the regenerator. A sidestream of MEA is withdrawn from the regenerator reboiler and cooled by heat exchange with the reclaimed solution returning to the reboiler. Sodium carbonate is added to the stream to liberate MEA. The sidestream is then cooled and filtered to remove precipitated sodium sulfate and sulfite and other degradation products. Purified MEA solution is returned to the reboiler. The reclaiming system is speculative at the present time and further development work is required before a final design can be prepared.[20]

(2) Deep Ocean Disposal of Recovered CO_2

Three methods of disposing of the recovered gaseous CO_2 from the MEA absorption/stripping process were considered (a) gaseous disposal via a deep ocean pipeline, (b) liquid disposal via a deep ocean pipeline and (c) deep ocean solid disposal.

(a) Gaseous disposal. In gaseous disposal, CO_2 recovered by the MEA process is compressed and fed to the deep ocean via a pipeline. A 500 meter injection depth was assumed even though ocean surface layers usually extend to a depth of only about 100 meters. The additional 400 meters was provided to allow for dissolution of the injected gas in the deep ocean. Any injected CO_2 which reaches the well mixed surface layers will eventually equilibrate with the atmosphere and thus reduce the effectiveness of disposal. Ocean depths of 500 meters are accessible at distances of about 100 miles from the shoreline of the northeast United States (approximate width of the continental shelf), therefore, a pipeline of about 100 miles in length was assumed (no allowance was made for inland distance).

For 90% recovery of the CO_2 generated by a 200 MWe coal fired power plant an 18 inch line was selected; for 50% recovery, a 14 inch line was selected. Most of the energy required by the CO_2 compressor is expended in overcoming the seawater static head between the CO_2 injection point and the surface of the ocean. A compressor discharge pressure (third stage) of about 800 psia is required.

A block diagram of the overall absorption/recovery/disposal system is shown in Figures 6 and 7.

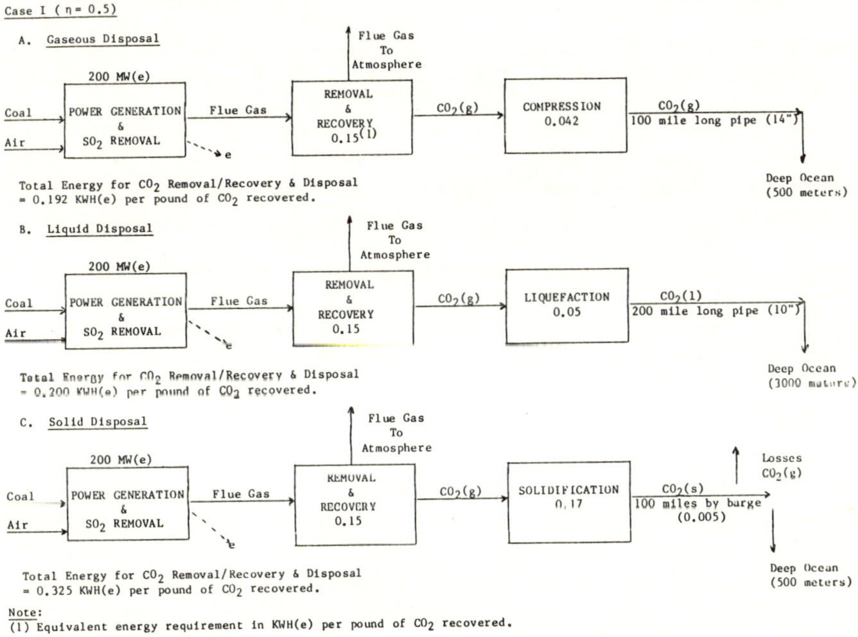

FIGURE 6. ENERGY REQUIREMENTS FOR MEA ABSORPTION AND DEEP OCEAN DISPOSAL OF FLUE GAS CO_2 FROM 200 MW(e) COAL FIRED POWER PLANT

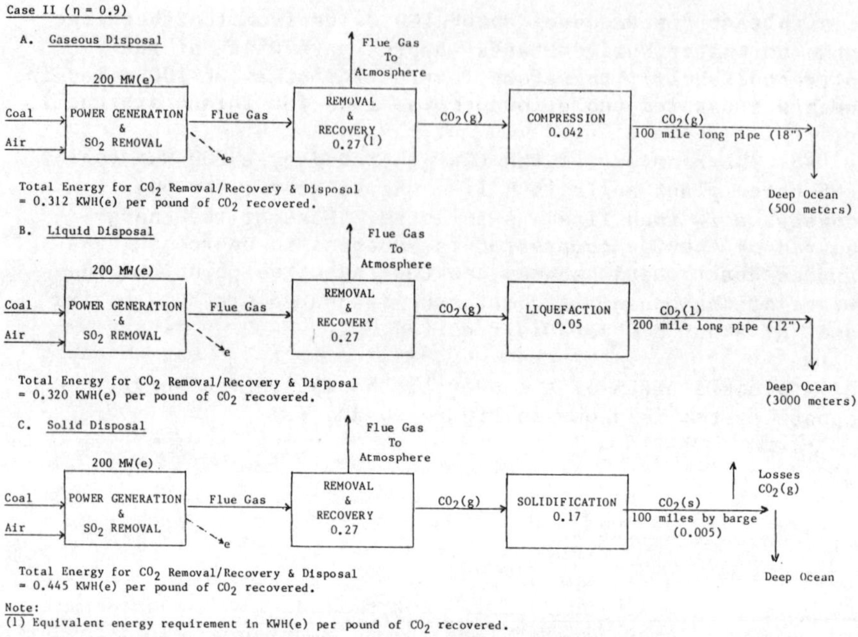

FIGURE 7. ENERGY REQUIREMENTS FOR MEA ABSORPTION AND DEEP OCEAN DISPOSAL OF FLUE GAS CO_2 FROM 200 MW(e) COAL FIRED POWER PLANT

(b) Liquid disposal. In order to produce liquid CO_2 product of the MEA system, the gaseous CO_2 must be compressed, cooled and dried. This is done in a three stage compression system equipped with coolers, water knockout drums, and a dryer. The gas is compressed to 920 psia in the third stage, passed through a cooler, and liquefied in a condenser at about $27^{o}C$[22] (the critical pressure and temperature of CO_2 are 1073 psia and $31^{o}C$). For deep ocean disposal, the liquefied CO_2 would then be piped to an appropriate location in the deep ocean.

In order to store CO_2 as a liquid pool in the deep ocean it must be injected to a depth where the density of CO_2, at the pressure corresponding to the seawater static head and the prevailing seawater temperature, is greater than the density of the seawater at the injection point. The specific gravity of CO_2 is approximately equal to 1 at pressures between 150 and 300 atmospheres and temperatures between 3 and $15^{o}C$. An injection depth of 3000 meters was selected which corresponds to a pressure of about 300 atmospheres; the temperature of the ocean at this depth is about $3^{o}C$. The specific gravity of CO_2 (liquid) at these conditions is 1.05.

Ocean depths of 3000 meters are accessible at distances of about 200 miles from the shoreline of the northeast United States and thus a pipeline of about 200 miles in length is required (no allowance is made for inland distance since the required offshore distance is quite variable with respect to the U.S. coastline). For 90% recovery of the CO_2 generated by a 200 MW(e) plant a 12 inch line was selected; for 50% recovery, a 10 inch line. For these line sizes, the liquid CO_2 will flow to the deep ocean under the influence of the CO_2 liquefaction pressure. Pumps are not required. In the subject evaluation it was assumed that the liquid would be cooled from about 27°C to the temperature of the deep ocean injection point during its journey to the deep ocean. If this is not the case, additional cooling of the liquid CO_2 will be required.

A block diagram of the overall absorption/recovery disposal system is shown in Figures 6 and 7.

(c) Solid disposal. In the production of commercial solid CO_2, the liquid CO_2 produced by compression and cooling, as described above, is expanded through a nozzle, forming "snow", and fed to a press. The vapors given off during expansion are used to precool the liquid carbon dioxide prior to expansion. The gas is then recompressed and reliquefied. The "snow" is compressed into 220 pound blocks and then quartered by band saws. The 55 pound blocks are wrapped in Kraft paper and either stored or shipped in insulated cars.[22] The density of the blocks is about 100 lb/ft^3. The temperature of saturated CO_2 ice at 1 atmosphere pressure is -78.5°C.

Since the density of the CO_2 blocks is much greater than that of seawater, they should readily sink to the ocean bottom when dropped into the surface waters of the ocean. The settling velocity of a 10 inch cube of solid CO_2 in seawater is roughly estimated at 5 ft/sec. For a disposal depth of 500 meters (the depth selected for gaseous CO_2 disposal) it would take about 5.4 minutes for a block to reach bottom; for a disposal depth of 3000 meters (the depth selected for liquid CO_2 disposal) it would take about 32 minutes. Based on the phase diagram for CO_2 and assuming the blocks are heated by the seawater to prevailing ocean temperatures, the final state of the blocks for both the 500 and 3000 meter disposal depths, is liquid. The liquid formed for the 3000 meter disposal depth will have a density greater than that of seawater and therefore should remain at the bottom of the ocean. For the 500 meter depth, the liquid will be less dense than seawater and would therefore tend to rise. Upon rising, some of the liquid CO_2 may dissolve in the seawater and some may flash into vapor which in turn will rise and dissolve. Depending upon the relative rates of heat and mass transfer and the rate of ascent of the CO_2 liquid droplets and gaseous bubbles, some CO_2 may reach the surface layers of the ocean. CO_2 reaching these layers may eventually end up in the atmosphere. Losses may also occur as the blocks descend to the ocean bottom as a result of gas and liquid formation and dissolution. Without experimental data or a detailed analysis of the physics of deep ocean disposal of solid CO_2 it is not possible to confidently predict the fate of the solid CO_2 blocks or accurately determine the extent of losses.

The energy required to transport the solid CO_2 blocks a distance of 100 miles offshore (500 meter disposal depth) by barge is estimated at 0.005 kWh(e)/lb CO_2 and for 200 miles (3000 meter depth) at 0.010 kWh(e)/lb CO_2. The energy to produce solid CO_2 blocks from gaseous CO_2 is about 0.17 kWh(e)/lb CO_2. Therefore, barging distance does not have a significant effect on the overall energy requirements of the system, assuming no CO_2 losses during transport, however, disposal costs for the 200 mile distance would be almost twice that of the 100 mile distance.

(3) Energy Requirements

Total system energy requirements for CO_2 recovery efficiencies of 50 and 90% and for gaseous, liquid and solid disposal are summarized in Figures 6 and 7, respectively. At 50% efficiency, gaseous or liquid disposal require the equivalent of about 0.20 kWh(e)/lb CO_2 recovered and at 90% efficiency about 0.32 kWh(e)/lb CO_2. For solid disposal at 50% recovery, assuming no sublimation losses during handling and transportation, about 0.33 kWh(e)/lb CO_2 are required, whereas, about 0.45 kWh(e)/lb CO_2 are needed at 90% recovery.

Assuming a power plant efficiency of 34% after sulfur dioxide removal, 0.49 kilowatt-hours of electricity is produced per pound of CO_2 generated. With no CO_2 removal, 2.04 pounds of CO_2 are discharged to the atmosphere per net kilowatt-hour of electricity produced. At 50% recovery and gaseous or liquid disposal, overall power plant efficiency would be reduced to 27% and for 90% recovery to 14%. The corresponding pounds of CO_2 discharged to the atmosphere per net kWh(e) generated are 1.3 and 0.5, respectively. For 50 and 90% recoveries and solid disposal, power plant efficiency would be reduced to 23% and 6%, respectively. The corresponding pounds of CO_2 discharged per net kWh(e) produced are 1.5 and 1.1.

(4) Investment and Power Generation Costs

Rough estimates of the required fixed capital investments[21] for 50% and 90% CO_2 recovery using MEA and for gaseous, liquid and solid disposal are summarized in Table 9.
Based on these estimates and estimates of the capital and operating costs of a 200 MWe coal fired power plant with flue gas desulfurization (FGD), electricity generation costs have been calculated and are included in Table 9 along with an estimate of the cost of generation without CO_2 control.

TABLE 9

COSTS OF GENERATING ELECTRICITY
WITH AND WITHOUT CO_2 CONTROL
FOR INITIAL POWER PLANT CAPACITY
OF 200-MW(e)

	SO_2 removal with no CO_2 control	SO_2 Removed Plus 50% CO_2 Control			SO_2 Removed Plus 90% CO_2 Control		
		Gaseous disposal	Liquid disposal	Solid disposal	Gaseous disposal	Liquid disposal	Solid disposal
Net capacity,(1) MW(e)	200	161	159	134	86	83	37
Capital investment MM$							
(a) Power plant(2)	160	160	160	160	160	160	160
(b) CO_2 control system(3)	--	152	216	46	194	290	68
Total	160	312	376	206	354	450	228
Operating costs,(4) MM$/year							
(a) Coal @ $30/ton	18.9	18.9	18.9	18.9	18.9	18.9	18.9
(b) O&M	4.8	5.8	5.8	7.2	5.8	5.8	7.2
(c) Barging costs(5,6)	—	—	—	5.0	—	—	5.0
(d) Capital charges @ 15% of Capital investment	24.0	46.8	56.4	30.9	53.1	67.5	34.2
Total	47.7	71.5	81.1	62.0	77.8	92.2	65.3
Electricity generation costs, Mills/kWh (revenue requirements)	30	56	64	58	113	139	221

(1) Electrical energy to drive the CO_2 control system is assumed to be obtained from the electrical output of the power plant, therefore: $e_{net} = e_{gross}(200\ MWe) - e_{control}$.

(2) Total capital investment @ $800/kW(e) including FGD.

(3) Total capital investment @ 1.2 times fixed capital investment.

(4) Based on 8000 hours of operation per year.

(5) Dry ice barging costs @ $15,000 per day; costs for transporting dry ice from power plant to barge are not included.

(6) Barging costs are based on a 100-mile barging distance (500-meter disposal depth); for a 200-mile distance (3000-meter disposal depth), barging costs are estimated at 9 million dollars per year.

Revenue requirements in mills/kWh were calculated on the basis of net electricity output assuming an initial plant capacity of 200 MWe (after SO_2 removal) and assuming that the electricity required to drive the CO_2 control system is supplied internally by the subject power plant. Thus, the net output of the power plant (e_{net}) is equal to gross ouput (e_{gross} = 200 MWe) minus the electricity consumed by the control system ($e_{control}$). A schematic of the overall system is presented in Figure 8.

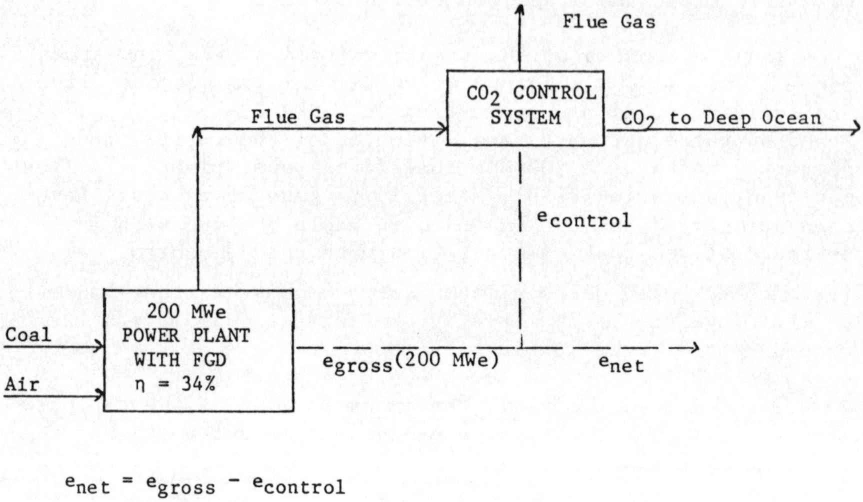

FIGURE 8. SCHEMATIC OF SELF-CONTAINED POWER PLANT WITH FGD AND CO_2 CONTROL

For a CO_2 removal efficiency of 50%, the costs of generating electricity for gaseous, liquid and solid disposal are approximately equal. The net power generated and yearly operating costs for gaseous and solid disposal are about the same, and therefore, so are power generation costs. Solid disposal requires more energy but less capital. For 50% removal of CO_2, these two factors balance one another and the cost of power generation for solid disposal turns out to be about the same as for gaseous and liquid disposal.

For 90% CO_2 removal, in comparison to 50%, the cost of power generation roughly doubles for both gaseous and liquid disposal whereas it increases by a factor of about 4 for solid disposal. This marked increase in generation costs for the solid disposal case is due to excessive control system energy requirements at the 90% CO_2 removal efficiency.

If the above analysis had been based on a net power plant output of 200 MWe rather than a gross output of 200 MWe the costs of power generation, particularly at the higher CO_2 removal efficiencies, may have been somewhat lower because of economies of scale.

(5) Discussion

Based on the above results, removal and recovery of flue gas CO_2 using MEA is technically and energetically feasible, however, economic penalties are significant. Disposal also appears to be feasible provided the effects of injected CO_2 on the deep ocean ecosystem are not prohibitive.

MEA absorption/stripping in combination with either gaseous or liquid disposal requires appreciably less energy than an MEA-solid disposal combination. On the other hand, the capital investments required by the former combinations far exceed those of the latter offsetting, to some degree, the economic benefits of the lower energy requirements. For gaseous or liquid disposal, the major portion of the capital investment is due to pipeline costs, which may be understated in the cost estimates.

A possible way to reduce the capital investment for a liquid disposal system is to barge liquid CO_2 to the required offshore distance (of about 200 miles) and then inject the liquid into the deep ocean via a pipe extending from the ocean surface to the injection point. With this arrangement the required length of injection piping would be substantially reduced (from about 330,000 meters to about 3000 meters) as well as the required pipe size. However, CO_2 would have to be shipped under pressure and below the critical temperature (31°C) to maintain it in the liquid state.

For direct gaseous or liquid deep ocean injection (via pipeline), essentially no losses of recovered CO_2 would be incurred during transport. For solid disposal by barging, losses during handling and transportation could be significant. For 10 inch cube blocks of about 55 pounds each, shipping losses (based on transport by truck) may be as high as 8 pounds during the first 24 hours in summer.[23] To reduce handling losses and improve handling logistics for inland power plants it may be better to pump the flue gas to the dock site and remove/recover and solidify at that point.

VII. CONCLUSIONS

1. The contribution by the United States to worldwide CO_2 emissions has been decreasing during the last 25 years and it is projected that it will decrease even further during the next 50 years. Therefore, a worldwide effort will be required to control the buildup of atmospheric CO_2.

2. Increased utilization of coal for producing heat, electric power and/or synthetic gaseous and liquid fuels will increase the rate of CO_2 emissions. The magnitude of this increase may or may not be significant depending upon the extent to which coal displaces oil and natural gas, and future energy use patterns.

3. Direct capture of CO_2 from the atmosphere or capture from the surface waters of the oceans does not appear to be feasible when using a fossil energy source to drive the CO_2 control system. More CO_2 is generated than is recovered.

4. For 60% to 90% absorption of the CO_2 contained in the flue gas of a coal fired power plant using seawater and deep ocean disposal, more CO_2 would be generated in supplying energy to the control system than would be recovered from the flue gas.

5. Removal and recovery of CO_2 from the stacks of fossil fuel plants by commercially available processes is feasible, e.g. by absorption/stripping with MEA. However, large quantities of energy are required which will substantially reduce the efficiency of power generation. For 50% CO_2 removal, the power plant efficiency decreases from 34% to 27 to 23% depending on whether gaseous, liquid or solid CO_2 is disposed of in deep ocean. For 90% CO_2 removal the efficiency decreases to between 15% and 6%.

6. Disposal of captured and recovered CO_2 will be a problem. Of the disposal/reuse options considered, deep ocean disposal appears to be a possibility. However, the costs of disposal are substantial. Also the effects of injected CO_2 on the ocean ecosystem have not as yet been assessed.

7. Of the processes considered, absorption by MEA is the least energy intensive. For a 50% CO_2 removal efficiency, the equivalent of 0.15 kWh(e)/lb captured CO_2 is required; and for 90% removal, 0.27 kWh(e)/lb CO_2. Gaseous or liquid disposal of the captured CO_2 requires an additional 0.05 kWh(e)/lb CO_2 whereas solid disposal requires about 0.18 kWh(e)/lb CO_2 (discounting transportation and handling losses). The required capital investments for gaseous and liquid pipeline disposal are substantially greater than for solid disposal. Low energy consumption and high capital investment have offsetting effects on the cost of power generation.

8. For effective control of atmospheric CO_2 by removal from flue gases, it would be advantageous and perhaps necessary to centralize the use of fossil fuels, such as in an all electric economy, since it would be almost impossible to

remove, recover and dispose of CO_2 from the numerous small sources, i.e., from the flues of home heating units, process plants and from automobile exhausts. In the U.S. about 30% of the total CO_2 released originates from central fossil fueled power plants. For direct removal of CO_2 from the atmosphere or from the surface layers of the ocean, centralized use of fossil fuels is not required. However, removal of CO_2 directly from the atmosphere or from the oceans does not appear to be feasible unless the energy required to drive the CO_2 system is supplied by a non-fossil energy source.

REFERENCES

1. R.M. Rotty, "Global CO_2 Production from Fossil Fuels and Cement, AD 1950-AD 2000", Proceedings of meeting held at University of Hawaii on Fate of Fossil Fuel CO_2 in the Ocean, sponsored by Office of Naval Research (Jan. 1976), Plenum Press, pp. 167-181 (Oct. 1977).

2. C.F. Baes Jr., H.E. Goeller, J.S. Olson and R.M. Rotty, "Carbon Dioxide and Climate: The Uncontrolled Experiment", American Scientist, 65, 310, (1977).

3. Statistical Abstract of the United States (1975), U.S. Department of Commerce.

4. A. Albanese and M. Steinberg, "Environmental Control Technology for Atmospheric Carbon Dioxide", BNL 50877, Brookhaven National Laboratory, (April 1978).

5. R.H. Perry and C.H. Chilton, Chemical Engineers Handbook, 5th Edition, McGraw Hill, 1973.

6. M. Steinberg and Vi-Duong Dang, "Use of Controlled Thermonuclear Reactor Fusion Power for the Production of Synthetic Methanol Fuel from Air and Water", BNL 20016R, Brookhaven National Laboratory, Upton, N.Y., (April 1975).

7. C. Marchetti, "On Geoengineering and the CO_2 Problem", Climatic Change 1, 59 (1977).

8. F.J. Dyson, "Can We Control the Carbon Dioxide in the Atmosphere", Energy 2, 287-291 (1977).

9. M. Steinberg, "Nuclear Power for the Production of Synthetic Fuels and Feedstocks", BNL 21330, Brookhaven National Laboratory, Upton, N.Y., (April 1976).

10. M. Steinberg, "Synthetic Carbonaceous Fuels and Feedstocks from Oxides of Carbon and Nuclear Power", BNL 22785, Brookhaven National Laboratory, Upton, N.Y., (March 1977).

11. F.C. Reisenfeld and A.L. Kohl, Gas Purification, 2nd Ed., Gulf Publishing Co., (1974).

12. J.P. Riley and G. Skirrow, Chemical Oceanography, Vol. 2, Editors, 2nd Ed., Academic Press, (1975).

13. T.K. Sherwood and R.L. Pigford, Absorption and Extraction, Editors, 2nd Ed., McGraw Hill, (1952).

14. T.B. Reed, "Efficiencies of Methanol Production from Gas, Coal, Waste or Wood", MIT, Lincoln Laboratory.

15. R. Detman, "Factored Estimates for Western Coal Commercial Concepts", C.F. Braun & Co., Interim Report, FE-2240-5, (October 1976).

16. R. Detman, "Factored Estimates for Eastern Coal Commercial Concepts", C.F. Braun & Co., Interim Report, FE-2240-31, (September 1978).

17. A.S. Albanese and M. Steinberg, "Environmental Control Technology for the Flash Hydropyrolysis of Coal", Progress Report No. 3, BNL 50983, Brookhaven National Laboratory, Upton, N.Y., (February 1979).

18. P.J. Rogoshewski, "Standards of Practice Manual for the Solvent Refined Coal Liquefaction Process", Hittman Associates, EPA-600/7-78-091, (June 1978).

19. A.W. Pappano et al., "Availability and Economics of CO_2 for Enhanced Oil Recovery in Appalachia", Dept of ChE, West Virginia University, TID-22787, (Aug. 1976).

20. W.M. Rump et al., "The Supply of Carbon Dioxide for Enhanced Oil Recovery", Pullman Kellog, Final Report, FE-2515-10, (September 1977).

21. R.M. Guthrie, Process Plant Estimating Evaluation and Control, Craftsman Book Company of America, 1974.

22. W.L. Faith, D.B. Keyes and R.L. Clark's 4th Edition, Chemicals, John Wiley & Sons, Inc. (1975).

23. E.L. Quinn and C.L. Jones, Carbon Dioxide, Reinhold Publishing Company, (1936).

24. W.D. Nordhaus, "Can We Control Carbon Dioxide?" IIASA-Working Paper WP-75-63, (June 1975).

DISCUSSION

Keeling:
What did you use to assemble your first table? Is the nuclear energy in units of thermal or electrical? What is the growth rate to 2000 A.D. assumed for total energy production? What is the growth rate assumed for nuclear to 2000? What are the rates from 2000 to 2020 A.D.?

Egberts:
The calculations were performed during 1979. The data gathering goes back to 1977. Nuclear energy is measured here using its fossil equivalent. The energy demand growth rate is about 2%/yr. before 2000 and 1%/yr. after 2000. Nuclear growth is about 50 Mio. tons of coal equivalent primary energy per annum.

Hekstra:
In the Netherlands we are debating the shift from gas to coal for electricity generation plants. One consequence would be that district heating using the waste heat of the electricity plants will no longer be possible, as gas fueled plants are mainly sited in or close to the residential areas, but coal fueled plants would have to be located about 10 km away. With nuclear fueled plants we would have to take even greater distances into account and therefore can completely forget the possibilities of cogeneration. The waste heat problem would increase tremendously.

Schaefer:
Switching from gas to coal will raise some new problems as mentioned. One will have to look for improved treatment of flue gases and should take into consideration the installation of small BWR and PWR units which only produce heat for district heating systems and industrial areas.

Keeling:
In expressing percentage of energy as space heating 39%, process heating 30%, light 1% etc., do you calculate in terms of how much thermal energy is required (i.e. primary energy)? For example in a pure coal economy would $\sim$1% be needed to produce the electricity to make the light?

Schaefer:
The given figures are the breakdown of final energy consumption = 100% which is about 2/3 of the primary energy consumption. So,

W. Bach, J. Pankrath, and J. Williams (eds.), Interactions of Energy and Climate, 552–555.

for example, light has a share of about 3% of primary energy consumption, since the efficiency of power generation accounts for some 33%.

Flohn:
As an atmospheric scientist, I am horrified by the great amount of waste heat lost in the generation and transport of energy. How large is the total amount of waste heat loss, and what means exist to reduce this loss to a minimum?

Schaefer:
The entire primary consumption is ultimately transferred to waste heat, no matter how the energy was used. Reducing waste heat requires a reduction in primary energy consumption.

Kellogg:
Earlier in the program Dr. Amory Lovins made a very emphatic point about the unsuitability of using electricity for the purpose of heating. He argued that this was a very inefficient way to produce heat. I would like to ask how you view this point, and what fraction of the large heating requirements in West Germany (75% of the total use of energy) is currently being met by electricity.

Schaefer:
In contrast to Dr. Lovins I think that the use of electricity for heating purposes is sometimes the only suitable way and often a better way of saving primary energy. About 25% of the total electricity consumption in West Germany is used for process heat and about 20% for home-heating purposes, hot water supply and cooking.

Quirk:
You have pointed out that renewable energy sources have large capital costs and may have adverse environmental impacts. On the other hand the first two subjects you discussed 1) avoiding undue consumption and 2) reducing end use demand, would save capital and would have beneficial environmental consequences. For this reason aren't these first two actions the most desirable way to make more energy available?

Schaefer:
Reducing end use demand is a very important possibility but not the only one. It does not create new energy sources in the same way that fasting is not a source of food.

Page:
The figures indicate that a process at 21% efficiency at the power station and a distributed efficiency of 16%, which displaces fossil fuel at 64% combustion efficiency will require 4 times as much primary fuel at perhaps 5 times the cost. The net CO_2 reduction is only 60% leaving 40% in the atmosphere. With a 15% efficiency one would need 7 times as much fossil fuel at 8 times the cost to get 70% of CO_2 produced by direct combustion to produce the equivalent amount of heat.

Baes:
If the primary use of electricity were for resistance heating, I would certainly agree that the option of CO_2 disposal becomes even less attractive than it already is. But for various reasons society may choose to increase its use of electricity (e.g. for transportation or for space heating with heat pumps). If so, then CO_2 disposal remains a last-resort option. If fossil fuel use is increasingly dispersed, this option will certainly not be available to us.

Hekstra:
1. In view of the very low energetic efficiency of dealing with CO_2 by dumping it into the oceans, you said, the ecological impact was at present the least to bother about. Not only would the pH be affected but also the considerable amounts of water pumped up and transferred would afflict the mainly planktonic organisms in the water body.
2. What is the significance of this study - other than purely scientific - for the policy of getting rid of CO_2 as long as we have much cheaper ways by simply saving energy?

Baes:
1. I think the ecological effects could be minimized by using fairly deep water (>1000 m) for disposal and by forming a negatively buoyant plume that would sink to deep water (ca. 4000 m) where there is little organic activity and where the great dilution of the stream would reduce the impact. While there would be some effects, it is difficult for me to imagine that they would compare with the extreme impacts that would have led mankind to such a disposal strategy.
2. I certainly agree that as long as we can meet our energy needs by conservation and other alternatives, we should not resort to this option. But I think it is by no means certain that we can meet our energy needs in the future without using more fossil fuels than may be climatically safe. Thus, the reason for this study was to consider the difficulties of the CO_2 disposal alternatives if such should ever become necessary.

Keeling:

In disposing of CO_2 in the oceans you need to take into account deeper currents, which are in general most intense near the continents. For the Eastern US, for example, the Gulf Stream is a problem. Even after disposing of CO_2 in deep water, after some one thousand years or so the concentration of CO_2 in the atmosphere will increase considerably if all available fossil fuel is consumed.

Baes:

I agree that ocean currents must be considered in selecting the location and the density excess at the disposal point; however, the substantial density differences that are produced by concentrated streams of carbon dioxide should produce effective disposal in many cases. I agree also that disposal of all the fossil carbon as CO_2 in the oceans would produce a substantial increase in the atmospheric content but much less than if it is injected into the atmosphere.

Kellogg:

The IIASA energy study suggests as one alternative the production of organic fuels (methane or methanol) from CO_2 taken out of the atmosphere and combined with water and heat (from nuclear reactors). This would conserve fossil fuel and recycle the present CO_2 in the atmosphere. How attractive do you consider this idea?

Baes:

This is equivalent to suggestions made by Steinberg in his paper but I would suggest that it would be much better to obtain the CO_2 from concentrated sources in order to avoid the high energy cost of concentrating CO_2 from its low atmospheric concentration of ca. 330 ppm.

Niehaus:

Autothermal coal gasification (and liquefaction) processes lead to an increase of specific CO_2-emissions since carbon is used to produce the energy and to produce hydrogen out of water. Autothermal gasification (using nuclear power) leads to somewhat higher emissions than burning coal directly since carbon is still needed to produce the hydrogen. However, if pure hydrogen produced by electrolysis or water splitting at higher temperatures (using a non-fossil source) is used for gasification it could reduce specific emissions by about 40% (27% for methanol).

INDEX

Absorption, of short-wave radiation, 344
Absorption recovery, from power plant flue gas, 539
Absorption-to-backscattering ratio, 288
Acid rain, 202
Accidental death, 426-428
Accidental injuries, 426-428
Active solar cooling, 86
Active solar process heat, 16
Aerodynamic roughness, of cities, 340-341; of suburban terrain, 341
Aerosols, anthropogenic, 282-293; causing warming/cooling, 288; effects on albedo, 287-293; effects on climate, 281-293; effects on vertical stability, 291-292; extent of industrial, 285-286; influence on heat balance, 283; optical characteristics of, 287-291; optical properties of, 283; radiative effects of, 283-293; regional trends, 283; stratospheric, 283; sulfate burden, 285
Aerosol concentrations, geographic distribution, 283; regional trends, 283
Aerosol spray cans, 244
Aggregation, excessive, 4
Agriculture, 55
Airborne fraction, 108-111, 135, 155-156, 159, 160, 162
Air conditioning, 55, 83
Aircraft measurements, 385
Arid land management, 461
Air quality problems, 61
Albedo, of cities, 346; urban-rural difference, 346
Albedo changes, caused by deforestation, 314
Albedo modification, effect on climate change, 317-325
Alternative energy sources, adoption of, 473-479; distribution of, 526; impact of, 524-526
Amelioration of CO_2 impacts, 224-225
Anaerobic digestion, 16, 30
Anomalous winters, 63
Antarctica, 124
Anthropogenic thermal effect, 392
Architectural education, 84
Assessment of risks, 422
Atmospheric heating, due to short-wave absorption, 345
Attenuation, of short-wave radiation, 345

Biogas, 10, 330, 492
Biogas plants, in India and China, 10
Biological effects, 244
Biomass, 492; conversion, 67; plantation, 18; productivity, 67, 306; substitution for fossil fuels, 145
Biosphere, as future sink for CO_2, 129-245; role in carbon cycle, 110, 121; role of, 108, 149-165
Biospheric conservation programs, 145
Biospheric response, to CO_2, 130
Biota growth factor, 154
Boundary layer, urban, 340-343
Bowen ratio, 332; for suburban terrain, 348
Box diffusion model, 114, 159
Buffer factor, 113, 122, 136, 155-156, 159
Building designs, 77
Building space utilisation, 79-80
Byrd Station, 124

Camp Century, 124
Canadian Climate Center, 53
Canopy layer, urban, 340-343
Capital asset loss, due to CO_2 growth, 444
Carbon, biospheric storage, 144;
Carbon, distribution in various reservoirs, 150; in living plants, 143; in recoverable fossil fuels, 197; produced by fossil fuel combustion, 134; storage in soils, 143; total dissolved inorganic in water, 136; uptake by soil, 144; withdrawal of, 155
Carbon cycle, 132, 149-165; modeling error in, 440; perturbed by man, 107-125; research on, 218-219; six reservoir model, 137
Carbon fluxes, 112, 129
Carbon storage, in deep soil, 131
Carbon transfer, from tropical land ecosystems to the atmosphere, 158
Carbon-14, 112-123; dilution of, 116; fluctuations of, 116; production of, 118; $^{14}C/C$ ratio, 108
Carbon dioxide, and the Climate Research and Assessment Program of the US, 214-227; biosphere as a sink for, 129-145; build up of, 275; collection of, 203; concentrated sources of, 498; control technology for, 521-549; costs of collecting from stack gas, 499; distribution between atmosphere and ocean, 156; evaluation of impacts, 225-234; in polar ice, 123-125; input and output functions for, 161; modeling surface air temperature changes for, 169-187; ocean as a sink for, 129-145; potential disposal sites for, 510-511; predicted increase in atmosphere of, 140-142; release of, 203-208, 328; retention time in ocean of, 501; scrubbing of, 498-500; seasonal and regional distribution of, 170; solubility in sea water of, 502-504; sources and sinks of, 23; transfer from atmosphere to ocean of, 135; unknown sinks of, 157
Carbon dioxide budget, 121-123
Carbon dioxide change, in the atmosphere, 125
Carbon dioxide concentration in the atmosphere, 204-208, 522; history of, 123-125; in 2025 AD, 208; prediction of future, 164; preindustrial, 130, 162; projected, 205
Carbon dioxide control, 526-529; energy requirements for, 528; steps for, 530
Carbon dioxide conversion, to useful products, 534
Carbon dioxide cost analysis, policy implications of, 458-459
Carbon dioxide disposal, by terrestrial burial, 533-534; in the deep ocean, 532-533; gaseous in ocean, 540-542; liquid in ocean, 542-543; solid in ocean, 543-544
Carbon dioxide doubling and surface air temperature change, annual average, 178; geographical distribution, 183-184; latitudinal variation, 181; mean tropospheric, 179; monthly mean, 180
Carbon dioxide emission, by source, 525; controls of, 523; for 15 IEA countries, 480; from energy supply, 473; from final energy forms, 474; from synthetic fuels, 526
Carbon dioxide exchange, in forests, 311-313
Carbon dioxide fertilzation, 113
Carbon dioxide impact, on the world economy, 441-442; remedial costs of, 445-449

Carbon dioxide increase, 108-111; and capital asset loss, 444; cost calculations for, 438-440; effected by market penetration time, 454; future atmospheric, 159-165
Carbon dioxide issue, 213-234
Carbon dioxide problem, 76, 522-524
Carbon dioxide production, from fossil fuels and cement, 522; industrial 135, 138-139
Carbon dioxide quadrupling and surface air temperature change, latitudinal and seasonal, 186
Carbon dioxide reduction strategy, discounted costs for, 448
Carbon dioxide removal, and deep ocean disposal, 534-538; from the atmosphere, 529-531; from the flue gas, 531-532, 538-547; net annual cost, 447; net asset losses, 447; sites, 528
Carbon dioxide research, on the amelioration of sectoral impacts, 224-225; on the effects on the environment, 221-223; on the effects on climate, 219-221; on societal costs, 223-224; program of US Department of Energy, 227 231
Carbon dioxide storage, at the South Pole, 515; in depleted gas and oil fields, 513-515; in plants and trees, 515, 534
Carbon dioxide system, 111-112
Carbon dioxide uptake, by the ocean, 119-120, 130; dynamic capacity of, 114-116; static capacity of, 113-114
Carbon dioxide values, of the Holocene, 125
Carbonate dissolution, 157
Catalytic cycles, 249
Charcoal, 84
Chlorine chemistry, stratospheric, 249
Chlorofluoromethanes, accumulation of, 245; effects on climate, 243-253; effect on IR absorption, 247-249; in stratosphere, 245; in troposphere, 245; release rate of, 245-246, 265
Circulation, thermally driven, 407,
Climate, and energy demand, 55-62; a hazard, 55,74; a resource, 55, 74, 81-82; determinant of energy demand, 53-62; effects of increasing CO_2 on, 219-221; human activity on, 270; influence on energy use, 55-56
Climate and energy conservation, 73-87
Climate ensembles, 401
Climate information, for energy systems management, 89-98
Climate model, inherent variability, 400
Climate-photochemical coupling, 273
Climate Program Act of the US, 215
Climate Program Office of the US, 214
Climate role, in affecting energy demand/supply, 53-69
Climate - sensitive activities, 55
Climate system, sensitivity of, 400
Climate variations, 89-90
Climatic anomalies, 54, 58-60
Climatic change, 54; cost comparisons, 456-458; costs of, 446; discounting the future costs of, 454; due to anthropogenic surface albedo modification, 317-325; equivalent costs of, 450; future, 277; impacts of, 440-445
Climatic effects, of aerosols, 281-293; of anthropogenic trace gases, 269-279; of chlorofluoromethanes, 243-253

Climatic feedback parameter, 276
Climatic impacts, amelioration of, 94-95; costs of, 461-468; of urbanisation, 339-352; of waste heat, 383-398, 399-413; on energy economy, 90; on energy systems, 90-93
Climatic Impact Assessment Program, 441
Climatic states, 401
Climatic variability, impact on agriculture, 61-62; impact on transportation, 62-63
Climatologists, status in world planning, 75
Coal, burning in air, 500; burning with pure oxygen, 500-501; fuel of the near future, 464; ultimately recoverable, 198
Coal conversion, 476
Coal exploitation, ultimate limit for, 134
Coal liquefaction, allothermal, 50; autothermal, 50
Coal resources, 194-196; global distribution of, 196
Collection, of CO_2, 497-501
Cold winters, 54
Colorado River Basin, 45
Combined heat and power systems, 76
CONAES (Committee on Nuclear and Alternative Energy Sources), 6
Conservation policy, 81
Conservation practices, 483-493
Conserver society, 12
Control technology, for atmospheric CO_2, 521-549
Conversions, gas to particle, 349
Cooking and energy use, 84
Cooling and rural food storage, 86
Cooling in stratosphere, 252
Cooling tower plumes, 387; structure of, 388
Costs, climatic, 457; environmental, 431; for scrubbing, 457; for the disposal of CO_2 from a floating power plant, 514; marginal, 432; of abatement, 431; of climatic change, 442-446; of electricity production, 431; of climatic impact, 458, 461-468; of natural variability of climate, 457; of remedial measures, 457; of the effect of uncertainty, 445; total discounted, 457
Cost-benefit analysis, 431-432
Cost calculations, for CO_2 increase, 438-440
Cost comparisons, for climatic change, 456-458
Cost-effectiveness analysis, 432
Critical surface albedo, 288
Crude oil, conventional, 192-194; global distribution, 194

Damping factor, 117
Deep ocean disposal, of recovered CO_2, 540-544
Deep sea, sedimentary carbonates in, 130
Deep water, bomb products in, 120
Deforestation, 23, 152, 275, 317, 440; and albedo changes, 314; and roughness parameter changes, 315; impacts of, 305-315; impacts on energy and water balance, 310
Deforested world, simulation of a, 313-315
Denitrification, 260-261
Desertification, 317
Design techniques, computer use, 83
Detrital carbon, 129
Developing countries, 9-10, 39, 77, 200
Diffusion model, 160
Diffusion process, 155

Dilution factor, dynamic, 114; for excess CO_2, 110; of ocean CO_2, 120-123; static, 113
Discounting, economic, 453-455; social, 453-455
Diseconomy, 14
Disposal, of CO_2 in ocean, 501-511
Distortions, in basic data, 3
DOE's Carbon Dioxide and Climate Research and Assessment Program, 214-227
Drought, 66, 91, 462
Dry ice, in sea water, 510
Dung, 10

Economic constraints, 133
Economic efficiency, 9
Economic impact, 92
Ecosystems, tropical, 157
Eddy diffusion coefficient, 155-156
Effect of urbanisation, 60
Effective penetration depth, 155, 159
Effective solvent, 532
Efficiency improvements, 4
Efficiency scenarios, 5-13
Electrical utilities, 76
Electricity, generation and distribution of, 485
Electricity production, costs of, 431
Emissions, of natural hydrocarbons, 279
Emission ratio, for CO/CO_2, 275
Energy, contained in fossil fuels, 198; conversion losses of, 485; end-use, 4; final, 41-42; lead times in, 20-23; needs and losses of, 485; primary, 41-42; secondary, 41-42; supply rate of, 368; total delivered, 13; useful, 41-42; what forms of, 13-14
Energy balance, radiative, 271; planetary, 319; surface, 333-335; urban and pre-urban, 347
Energy balance models, 173, 175-176
Energy balance scheme, for the Federal Republic of Germany, 485
Energy Centers, impact on local and regional climate, 394-398
Energy carriers, 475
Energy conservation, and climate, 73-87; and window systems, 83; in buildings, 83; in human settlements, 81-82; problems and constraints of, 492-493; reasons for, 484-486; role of building mass and color in, 83; ways of, 486-492
Energy conservation city, 77-78
Energy consumption, commercial, 34; for different source categories, 371-372; global annual, 364; per capita, 35; regional, 56; unnecessary, 486-487
Energy conversion, 483-493; biomass, 330-333; decentralized, 329-331; photovoltaic, 328-330; solar thermal electricity, 333-336
Energy conversion systems, impacts of, 327-336
Energy costs, 91; of conversion to dry ice, 498
Energy crops, 330
Energy demand, 1-13; and climate, 55-62; by world regions, 36; final for the Federal Republic of Germany, 486; forecasts of, 7-8; future, 328; global, 39-42, 203; high supply scenario, 33-52; of human settlements, 76; price elasticity of, 2; reduction of, 487-488
Energy demand and supply, role in climate, 53-69
Energy economy models, Impact, 37-38; Macro, 37-38; Medee 2, 37-39; Message, 37-39, 48
Energy efficiency, at end point,

4; improvement of, 488-489; in Canada, France, Switzerland, 7; in Denmark, Germany, Sweden, United Kingdom, U.S.A., 6; increase of primary, 4, 483-493
Energy flux densities, 364-366
Energy futures, efficient, 1-25
Energy - GDP elasticities, 42, 44
Energy growth, for U.S. and world, 526
Energy intensiveness, 41, 43
Energy needs, of the ECE region, 9
Energy planner, 54
Energy plantations, 331
Energy policy, minimum climatic risk, 25
Energy Policy Office of the US President, 94
Energy production and demand, 89; prospects of, 96-98
Energy productivity, 2, 20
Energy projections, primary, 4
Energy recovery, 489-490
Energy requirements, by CO_2 control system, 528; for CO_2 disposal in ocean, 544; for MEA absorption and deep ocean disposal, 541-542
Energy resources, 42-48
Energy savings, 61, 493
Energy sectors, 485
Energy services, 41
Energy scenarios, future global, 37-39; construction of, 199-201
Energy shortages, 54
Energy shrinkage, 9
Energy sources, renewable, 76, 328, 491-492; biological, 330
Energy storage, 69, 81
Energy strategies, 19, long-term, 159
Energy supply, 63-69; a source of carbon dioxide, 473; by sectors in the Upper Rhine Region, 368
Energy supply scenarios, 48-50
Energy supply structure, of IEA countries, 477
Energy systems, comparison of risks and benefits of, 421-433; conventional, 68-69; man-days lost in, 430; vulnerability to climatic change, 54
Energy systems management, through climate information, 89-98
Energy systems program, of the International Institute for Applied Systems Analysis, 37
Energy transfer, radiative, 271
Energy usage, 464
Energy use, components of urban, 76-77; ecological impact of, 484; per capita, 202
Enhanced oil recovery, 498
Entrainment/detrainment mechanism, 394
Environment, effects of increasing CO_2 on, 221-223
Epidemiological studies, 429
Equation, of motion, 377; of continuity, 377
Euler backward method, 378
Eulerian frames, 384
Exchange models, of atmosphere-ocean, 157
Extrapolative models, 2

Farm and forestry residues, 16
Fatal diseases, 426-428
Fatalities, frequency of events, 425
Feedback, ice-albedo, 323
Fermentation, 16
Fertilizers, 270
Fertilizer plant, 10
Firewood, 10
Floating power stations, 511-513
Fluid fuels, 16
Food and agriculture, impact on, 461
Food demand, 464,
Forecasting, future world energy demand, 132

Ford Foundation, 5
Forests, climate related properties, 307-311, 331; dynamic properties, 310-311; energy exchanges, 309-310; hydrological pecularities, 308-309; impact on energy balance, 332; radiation and thermal characteristics, 307-308
Fossil fuels, alternatives, 209; combustion of, 23-24; consumption of, 51, 108; energy and carbon in, 198; growth in consumption, 145; impact on climate, 328; prediction of future consumption, 132-135; recovery of, 43-48; unconventional, 195-197
Fossil fuel use, constraints on, 191-209; environmental constraints, 202-208; fuel-demand constraints, 197-202; impact of US emissions, 524; reduction of, 473-479; resource constraints, 192-199
Fossil fuel substitution, in OECD countries, 476-479
Fossil input, 111
Fuels, commercial, 10; non-commerical, 10
Fuelwood, 84

Gas chromatographs, 258
General circulating models, 169, 170, 173, 176-187, 400-401; experiments with, 185; predictability of, 180; with idealized geography, swamp ocean without seasons, 177; with realistic geography, climatological ocean with seasons, 177-185; with realistic geography, mixed-layer ocean with seasons, 185-187
Geochemical models, 149, 154
GEOSECS, ^{14}C and ^{3}H data, 120
Geostrophic wind, 378
Global carbon cycle, possibilities to balance the, 157-158
Global climate, 23
Global economy, 89
Global energy study, 35
Global forest area, 312
Global weather patterns, 97
Greenhouse effect, 282, 496, 523
Greenhouse gases, other, 205
Greenland, 124
Growth, primary energy, 21; real economic, 21
Growth rate, of GDP, 40; of population, energy and economy, 201
Gross national pollution, 286

Hadley circulation, 321
Hamilton concept, 20
Hard solar technologies, 18
Health effects, due to energy use, 431
Heat emission, anthropogenic, 347; impacts on climate, 383-398
Heat energy balances, 489
Heat fluxes, anthropogenic and natural, 374
Heat island, 344
Heat pumps, 491
Heat recovery, from domestic sewage water, 490
Heat release, impact on regional weather and climate, 366
Heat source inventory, 389
Heating, of earth-troposphere system, 277
Heating degree days, 94
Heating rate, due to CFM in tropical atmosphere, 248
Heating systems, floor, wall, ceiling, 490
Heavy crude oil, 197-198
Helium ionization detectors, 258
High consumption city, 80
High rise buildings, 77
Human capital approach, 431
Human influences, on global

warming, 463-464
Humidity island, 344
Humus oxidation, 152
Hydrogasification, 330
Hydrogen to carbon ratio, 192
Hydrogenation, 330
Hydrocarbons, 270-273
Hydrological characteristics, 333
Hydrolysis, 16
Hydropower, 16, 65-66, 91
Hypotheticality, 422

Ice fog, 344
Imaginery index of refraction, 288
Impacts, due to deforestation, 305-315; local/regional, 329; micrometeorological, 330-334; of nitrogen fertilization, 257-266
Impact of climate, on energy demand, 59
Impact on society, of climate variations, 90
Impact of urban areas, on humidity, 343
Industrialization, 389
Industrial future, 81
Infrastructure costs, 78
Input/output methodology, 430
Institute for Energy Analysis, 20, 191
Interactions, between climate and photochemistry, 270-273; stratospheric-tropospheric, 273
International economy, 97
International Energy Agency, 476
International Institute for Applied Systems Analysis, 20, 33
Inverse price elasticity, of energy demand, 3
Investment and power generation costs for CO_2 disposal, 545-546
Irrigation, 90, 332-333

Jefferson concept, 20
Jet zone, in ocean, 504

Land biosphere, 129
Land clearing, 130
Land ecosystems, 150-157
Land-use change, irreversible, 339
Langrangian frames, 384
Latent waste heat, 367
Lifestyle, changes in, 12
Linear city, 78
Linear optimization model, 477
Liquefaction pressure of CO_2, 502
Liquid fuel supply, 49
Litter, 153
Living biomass, 129
Local impacts from point sources, 386-388
Logistic function, 134
Low frequency risks, 423
Lung cancer, 430

Man's impact, local versus distant, 75-76; on climate, 60-61
Market penetration analysis, 34
Market penetration times, 455-456
Market penetration time, effect on CO_2 growth, 454
MEA absorption/stripping, 538-540
MEA process, 499
Megalopolis, 400
Mesoclimate, 61
Meteorological instruments, for aircraft measurements, 385
Microclimate, 63
Microhydropower, 10
Micro-'oasis' effect, 348
Mixed layer, 155
Mixing length, Blackadar's, 378
Models, box diffusion, 138; for cooling tower plumes, 384; general circulation, 169-170, 173, 176-187, 400-401; mesoscale, 334, 376; of the carbon cycle, 137,

radiative-convective, 171-175, 244, 276
Model response, seasonal differences, 405
Modeling, photochemical, 270
Modeling, surface temperature changes due to CO_2, 169-187
Moisture release, anthropogenic, 343
Municipal wastes, 16

National Energy Act of US, 20
Natural gas, conventional, 193-194; global distribution, 194; shortage of, 92; unconventional, 197; ultimately recoverable conventional, 198
Natural Gas Policy Act of US, 93
Natural ventilation, 82
Negatively buoyant plume, 502; the behaviour of, 504-508
Net radiation, 375
Nitric oxide, 270-273
Nitrification, 260-261
Nitrogen, amount fixed, 260
Nitrogen fertilization, impact of, 257-266
Nitrous oxide, anthropogenic sources of, 262-264; atmospheric distribution of, 258-259; effect of doubling on ozone, 265; effects on stratospheric ozone, 264-266; future increases of, 263; greenhouse effect of, 264; release rates of, 261; residence time of, 258-259; sources and sinks of, 259-262
Normalized response, 403
Nuclear energy, 45-48, 328
Nuclear power stations, 14
Nuclei, Aitken condensation, 349-350; cloud condensation, 349; seeding of, 350; ultra-giant, 350
Numerical model, for the simulation of undisturbed climate, 389

Occupational and public health effects, 426-428
Oceans, as future sinks for CO_2, 129-145
Ocean mixing, rate of, 119-120
Ocean sediments, 502
Ocean thermal electric conversion, 18, 68, 511
Ocean uptake, of CO_2, 113-114, 130, 155-157, 159
OECD countries, 9
Operational efficiency, 61
Optimized energy supply, IEA countries-base case, 478; forced oil saving, 479
Organic carbon, stored in land ecosystems, 151
Outdoor space utilization, 80
Ozone, stratospheric, 244-246; tropospheric, 270-273
Ozone depletion, climatic effects of, 249-253; effect on tropospheric temperature, 252-253; total effect on surface temperature, 253; vertical profile of, 251;
Ozone layer, depletion of, 244-253

Passive cooling techniques, 86
Peat bogs, 153
Petroleum, ultimately recoverable conventional, 198
Photochemical decomposition, in stratosphere, 260
Photolysis, of chlorine atoms, 244
Photosynthesis, 130; acceleration of, 154; stimulation by increased CO_2, 154-155
Planetary boundary layer, impact of heat emission, 384; winds in, 340
Plume zone, 504

Population, growth rates, 39; historical and projected, 38; increase of, 464
Portable liquids, 13
Power plant efficiency, affected by CO_2 control, 528; overall, 531
Power towers, 18, 333
Prandtl approach, 377
Precipitation, enhanced by large cities, 349; patterns of, 66, 408-410; response to prescribed change, 408-410
Precipitable water vapor, 323
Prediction, of future fossil fuel consumption, 132-135; of future response of biosphere to industrial CO_2, 141-145; of future response of oceans to industrial CO_2, 135-141
Preventive measures, 496
Primary energy, by region, 50; market shares for the FRG, 34-35
Probabilities, of global temperature change, 439
Productive forests, 312
Profile, vertical of CFM, 248
Progress, technological, 4
Project METROMEX, 349
Promotion, of energy, 3
Propellants, 270
Pyrolysis, 10, 16, 330

Quasi-geostrophic atmospheric model, 176

Radiation, shortwave, 344
Radiation budget, infrared, 244
Radiative-convective models, 171-175
Radiative fluxes, in a city-atmosphere system, 345
Radiative warming, 172
Radioactive emissions, from coal-fired plants, 429
Rankine engines, 16
Reaction paths, of stratospheric chlorine, 251
Reforestation, 23, 154
Refrigerants, 244, 270
Refrigeration, 55
Regional impacts, from heat release, 389-393
Renewable energy sources, 14-20, 45-48, 54, 64-68, 86-87
Renewable fuels, 144
Research, on the effects of climate change and CO_2 on the environment, 221-223; on the effects of increasing CO_2, 219-221; on the global carbon cycle, 218-219; on sectoral CO_2-induced impacts, 224-225; on societal costs, 223-224
Risks, acceptable levels of, 423; occupational and public, 432; of climatic change, 2
Risk averse costs, 451-453
Risk aversion, and capital loss rate, 452; and capital asset losses, 452
Risk and benefits, for different energy systems, 421-433; of coal, 429; of gas, 429; of light water reactors, 430; of oil, 429; of solar energy, 430
Risk benefit analysis, for CO_2-induced climatic change, 437-459
Risk evaluation, technical approaches of, 423
Roughness parameter, 332; changes caused by deforestation, 315

Saturations, in energy markets, 3
Scattering, of short-wave radiation, 344
Scenarios, for the roles of land ecosystems and the ocean, 163
Sea breeze, 352
Secondary cloud formation, 393

Secondary law efficiency, 13
Sensible heat flux, 367
Sensitivity, of statistical dynamic model, 319
Sensitivity parameter, 172
Severe heat, 462
Shale oil, exploitation of, 134, 195, 198
Shallow water approximation, 377
Shelter belts, 82
Significance criterion, 403
Simulation modeling, mesoscale, 374-380
Simulation studies, of climatic impacts of waste heat, 399-413
Snow removal, costs of, 63
Societal costs/benefits, research on the CO_2 effects of, 223-224
Soft technologies, 16-20
Soil moisture, 410-411
Soil organic matter, 153
Solar energy, 45-48, 328; active and passive, 66-67
Solar collectors, 491
Solar cooking, 84
Solar energy conversion, 328
Solar heat, 10
Solar heating and cooling, 16
Solar ponds, 16
Solar radiation, flux density of, 375; global mean, 364
Solar space satellites, 18
Solar thermal energy conversion (STEC), 333-336
Solid waste products, 79
Southern Oscillation, 110, 130
Space heating, 55
Spacing of cooling towers, 397
Stack gas, 499
Storage of carbon, biospheric, 143
Storage, of CO_2, 513-517
Storage of food, 86
Strategies, for a viable future, 465-468
Stratospheric ozone, decrease of, 273; effects of N_2O on, 264-266; reduction in, 273
Subsidies, to make energy, 3
Subsistence agriculture, 152
Substitutability, of resources, 13
Suess effect, 107, 116
Sulphate particles, 349
Surface energy balance, 402
Surface inversion, 405
Surface roughness, 333
Surface water flux, 410
Synoptic fronts, 352

Tar sands, 197-198
Technical fixes, 4-13, 202-203
Technologies, optimal mix of, 477
Temperate forests, increase of, 160; regrowth of, 154
Temperature, ground surface, 404-405, at 1.5 km, 405-406
Temperature effects, of N_2O, 264
Temperature rise, and fossil fuel growth rate, 439
Terrestrial sinks, miscellaneous, 153-154, 160
Thermal asymmetry, 390-392
Thermal conductivity detectors, 258
Thermal forcing, 376
Thermal impacts, 77, of man-made heat emission, 384
Thermal plumes, 384, 386-388
Thermal pollution, experimental design for, 401-403; GCM experiments of, 400-413; response of model to, 401
Thermal power, 394, 397
Thermal storage, 83
Thermogasification, 16
Total enthalpy, 395
Total heat, 13
Total fossil fuel resources, 134
Trace elements, 429
Transportation, 55
Trace gases, climatic effects of, 269-279; future change

scenario of, 273-276; global alterations of, 274; radiatively and chemically active, 269-279
Traverses, of temperature, humidity and wind, 387
Tree planting program, 497
Tree rings, 108
Tropical forests, 312
Tropical forest area, reduction of, 152
Tropospheric chemistry, 278
Tropospheric ozone, increase of, 273; production of, 271

Ultimate resources, 133
Unemployment, 90, 92
Uncertainties, in the projections of CO_2, 438
Upper Rhine Region, impact of waste heat, 363-380
Upper Rhine Valley, 389
Urban, albedo, 345; boundary layer, 340-343; breeze system, 342; canopy layer, 340-343; energy balance, 348; evapotranspiration, 343, 348; haze layer, 350; heat island, 340-349; latent heat flux, 348; mixed layer, 341; net radiation, 348; plume, 341, 346; precipitation modification, 351; run-off, 342; sensible heat flux, 348; thunderstorms, 351; turbulence, 340
Urban design, effects on materials and energy resources, 77; effects on the consumption of materials, 77
Urban heat island, 60 , 384
Urban impacts, on airflow, 340-342; on clouds and precipitation, 349-351; on energy balance, 348; on large-scale climates, 351-352; on net all-wave radiation, 346; on radiation balance, 344-347;on surface emissivity, 346; on water balance, 342-344
Urban metabolism, 339
Urban shape and energy use, 78-79
Urbanism, recycling in, 79
Urbanization, 349, 389; climatic impacts of, 339-352; effects in the global albedo, 351; impact on surface long-wave radiation balance, 346; waterproofing effects of, 348
Urban-rural roughness, differences in, 341
US Climate Program, 96, 215
US Department of Energy, 6, 97, 213
US Department of Energy Climate Program, 97
US National Climate Program Office, 214
US National Program on CO_2, 215-227
Utility function, 450

Vertical dynamics, 407
Vertical flux, of heated air, 394
Vertical soundings, 389
Vertical velocity, in the atmosphere, 397; at 3 km, 407-408

Warming, due to CO_2, 464; due to doubling of CO_2, 208; of surface by CFMs, 247-249; of surface and troposphere, 273; tropopause by CFM, 247-249
Waste heat, categories of, 368-369; climatic impacts of, 399-413; components of, 366-368; contribution of radiation to, 367; flux density of, 373; future emission of 374; globally averaged response to, 411-413; impact on the Upper Rhine Region,

363-380; inventory for the Upper Rhine Region, 366-368; recovery of, 490; regional response to, 403-411; release rate of, 328, 368; temporal variation of, 370-373
Waste Heat Project, in the Upper Rhine Region, 366
Waste recycling systems, 79
Water balance, of an urban building-air volume, 342-343
Water heating, 55
Waterproofing, of the urban surface, 342
Wave energy, 68
Weathering of carbonates, 153
Wind energy, 10, 16, 67, 82-83
Wind tunnels, 82
Woodburners, 16
World Energy Conference, 192
World population, 199

Zonally-average model, 175
Zone of flow establishment, 504